AF782885

OXYGEN TRANSPORT
TO TISSUE XXVI

ADVANCES IN EXPERIMENTAL MEDICINE AND BIOLOGY

Editorial Board:

NATHAN BACK, *State University of New York at Buffalo*

IRUN R. COHEN, *The Weizmann Institute of Science*

DAVID KRITCHEVSKY, *Wistar Institute*

ABEL LAJTHA, N.S. Kline *Institute for Psychiatric Research*

RODOLFO PAOLETTI, *University of Milan*

Recent Volumes in this Series

Volume 558
DEFECTS OF SECRETION IN CYSTIC FIBROSIS
Edited by Carsten Schultz

Volume 559
CELL VOLUME SIGNALING
Edited by Peter Lauf

Volume 560
MECHANIMS OF LYMPHOCYTE ACTIVATION AND IMMUNE REGULATION X: INNATE IMMUNITY
Edited by Sudhir Gupta, William E. Paul, and Ralph Steinman

Volume 561
CHEMISTRY AND SAFETY OF ACRYLAMIDE IN FOOD
Edited by Mendel Friedman and Don Mottram

Volume 562
CHOLINGERGIC MECHANSIMS
Edited by José Gonzalez-Ros

Volume 563
UPDATES IN PATHOLOGY
Edited by David C. Chhieng and Gene P. Siegal

Volume 564
GLYCOBIOLOGY AND MEDICINE
Edited by John S. Axford

Volume 565
SLIDING FILAMENT MECHANISM IN MUSCLE CONTRACTION: FIFTY YEARS OF RESEARCH
Edited by Haruo Sugi

Volume 566
OXYGEN TRANSPORT TO TISSUE XXVI
Edited by Paul Okunieff, Jacqueline Williams, and Yuhchyau Chen

---

A Continuation Order Plan is available for this series. A continuation order will bring delivery of each new volume immediately upon publication. Volumes are billed only upon actual shipment. For further information please contact the publisher.

Paul Okunieff
Jacqueline Williams
Yuhchyau Chen
(Eds.)

Oxygen Transport to Tissue XXVI

Springer

Paul Okunieff
Department of Radiation Oncology
University of Rochester
Rochester, NY 14642
USA

Jacqueline Williams
Department of Radiation Oncology
University of Rochester
Rochester, NY 14642
USA

Yuhchyau Chen
Department of Radiation Oncology
University of Rochester
Rochester, NY 14642
USA

Proceedings of the Thirty-first Scientific Meeting of the International Society on Oxygen Transport to Tissue (ISOTT), held August 16-20, 2003 at the University of Rochester in Rochester, NY, USA.

Library of Congress Control Number: 2005923473

ISBN-10: 0-387-25062-X
ISBN-13: 978-0387-25062-5

Printed on acid-free paper.

© 2005 Springer Science+Business Media, LLC.
All rights reserved. This work may not be translated or copied in whole or in part without the written permission of the publisher (Springer Science+Business Media, LLC, 233 Spring Street, New York, NY 10013, USA), except for brief excerpts in connection with reviews or scholarly analysis. Use in connection with any form of information storage and retrieval, electronic adaptation, computer software, or by similar or dissimilar methodology now known or hereafter developed is forbidden.
The use in this publication of trade names, trademarks, service marks and similar terms, even if they are not identified as such, is not to be taken as an expression of opinion as to whether or not they are subject to proprietary rights.

Printed in the United States of America.

9 8 7 6 5 4 3 2

springer.com

INTERNATIONAL SOCIETY ON OXYGEN TRANSPORT TO TISSUE

2003

Officers

President	Paul Okunieff
Past-President	Maureen Thorniley
President-Elect (2004)	Giuseppe Cicco
Secretary	Oliver Thews
Treasurer	Peter Keipert
Chairperson, Knisely Award Committee	Duane F. Bruley

Executive Committee

Clare E. Elwell
David J. Maguire
Roland N. Pittman
Eiji Takahashi
Artur Fournell
Avraham Mayevsky
Harold M. Swartz
Martin P. Wolf

Local Committee

Yuhchyau Chen
Amy K. Huser
Ann Muhs
George Uschold
Bruce M. Fenton
Wendy Malorzo
Paul Okunieff
Jacqueline P. Williams

Technical Editors

Amy K. Huser
Laraine Visser

Volume Administrator

Amy K. Huser

PREFACE

The International Society of Oxygen Transport to Tissue (ISOTT) was founded in 1973 to provide a forum for bioengineers, basic scientists, physiologists, and physicians to discuss new data, original theories, new interpretations of old data, and new technologies for the measurement of oxygen. At each annual meeting all posters are presented orally along with plenary lectures, and all presentations are given in a general session attended by everyone. Each meeting has had a specific focus, ranging from neonatology to physical chemistry to cancer biology. The Society has helped to build many careers, through opportunities to meet leaders in the field, and through awards made to young physicians and scientists. The Society also, through cross fertilization of ideas and scientific comradery, has inspired many breakthroughs in clinical medicine that now benefit mankind. I find myself president of the society after having been a winner of the Melvin Knisely Award for young scientists, in 1991.

The 2003 meeting emphasized the role of oxygen and oxygen measurement in tumor growth, metastasis, physiology, and treatment resistance. Additionally, however, completely novel approaches to measurement of tissue oxygen were presented (notably work by Dr. Takahashi) and molecular methods for estimating tissue oxygen were evaluated. Papers discussing other aspects of oxygen measurement and pathophysiology were presented including *in vivo* ESR spectroscopy (notably including Dr. Swartz and colleagues), exercise physiology, organ transplant outcome (discussed by Dr. Cicco, our 2004 president), circulatory physiology, and cerebral oxygenation (notably including Dr. Chance).

The 31st ISOTT Meeting was held in Rochester, New York from Augusts 16 to 20, 2003. There were 77 participants from all 5 continents. The presentations were chosen from the submitted abstracts and by invitation. Presentations included 7 plenary lectures, and 61 posters each with short oral presentation. Fifty-two presentation papers were found suitable for publication after major or minor modifications, sometimes requiring additional experiments or new data submission. All papers had at least two reviews from external reviewers or by ISOTT members chosen by the editors.

The Editors and organizers wish to thank our many industry supporters: Amgen, Applied Biosystems, AstraZeneca, Axentis, BrainLAB, Daiichi, Eli Lilly, Kyphon, McNeil, MedImmune, MGI Pharma, and Varian for their unrestricted educational grants. We would also like to thank our industry exhibitors for their time and support: AstraZeneca, Aventis, Biospherix, Biotrace, Bruker Biospin, and ISS.

The meeting would not have been possible without the support of many faculty and staff of the Department of Radiation Oncology at the University of Rochester Medical Center. These include Dr. Bruce Fenton for his scientific expertise, and Amy Huser, Wendy Malorzo, Ann Muhs, and Dr. George Uschold for organization of the meeting.

The task of editing the book would not have been possible without the efforts of Amy Huser, with the additional support of Laraine Visser.

The Editors congratulate Dr. Mohammad Nadeem Khan for the honor of being selected as the 2003 Melvin Knisely Award winner, Dr. James Lee for the honor of being selected for the 2003 Lubbers Award winner, and Dr. Mark Noble for the honor of being selected for the 2003 Kovach Lectureship. Finally, I would like to thank the Society for helping to build my career and for allowing me the honor of serving as its president.

On behalf of the Editors,
Paul Okunieff

CONTENTS

CEREBRAL OXYGENATION DURING REPETITIVE APNEA IN NEWBORN PIGLETS

Gregory Schears, Jennifer Creed, Tatiana Zaitseva, Steven Schultz, David F. Wilson, and Anna Pastuszko

Abstract: This study examined the effect of repetitive apnea on brain oxygen pressure in newborn piglets. Each animal was given 10 episodes of apnea, initiated by disconnecting them from the ventilator and completed by reconnecting them to the ventilation circuit. The apneic episodes were ended 30 sec after the heart rate reached the bradycardic threshold of 60 beats per min. The oxygen pressure in the microvasculature of the cortex was measured by oxygen-dependent quenching of the phosphorescence. In all experiments, the blood pressure, body temperature, and heart rate were continuously monitored. Arterial blood samples were taken throughout the experiment and the blood pH, PaO_2 and $PaCO_2$ were measured.

During pre-apnea, cortical oxygen was 55.1 ± 6.4 (SEM, n = 7) mm Hg and decreased during each apnea to 8.1 ± 2.8 mm Hg. However, the values of cortical oxygen varied during recovery periods. Maximal oxygen levels during recovery from the first two apneic episodes were 76.8 ± 12 mm Hg and 69.6 ± 9 mm Hg, respectively, values higher than pre-apnea. Cortical oxygen pressure then progressively decreased following consequent apnea.

In conclusion, the data show that repetitive apnea caused a progressive decrease in cortical oxygen levels in the brain of newborn piglets. This deficit in brain oxygenation can be at least partly responsible for the neurological side effects of repetitive apnea.

1. INTRODUCTION

Perinatal asphyxia, with attendant bradycardia and hypotension, is an hypoxic/ischemic condition, which is followed by reperfusion upon resuscitation. Acute asphyxia events, like apnea, are commonly associated with prematurity in the perinatal period. Apneic episodes longer than 15 sec and accompanied by hypoxia or bradycardia have been reported to occur commonly in infants with gestational ages less than 34 weeks.[1] Apneic episodes requiring ventilatory support or pharmacological intervention occur in at least 50% of surviving neonates weighing less than 1500 g at birth.[2] It is known that neonatal apnea may trigger cellular and molecular changes that can lead to metabolic disturbance and/or cell death. However, the cellular and molecular mechanisms of apnea are still poorly understood, due in part to the lack of reliable measurements of brain oxygenation. In the present study, a minimally invasive optical method, oxygen-dependent quenching of phosphorescence, was used to measure oxygenation of the brain's cortex. This method has been used previously for measuring oxygen within the brain.[3-7] Phosphorescence quenching directly measures the free oxygen dissolved in the blood plasma in the microcirculation and these measurements are not affected by pH in the physiological range (pH 6-8) or by alterations in cellular metabolism.

2. METHODS

2.1. Animal Model

Seven newborn piglets, age 3-5 days, were used for all studies. This animal model was chosen because the brain of the newborn piglets is comparable, from a maturational standpoint, to a full-term human newborn. Anesthesia was induced with halothane (Halocarbon Laboratories, Augusta, SC; 4% mixed with 96% oxygen), and 1.5% lidocaine-HCl (Abbott Laboratories, North Chicago, IL) was used as a local anesthetic. Halothane was withdrawn entirely after the tracheotomy, and pancuronium was used to induce respiratory paralysis (Gensia Pharmaceuticals, Irvine, CA; 1.5 mg/kg). Fentanyl citrate (Elkins-Linn, Inc., Cherry Hill, NJ) was intravenously injected at 30 μg/kg, and the animals were mechanically ventilated with a mixture of oxygen and 0.5% isoflurane (Baxter Healthcare Corp., Dearfield, IL). The femoral artery and femoral vein were then cannulated and the piglet was maintained on a D_5LR infusion with 10 mg/kg/hr of fentanyl citrate throughout the experiments. The head was placed in a Kopf stereotaxic holder and an incision was made along the circumference of the scalp. The scalp was removed to expose the skull and a hole approximately 5 mm in diameter was made in the skull over one parietal hemisphere for measuring the oxygen pressure. In all experiments, the blood pressure, body temperature, and respiratory rate were monitored. The blood pH, $PaCO_2$, and PaO_2 were measured using a Chiron/Diagnostics Rapidlab 800 blood gas machine.

2.2. Repetitive Apnea Model

Each animal underwent 10 episodes of apnea and recovery. Apnea was initiated by disconnecting the animal from the ventilator and completed by reconnecting it to the ventilator. The apneic episodes were terminated 30 sec after the heart rate reached the bradycardic threshold of 60 beats per minute. This simulates the delayed response time to resuscitation. The normal heart rate range for a piglet is between 140-190, which is similar to that of human infants. The piglets had a recovery period of 10 minutes on mechanical ventilation between periods of apnea.

2.3. Measurement of Cortical Oxygen

Cortical oxygen pressure was measured by the oxygen-dependent quenching of phosphorescence.[3, 4, 8] This is a non-invasive optical method in which an oxygen sensitive phosphor (Oxyphor G2) was injected IV at approximately 2 mg/kg. Oxyphor G2 is the second generation glutamate dendrimer of Pd-tetracarboxyphenyl tetrabenzoporphyrin.[9] The measurements were made with a frequency domain phosphorometer (PMOD 2000, Oxygen Enterprises, Ltd.). The excitation light (635 nm) was carried to the tissue through one branch of a bifurcated light guide and it illuminated an approximately 5 mm diameter area. The phosphorescence (790 nm maximum) was returned through the second branch of the bifurcated light guide, filtered to remove the excitation light, and measured. The phosphorescence lifetime was determined from the phase relationship relative to the excitation light measured at a phase shift of 25 ± 2 degrees. The calculated lifetime is approximately the mean for the phosphor in the tissue microvasculature. The lifetime at zero oxygen and quenching constant for oxygen at 38°C are 255 μsec and 280 mm Hg^{-1} sec^{-1}, respectively.[10]

2.4. Statistical Evaluation

All values are expressed as means for n experiments ± SEM. Statistical significance was determined using paired *t* test analysis or one-way analysis of variance with repeated measures by Wilcoxon signed-rank test. $P < .05$ was considered statistically significant.

Table 1. Physiological parameters in newborn piglets during repetitive apnea and recovery periods.

	PaO_2 (mm Hg)	**$PaCO_2$ (mm Hg)**	**pH**	**BP (mm Hg)**
Control	81.6 ± 4.9	32.9 ± 1.7	7.49 ± 0.02	90.6 ± 4.5
Apnea 1	96.5 ± 7.2	32.1 ± 2.8	7.40 ± 0.03[a]	93.1 ± 5.8
Apnea 2	88.8 ± 7.4	33.8 ± 4.1	7.33 ± 0.05[a]	100.1 ± 5.6
Apnea 3	96.3 ± 4.8	32.7 ± 3.0	7.27 ± 0.04[b]	96.3 ± 6.0
Apnea 4	96.8 ± 4.0	31.7 ± 2.9	7.24 ± 0.05[b]	98.3 ± 4.8
Apnea 5	99.4 ± 2.3	31.4 ± 2.7	7.25 ± 0.05[b]	88.4 ± 4.0
Apnea 6	102.7 ± 5.1	30.1 ± 3.0	7.23 ± 0.05[b]	90.0 ± 4.4
Apnea 7	86.11 ± 5.1	29.9 ± 2.9	7.22 ± 0.04[b]	89.8 ± 4.7
Apnea 8	102.1 ± 3.2	30.2 ± 3.1	7.22 ± 0.05[b]	87.8 ± 5.7
Apnea 9	104.7 ± 2.7	30.7 ± 2.9	7.19 ± 0.03[b]	86.6 ± 4.0
Apnea 10	98.5 ± 5.9	29.4 ± 2.4	7.20 ± 0.04[b]	81.3 ± 5.3
Recovery, 1 hr	80.4 ± 6.0	32.5 ± 3.9	7.37 ± 0.03[a]	82.2 ± 3.4
Recovery, 2 hr	78.8 ± 5.8	36.4 ± 2.5	7.41 ± 0.03[a]	80.8 ± 4.3

The values are the means ± SEM for n = 7 experiments. [a]$P < .05$; [b]$P < .001$ for difference from pre-apnea conditions as determined by one-way analysis of variance with repeated measures by Wilcoxon signed-rank test.

3. RESULTS

3.1. Physiological Parameters of the Piglets During Repetitive Apnea

The effects of repetitive apnea on the physiological parameters of newborn piglets are shown in Table 1. As can be seen, during 10 episodes of apnea, there was no significant difference in $PaCO_2$, PaO_2, and blood pressure. However, pH decreased progressively with increasing numbers of apnea, and at the end of the last apnea was 7.2 ± 0.04 ($P < .001$).

3.2. Cortical Oxygen Pressure During Repetitive Apnea/Post-Apneic Recovery

Cortical oxygen pressure was measured at 5 sec intervals throughout the apnea and post-apneic recovery. The time course for an individual experiment of 10 apneic episodes is presented in Figure 1. Three illustrative apnea/recovery periods were selected (apnea 1, 5, and 10) and are presented in Figure 2. The selection of only 3 periods makes it possible to see more clearly the differences in the time courses with increasing number of apneic episodes. Both the pre- and post-apneic cortical oxygen pressures decrease with increasing numbers of apnea. The values for the pre-apneic cortical oxygen pressure and the maximum in the post-apneic recovery have been summarized (n = 7) and presented in Figure 3. During repetitive apnea, the cortical oxygen decreased to the same value (about 8 mm Hg), a value that includes some contribution of surface vessels that are not deoxygenated. The maximum that occurs during the post-apneic recovery decreases significantly with increasing numbers of apneic episodes, falling from about 76 mm Hg to 50 mm Hg by the apnea 6 period, and then remaining nearly constant. When the data are evaluated by paired *t* test, the first and second apneas have an "overshoot" in the oxygen levels, indicating a reactive hyperemia. This overshoot is markedly attenuated with increasing numbers of apneic episodes, consistent with the lack of full recovery.

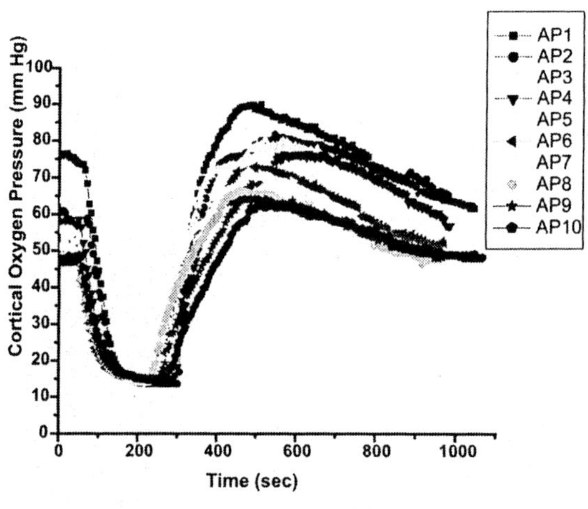

Figure 1. Changes in cortical oxygen pressure during repetitive apnea and recovery periods. Each individual curve represents the change in cortical oxygen pressure that occurred during one individual apnea in an experiment with a set of 10 apneic episodes.

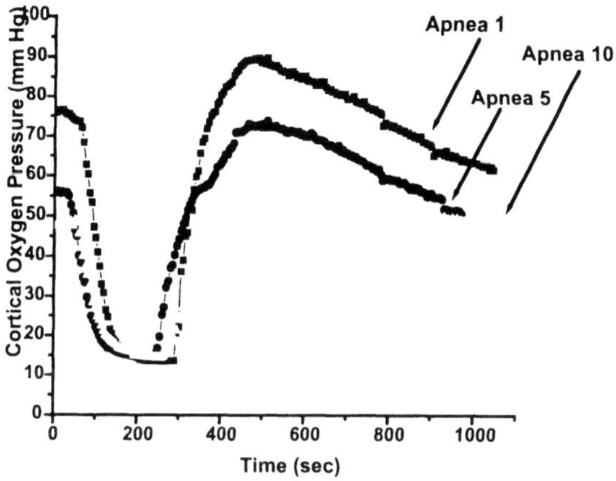

Figure 2. Changes in cortical oxygen pressure during repetitive apnea and recovery periods. The data for three illustrative apnea/recovery periods (apnea 1, 5, and 10 from Fig. 1) are shown. The results are representative of the progression of changes in the apnea that occurred during the experiment.

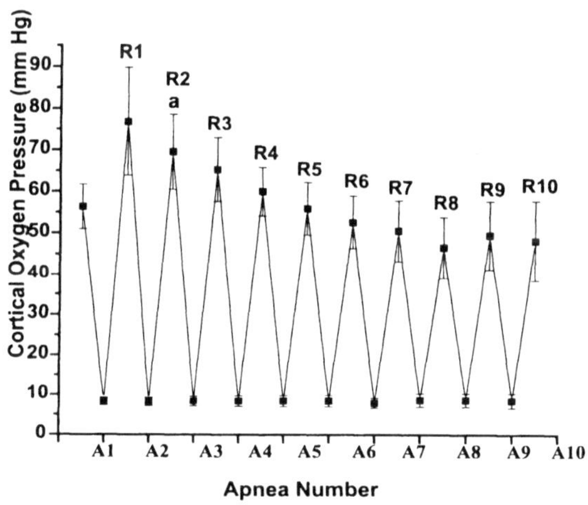

Figure 3. Cortical oxygen pressure as a function of repeated numbers of apneic episodes. Maxima (R1-R10) represent the peak oxygen values during recovery periods and the minima (A1-A10) represent the lowest oxygen values during apneic periods. The values are the means ± SEM for n = 7 experiments. [a]$P < .05$ for difference from pre-apnea conditions as determined by paired *t* test analysis.

4. DISCUSSION

There is a lack of information about the changes in brain oxygenation after cessation of breathing in an infant and during recovery from an apneic episode. The relationship between measured arterial oxygen pressure and oxygen pressure in the brain is dependent on many factors, including the cause of the alteration and its duration. As a result, the duration of an apneic episode necessary to cause significant hypoxia in the brain and therefore potentially to affect later neurodevelopment prognosis is not yet understood.[11,12]

The data presented show the effects of repetitive apnea on cortical oxygen pressure. Cortical oxygen pressure was measured by oxygen-dependent quenching of phosphorescence. This method allows quantitative measurement of oxygen distribution in the veins and capillaries of the brain cortex *in vivo* and, therefore, correlation of tissue oxygenation with neuronal function. Each animal underwent 10 episodes of apnea and recovery. The number was based on our experience that 10 episodes produces measurable brain injury and yet the cardiovascular function is sufficiently retained. This experimental model was chosen because the cumulative effect of brief but repetitive asphyxial events, as may occur in the human newborn with severe apnea, may have clinical significance for later neurological function. Study of bicuculline-induced seizures in rats[13] revealed that systemic and cerebral vascular responses and associated changes in cerebral oxygenation were better maintained during long-duration seizures than during shorter, repeated seizures. In a gerbil model of repeated ischemic events secondary to bilateral carotid occlusion, a series of three 5-min occlusions at various intervals produced more cerebral edema and neuronal loss than a single 15-minute occlusion.[14] In fetal sheep, 3 episodes of 10-minute ischemia were more deleterious than an isolated 30-minute ischemia,[15] with a particular increase in neuronal loss in the striatum after multiple insults.

Our results show that, following the first few apneas, an overshoot in brain oxygenation occurred during post-apneic recovery. This is consistent with other studies showing that resuscitation after mild to moderate ischemia is accompanied by an initial period of hyperemia.[16] This process, triggered by the primary insult, is possibly related to accumulation of vasoactive agents such as adenosine and reactive oxygen metabolites. In a severe insult, hyperemia is followed by hypoperfusion,[17-19] endothelial injury and swelling,[20, 21] granulocyte plugging of microvessels,[22] and intravascular clotting.[23] Steady decreases in cortical oxygen and pH following increasing numbers of apneic episodes suggest a lack of full metabolic recovery between apnea, particularly in the brain. These biochemical alterations, such as acid-base imbalance, accumulated exposure to reactive oxygen species, and general metabolic alterations that result from repetitive apnea ultimately result in short- and long-term neuronal damage.

In conclusion, we have shown that, in newborn piglets, repetitive apnea cause a progressive decrease in cortical oxygenation combined with an imbalance in brain metabolism. The progressive nature of these disturbances is consistent with repetitive apnea being more likely to produce long-term injury than an equal number of individual events.

5. ACKNOWLEDGEMENTS

This work was supported by grants NS-31465 and HD041484.

REFERENCES

1. K. Barrington, and N. Finer, The natural history of the appearance of apnea of prematurity, *Pediatr. Res.* **29**, 372-375 (1991).
2. R. J. Martin, M. B. Miller, and W. A. Carlo, Pathogenesis of apnea in preterm infants, *J. Pediatr.* **109**, 733-741 (1986).
3. A. Pastuszko, S. N. Lajevardi, J. Chen, O. Tammela, D. F. Wilson, and M. Delivoria-Papadopoulos, Effects of graded levels of tissue oxygen pressure on dopamine metabolism in striatum of newborn piglets, *J. Neurochem.* **60**, 161-166 (1993).
4. A. Pastuszko, Metabolic responses of the dopaminergic system during hypoxia in newborn brain, *Biochem. Med. Metab. Biol.* **51**, 1-15 (1994).
5. M. Yonetani, Ch-Ch. Huang, N. Lajevardi, A. Pastuszko, M. Delivoria-Papadopoulos, and D. F. Wilson, Effect of hemorrhagic hypotension on extracellular level of dopamine, cortical oxygen pressure and blood flow in brain of newborn piglets, *Neurosci. Lett.* **180**, 247-252 (1994).
6. M. Olano, D. Song, S. Murphy, D. F. Wilson, and A. Pastuszko, Relationships of dopamine, cortical oxygen pressure, and hydroxyl radicals in brain of newborn piglets during hypoxia and posthypoxic recovery, *J. Neurochem.* **65**, 1205-1212 (1995).
7. W. M. DeCampli, G. Schears, R. Myung, S. Schultz, J. Creed, A. Pastuszko, and D. F. Wilson, Tissue oxygen tension during regional low flow perfusion in neonates, *J. Thorac. Cardiovasc. Surg.* **125**(3 Pt 1), 472-480 (2003).
8. S. A. Vinogradov, M. A. Fernandez-Seara, B. W. Dugan, and D. F. Wilson, Frequency domain instrument for measuring phosphorescence lifetime distributions in heterogeneous samples, *Rev. Sci. Inst.* **72**(8), 3396-3306 (2001).
9. I. B. Rietveld, E. Kim, and S. A. Vinogradov, Dendrimers with tetrabenzoporphyrin cores: near infrared phosphors for in vivo oxygen imaging, *Tetrahedron* **59**(22), 3821-3831 (2003).
10. I. Dunphy, S. A. Vinogradov, and D. F. Wilson, Oxyphor R2 and G2: Phosphors for measuring oxygen by oxygen dependent quenching of phosphorescence, *Analy. Biochem.* **310**, 191-198 (2002).
11. R. A. Jones, and D. Lukeman, Apnea of immaturity. 2. Mortality and handicap, *Arch. Dis. Child* **57**, 766-768 (1982).
12. G. A. Levitt, A. Mushin, S. Bellman, and D. R. Harvey, Outcome of preterm infants who suffered neonatal apneic attack, *Early Human Dev.* **16**, 235-243 (1988).
13. N. R. Kreisman, T. J. Sick, and M. Rosenthal, Importance of vascular responses in determining cortical oxygenation during recurrent paroxysmal events of varying duration and frequency of repetition, *J. Cereb. Blood Flow Metab.* **3**, 330-338 (1983).
14. S. Tomida, T. S. Nowak, K. Vass, J. M. Lohr, and I. Klatzo, Experimental model of repetitive ischemic attacks in the gerbil, *J. Cereb. Blood Flow Metab.* **7**, 773-782 (1987).
15. E. C. Mallard, C. E. Williams, A. J. Gunn, M. I. Gunning, and P. D. Gluckman, Frequent episodes of brief ischemia sensitize the fetal sheep brain to neuronal loss and induce striatal injury, *Pediatr. Res.* **33**(1), 61-65 (1993).
16. V. Fellman, and K. O. Raivio, Reperfusion injury as the mechanism of brain damage after perinatal asphyxia, *Pediatr. Res.* **41**, 599-606 (1997).
17. C. Palmer, Hypoxic-ischemic encephalopathy. Therapeutic approaches against microvascular injury, and role of neutrophils, PAF, and free radicals, *Clin. Perinatol.* **22**, 481-517 (1995).
18. A. A. Rosenberg, E. Murdaugh, and C. W. White, The role of oxygen free radicals in postasphyxia cerebral hypoperfusion in newborn lambs, *Pediatr. Res.* **26**, 215-219 (1989).
19. B. R. Karlsson, B. Grögaard, B. Gerdin, and P. A. Steen, The severity of postischemic hypoperfusion increases with duration of cerebral ischemia in rats, *Acta Anaesthesiol. Scand.* **38**, 248-253 (1994).
20. R. Pluta, A. S. Lossinsky, H. M. Wisniewski, and M. J. Mossakowski, Early blood-brain barrier changes in the rat following transient complete cerebral ischemia induced by cardiac arrest, *Brain Res.* **633**, 41-52 (1994).
21. C. K. Petito, W. A. Pulsinelli, G. Jacobson, and F. Plum, Edema and vascular permeability in cerebral ischemia: Comparison between ischemic neuronal damage and infarction, *J. Neuropathol. Exp. Neurol.* **41**, 423-436 (1982).
22. R. L. Zhang, M. Chopp, H. Chen, and J. H. Garcia, Temporal profile of ischemic tissue damage, neutrophil response and vascular plugging following permanent and transient middle cerebral artery occlusion in the rat, *J. Neurol. Sci.* **125**, 3-10 (1994).
23. W. S. Thomas, E. Mori, B. R. Copeland, J. Q. Yu, J. H. Morrissey, and G. J. del Zoppo, Tissue factor contributes to microvascular defects after focal cerebral ischemia, *Stroke* **24**, 847-854 (1993).

OPTIMAL DETERMINATION OF DETECTOR PLACEMENT IN CEREBRAL NIR SPECTROSCOPY OF NEONATES USING CHEMOMETRIC TECHNIQUES

Terence S. Leung, Clare E. Elwell, and David T. Delpy

Abstract: This paper investigates the optimal placement of NIRS optodes in order to maximise the detection of haemoglobin changes in cortical grey matter resulting from an evoked response in neonates. The analysis is based upon predictions of optical signal at the surface of the head, using a Finite Element based model of light diffusion in tissue. Using the generated intensity data, the combination of optode positions, which maximise the signal from cortical grey matter whilst minimising that from surface tissue or cerebral white matter, is determined using a Chemometric statistical analysis. The neonatal head is modelled as a 2 dimensional circle with 3 layers corresponding to the skin/scalp, and grey and white matter. A wide range of absorption coefficients for each layer is simulated, based upon physiologically reasonable values for parameters. Surface intensity at 10 different optode positions have been generated for a total of 31,250 combinations of these variables for the 3 layers. It was found that with 3 optodes at 5, 15, and 50 mm apart from the source, the smallest root-mean-square error between the estimated and modelled values can be obtained. Increasing the number of optodes further does not improve the performance.

1. INTRODUCTION

Non-invasive near infrared (NIR) spectroscopy measurement is widely used in investigating the cerebral oxygenation and haemodynamics in humans. However, it is known that the measurements obtained from the surface of a human head are not totally comprised of intracerebral signals, but also contain extracerebral signals from the skin and skull. It has also been shown that the mean interrogation depth of photons increases with increasing optode spacing,[1] resulting in different signal contributions from skin, skull, and cerebral grey and white matter as a function of optode spacing. As illustrated schematically in Figure 1, detector 1 predominantly measures the attenuation (A_1) through skin/skull, detector 2 measures A_2, which includes some signals from the grey matter, and detector 3 measures A_3, which further includes some white matter contribution. Signals purely from the brain are clinically useful, and several schemes have been proposed to minimise the contamination from extracerebral tissues by subtraction of signals measured at different optode spacings.[2] Generalising this methodology, we investigated the optimal number and positions of detectors on a neonatal head to maximise the detection of total haemoglobin concentration change in the cortical grey matter where evoked responses occur. To achieve this aim through experiment would require measurements of optical signals (e.g. intensity or mean time) at all points on the surface whilst systematically changing the optical properties (i.e. absorption (μ_a) and scattering (μ_s') coefficient) of each tissue layer. With prior knowledge of the actual grey matter haemoglobin concentration which determines μ_a, the optimal number and positions of detectors can then be determined by a chemometric technique known as Partial Least Square (PLS). In practice, such experimental measurements are not possible in human or animal subjects, and using experimental phantoms would be difficult and require a very large number of studies. This paper therefore presents a computer simulation to tackle this issue.

2. THEORY

2.1. Maximising Intracerebral Signals by the Weighted Sum of Surface Measurements

The attenuation caused by absorption at a particular layer can be approximated by the partial differential pathlength[3] (PDP), which is the mean pathlength photons travel in that layer. The attenuation measured by detectors 1-3 in the example shown in Figure 1A can be written as:

$$\Delta A_n(\lambda) = L_n^{skin}\Delta\mu_a^{skin}(\lambda) + L_n^{grey}\Delta\mu_a^{grey}(\lambda) + L_n^{white}\Delta\mu_a^{white}(\lambda) \qquad (1)$$

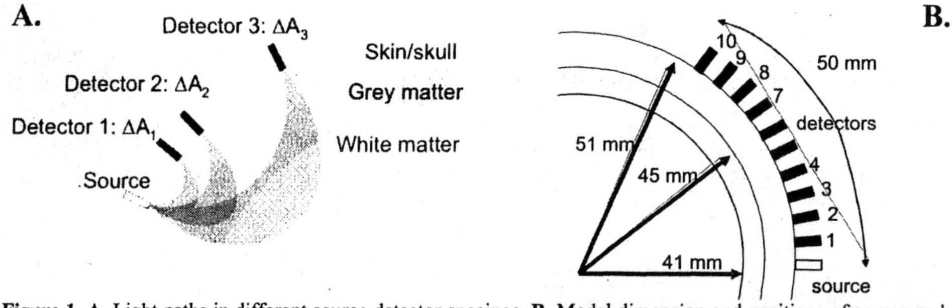

Figure 1. A. Light paths in different source detector spacings. **B.** Model dimension and positions of source and detectors.

where $\Delta\mu_a(\lambda)$ is the change in μ_a, $\Delta A_n(\lambda)$ and L_n are the change in attenuation, and the PDP is measured by the detector n. In the example of Figure 1, $L_1^{grey} = L_1^{white} = L_2^{white} = 0$. Generally speaking, there exists a weighted sum of $\Delta A_n(\lambda)$ that maximises $\Delta\mu_a(\lambda)$ from a particular layer (the grey matter, in this case),

$$\Delta\mu_a^{grey}(\lambda) \approx \sum_{n=1}^{N} w_n^{\lambda}\Delta A_n(\lambda) \tag{2}$$

where w_n^{λ} is the weighting given to $\Delta A_n(\lambda)$ measured by detector n at wavelength λ (w_n^{λ} can also be expressed as a function of PDP in certain layers), and N is the number of detectors. In this work, the total haemoglobin concentration in the grey matter is the target, which can be written as (in this case, using two wavelengths):

$$\Delta[HbT]^{grey} = \Delta[HbO_2]^{grey} + \Delta[HHb]^{grey} = (k_{11} + k_{12})\Delta\mu_a^{grey}(\lambda_1) + (k_{21} + k_{22})\Delta\mu_a^{grey}(\lambda_2) \tag{3}$$

where $\Delta[HbT]^{grey}$, $\Delta[HbO_2]^{grey}$, $\Delta[HHb]^{grey}$ correspond to changes in the total, oxy- and deoxy-haemoglobin concentrations in the grey matter, and k_{ij} is the element of the inverse of the matrix containing the specific absorption coefficients of HHb and HbO_2, respectively. Substituting (2) into (3), the following can be written:

$$\Delta[HbT]^{grey} = \sum_{n=1}^{N}\left[b_n^{\lambda_1}\Delta A_n(\lambda_1) + b_n^{\lambda_2}\Delta A_n(\lambda_2)\right] \tag{4}$$

where $\quad b_n^{\lambda_1} = (k_{11} + k_{12})w_n^{\lambda_1} \quad$ and $\quad b_n^{\lambda_2} = (k_{21} + k_{22})w_n^{\lambda_2}$.

2.2. Modelling Light Transport

Light transport has been modelled by the diffusion theory implemented in a finite element method (FEM),[4] which has been verified against Monte Carlo and phantom experiments. This light transport modelling is part of a software package known as Time-resolved Optical Absorption and Scattering Tomography (TOAST) developed to perform light transport modelling and tomographic image reconstruction. TOAST provides fast computation, which is crucial in this study because of the need to generate thousands of datasets. The neonatal head is modelled as a 2 dimensional circle with 3 layers corresponding to the skin/scalp (radius = 51 mm), grey matter (radius = 45 mm), and white matter (radius = 41 mm). The FEM mesh consists of 25,327 nodes and 12,474 elements. Because of signal:noise considerations, practical NIRS measurements are typically limited to optode spacings of about 50mm, and this limitation has been included in the model. The resulting locations of the source and 10 detectors are shown in Figure 1B. The first detector is 5 mm from the source, and subsequent detectors are 5 mm apart.

2.3. Physiological Changes in Each Layer

A wide range of μ_a for each layer is simulated, based upon physiologically reasonable values for parameters such as the total haemoglobin concentration (HbT, in

Table 1. Summary of modelled parameter variations.

Layer	Parameters	Range (min:step:max)
Skin/scalp (layer 1)	HbT conc (μM)	30:10:70
	Tissue Oxygenation (%)	50:10:90
Grey Matter (layer 2)	HbT conc (μM)	60:10:100
	Tissue Oxygenation (%)	50:10:90
White Matter (layer 3)	HbT conc (μM)	20:10:60
	Tissue Oxygenation (%)	50:10:90

μM), tissue blood oxygenation (*TO*, in fractional %), water content (*W*, in fractional %), and background absorption (*B*, in mm^{-1}). The variations in parameters are summarised in Table 1. The μ_a of each layer is calculated as follows:

$$\mu_a(\lambda) = \varepsilon_{HHb}(\lambda) HbT(1-TO) + \varepsilon_{HbO2}(\lambda) HbT \cdot TO + \mu_{a,H2O}(\lambda) W + B \tag{5}$$

where $\varepsilon_{HHb}(\lambda)$, $\varepsilon_{HbO2}(\lambda)$ are the specific absorption coefficients of HHb and HbO_2, $\mu_{a,H2O}(\lambda)$ is the μ_a of 100% water, and B is a wavelength independent background absorption. The transport scattering coefficient (μ_s') and background absorption are fixed at 1 mm^{-1} (λ = 780 nm) and 0.003 mm^{-1} (at all wavelengths), respectively, for all three layers, while the tissue water contents are set at 70% in skin/scalp and 80% in the grey and white matter. A summary of variations of modelled parameters is given in Table 1. The wavelength dependency of μ_s' is also included[5]; for example, when μ_s' *(780 nm)* is 1 mm^{-1}, μ_s' *(820nm)* is scaled to 0.9534 mm^{-1}.

2.4. Optimization by Chemometric Techniques

Given a target and a set of measurements that are influenced by other factors, chemometric techniques such as PLS seek to optimise the estimation of the target by weighting the measurements optimally. Equation (4) can be re-written in matrix form as:

$$\mathbf{y}=\mathbf{Xb} \tag{6}$$

where **y** is the target, **X** a set of measurements, and **b** the optimal weightings. In our case, **y** is the total haemoglobin concentration in the grey matter (i.e. $\Delta[HbT]^{grey}$), and **X** is a set of measurements (e.g. $\Delta A_n(\lambda)$) from 10 (or fewer) detectors at two wavelengths, i.e.

$$\mathbf{X} = \begin{bmatrix} \Delta A_1(\lambda_1,t_1) & \cdots & \Delta A_N(\lambda_1,t_1) & \Delta A_1(\lambda_2,t_1) & \cdots & \Delta A_N(\lambda_2,t_1) \\ \vdots & & \vdots & \vdots & & \vdots \\ \Delta A_1(\lambda_1,t_M) & \cdots & \Delta A_N(\lambda_1,t_M) & \Delta A_1(\lambda_2,t_M) & \cdots & \Delta A_N(\lambda_2,t_M) \end{bmatrix}$$

$$\mathbf{y} = \begin{bmatrix} \Delta[HbT]^{grey}(t_1) \\ \vdots \\ \Delta[HbT]^{grey}(t_M) \end{bmatrix} \qquad \mathbf{b}^T = [\mathbf{b}^{\lambda_1} \quad \mathbf{b}^{\lambda_2}] = [b_1^{\lambda_1} \quad \cdots \quad b_N^{\lambda_1} \quad b_1^{\lambda_2} \quad \cdots \quad b_N^{\lambda_2}] \tag{7}$$

In PLS, the optimal weighting, **b**, can be considered as a projection of the original measurements on to a new plane with a given number of principle components. This approach captures the largest amount of variance in **X** and at the same time optimises the correlation between **X** and **y**. More details can be found in the literature.[6]

3. METHODS

3.1. Data Generation

As shown in Table 1, the values of six variables were changed one by one with 5 steps between the minimum and maximum values in the range, resulting in 15,625 ($\equiv 5^6$) sets of data. Each data set was generated by TOAST and contained estimates of intensity and mean time at the 10 detector positions shown in Figure 1B. All NIRS instruments are equipped with at least 2 laser sources with wavelengths at both sides of the isobestic point (800 nm). We therefore also generated the intensity and mean time data at two wavelengths (780 & 820 nm) to simulate the practical situation. In this initial study, only intensity data were included in the analysis, and the intensity measurements (I) were converted to attenuations by the conventional formulation: $A = -log_e(I)$.

3.2. Calibration with Partial Least Square

The data $[HbT]^{grey}$ and $A_n(\lambda)$ were scaled to zero mean and unit variance. These data can thus be considered as changes from their mean values, i.e. $\Delta[HbT]^{grey}$ and $\Delta A_n(\lambda)$. The aim of this work is to find the best combination of detectors given, say, P detectors, e.g. if 3 detectors are to be chosen out of the 10 detectors in Figure 1B, the possible combinations will include detectors 1, 2, 3; detectors 1, 2, 4; detectors 1, 2, 5 and so on. The total possible number of combinations can be found by the binomial coefficient, ${}_PC_N = P!/[N!\ (P-N)!]$, in which N elements are selected from P elements. For this example, the total number of combinations is ${}_{10}C_3 = 120$. For each combination of detectors, the corresponding measurements, $\Delta A_n(\lambda)$ and $\Delta[HbT]^{grey}$, were used as **X** and **y**, respectively, in the basic model in Eq. (6). Using PLS with 3 principle components, the weighting of **b** was found. Subsequently, the root-mean-square (RMS) value between the actual and predicted $\Delta[HbT]^{grey}$ was calculated and used as a performance measure. Analyses were carried out based on $P = 10$ and $N = 1, 2, 3, 4, 6$, and 10. It should be noted that for 1 detector, the analysis was performed using multiple linear regression instead of PLS.

4. RESULTS

Table 2 summarises the results in terms of minimum RMS errors obtained by the optimal combination of N detectors out of a total of 10 (P) detectors. With only one detector, which is a common setup for many commercial NIRS instruments, the RMS error is the largest at 12.9 μM (real concentration varying from 60 to 100 μM). With two and three detectors, the RMS errors have been reduced to 4.0 and 1.6 μM, respectively.

Table 2. Minimum RMS errors obtained by the optimal combination of *N* detectors.

Number of detectors (*N*)	Optimal position of detectors (refer to Figure 2)	Minimum RMS error (μM)
1	8	12.9
2	1, 2	4.0
3	1, 3, 10	1.6
4	1, 2, 3, 10	1.6
6	1, 2, 4, 9, 10	1.6
10	1, 2, 3, 4, 5, 6, 7, 8, 9, 10	1.8

With more detectors for N = 4, 6, 10, the RMS errors are not reduced further. The weightings $\mathbf{b}^{\lambda 1}$ [λ_1=780 nm, refers to Eq. (7)] for N = 2, 3, 4, 6, 10 are shown in Figure 2. The weightings $\mathbf{b}^{\lambda 2}$ (λ_2=820 nm) have similar relative values and are not shown.

5. DISCUSSION AND CONCLUSIONS

For neonatal measurements, when only a single detector is available, the optimal position of this detector is 40 mm from the source (detector 8), in agreement with the literature.[7] Figure 2 shows the relative signal weightings that resulted from PLS analysis for 2, 3, 4, 6, and 10 detectors and provides some intuitive insight into the logic behind the calibration. The closest detectors pick up more photons that have travelled through the skin/skull, whose contribution is to be minimised, and hence have a negative weighting. A relatively higher proportion of photons measured over the next few detectors also pass through the grey matter and hence have a rather larger positive weighting. Detectors further away receive photons that have travelled through the white matter (in addition to the first two layers), whose contribution is also minimised by negative weightings. In an N layer model, the absorption change in a particular layer can be calculated if N number of surface measurements are available. As briefly discussed in section 2.1., this is because one can always express the absorption change in a particular

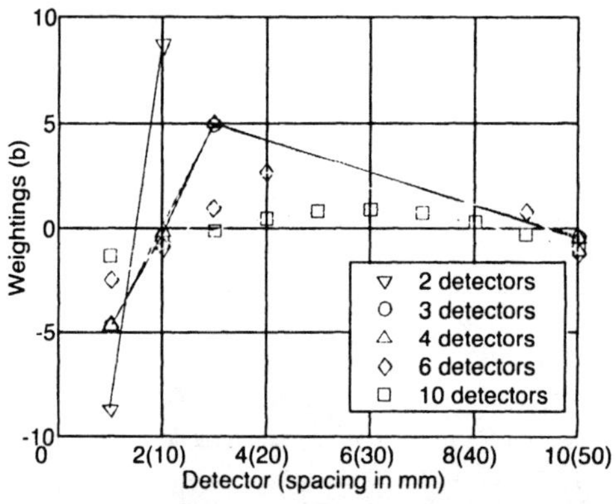

Figure 2. The optimal weighting $\mathbf{b}^{\lambda 1}$ for different number of detectors at 780 nm.

layer with the weighted sum of *N* surface measurements. This is, however, only true if we limit our discussion of the effects of inhomogenuity in terms of PDP,[3] which is a linear approximation. In a more realistic non-linear diffusion theory based model such as the one simulated by the TOAST software, one should not expect the weighted sum to be totally due to absorption change in one particular layer alone. When at least two wavelengths are available, the total haemoglobin concentration change can be obtained as described in Eq. (4). The minimum RMS error is obtained with 3 or more detectors, which agrees with the above discussion that at least 3 (or *N*) detectors are needed for a 3 (or *N*) layer model, although it is interesting that virtually no improvement in RMS error results from inclusion of further detectors. Analyses based on the mean time measurements also show similar results: that 3 detectors are sufficient to minimise the RMS error and that additional detectors do not further improve the performance. Due to the constraint of space, more details cannot be discussed here. The simplified model of the head used here does not include a cerebrospinal fluid (CSF) layer, which is known to affect light distribution. This is justified in the case of the neonatal head where the CSF layer is thin (0.5 – 1.0 mm) and the scattering coefficient of the white matter in the neonatal brain is lower than that in an adult brain because of reduced myelination. However, it is anticipated that the presence of CSF and local thickness variation, as would be required in a more realistic adult brain model, would change the way light propagates, and in this instance a hybrid diffusion/radiosity model[8] would need to be employed. In this case, the optimal number of detectors that results in the minimum prediction error would also depend on the underlying number of layers, and the optimal locations of the detectors may also be significantly different. However, the general analysis methodology described here should still apply.

6. ACKNOWLEDGEMENTS

The authors thank Dr. A. Gibson, Dr. J. Riley, and Prof. S. Arridge for their help in running the simulations and useful comments. This work was funded by the Wellcome Trust, EPSRC, and Hamamatsu Photonics KK.

REFERENCES

1. G. H. Weiss, R. Nossal, and R. F. Bonner, Statistics of penetration depth of photons re-emitted from irradiated tissue, *J. Mod. Opt.* **36**, 349-359 (1989).
2. P. W. McCormick, M. Stewart, M. G. Goetting, M. Dujovny, G. Lewis, and J. I. Ausman, Noninvasive cerebral optical spectroscopy for monitoring cerebral oxygen delivery and hemodynamics, *Crit. Care Med.* **19**, 89-97 (1991).
3. M. Hiraoka, M. Firbank, M. Essenpreis, M. Cope, S. R. Arridge, P. van der Zee, and D. T. Delpy, A Monte Carlo investigation of optical pathlength in inhomogeneous tissue and its application to near-infrared spectroscopy, *Phys. Med. Biol.* **38**, 1859-1876 (1993).
4. S. R. Arridge, M. Schweiger, M. Hiraoka, and D. T. Delpy, A finite element approach for modelling photon transport in tissue, *Med. Phys.* **20**, 299-309 (1993).
5. S. J. Matcher, M. Cope, and D. T. Delpy, In vivo measurements of the wavelength dependence of tissue-scattering coefficients between 760 and 900 nm measured with time-resolved spectroscopy, *App. Opt.* **36**, 386-396 (1997).
6. H. Martens, and T. Naes, *Multivariate Calibration*, (John Wiley & Sons Ltd, UK, 1991).
7. S. Kohri, Y. Hoshi, M. Tamura, C. Kato, Y. Kuge, and N. Tamaki, Quantitative evaluation of the relative contribution ratio of cerebral tissue to near-infrared signals in the adult human head: a preliminary study, *Physiol. Meas.* **23**, 301-312 (2002).
8. M. Firbank, S. R. Arridge, M. Schweiger, and D. T. Delpy, An investigation of light transport through scattering bodies with non-scattering regions, *Phys. Med. Biol.* **41**, 767-783 (1996).

SIMULATION STUDY OF BREAST TISSUE HEMODYNAMICS DURING PRESSURE PERTURBATION

Shoko Nioka, Simon Wen, Jun Zhang, Juan Du, Xavior Intes, Zhongyao Zhao, and Britton Chance

Abstract: We simulated the effects of compression of the breast on blood volume and tissue oxygenation. We sought to answer the question: how does the compression during breast examination impact on the circulatory systems of the breast tissue, namely blood flow, blood pooling, and oxygen concentration? We assumed that the blood was distributed in two compartments, arterial and venous. All the parameters were expressed with oxy- and deoxyhemoglobin quantities and were measured with a non-invasive method, Near Infrared Spectroscopy (NIRS). The simulated data showed that the blood volume pool in the breast decreased due to lower arterial flow and higher venous outflow, as the breast was squeezed under 100 cm H_2O with a 10 cm diameter probe (or 78 cm^2). The blood volume was reversed when the pressure was released. The breast venous oxygen saturation dropped, but overall tissue saturation (presenting NIRS signal, volume weighted average saturation) was increased. The results showed that simulation can be used to obtain venous and average oxygen saturation as well as blood flow in compressed breast tissues.

1. INTRODUCTION

While breast examination for cancer prevention has been a routine practice, and external compression is always associated with it, effects of compression on the breast have not been fully investigated. The effects of compression of the breast have been studied regarding the mechanoreceptivity,[1] anatomical point of view,[2] relation to milk ejection,[3] as the diagnostic tool,[4] and in the optical property.[5] The effect of compression on blood volume and oxygenation is of importance since it is the signature of cancer.[6] While we did not find many previous studies on breast tissue, one study observed that blood volume in a leg was reduced, and the study simulated an increase in the venous flow due to substantial pressure.[7]

For this article, we simulated the relationships between compression and blood flow, and between blood reduction and oxygenation. With the simulation, we were able to calculate blood flow (arterial and venous) through the compression. This leads to a potential use of the simulation. The following equations allowed us to calculate blood flow, blood pooling, and oxygen concentration using hemoglobin kinetics data. These data sets can be obtained by a non-invasive procedure, Near Infrared Spectroscopy and Imaging, during a breast examination.

2. SIMULATION OF HEMOGLOBIN KINETICS

2.1. Simulation Equations

We established relationships between blood flow and tissue blood volume (BV) and oxygenation information using oxy- and deoxyhemoglobin quantities described as $[HBO_2]$ and [HB], respectively. What we gained from the NIRS signal were $[HBO_2]$ and [HB] and their changes. Addition of the quantities yielded blood volume, while subtraction yielded oxygenation, respectively, as follows:

$$[HBO_2] + [HB] = \text{Blood Volume or t} [HB] \tag{1}$$

$$[HBO_2] - [HB] = \text{Oxygenation} \tag{2}$$

In particular, changes of oxygenation occurring during compression are expressed as $d[HBO_2] - d[HB]$, which is used to determine whether the tissue was oxygenated or deoxygenated.

Using a two compartment model, these two blood hemoglobin forms exist either in artery or venous vasculature. The following is a modified Fick's equation describing the relationships of blood flow and tissue blood information:

$$d\lfloor HBO_2 \rfloor / dt = Q_a - Q_v (SvO_2) - M \tag{3}$$

where Qa and Qv are input (arterial) and output (venous) blood flows, expressed as velocity of hemoglobin mass mg/100 g tissue/sec. SvO_2 is saturation of hemoglobin with O_2 in the venous blood:

$$S_v O_2 = {[HBO_2]_v} \Big/ {[HBO_2]_v + [HB]_v} \tag{4}$$

M is the metabolic rate of the breast tissue, or oxygen consumption expressed as hemoglobin oxygen carrying capacity mg/100g tissue/sec.

We assume that the arterial blood consists of 100% HBO_2, or $SaO_2 = 1$. The following equation explains that oxygen is extracted from HBO_2 with the rate of M, and HBO_2 is converted to HB at the same time:

$$d[HB]/dt = M - Q_v(1 - S_vO_2) \tag{5}$$

With Eqs. (3)-(5), we come to the solutions of Q_a and Q_v separately, as follows:

$$Q_a = \frac{d}{dt}[HBO_2] - \frac{[HBO_2]-[HBO_2]_a}{[HB]}\frac{d}{dt}[HB] + \frac{[HBO_2]-[HB]-[HBO_2]_a}{\{HB]}M \tag{6}$$

$$Q_v = \frac{[HBO_2]+[HB]-[HBO_2]_a}{[HB]}\overline{M} - \frac{d}{dt}[HB] \tag{7}$$

Eqs. (6)-(7) show that if we measure $[HBO_2]$, $[HB]$, and $[HBO_2]_a$, as well as $d[HBO_2]$ and $d[HB]$, then we can calculate the inflow and outflow of blood to the breast tissue separately.

We assumed the breast tissue hemoglobin was approximately 30 mg/100 g tissue, tissue hemoglobin saturation to be near 75% and venous blood saturation to be 66%. We assumed an arterial venous blood volume ratio of 20%/80%. The initial values of blood flow were estimated as 32.5 mg total hemoglobin/100 g tissue/sec.

2.2. Model of Breast Compression

In our model, 100 cm H_2O of pressure (equivalent to 87 mm Hg) was applied to the breast from above with a circular probe whose diameter is 10 cm, while the female subject lay in a supine position. In the breast tissue volume located under the probe, arterial flow (Q_a) and venous flow (Q_v) were altered by the pressure as indicated in Figure 1.

We assumed that a healthy woman has an arterial pressure that fluctuates between approximately 120 mm Hg and 60 mm Hg, as systolic and diastolic pressures, respectively. These pressures translated into approximately 160 cm H_2O and 80 cm H_2O in the vessels below the probe, and the external compression of 100 cm H_2O in the 10 cm

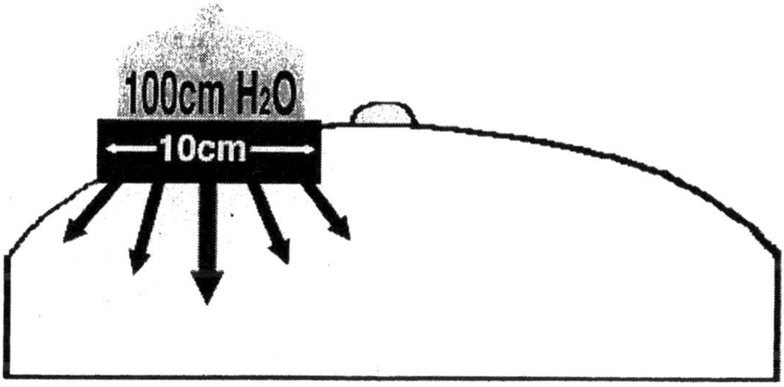

Figure 1. Illustration of compression of a human breast model. A pressure of 100 cm H_2O pressed the breast with a circular probe that has a 10 cm diameter.

diameter probe area altered a small quantity in Qa. When the arterial pressure was near diastolic pressure of 80 to 100 cm H_2O, there may have been some alterations of Qa, maximally by 25%. We assumed the arterial blood volume was affected, so the BV of the arterial vessels was maintained. Secondly, we assumed that venous blood volume decreased since the compliance of the venous vessels are much more than those of the artery.[8] This results in a higher venous flow (Qv) with the 100 cm H_2O compression in the area under the probe. On the other hand, when the compression is released, the venous vessels inflated to the normal pressure by having less outflow of (Qv) for a short time, while arterial flow and blood volume was recovered.

3. RESULTS

3.1. Blood Volume and Oxygenation Changes Due to Compression

We simulated expected changes in Qa and Qv when the human breast is pressed with 100 cm H_2O of pressure using a 10 cm diameter probe (Figure 2, left). Our simulation resulted in the following changes in blood volume and oxygenation (Figure 2, right): the blood volume was decreased during the pressing time, and oxygenation was decreased.

Interestingly, when the tissue saturation (S_tO_2; overall saturation) and the venous blood saturation (S_vO_2) were plotted, oxygenation levels in the venous blood indeed were decreased; however, overall tissue saturation (S_tO_2) was increased to 80% from 75% (Figure 3). Venous blood was desaturated due to a reduced overall blood flow, but tissue had relatively higher remaining [HBO_2] than [HB] due to the fact that the arterial blood volume increased its ratio to that of venous blood. Further, since arterial saturation was 100%, it contributed more to the average saturation. The result suggests that the pressure would alter the oxygenation values of breast tissue, and since the pressure controls the remaining blood flow to the tissue, the blood volume and oxygenation will depend upon the pressure. Therefore, monitoring the pressure is essential in finding the tissue BV and oxygenation in the examination process of the breast.

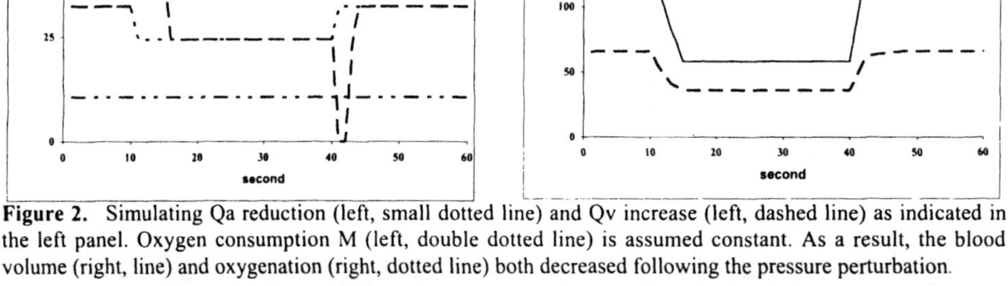

Figure 2. Simulating Qa reduction (left, small dotted line) and Qv increase (left, dashed line) as indicated in the left panel. Oxygen consumption M (left, double dotted line) is assumed constant. As a result, the blood volume (right, line) and oxygenation (right, dotted line) both decreased following the pressure perturbation.

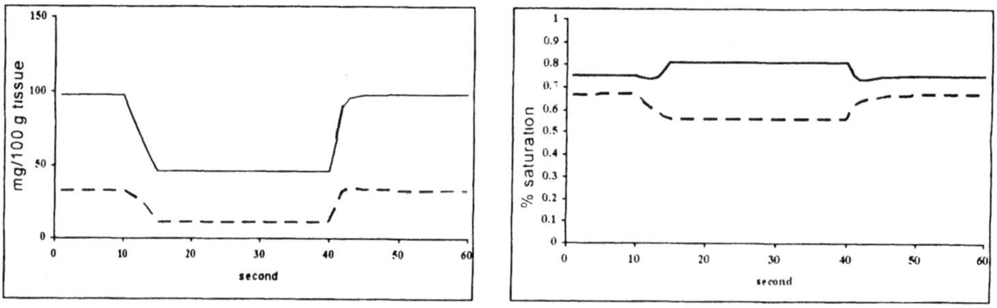

Figure 3. The left panel represents [HB] (dashed line) and [HBO_2] (solid line) reductions during compression of the breast, while the right panel shows the increase of S_tO_2 (tissue saturation, solid line) and the decrease of S_vO_2 (venous saturation, dashed line).

3.2. Calculation of Blood Flow During the Compression

We then calculated the blood flows, Qa and Qv, using Eqs. (6)-(7) with the data we obtained from the simulation of human breast tissue. Using the results of the simulation of hemodynamic response, [HB] and [HBO_2], we were able to recover the same blood flow patterns input initially (both Qa and Qv). This implies that NIRS data of [HB] and [HBO2] can be used to calculate both arterial and venous blood flows.

4. DISCUSSION

We have shown simulation of the external compression effect on the breast hemodynamic response. In the simulation, we expressed all the hemodynamic parameters with hemoglobin. The purpose of expressing with all the hemoglobin forms is that, with this, we can interpret data from NIR spectroscopy and imaging. There have been studies of blood volume and oxygen concentration,[8-13] and blood flow in breast cancers[14-15] using NIRS, as well as oxygen measurement using O_2 electrodes.[6] In addition, we can obtain information about inflow and outflow rate, (i.e., arterial and venous flow) as well as tissue oxygenation and venous oxygenation by solving the inverse problem listed as Eqs. (6)-(7). The results of the simulation showed that, with mild compression (100 cm H_2O per 10 cm diameter area), the breast blood volume can be reduced. Furthermore, the compression can change the oxygen concentration of the vasculature, especially venous blood saturation; therefore, it may affect the breast cells metabolically if the effect is prolonged.

The simulated data shows both oxygenation and deoxygenation in several compartments in the compressed breast, implying complexity. First, we have seen reduction of oxygen from the lost blood, expressed as "decreased oxygenation," and this can be interpreted simply as more [HBO_2] was lost than [HB] in Eq. (2). Second, in the venous compartment, oxygen concentration is reduced due to the slight reduction of arterial flow into the breast. Lastly, when we look at overall oxygen concentration in the breast tissue blood, as the NIR spectra can reveal, saturation increased, since venous contribution is reduced, making the average oxygen saturation higher as an average of the whole breast tissue.

In summary, we have simulated the effect of compression on blood pooling and oxygen concentration, and both parameters were greatly influenced by mild compression.

5. ACKNOWLEDGEMENTS

This research was funded by NIH grant CA 87046 and CA 72895.

REFERENCES

1. J. K. Terzis, M. P. Vincent, L. M. Wilkins, K. Rutledge, and L. M. Deane, Breast sensitivity: a neurophysiological appraisal in the normal breast, *Ann. Plastic Surg.* **19**, 318-322 (1987).
2. R. Novak, Transformation of the female breast during compression at mammography with special reference to the importance for localization of a lesion, *Acta Radiol.* **371**, 41-47 (1988).
3. N. L Sala, E. C. Luther, J. C. Arballo, and J. C. Funes, Roles of temperature, pressure, and touch in reflex milk ejection in lactating women, *J. Appl. Physiol.* **37**, 840-843 (1974).
4. W. A. Chilcote, G. A. Davis, P. Suchy, and D. M. Paushter, Breast specimen radiography: evaluation of a compression device, *Radiol.* **168**, 425-427 (1988).
5. J. Shudong, B. W. Pogue, and K. D. Paulsen, In vivo near-infrared spectral detection of pressure-induced changes in breast tissue, *Optics Lett.* **28**, 1212-1214 (2002).
6. P. Vaupel, S. Briest, and M. Höckel, Hypoxia in breast cancer: pathogenesis, characterization and biological/therapeutic implications, *Wiener Medizinische Wochenschrift* **152**, 334-342 (2002).
7. G. Dai, J. P. Gertler, and R. D. Kamm, The effects of extemal compression on venous blood flow and tissue deformation in the lower leg, *J. Biomech. Eng.* **121**, 557-564 (1999).
8. A. C. Guyton, G. T. Armstrong, and P. L. Chipley, Pressure-Volume curves of the entire arterial and venous systems in the living animal, *Am. J. Physiol.* **184**, 253-259 (1956).
9. V. Quaresima, S. Matcher, and M. Ferrari, Identification and quantification of intrinsic optical contrast for near-infrared mammography, *Photochem. Photobiol.* **67**(1), 4-14 (1998).
10. A. E. Cerussi, D. Jakubowski, N. Shah, F. Bevilacqua, R. Lanning, A. Berger, D. Hsiang, J. Butler, R. F. Holocombe, and B. J. Tromberg, Spectroscopy enhances the information content of optical mammography, *J. Biomed. Opt.* **7**, 60-71 (2002).
11. B. W. Pogue, S. P. Poplack, T. O. McBride, W. A. Wells, K. S. Osterberg, U. L. Osterberg, and K. D. Paulsen, Quantitative hemoglobin tomography with diffuse near infrared spectroscopy: pilot results in the breast, *Radiol.* **218**, 261-266 (2001).
12. J. Zhang, Y. Lin, S. Nioka, N. O'Connor, B. Czemiecki, E. F. Conant, and B. Chance, Application of LED imager for breast cancer diagnosis, *Proc. SPIE* **4916**, 30-36 (2002).
13. T. Durduran, R. Choe, J. P. Culver, L. Zubkov, M. J. Holboke, J. Giammarco, B. Chanee, and A. G. Yodh, Bulk optical properties of healthy female breast tissue, *Phys. Med Biol.* **47**, 2847-2861 (2002).
14. S. Nioka, S. Yung, M. Shnall, S. Zhao, S. Orel, C. Xie, B. Chance, and L. Solin, Optical imaging of breast tumor by means of continuous waves, *Adv. Exp. Med. Biol.* **411**, 227-232 (1997).
15. S. Nioka, S. B. Colak, X. Li, and B. Chance, Breast tumor images of hemodynamic information using a contrast agent with back projection and FFT enhancement, OSA TOPS, *Adv. Opt. Imag. Photon Migration* **21**, 266-270 (1998).
16. X. Intes, J. Ripoll, Y. Chen, S. Nioka, A. Yodh, and B. Chance, In vivo continuous-wave optical breast imaging enhanced with indocyanine green, *Med. Phys.* **30**, 1039-1047 (2003).

MICROVASCULAR OXYGENATION AND OXIDATIVE STRESS DURING POSTISCHEMIC REPERFUSION

PO_2, ROS, and NO during reperfusion

Silvia Bertuglia and Andrea Giusti

Abstract: Increased formation of ROS on reperfusion after ischemia underlies ischemia reperfusion (I/R) damage. We measured, in real time, both oxygen tension in microvessels and tissue and oxidant stress during postischemic reperfusion in hamster cheek pouch microcirculation. We measured PO_2 by using phosphorescence quenching microscopy and oxygen radical species (ROS) production in the systemic blood. We evaluated the effects of a NOS inhibitor (L-NMMA) and superoxide dismutase (SOD) on the oxidative stress during reperfusion. Microvascular injury was assessed by measuring diameter change, the perfused capillary length (PCL), and leukocyte adhesion.

Our findings demonstrate that early reperfusion is characterized by low concentration of oxygen linked to increased production of ROS. After this initial transience in arterioles, the oxygen tension and production of ROS return to normal after reperfusion, while the blood flow and capillary perfusion decrease. The early increased ROS production, in turn, may impair oxygen consumption by endothelial cells, thus further promoting activation of oxygen to ROS. This event is substantiated by the finding that treatment with SOD maintains ROS at normal levels, which, in turn, should be effective to increase the production of endothelial NO. Conversely, a decrease in NO levels led to decreased ROS production during early reperfusion, which increased later during reperfusion, ultimately causing vasoconstriction and greatly increasing venular leukocyte adhesion on postcapillary venules during hypoxic conditions. Therefore, low-flow hypoxia is primarily responsible for vascular endothelial damage during reperfusion through changes in ROS and NO production.

1. INTRODUCTION

The excessive generation of reactive oxygen species (ROS) has been implicated as a final common pathway for the development of endothelial dysfunction by ischemia reperfusion (I/R).[1] Upon reperfusion, there is an overproduction of ROS that leads to the generation of products of lipid peroxidation. ROS may be responsible for the attenuated nitric oxide (NO) formation by the vascular endothelium during I/R.[2] NO limits injury to target molecules and tissue during events associated with excess production of ROS including reduction of I/R injury. Protection from ROS comes from superoxide dismutase (SOD) that is rapidly depleted by the high amount of ROS produced by the reentry of oxygenated blood into the ischemic tissue.[3, 4]

The overall picture of the inflammatory process has been complicated by the potential pathogenetic contribution of reactive nitrogen species. NO is capable of interacting with O_2, redox metals, especially iron-sulphur centers in protein, and superoxide to form the highly reactive peroxynitrite.[5, 6] This can subsequently be converted to perhydroxy nitrite and ultimately hydroxyl radicals. The reaction with superoxide is the most rapid reaction for NO. A recent development has been the concept that NO represents a third molecule, in addition to O_2 and CO_2, carried away by hemoglobin that could affect cardiovascular physiology significantly.[6] Therefore, it is reasonable to suggest that changes in PO_2 can affect significantly ROS and NO formation not only during baseline, but also during reperfusion.

Our aim was to measure oxygen tension in arterioles and lipoperoxide formation in the systemic blood during baseline and reperfusion, and to determine whether ROS and NO contribute to microcirculation damage during reperfusion, using a hamster model.[2, 7] We measured oxygen tension during baseline and reperfusion by using phosphorescence quenching microscopy that allows instantaneous online PO_2 analysis and rapid sequences of measurements in the microvessels and in the tissue.[2, 8, 9] Furthermore, we evaluated the ability of SOD and a conventional NO donor (2,2'-hydroxynitrosohydrazino -bis-etanamine, DETA-NO) to inhibit ROS formation during I/R.[10] The microvascular damage that occurs during I/R was quantified by changes in blood flow and perfused capillary length, namely the capillaries perfused by blood.[2, 7]

2. METHODS

Male Syrian hamsters (80-100 g, Charles River, Italy) were anesthetized by intraperitoneal injections of sodium pentobarbital, 50 mg/kg body wt. The animals were tracheotomized and the right carotid artery and femoral vein were cannulated to measure blood pressure, to inject the phosphorescence probes and supplementary doses of anesthetic. Animal handling and care were provided following the procedures outlined in the Guide for the Care and Use in the laboratories of the Italian Research Council.

We used four groups of animals subjected to I/R. The first group (I/R, n = 10) was used as the control group and subjected to PO_2 measurements. The second group (SOD, n = 7) was injected with a bolus of 9.0 mg/kg SOD from bovine erythrocytes dissolved in 0.9% saline (Sigma, St. Louis, Mo, USA) and administered 30 min before I/R, and 0.15 mg/kg/min were infused intravenously during reperfusion. The third group (DN, n = 5) was treated with a NO donor, 2,2'-hydroxynitrosohydrazino bis-etanamine DETA-NO (Inalco, Spa, Milan, Italy) (0.5 mg/kg). DETA-NO was dissolved in 0.5 ml of saline and administered intravenously over 30 min into the femoral vein 30 min before I/R and during the period of reperfusion. Atraumatic microvascular clips were placed on the proximal part of the cheek pouch to achieve complete ischemia for 30 min. The clamp was then removed and the microcirculation was observed for 30 min.

The hamster cheek pouch was prepared as previously reported.[2, 7] A trans-illumination technique (Orthoplan, Leica Microsystem GmbH, Wetzlar, Germany) was used. All selected fields were recorded by a video camera (COHU Inc. San Diego, CA, USA) displayed on a monitor and transferred to a video recorder. Video images were videotaped and microvascular measurements were made off-line by a computer-assisted imaging software system (MIP Image, CNR, Institute Clinical Physiology, Pisa, Italy). The hamster's body temperature and cheek pouch temperature were maintained at 37°C with circulating warm water. An intravenous injection of pentobarbital (300 mg/kg) was used as the method of euthanasia.

The oxygen measurements, based on oxygen-dependent quenching of phosphorescence emitted by palladium meso-tetra-(4-carboxyphenyl)-porphyrin after pulse excitation, was previously described.[2] Animals received a slow intravenous injection of 15 mg/kg at a concentration of 10 mg/ml of porphyrin (Porphyrin Products, Inc. Logan, UT) bound to albumin. Phosphorescence was excited by flashes (10 flashes/s) generated by a 45-W xenon strobe arc (EG&G Electro Optics, Salem, MA). An optical window of 15 x 20μm placed longitudinally along the microvessel allowed for location of the measuring sites. Filters of 420 and 630 nm were used for porphyrin excitation and phosphorescence emission, respectively. Phosphorescence signals were captured by a photomultiplier (EMI, 9855B, Knott Elektronik, Munich, Germany). The decay curves were averaged, visualized, and saved by a digital oscilloscope (Hitachi Oscilloscope V-1065, 100 MHz, Hitachi, Denshi). Decay time constants were determined by a computer fitting the averaged decay curves to a single exponential using the Stern-Volmer equation.[2, 8, 9]

To measure plasma hydroperoxides, the analytic method d-ROMs (Diacron s.r.l., Parma, Italy) was used as previously reported.[2] Ten μl of a chromogenic substance and 1 ml of the kit buffer were mixed with 10 μl of blood for 1 min at 37°C. The results were expressed in arbitrary units (au = 0.08 mg/100 ml H_2O_2). Measurements of PO_2 were done and, contemporaneously, blood samples were taken from the cannulated carotid artery at baseline, 5, 15, and 30 min of reperfusion for each hamster.

PCL, defined as the total length of capillary segments that have at least one RBC passing through them in a 30-s period, was assessed in a region of ~0.5 mm^2. Microvessel diameters were measured by an image shearing system (Digital Image Shearing Monitor Mod 907, IPM). RBC velocity was determined by using the dual slit cross correlation (velocity tracker Mod 102 B, IPM, San Diego, CA, USA). The blood flow was measured with the measured parameters: $Q = V \pi (D/2)^2$, where V is the velocity, and D is the diameter of the microvessels. The measured centerline velocity was corrected according to vessel size so that the mean RBC velocity could be obtained.

Mean arterial blood pressure (MAP) and heart rate (HR) were measured by a Gould Windograf recorder (Mod. 13-6615-10S, Gould Inc., Ohio, USA).

Data were expressed as mean ± SD. When two groups were compared, the unpaired Student *t* test was used. Comparisons between multiple groups were conducted using one-way analysis of variance followed by a Tukey-Kramer test to compare all treatment groups. The statistical significance was determined at $P < .05$.

Table 1. Changes of diameter, RBC velocity and blood flow in arterioles, lipid peroxides and perfused capillary length (PCL) after 5 and 30 min of reperfusion (5R, 30R) in the groups treated with saline (I/R), SOD and DETA-NO (DN), respectively.

	I/R	SOD	DN
Arteriole (μm)	31 ± 12	31 ± 12	31 ± 12
5R	23 ± 7*	30 ± 4°	32 ± 3°
30R	16 ± 5*	32 ± 7°	35 ± 4°
RBC Velocity (mm/s)	2.0 ± 0.7	2.0 ±0.7	2.0 ± 0.7
5R	1.2 ± 0.7*	1.8 ±0.3°	1.9 ± 0.3°
30R	1.0 ± 0.7*	2.1 ±0.3°	2.0 ± 0.3°
Flow (nl/s)	15 ± 9	15 ± 9	15 ± 9
5R	10 ± 4*	14 ± 7°	16 ± 3°
30R	3 ± 2*	16 ± 5°	15 ± 5°
Lipid peroxides mg/100 ml H_2O_2	20 ± 3	22 ± 5	20 ± 3
5R	28 ± 2*	20 ± 3°	21 ± 4°
30R	22 ± 2	20 ± 3	25 ± 3°°
PCL %	55 ± 5*	97 ± 3	95 ± 3

Values are means ± SD;**P* < .001, vs. baseline; °*P* < .05 vs. I/R; °°*P* < .05 vs. I/R and SOD groups.

3. RESULTS

The systemic parameters (MAP: 90 ± 5 mmHg; heart rate: 250 ± 30 beats/min, PaO_2: 98 ± 12 mm Hg; hematocrit: 47% ± 5%) were not significantly different during baseline and after I/R in the I/R and SOD groups ($P < .05$). In the DN group, MAP and heart rate were significantly different compared with the I/R and SOD groups (80 ± 7 mm Hg, 260 ± 30 beats/min, $P < .01$). PO_2 values for arterioles (diameter: 31 ± 12 μm) and interstitial tissue are reported in Figure 1. Lipid peroxide concentration, diameter, RBC velocity and flow changes in arterioles are reported in Table 1. In the I/R group PCL decreased significantly when compared with the baseline (baseline: 9,500 ± 450 μm, $P < .01$).

4. DISCUSSION

Our data provide evidence that lipid peroxides in the systemic blood of hamsters increased significantly within 5 minutes of reperfusion and returned to baseline values when PO_2 recovered. PO_2 did not recover rapidly in the blood and tissue but only after 30 minutes of reperfusion. Moreover, the arterial and capillary flow decreased significantly after 30 minutes of reperfusion. Both SOD and the NO donor maintained ROS formed during reperfusion at normal levels. However, ROS production increased after 30 minutes of reperfusion with DETA-NO. Both SOD and NO are correlated with vasodilatory effects and normal microvascular blood flow during reperfusion. These results indicate that SOD as well as NO inhibit lipid peroxide formation during hypoxic conditions thus preventing vasoconstriction during reperfusion.

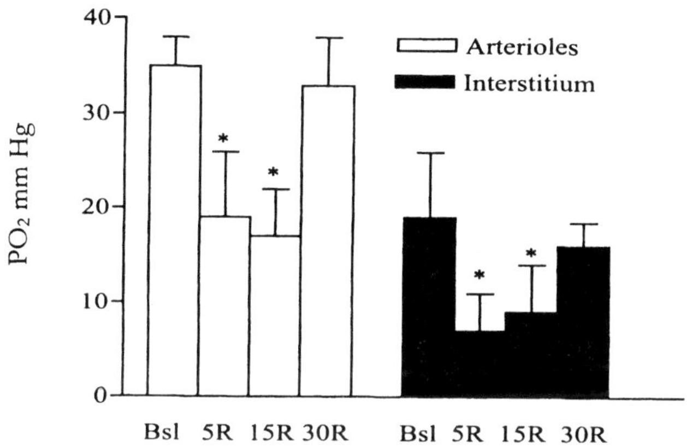

Figure 1. PO_2 measurements were made in arterioles (diameter: 31 ± 12 μm) and interstitial tissue at baseline (Bsl), 5 (5R), 15 (15R), and 30 (30R) min of reperfusion. Data are expressed as means ± SD. Each point represents an average of at least 25 measurements. *$P < .05$ vs. baseline.

It has long been assumed that overproduction of ROS during reperfusion was related to over-oxygenation, while hypoxia in the early reperfusion is contemporaneous with increased lipid peroxidation that could impair oxygen consumption by endothelial cells. At the beginning of reperfusion the reduced oxygen availability, caused also by the oxygen consumption of upstream arterioles and tissue, may alter biochemical activity of vascular endothelial cells in proportion to hypoxia thus promoting activation of O_2 to reactive species.[2] The hypoxic conditions during reperfusion determine competition between oxygen consumption by tissue and vascular endothelial cells.[6] Tissue PO_2 recovered after 30 minutes of reperfusion, notwithstanding that capillary perfusion decreased significantly. This paradox could be explained by the large drop in PO_2 between arterioles and capillaries that allows the transport of oxygen to tissue from arterioles.[2, 9, 11]

Our findings indicated that SOD and/or NO are required for vasodilation during reperfusion. One of the earliest observed abnormalities in reperfusion-induced injury is vascular endothelial dysfunction manifested as loss of NO-dependent vasodilation. This phenomenon occurs within 2-5 minutes of reperfusion, as shown also in ischemic reperfused rat hearts.[12] Our findings also indicate that vasoconstriction during reperfusion may be dependent on the altered balance of ROS and NO. In agreement with this hypothesis, Beckman et al.[3] demonstrated that superoxide inactivates NO, resulting in generation of peroxynitrite, and Whorton et al.[13] showed that NO formation in cultured endothelial cells was decreased when the PO_2 of the medium was lowered.

Collectively these findings suggest that the local regulation of vasomotor tone for the purpose of oxygen delivery is very complex since NO is oxidized by both oxygen and superoxide anion. DETA-NO promoted arteriolar vasodilation to hypoxic tissue and therefore rapidly re-established capillary perfusion, which was improved relative to that found in the ischemia reperfusion group with no treatment. Coronary artery disease has been associated with impaired NO production with augmented responses to several vasoconstrictors.[13] Therefore, the increased NO availability with DETA-NO may prevent vasoconstriction and protect arterioles during vasospasm in the reperfusion period. In agreement with our data, Edmunds et al.[14] showed that the increase in femoral vascular conductance evoked by systemic hypoxia is attenuated by NO synthesis blockade by L-NAME, but restored when baseline vasodilation is restored by infusion of the NO donor. The potential importance of the vasodilator properties of NO during hypoxic conditions can be considered by appreciating that the main determinant of tissue recovery is to restore microvascular perfusion rather than normal oxygenation.[6]

Moreover, SOD protects NO bioactivity. Hypoxia has a significant modulating effect on NO synthesis and it induces vascular relaxation through changes in superoxide production and cGMP formation in microvascular coronary endothelial cells at baseline and upon reoxygenation.[1] However, it is interesting to note that many experimenters observed that while L-NMMA determined the reduction of blood flow, the delivery of NO gas had no effect on peripheral blood flow.[12] As shown in previous results, L-NMMA caused an increase in ROS, aggravated vasoconstriction, and reduced capillary perfusion.[2] However, our findings suggest that the mechanism responsible for vasoconstriction during reperfusion is dependent on either NO and/or SOD and their decomposition due to oxidative stress. The increase in the NO level with the NO donor results in increased microvascular blood flow and decreased production of ROS during reperfusion. However, DETA-NO produced a slight but significant increase in the ROS level after 30 min of reperfusion. Yamamoto and Bing[10] showed also that oxidation products of NO (NO_2^- and NO_3^-) in arterial blood rose significantly after treatment with NO donors.[37] There is production of non-enzymatic NO during postischemic reperfusion nitrous acid and nitrogen oxides (HNO_2, N_2O_3, NO, NO_2). Such reactive nitrogen species are formed when fluxes of NO are enhanced. In addition, peroxynitrite has been detected in inflammation during treatment with NO donors. In view of these findings, treatment with NO donors results in the delivery of high local concentrations of NO and its metabolites in microcirculation during reperfusion. Therefore, it appears that basal blood flow is substantially regulated by the local level of NO, whereas the impact of oxidative stress and nitrosative stress on regulation of vascular function occurring during baseline and hypoxia remains unclear.

In conclusion, we found that early reperfusion is characterized by low PO_2 contemporaneous to increased production of lipid peroxides in the systemic blood. Our data show that the mechanism by which hypoxia during reperfusion elicits changes in vascular tone and blood flow during reperfusion is a function of both SOD and NO. DETA-NO inhibited ROS production during early reperfusion which, however, increased later during reperfusion. We hypothesize that the major determinant of the recovery of blood flow during reperfusion is dependent on the level of NO and its interaction with ROS.

REFERENCES

1. B. Haliwell, and J. M. C. Gutteridge, Free radicals in biology and medicine, 3rd ed. Oxford University Press (1993).
2. S. Bertuglia, and A. Giusti, Microvascular oxygenation, oxidative stress, nitric oxide suppression and superoxide dismutase during postischemic reperfusion, *Am. J. Physiol.* **278,** H1064-1071 (2003).

3. J. S. Beckman, and W. H. Koppenol, Nitric oxide, superoxide and peroxynitrite: the good, the bad and the ugly, *Am. J. Physiol.* **271,** C1424-1437 (1996).
4. J. S. Beckman, T.W. J. Beckman, P. Chen, P. A. Marshall, and B. A. Freeman, Apparent hydroxyl radical production by peroxinitrite, implications for endothelial injury from NO and superoxide, *Proc. Nat. Acad. Sci. USA* **87,** 1620-1624 (1990).
5. L. Kenneth, C. J. Giovanelli, and S. Kaufman, Characteristics of the NO synthase-catalyzed conversion of arginine to N-hydroxyarginine, the first oxygenation step in the enzymic synthesis of NO, *J. Biol. Chem.* **270,**1721–1728 (1995).
6. M. T. Gladwin, J. R. Lancaster, B. A. Freeman, and A. N. Schechter, NO's reactions with hemoglobin a view through the SNO-storm, *Nature Med.* **9,** 496-500 (2003).
7. S. Bertuglia, A. Giusti, S. Fedele, and E. Picano, Glucose-insulin-potassium treatment in combination with dipyridamole inhibits ischemia-reperfusion-induced damage, *Diabetologia* **44,** 2165-2170 (2001).
8. A. Golub, A. S. Popel, L. Zheng, and R. N. Pittman, Analysis of phosphorescence in heterogeneous systems using distributions of quencher concentration, *Biophys. J.* **73,** 452-465 (1999).
9. M. Intaglietta, P. C. Johnson, and R. M. Winslow, Microvascular and tissue oxygen distribution, *Cardiovasc. Res.* **32,** 632-643 (1996).
10. T. Yamamoto, and R. J. Bing, NO donors, *Proc. Soc. Exp. Biol. Med.* **225,** 200-206 (2000).
11. A. S. Popel, R. N. Pittman, and M. L. Ellsworth, Rate of oxygen loss from arterioles is an order of magnitude higher than expected, *Am. J. Physiol.* **256,** H921-924 (1989).
12. P. S. Tsao, and A. M. Lefer, Time course and mechanism of endothelial dysfunction in isolated ischemic and hypoxic perfused rat hearts, *Am. J. Physiol.* **259,** H1660-1666 (1990).
13. A. R. Whorton, D. B. Simonds, and C. A. Piantadosi, Regulation of NO synthesis by oxygen in vascular endothelial cells, *Am. J. Physiol.* **272,** L1161-1166, (1997).
14. N. J. Edmunds, S. Moncada, and J. M. Marshall, Does nitric oxide allow endothelial cells to sense hypoxia and mediate hypoxic vasodilation? In vivo and in vitro studies, *J. Physiol.* **546,** 521-577 (2003).

HYPOXIA, TUMOR ENDOTHELIUM, AND TARGETS FOR THERAPY

Beverly A. Teicher

Abstract: Hypoxia is a well-recognized feature of human solid tumors. It is also well recognized, by both physicians and investigators, that malignant disease in various organs/tissues in the same patient, or the same tumor cells implanted in different sites or organs in the preclinical host, have different levels of hypoxia and different levels of response to systemic therapies. Over the past 10 years, it has been established that normal cells involved in the malignant disease process can be important targets for therapeutic attack. A prime example of 'normal' cells that have come to the fore as anticancer therapeutic targets is endothelial cells.

The field of antiangiogenic therapies was fueled by the early hypothesis which held that angiogenesis was the same no matter where it occurred. The corollary to this hypothesis was that models of normal embryo development, as well as models working with mature well-differentiated endothelial cells in culture, would be sufficient and satisfactory models for tumor endothelial cells. However, the current hypothesis is that angiogenesis occurring during malignant disease is abnormal, and that therapeutic targets identified by studying endothelial cells isolated from fresh samples of human cancers will be most relevant in developing therapeutic agents to treat human malignant disease.

1. INTRODUCTION

Hypoxia is a well-recognized feature of human solid tumors.[1-6] It is also well recognized by physicians and investigators that malignant disease in various organs/tissues in the patient or the same tumor cells implanted in different sites or organs in the preclinical host have different levels of hypoxia and different levels of response to systemic therapies.[7, 8] It has been established over the past 10 years that normal cells involved in the malignant disease process can be important targets for therapeutic attack. A prime example of 'normal' cells that have come to the fore as anticancer therapeutic targets is endothelial cells.

Angiogenesis as an Anticancer Target

Early hypothesis	Current hypothesis
Angiogenesis is a normal process, wherever it occurs. Angiogenesis during wound healing or embryo development is the same as angiogenesis in malignant disease. Targets identified by studying normal or neoangiogenesis will apply to malignant disease	Angiogenesis during malignant disease is abnormal. Targets identified by studying endothelial cells isolated from fresh samples of human cancers will be most relevant for developing therapeutic agents to treat human malignant disease.

Figure 1. The evolution of a hypothesis: angiogenesis as an anticancer target.

2. ENDOTHELIAL CELL TARGETS

The field of antiangiogenic therapies has moved very quickly from laboratory discoveries into the clinic. As with other areas of science, the rapidity of the development of the antiangiogenic field was fueled by the availability of models and the identification of therapeutic targets. The field was also fueled by an early hypothesis, which held that angiogenesis was the same no matter where it occurred. Therefore, angiogenesis during embryo development or wound healing was the same as angiogenesis during the growth of malignant disease. The corollary to this hypothesis was that models of normal embryo development and models working with mature well-differentiated endothelial cells in culture would be sufficient and satisfactory models for tumor endothelial cells. This hypothesis also held that because endothelial cells involved in malignant disease were normal, these cells would be less susceptible to developing drug resistance because they were genetically stable.

As Figure 1 outlines, the current hypothesis is that angiogenesis occurring during malignant disease is abnormal and that therapeutic targets identified by studying endothelial cells isolated from fresh samples of human cancers will be most relevant for developing therapeutic agents to treat human malignant disease. To accomplish this, fresh samples of human colon carcinoma, normal colon mucosa, breast carcinoma, normal breast tissue, brain tumor, and normal brain were obtained (Fig. 2). The tissues were disaggregated and the endothelial cells were isolated using selection with an antibody to P1H12 linked to a magnetic bead.[9] The RNA from the endothelial cells was collected and subjected to SAGE (Serial Analysis of Gene Expression) analysis. This methodology allows elucidation of the mRNA transcripts in the cells at the time of RNA isolation, providing the identity of the transcript and the relative abundance of each transcript. Thus far, SAGE-derived transcript libraries have been generated for endothelial cells isolated from 7 fresh human tumor specimens and 5 specimens of corresponding normal tissues.

The first bioinformatics analysis was to compare the genes/mRNA expressed in each of the 3 tumor types with the genes/mRNA expressed in each corresponding normal tissue. In each case, a similar pattern emerged. The vast majority of the genes/mRNA

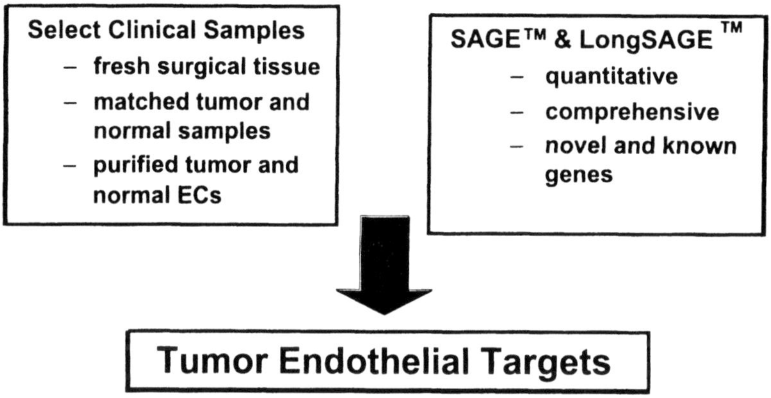

Figure 2. Fresh samples of human colon carcinoma, normal colon mucosa, breast carcinoma, normal breast tissue, brain tumor, and normal brain, were obtained. Tissues were disaggregated, endothelial cells were isolated, and the RNA from the endothelial cells was collected and subjected to SAGE analysis.

expressed by the tumor endothelial cells was very similar to the genes/mRNA expressed by the endothelial cells from the corresponding normal tissue. However, there was a small subpopulation of genes/mRNA that was expressed at much higher levels by the tumor endothelial cells and a different, small subpopulation of genes/mRNA that was expressed at much higher levels by the normal endothelial cells. Generally, the tumor endothelial cells appeared to be expressing at least a partial 'malignant phenotype.' The tumor endothelial cells appeared to be relatively de-differentiated or immature relative to the corresponding normal endothelial cells.

The second bioinformatics analysis was to compare the genes/mRNA that were expressed at high levels by the tumor endothelial cells from the colon carcinoma, breast cancer, and brain tumors with each other. Venn diagrams were developed for the subpopulations of genes that by the Chi square test had >99% confidence of being over-expressed in the tumor endothelial cells compared with the corresponding normal endothelial cells. The genes/mRNA that fulfilled these criteria included 280 genes from the colon carcinoma, 109 genes from the breast carcinomas, and 111 genes from the brain tumors (Fig. 3). The number of genes that were over-expressed in endothelial cells from both breast cancer and brain cancers was 22, from brain cancers and colon cancer was 24, and from breast cancer and colon cancer was 30. Thus, there is a high degree of organ/tissue specificity in the endothelium and there is a high degree of heterogeneity among tumor endothelium. When the highly over-expressed genes from the endothelial cell libraries were compared, there were only 12 genes that were highly over-expressed in all three tumor types. Based on these findings, it may be less likely that therapeutic antiangiogenesis targets that can be identified would be universally applicable. It may be more likely that antiangiogenic therapeutic targets can be found that will apply to major tumor categories.

The third bioinformatics analysis was to compare the expression of the genes/mRNA that were highly up-regulated in tumor endothelial cells with genes/mRNA expressed in cells commonly used as model systems in the angiogenesis and antiangiogenesis fields. The cells whose gene/mRNA expression was examined included HUVEC (human

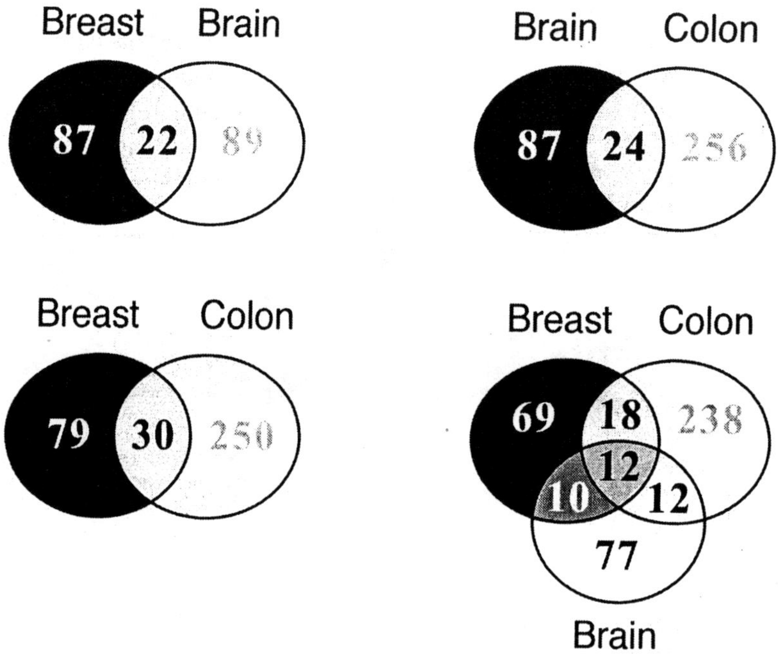

Figure 3. Venn diagram depicting tumor endothelial gene expression overlap (known and novel genes).

umbilical vein endothelial cells), HMVEC (human microvascular endothelial cells), and EPC (human endothelial precursor cells).[10] SAGE libraries were available for several cell culture conditions, including VEGF-stimulated and unstimulated cells. Many of the genes/mRNA expressed at high levels in the tumor endothelial cells isolated from fresh human tumor specimens were either not expressed or were expressed at very low levels in HUVEC and HMVEC under stimulated and unstimulated conditions.

The EPC developed by driving AC133+/CD34+ human bone marrow progenitor cells toward an endothelial cell differentiation in cell culture were a better model for tumor endothelial cells than were HUVEC and HMVEC.[10] Analysis of several cell surface markers by flow cytometry showed that EPC, HUVEC, and HMVEC had similar expression of P1H12, VEGFR2, and endoglin, but that EPC had much lower expression of ICAM1, ICAM2, VCAM1, and thrombomodulin than did HUVEC and HMVEC. The EPC generated can form tubes/networks on Matrigel, migrate through porous membranes, and invade through thin layers of Matrigel similarly to HUVEC and HMVEC. However, in a co-culture assay using human SKOV3 ovarian cancer cell clusters in collagen as a stimulus for invasion through Matrigel, EPC were able to invade into the malignant cell cluster while HMVEC were not able to invade the malignant cell cluster. *In vivo*, a Matrigel plug assay where EPC were suspended in the Matrigel allowed tube/network formation by EPC to be carried out in a murine host.

EPC appear to represent a more immature endothelial cell or a more de-differentiated endothelial cell than do HUVEC and HMVEC, and thus provide a more accurate mimic of tumor endothelial cells. EPC function well in cell-based assays, including proliferation, tube formation, migration, and invasion. EPC from several donors express targets identified by studying tumor endothelial cells and thus may represent an improved or second generation model cell system that can be used to study and screen potential antiangiogenic therapeutics. EPC may be a better model of human tumor endothelial cells than HUVEC and HMVEC and, thus, may provide an improved cell-based model for second generation antineoplastic, antiangiogenic drug discovery.

3. *IN VIVO* MODELS

Many of the models that have traditionally been used in the discovery and development of antiangiogenic therapeutics for cancer have been rodent models of neo-angiogenesis. These models generally involve the implantation of a single angiogenic factor-driven stimulus in a rodent and the observation of neovascular growth toward or into the stimulus. These models are a low hurdle for potential antiangiogenic agents, and neoangiogenesis in these models can frequently be completely inhibited.

The most common tumor models used in the field are transplantable syngeneic tumors and human tumor xenografts.[11] However, with both of these types of models, the angiogenic molecular targets and the angiogenic processes are all rodent. There are several approaches to growing vasculature that are, at least in part, human in a rodent host. These include producing bone marrow ablation in a group of recipient animals and rescuing them with human bone marrow. When human tumor xenografts are grown in these animals, the vasculature of the tumors is in part made up of human endothelial cells. Secondly, human foreskin can be xenografted onto Scid mice. After the skin graft

In Vivo Models

Neoangiogenesis assays:
- **Corneal micropocket - rat or mouse**
- **Matrigel plug**
- **Foam sponge**

Tumor models:
- **Immunocompetent, syngeneic mouse tumors**
- **Transfected (or infected) syngeneic marrow transplant**
- **Allogeneic marrow transplant into -**
  - **immunocompetent mice/murine tumor**
  - **SCID mice/human tumor xenograft**
- **Human foreskin xenograft - SCID mice/human tumor xenograft**
- **Transgenic mouse with target expression in tumor vessels**
- **Endpoints: tumor response & vessel counts**

Figure 4. *In vivo* models for antiangiogenic therapies.

has established, a human tumor can be implanted in the grafted skin and the vasculature of the tumor nodule that grows will be in part made up of human endothelial cells. These points are important because it is often the compound or treatment agent that performs best in these rodent-based efficacy models that is selected to be the clinical candidate. Every effort should be made to select a compound optimized to the human target and not to the mouse or rat homolog. The ideal model might be a host animal that has had the mouse (or rat) gene for the molecular target knocked out and has had the human gene for the molecular target genetically engineered in to be expressed in the vasculature.

It can be demonstrated in cell culture systems that malignant cells maintained in a hypoxic atmosphere secrete angiogenic growth factors into the medium.[12-15] These same angiogenic growth factors can be measured in circulating blood of animals bearing human tumor xenografts. The protein kinase C inhibitor, LY317615, was selected as an antiangiogenic agent for clinical development. The acyclic bisindolylmaleimide, LY317615, was identified as a potent selective inhibitor of protein kinase Cß. This compound, LY317615, was tested in culture and *in vivo* in the rat corneal micropocket and in the SW2 small cell lung carcinoma human tumor xenograft. In cell culture, LY317615 was a more potent inhibitor of VEGF-stimulated HUVEC proliferation (IC_{50} = 150 nM, 72 h) than of human SW2 small cell lung carcinoma cell proliferation (IC_{50} = 3.5 μM, 72 h). In addition, this compound decreased the secretion of VEGF into the medium by tumor cells. Using the *in vivo* neoangiogenesis corneal micropocket assay in rats, treatment of the animals with LY317615 appeared to completely block blood vessel growth and, when administered orally twice daily for 10 days, LY317615 markedly decreased the neoangiogenesis induced by VEGF or bFGF. To assess antitumor efficacy, LY317615 was administered orally twice daily on days 14-30 post-tumor implantation to nude mice bearing SW2 xenograft tumors. The number of countable intratumoral vessels was decreased in a dose-dependent manner, reaching as low as one-quarter the number in the control tumors. The decrease in intratumoral vessels was paralleled by increases in tumor growth delay. Treatment of the tumor-bearing animals with paclitaxel or carboplatin followed by treatment with LY317615 resulted in a 2.5- to 3.0-fold increase in tumor growth delay compared with the standard chemotherapeutic agents alone.

In mice bearing human tumor xenografts, such as the SW2 small cell lung carcinoma growing subcutaneously in a hind leg of a nude mouse, treatment with LY317615 (30 mg/kg orally twice per day on days 14-30 post-tumor implantation) resulted in a decrease in VEGF in circulating blood (Fig. 5A). The decreased levels of VEGF became significant after 5 days of treatment. Surprisingly, the VEGF levels stayed 3- to 4-fold lower in the treated animals than the VEGF levels in the controls for the duration of the experiment until some animals were removed from the study due to tumor size. On the other hand, the tumor volumes in the LY317615-treated animals did not differ from those of the control animals until after day 30 (Fig. 5B). Tumors from these animals were collected and intratumoral vessels were counted. The tumors collected after 9 days of LY317165 administration and were stained for CD31. The mean number of intratumoral vessels present in the control SW2 tumors was 20.4 ± 8.3 and the mean number of intratumoral vessels present in the LY317615-treated tumors was 12.2 ± 8.6. While there was a 40% decrease in the number of countable intratumoral vessels, this decrease was not sufficient to result in a measurable decrease in tumor growth. LY317615 represents a new approach to antiangiogenic therapy in cancer, blocking multiple growth factor signaling pathways in endothelial cells with a single agent. LY317615 is now in Phase II clinical testing.

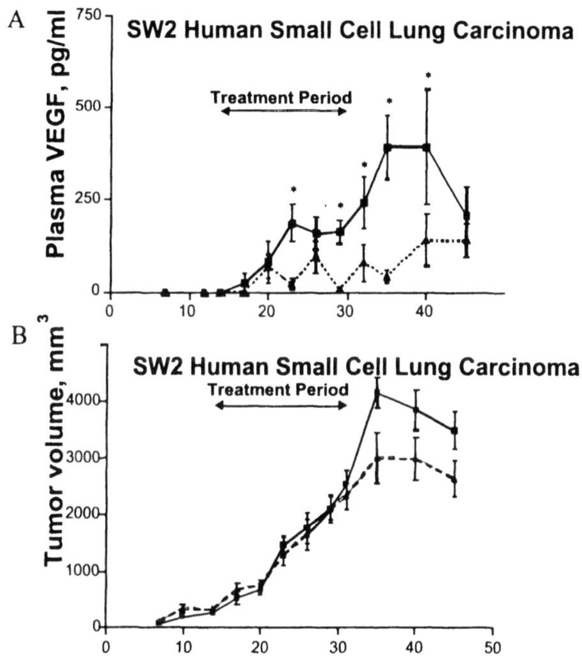

Figure 5. Effect of LY317615 administration on SW2 xenograft tumors on: **A**. circulating levels of VEGF; **B**. tumor volume.

4. REFERENCES

1. C. M. Doll, M. Milosevic, M. Pintilie, R. P. Hill, and A. W. Fyles, Estimating hypoxic status in human tumors: A simulation using Eppendorf oxygen probe data in cervical cancer patients, *Int. J. Radiat. Oncol. Biol. Phys.* **55**(5), 1239-1246 (2003).
2. A. Fyles, M. Milosevic, D. Hedley, M. Pintilie, W. Levin, L. Manchul, and R. P. Hill, Tumor hypoxia has independent predictor impact only in patients with node-negative cervix cancer, *J. Clin. Oncol.* **20**(3), 680-687 (2002).
3. J.-Y. Wang, K.-Y. Chen, J.-T. Wang, J.-H. Chen, J.-W. Lin, H.-C. Wang, L.-N. Lee, and P.-C. Yang, Outcome and prognostic factors for patients with non-small cell lung cancer and severe radiation pneumonitis, *Int. J. Radiat. Oncol. Biol. Phys.* **54**(3), 735-741 (2002).
4. J. Dunst, T. Kuhnt, H. G. Strauss, U. Krause, T. Pelz, H. Koelbl, and G. Haensgen, Anemia in cervical cancers: impact on survival, patterns of relapse, and association with hypoxia and angiogenesis, *Int. J. Radiat. Oncol. Biol. Phys.* **56**(3), 778-787 (2003).
5. B. Movsas, J. D. Chapman, A. L. Hanlon, E. M. Horwitz, W. H. Pinover, R. E. Greenberg, C. Stobbe, and G. E. Hanks, Hypoxia in human prostate carcinoma: An Eppendorf PO2 study, *Am. J. Clin. Oncol.* **24**(5), 458-461 (2001).
6. P. Subarsky, and R. P. Hill, The hypoxic tumor microenvironment and metastatic progression, *Clin. Exp. Metastasis* **20**, 237-250 (2003).
7. S. A. Holden, Y. Emi, Y. Kakeji, D. Northey, and B. A. Teicher, Host distribution and response to antitumor alkylating agents of EMT-6 tumor cells from subcutaneous tumor implants, *Cancer Chemother. Pharmacol.* **40**, 87-93 (1997).
8. B. A. Teicher, G. Ara, S. R. Keyes, R. S. Herbst, and E. Frei III, Acute in vivo resistance in high-dose therapy, *Clin. Cancer Res.* **4**, 483-491 (1998).

9. B. St. Croix, C. Rago, V. Velculescu, G. Traverso, K. E. Romans, E. Montgomery, A. Lai, G. J. Riggins, C. Lengauer, B. Vogelstein, and K. W. Kinzler, Genes expressed in human tumor endothelium, *Science* **289**(5482), 1197-1202 (2000).
10. R. G. Bagley, J. Walter-Yohrling, X. Cao, W. Weber, B. Simons, B. P. Cook, S. D. Chartrand, C. Wang, S. L. Madden, and B. A. Teicher, Endothelial precursor cells as a model of tumor endothelium: characterization and comparison with mature endothelial cells, *Cancer Res.* **63**(18), 5866-5873 (2003).
11. Y. Kakeji, and B. A. Teicher, Preclinical studies of the combination of angiogenic inhibitors with cytotoxic agents, *Invest. New Drugs* **15**, 39-48 (1997).
12. K. A. Keyes, L. Mann, K. Cox, P. Treadway, P. Iversen, Y.-F. Chen, and B. A. Teicher, Circulating angiogenic growth factor levels in mice bearing human tumors using Luminex multiplex technology, *Cancer Chemother. Pharmacol.* **51**, 321-327 (2003).
13. K. Keyes, K. Cox, P. Treadway, L. Mann, C. Shih, M. M. Faul, and B. A. Teicher, An in vitro tumor model: analysis of angiogenic factor expression after chemotherapy, *Cancer Res.* **62**, 5597-5602 (2002).
14. B. A. Teicher, K. Menon, E. Alvarez, E. Galbreath, C. Shih, and M. M. Faul, Antiangiogenic and antitumor effects of a protein Kinase Cb inhibitor in human T98G glioblastoma multiforme xenografts, *Clin. Cancer Res.* **7**, 634-640 (2001).
15. K. A. Keyes, L. Mann, M. Sherman, E. Galbreath, L. Schirtzinger, D. Ballard, Y.-F. Chen, P. Iversen, and B. A. Teicher, LY317615 decreases plasma VEGF levels in human tumor xenograft bearing mice, *Cancer Chemother. Pharmacol.* **53**(2), 133-140 (2003).

GENETIC OXYGEN SENSOR:
GFP as an indicator of intracellular oxygenation

Eiji Takahashi, Tomohiro Takano, Aya Numata, Natsuho Hayashi, Satoshi Okano, Osamu Nakajima, Yasutomo Nomura, and Michihiko Sato

Abstract: We report in this article a new method for *in vivo* oxygen measurement using green fluorescence protein (GFP). COS7 cells were transiently transfected with an expression vector, pCMX-GFP, using a polyethylenimine reagent and cultured for 48 hrs. After exposure of the cell to anoxic gas ($O_2 < .001\%$), a 1 min illumination of the cell to strong 470-490 nm light evoked a significant red fluorescence (excitation 520-550 nm, emission >580 nm) that had been negligible before the photoactivation. This red shift of (green) GFP fluorescence was never observed in normoxia. We then examined the validity of this method in transgenic mice in which GFP is stably expressed (green mice). All the ventricular myocytes isolated from the green mice showed significant green fluorescence, although the intensity was ~1/200 of the transiently GFP-expressing COS7 cells. The photoactivation in anoxia increased the red fluorescence in these cells, but the magnitude was much smaller than expected. In summary, GFP can be used as an *in situ* probe for hypoxia. In GFP-expressing transgenic animals, *in vivo* imaging of anoxic loci with a submicron spatial resolution may be possible.

1. INTRODUCTION

In vivo measurement of oxygen concentration in various tissues is crucial for understanding the complex interaction between cellular oxygen supply and mitochondrial oxidative metabolism.[1, 2] Importantly, these oxygen measurements require a subcellular spatial resolution because significant oxygen concentration gradients may be established from the capillary blood to mitochondria, particularly those located near the center of cell.[3, 4] Previously, we utilized myoglobin in the rat ventricular myocyte as an intrinsic *in situ* oxygen probe.[3, 5] Although spectrophotometric determination of myoglobin oxygen saturation in isolated single cardiomyocytes provides a spatial resolution sufficient to visualize intracellular radial gradients of oxygen concentration, this approach is only applicable to red muscles with abundant myoglobin.

In 1997, Elowitz et al.[6] reported that, in *E. coli* transformed with a plasmid which expresses a green fluorescence protein (GFP, a fluorescent protein originally isolated from the *Aequorea victoria*), GFP fluorescence may turn red after a brief (2-60 sec) illumination of 475-495 nm light when the bacteria are sealed between microscope slides and left for a prolonged time (~25 min). The red shift of GFP fluorescence was also demonstrated in the transformed *E. coli* when photoactivation was conducted after an oxygen scavenger was added to the suspension medium. These observations suggest the possibility that GFP might be used as an *in situ* oxygen probe if this molecule is efficiently expressed in cells.

2. GFP AS AN OXYGEN INDICATOR IN CULTURED CELLS

To test whether the red shift under anoxic atmosphere can be demonstrated in mammalian cells, cultured COS7 cells were transiently transfected with an expression vector, pCMX-SAH/Y145F (pCMX-humanized GFP-S65A/Y145F), using a polyethylenimine (jetPEI, PolyPlus-transfection) and cultured for 48 hrs. Cultured cells were placed in an airtight measuring cuvette and superfused with a humidified gas with various oxygen concentrations at 36.5°C. A 16-bit CCD camera (SV512, PixelVision) captured the green and red fluorescence images of the cells by using NIBA (excitation 470-490 nm, emission 515-550 nm) and WIG (excitation 520-550 nm, emission >580 nm) filter cubes in inverted microscope (IX70, Olympus), respectively. Photoactivation was conducted by illuminating the cells with 470-490 nm light (neutral density filter removed) for 1 min.

In normoxic cells, red fluorescence was almost absent (<0.6% of the green fluorescence) irrespective of photoactivation. In contrast, in anoxic cells (Table 1), photoactivation significantly increased the red fluorescence to 23% and 80% of the initial green fluorescence level at extracellular oxygen concentrations of 1% and <0.001%, respectively (Fig. 1). After photoactivation, the green fluorescence decreased to 40% of the initial level (Table 1), suggesting a shift of GFP fluorescence spectrum as demonstrated by Elowitz et al.[6]

Table 1. Fluorescence intensities in anoxic atmosphere

		Green fluorescence		Red fluorescence	
		Before PA	After PA	Before PA	After PA
COS7	Transient transfection	~300,000	~120,000	~1,800	~240,000
Single	WT	44	38	250	249
cardiomyocytes	TG	1624	1233	381	524

Fluorescence values are in arbitrary units. PA: photo activation. WT: wild-type. TG: transgenic. Green and red fluorescences were determined for excitation 470-490 nm/emission 515-550 nm, and excitation 520-550 nm/emission >580 nm, respectively. Fluorescence intensities of WT single cardiomyocytes represent autofluorescence.

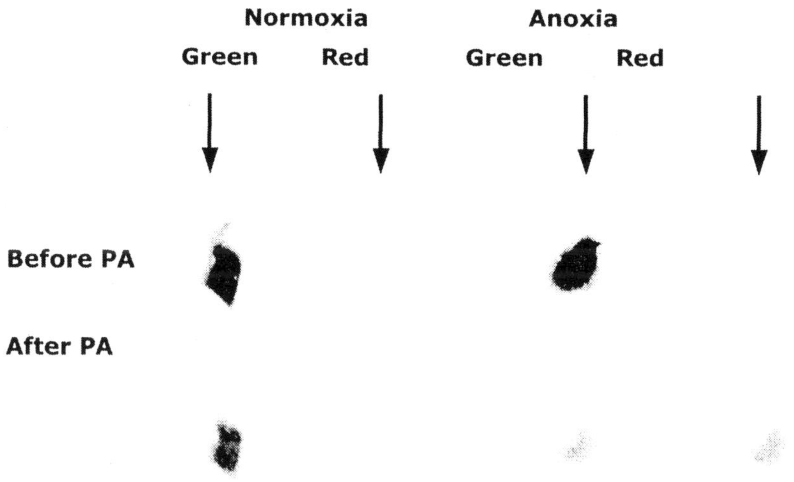

Figure 1. Changes in fluorescence before and after a brief photoactivation (PA) by a blue light (470-490 nm, 1 min) in GFP-expressing COS7 cells. Cell suspension was superfused with 15% O_2 (normoxia) and/or < 0.001% O_2 (anoxia), respectively.

3. GFP AS AN OXYGEN INDICATOR IN CARDIOMYOCYTES ISOLATED FROM A TRANSGENIC MOUSE

We further examined the validity of this technique for anoxia detection in transgenic (TG) mice that stably express the enhanced GFP (EGFP, Clontech). Single ventricular myocytes were isolated from the TG mouse using the collagenase digestion method. All of the ventricular myocytes isolated from the green mice showed significant green fluorescence, although the intensity was only ~1/200 of the transiently GFP-expressing COS7 cells (Table 1). Thus, we had to prolong exposure duration of the CCD camera to detect such weak GFP fluorescence. As expected, photoactivation of single cardiomyocytes for 1 min increased the red fluorescence while decreasing the green fluorescence, but the magnitude of the red shift was very small (Table 1). This appeared to arise from strong red autofluorescence in cardiomyocytes (evident in wild type cardiomyocytes, Table 1) that may have masked the red shift of GFP fluorescence.

To exclude effects of autofluorescence, we calculated the difference between red fluorescence images before and after a photoacivation (ΔF_{535}). As shown in Figure 2, photoactivation for 1 min in anoxic atmosphere produced significant augmentation of the red fluorescence in single cardiomyocytes isolated from the transgenic mouse. For more precise quantification of the red shift, each ΔF_{535} image was normalized to its green fluorescence ($\Delta F_{535}/F_{480}$). Figure 3 demonstrates appearance of the red fluorescence in

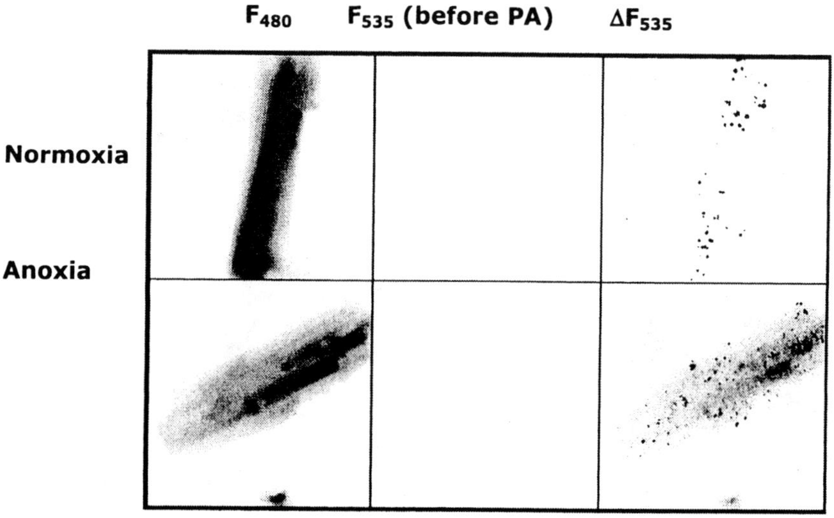

Figure 2. Red shift of GFP fluorescence after a brief photo activation (PA) in single cardiomyocytes isolated from GFP-expressing transgenic mice. F_{480} and F_{535} represent green and red fluorescences, respectively. ΔF_{535} is the difference in red fluorescence images before and after photoactivation.

anoxia. Photoactivation in anoxia converted approximately 35% of GFP to red fluorescent GFP that was quite distinct from cardiomyocytes photoactivated in normoxia (15% oxygen superfusion).

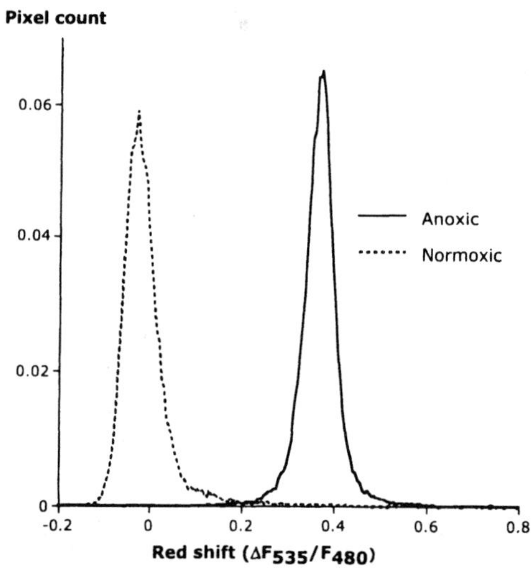

Figure 3. Red shift of GFP fluorescence after a brief photoactivation in single cardiomyocytes isolated from GFP-expressing transgenic mice. F_{480} and F_{535} represent green and red fluorescences, respectively. ΔF_{535} is the difference in red fluorescence images, before and after photoactivation. Pixel count was normalized to the cell area.

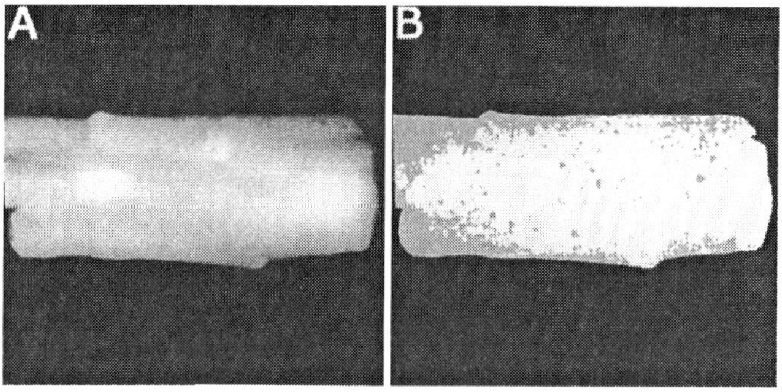

Figure 4. Representative data demonstrating the hypoxic core in a single cardiomyocyte with elevated oxygen consumption. A single 1.0 μM CCCP-treated cell was superfused with 3% O_2 gas. **A**. GFP fluorescence (green fluorescence). **B**. Reconstructed image of the same cell in which regions with the red shift of GFP fluorescence (i.e. anoxic regions) are indicated in white.

4. VISUALIZATION OF RADIAL OXYGEN CONCENTRATION GRADIENTS IN SINGLE CARDIOMYOCYTES USING GFP AS AN OXYGEN INDICATOR

Previously, we demonstrated that significant radial gradients of oxygen concentration may be established within a single cardiomyocyte if mitochondrial oxygen consumption was moderately elevated.[3] This was demonstrated again with the new technique. Mitochondrial respiration was stimulated with an uncoupler of oxidative phosphorylation (carbonyl cyanide *m*-chlorophenylhydrazon, CCCP) while superfusing the cell with a hypoxic gas (3% O_2). In the steady state, photoactivation was conducted for 1 min, fluorescence images were captured, and $\Delta F_{535}/F_{480}$ image was reconstructed. As shown in Figure 4, heterogeneities in the red shift of the GFP fluorescence were demonstrated in the intracellular space. Furthermore, distribution of the red shift appeared to depend upon the shape of the cell. These results are compatible with our previous findings of an hypoxic core in actively respiring single cardiomyocytes,[1, 3] and suggest that, at elevated oxygen consumption, intracellular diffusion of oxygen may limit oxygen availability at the mitochondria, particularly those located in the cell core.

5. DISCUSSION

Green fluorescent protein can be used as an *in situ* probe for hypoxia/anoxia with such a high spatial resolution that imaging of intracellular heterogeneity in oxygen concentration may be possible in single cardiomyocytes. One of the issues in applying this technique in single cardiomyocytes isolated from the transgenic mouse was the relatively low level of GFP expression in these cells. To detect such weak GFP fluorescence in single cardiomyocytes, sensitivity of the optical measurement had to be increased. This also increased the signal arising from non-specific artifactual red fluorescence that could be otherwise neglected in the case of GFP-expressing cultured COS7 cells. Thus, the present technique might preferably be used in organs, where GFP fluorescence signal may be much stronger, rather than isolated single cells.

Potentially, the present genetic technique for detection of hypoxia/anoxia is very powerful and may allow for various new applications. For example, site-directed expression of GFP is possible by using subcellular localizing vectors. In this case, one could specifically determine oxygenation status of target subcellular fractions, such as plasma membrane, mitochondria, nucleus, and so on.

6. ACKNOWLEDGMENTS

Part of this work was supported by JSPS KAKENHI 15390061.

REFERENCES

1. E. Takahashi, and K. Asano, Mitochondrial respiratory control compensates for slow oxygen diffusion in cardiomyocytes, *Am. J. Physiol.* **283**, H871-H878 (2002).
2. E. Takahashi, H. Endoh, and K. Doi, Intracellular gradients of O_2 supply to mitochondria in actively respiring single cardiomyocyte of rats, *Am. J. Physiol.* **276**, H718-H724 (1999).
3. E. Takahashi, K. Sato, H. Endoh, Z. L. Xu, and K. Doi, Direct observation of radial intracellular Po_2 gradients in a single cardiomyocyte of the rat, *Am. J. Physiol.* **275**, H225-H233 (1998).
4. K. Groebe, An easy-to-use model for O_2 supply to red muscle. Validity of assumptions, sensitivity to errors in data. *Biophys. J.* **68**, 1246-1269 (1995).
5. E. Takahashi, and K. Doi, Visualization of oxygen level inside a single cardiac myocyte, *Am. J. Physiol.* **268**, H2561-H2568 (1995).
6. M. B. Elowitz, M. G. Surette, P.-E. Wolf, J. Stock, and S. Leibler, Photoactivation turns green fluorescent protein red, *Curr. Biol.* **7**, 809-812 (1997).

EFFECTS OF INSULIN AND CATECHOLAMINES ON INOTROPY AND OXYGEN UPTAKE

Best perfusate for skeletal muscles

Hisaharu Kohzuki, and Hidemi Fujino

Abstract: The purpose of this study is to identify the best perfusate after blood for maintaining skeletal muscle inotropy, muscle peak oxygen consumption (peak VO_2), and oxygen consumption at rest (resting VO_2) in *in situ* isolated canine gastrocnemius-plantaris muscle. Rejuvenated red cells suspended in perfusate at hematocrit 30% and 45%, perfusate contained insulin (100 $\mu U \cdot ml^{-1}$), adrenalin (0.3 and 3 $ng \cdot ml^{-1}$), and noradrenaline (3 $ng \cdot ml^{-1}$). Insulin significantly augmented resting VO_2 and contracting muscle peak VO_2, and developed isometric twitch tension at 4 Hz, compared with control. Insulin-induced increase in resting muscle VO_2 was abrogated by catecholamines. In addition to insulin and catecholamines, the developed twitch tension increased significantly by 178% with the accompanied increase in flow rate. O_2 cost (peak VO_2 / tension) significantly decreased by 52%. The developed tension did not correlate with O_2 delivery but with flow rate and peak VO_2 of contracting muscle. We successfully identified the characteristics of the best perfusate after blood. Our results suggest that the positive inotropy by insulin and catecholamines is attributed partly to an O_2 delivery-independent increase in flow to contracting muscle and redistribution of flow within the contracting muscle, which suffered from low perfusion by perfusate containing rejuvenated red cells.

1. INTRODUCTION

We have previously reported that epinephrine and norepinephrine are required for the development of normal twitch tension in *in situ* canine gastrocnemius muscle perfused with rejuvenated red blood cell-containing perfusate.[1, 2] Our preparations also showed weak autoregulation of blood flow rate and perfusion pressure.[3] Dawson et al.[4] have recently reported that insulin increases blood flow and vascular recruitment in the rat hind-leg muscle, suggesting that insulin can switch flow from non-nutritive to the nutritive route in the muscle.[5] We can expect a positive inotropic effect for insulin on muscle perfused with red blood cell-containing perfusate by increasing flow distribution within the muscle.

There is convincing evidence that epinephrine potentiates muscle twitch in skeletal muscles consisting mostly of fast-twitch muscle fibers.[6] Although the mechanism underlying the inotropic effects of epinephrine and β-adrenergic agonists is not well understood, the inotropic effects of the latter are observed solely in fast skeletal muscles.[7] It is an intriguing issue whether the positive inotropy increases peak VO_2 during contractions and O_2 cost (peak VO_2/developed twitch tension) or not. O_2 cost can be thought to be inversely related to economy that is proportional with myofilament length and inversely proportional with actomyosin ATPase, sarcoplasmic reticulum (SR) Ca^{2+}-ATPase, and the number of Ca^{2+} transporting sites in the SR.[8] The present experimental study was designed to identify the characteristics of the best perfusate after blood and clarify the effects of insulin and catecholamines on VO_2 and O_2 cost in *in situ* canine muscle perfused with perfusion medium solution containing rejuvenated red blood cells.

2. MATERIALS AND METHODS

Eighty mongrel dogs (weight, 4.3 to 13.2 kg) were anesthetized with pentobarbital sodium (30 mg•kg^{-1}, i.v.), with additional doses given as necessary. The animals were artificially ventilated with room air through a cuffed endotracheal tube by a respirator. The left gastrocnemius-flexor digitorum superficialis muscle group was surgically prepared using the method described previously by our laboratory.[1] Briefly, the achilles tendon was severed close to the calcaneus and attached to a force transducer (model WBS-20K, Showa, Tokyo, Japan). The length of the muscle was adjusted at 10 g force•g muscle^{-1}. The sciatic nerve was sectioned and placed on stimulating electrodes.

The method used for perfusion was almost the same as previously reported.[1] Briefly, an arterial cannula, which was connected to a dual-tube heat exchanger (37°C), was inserted into the popliteal artery. After cannulation, the muscle was perfused for 30 min under resting conditions using a peristaltic pump. The muscle was then stimulated via the sciatic nerve (4 V, 0.2 ms duration) to contract isometrically at a rate of 4 Hz for 3 min and, simultaneously the pump speed was gradually increased. Flow was determined by weighing the venous effluent collected over a 5-10 s period in a tarred glass vessel. Packed red blood cells were prepared by the method described previously. The hematocrit (Ht) of the suspension was adjusted to 30% or 45%.

The washing solution used in the present experiment was passed through a filter of 1-μm pore size (Advantec, Tokyo). The base excess of the perfusate was adjusted to zero with 7% bicarbonate. Perfusate-gas equilibration was performed using a bubble oxygenator for 15 min in 1,500 ml perfusate at 37°C. The gas phase above the equilibrated blood in the reservoir was replaced by N_2 gas. The blood gas values of the equilibrated blood in the reservoirs at 4°C did not change throughout the experiment. The perfusion blood was not recirculated.

Venous effluents were collected at rest and at 3 min of contraction time. Arterial samples were drawn directly from the reservoir. Oxygen saturation of hemoglobin (SO_2) of all samples was measured with an OSM-2 (Radiometer, Copenhagen, Denmark) and their pH, PCO_2, and PO_2 were measured with a blood-gas analyzer (BMS-Mk2, Radiometer). VO_2 was calculated using the Fick principle. The flow rate, O_2 delivery, resting VO_2, peak VO_2, and muscle tension developed were normalized for 100 g wet weight of muscle. Fatigue rate (%) was calculated and represented the difference between peak tension and tension at three minutes, divided by peak tension.

We performed five series of experiments. *Series 1 (control, C):* The perfusate was prepared at Ht 45% without insulin or catecholamines. *Series 2 (Insulin run, I):* A perfusate containing insulin (Humulin R, Shionogi, Osaka) was prepared with an insulin concentration of 100 $\mu U \cdot ml^{-1}$ perfusate at Ht 45%. *Series 3 (Low-dose epinephrine run, NE):* In these studies, we prepared a perfusate containing epinephrine (Bosmin, Daiichi Pharmaceutical, Tokyo) at a concentration of 0.3 $ng \cdot ml^{-1}$, norepinephrine (Nor-epinephrine, Sankyo Pharmaceutical, Tokyo) at 3 $ng \cdot ml^{-1}$, and insulin at 100 μU/ml at Ht 45%. *Series 4 (Low-dose epinephrine and low-Ht run, NE-L):* The perfusate was similar to that used in the Low-dose epinephrine run except Ht was set at 30%. *Series 5 (High-dose epinephrine and low-Ht run, E-L):* The perfusate contained epinephrine at a concentration of 3 $ng \cdot ml^{-1}$, norepinephrine at 3 $ng \cdot ml^{-1}$, and insulin at 100 $\mu U \cdot ml^{-1}$, with Ht at 30%. All drugs were added to the perfusate just before pump perfusion was started. All experimental data are presented as mean ± SD. A one-way analysis of variance (ANOVA) and repeated measures ANOVA were performed using Statistica (StatSoft, Tulsa, OK) software. Duncan's *post hoc* test was used to identify differences between groups at $P < .05$.

3. RESULTS

Muscle flow rate was regulated by keeping the perfusion pressure adjusted to 100 ± 5, 106 ± 7, and 100 ± 5 mm Hg in C, I and NE, respectively. There was no significant difference in the perfusion pressure among the five series. Muscle flow rate in NE-L and E-L was regulated by adjusting the perfusion pressure to 77 ± 5 and 82 ± 4 mm Hg, respectively. The perfusion pressure was not significantly different between these two series, but it was significantly lower than other series ($P < .05$). Resting muscle flow was significantly higher in I of insulin run than C (Table 1, $P < .05$). Contracting muscle

Table 1. O_2 supply values for resting and contracting gastrocnemius muscle perfused with pump at constant perfusion pressure.

At Rest	C	I	NE	NE-L	E-L
Flow rate	7.8 ± 2.5	10.51 ± 3.3*	11.2 ± 2.2	11.8 ± 2.7	11.6 ± 4.3
O_2 delivery	1.88 ± 0.59	2.56 ± 0.74$^{*\,b}$	2.54 ± 0.56*	1.85 ± 0.43^d	1.77 ± 0.66
Resting VO_2/flow	4.76 ± 1.74	6.91 ± 3.32^f	3.57 ± 1.33	3.87 ± 1.43	3.51 ± 1.39
Contraction					
Flow rate	87.6 ± 33.1	76.8 ± 25.3^a	114.0 ± 26.5	112.2 ± 24.4^e	159.7 ± 48.4*
O_2 delivery	21.15 ± 8.03	18.68 ± 5.83^c	25.58 ± 5.53	17.9 ± 3.84^c	24.19 ± 7.03
Peak VO_2/flow	9.79 ± 3.79	13.74 ± 2.88$^{*\,b}$	11.96 ± 1.30	10.53 ± 1.29*	7.16 ± 1.84^f

Values are means ± SD. O_2 delivery = $[O_2] \times$ flow, $ml \cdot min^{-1} \cdot 100g^{-1}$; flow, $ml \cdot min^{-1} \cdot 100g^{-1}$; VO_2/flow, $ml \cdot ml^{-1} \cdot 10^{-2}$. *$P < .05$, compared with C. $^a P < .05$, between I and NE, and NE-L and E-L. $^b P < .05$, between I, and NE-L and E-L. $^c P < .05$, between NE-L, and NE and E-L. $^d P < .05$, between NE-L and NE. $^e P < .05$, between NE-L and E-L. $^f P < .05$, compared with others.

flow was successfully increased through experimental protocols of C to E-L (Table 1), in which we attempted developing muscle tension to a similar value observed in muscle perfused with blood. According to our low-Ht run, O_2 delivery did not increase through experimental protocols of C to E-L (Table 1).

Resting muscle VO_2 for insulin run of I was significantly higher than other series. The addition of catecholamines to the perfusate in NE abrogated an insulin-induced increase in VO_2 observed in I. Resting muscle VO_2 (0.35-0.44 $ml \cdot min^{-1} \cdot 100g^{-1}$) was not significantly different among the different series of experiments except for I. The resting muscle VO_2 was similar to the value reported previously in *in situ* gastrocnemius muscle perfused with blood.[9]

The highest value of twitch tension at the third-minute of contraction time was noted in E-L, which increased by 178% compared with that observed in the control. Tension of gastrocnemius perfused with a perfusion medium solution containing red cells was equivalent to that observed in *in situ* gastrocnemius muscle perfused with blood. The fatigue rate (%) was 27 ± 13 for C, 19 ± 13 for I, 10 ± 13 for NE, 10 ± 8 for NE-L, and 9 ± 10 for E-L. The values of series 3, 4, and 5 were significantly lower than that of the control ($P < .05$, each). The highest value of peak VO_2 was observed in NE, though it was not significantly different from that in NE-L and E-L (Fig. 1A, $P > .05$). Peak VO_2 in I, but not in E-L, was significantly higher than the control. The O_2 cost was lower in I to E-L, relative to the control; the lowest value was observed in E-L ($P < .05$), while that of I was not different from the control (Fig. 1B). Insulin significantly increased peak VO_2/flow compared with the control (Table 1, $P < .05$). Peak VO_2/flow of the control was significantly different from those of I and E-L.

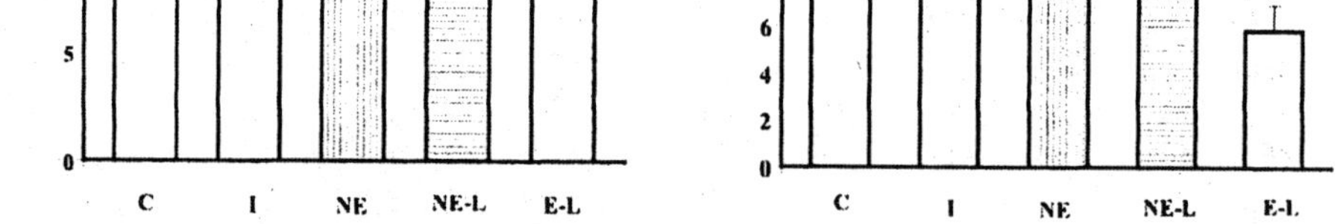

Figure 1. A: Peak oxygen consumption (peak VO_2) at the third minute of contraction time in *in situ* gastrocnemius muscle perfused with perfusate containing insulin and catecholamines at constant perfusion pressure in the five series of experiments. Values are mean ± SD. * $P < .05$, compared with the control (C), # $P < .05$. **B**: O_2 cost (peak VO_2/tension at the third min) in *in situ* gastrocnemius muscle perfused with perfusate containing insulin and catecholamines at constant perfusion pressure among the five series of experiments. Values are mean ± SD. *$P < .05$, compared with the control (C), # $P < .05$. For definition of different series, see Materials and Methods section.

4. DISCUSSION

The major findings of the present study were the following. Firstly, we can identify the best perfusate used for *in situ* canine muscle instead of blood. The positive inotropic effects of insulin and catecholamines were observed in *in situ* muscle perfused with perfusion medium mixed with rejuvenated red cells. Secondly, low O_2 cost by insulin- and catecholamine-perfused muscle developed a tension equivalent to that generated by muscle perfused with a perfusion medium free of these agents. Thirdly, the sole addition of insulin was able to augment the developed tension.

Several studies[10, 11] using epinephrine or related amines have confirmed the inotropic effects of epinephrine, similar to our results reported in the present study. It has been suggested that the inotropic effects of epinephrine are dependent on β-adrenergic receptors.[6] The inotropic effects of β-agonists are observed solely in skeletal muscles consisting mostly of fast-twitch muscle fibers. In energetic points, the O_2 cost of fast twitch muscle augmented by epinephrine expects higher O_2 cost than slow twitch muscle because fast twitch muscle has higher actomyosin ATPase activity.[12] In fact, Richter et al.[11] reported that epinephrine increased O_2 uptake and the proportional positive inotropic effect. In the present study, O_2 cost decreased in muscles perfused with insulin and catecholamines. Thus, other factors should be considered in relation to the positive inotropy. Holmberg and Waldeck[13] reported that insulin partly restored the twitch tension of both the soleus, a slow-contracting and the extensor digitorum longus, a fast-contracting muscle when depressed by KCl, although insulin had a suppressing effect on the force of contraction of the soleus under normal conditions. They discussed that this effect of restoration may be due to the ability of insulin to stimulate the transport of Na^+ and K^+. Furthermore, both epinephrine and insulin can increase active Na^+-K^+ transport.[14] The role of Na^+-K^+ ATPase on muscle force is complex and still unclear.

Crow and Kushmerick[15, 16] suggested that myosin light-chain phosphorylation is associated with a decrease in the energy cost for contraction in fast twitch mouse muscle. They reported that myosin light-chain phosphorylation was associated with a decrease in the total isometric high-energy phosphate-splitting. However, the exact contribution of actomyosin ATPase to the total energy cost could not be established in their study. To our knowledge, the relationship between insulin and catecholamines and myosin light-chain phosphorylation in skeletal muscles has not yet been determined. Since we previously indicated that the regulatory mechanism(s) responsible for the staircase phenomenon could be modulated by limited O_2 availability,[1] the present findings of the effects of catecholamines and insulin on the staircase phenomenon with increased peak VO_2 may be explained by the mechanism reported by Crow and Kushmerick.[15]

Barclay[17] reported that hyperperfusion decreased fatigue by a mechanism independent of increased O_2 and substrate delivery in canine gastrocnemius-plantaris *in situ* preparations stimulated at 5 Hz. It was suggested that hyperperfusion could prevent any flow maldistribution and thus alter the cellular environment with subsequent changes in tissue function and metabolism. Our results indicated that the fatigue rate during perfusion with insulin and catecholamines was significantly lower than the control. The flow rate was significantly different from the control only during perfusion with insulin and catecholamines (series 5). Thus, hyperperfusion does not seem to be the only factor responsible for the augmented, developed tension at the third-minute of contraction time.

In conclusion, we successfully identified the characteristics of the best perfusate after blood and quantified the effects of insulin and catecholamines on skeletal muscle inotropy, muscle peak VO_2, and resting VO_2 in the *in situ* isolated canine gastrocnemius-flexor digitorum superficialis muscle group using a perfusate containing rejuvenated red blood cells at constant perfusion pressure in resting and isometrically contracting at 4 Hz. Insulin significantly augmented resting VO_2 and contracting muscle peak VO_2, and developed twitch tension, compared with control. An insulin-induced increase in resting muscle VO_2 was abrogated by catecholamines. In addition to insulin and catecholamines, the developed twitch tension and flow increased significantly by 178% and 82%, and O_2 cost (peak VO_2/tension) significantly decreased by 52%. Improvement of O_2 cost was not observed for insulin perfusion. The developed tension did not correlate with O_2 delivery, but rather with flow rate and peak VO_2 of contracting muscle. Our results suggest that the mechanisms underlying positive inotropy induced by insulin and catecholamines are different from each other, and thus insulin does not effect O_2 delivery-independent increases in flow to increase muscle force.

REFERENCES

1. H. Kohzuki, Y. Enoki, K. Matsumura, S. Sakata, and S. Shimizu, Flow-dependent influence of high-O_2-affinity erythrocytes on peak VO_2 in exercising muscle in situ, *J. Appl. Physiol.* **80**(3), 832-838 (1996).
2. H. Kohzuki, Y. Enoki, S. Sakata, S. Shimizu, and K. Matsumura, Influence of high O_2 affinity red cells on developed tension-VO_2 relation in skeletal muscle, (*in Japanese*) *J. Physiol. Soc. Japan* **53**(Suppl), 13 (1990).
3. H. Kohzuki, Y. Enoki, S. Shimizu, and S. Sakata, High blood O_2 affinity and relationship of O_2 uptake and delivery in resting muscle, *Respir. Physiol.* **92**(2), 197-208 (1993).
4. D. Dawson, M. A. Vincent, E. J. Barrett, S. Kaul, A. Clark, H. Leong-Poi, and J. R. Lindner, Vascular recruitment in skeletal muscle during exercise and hyperinsulinemia assessed by contract ultrasound, *Am. J. Physiol. Endocrinol. Metab.* **282**(3), E714-E720 (2002).
5. M. G. Clark, S. Rattigan, L. H. Clerk, M. A. Vincent, A. D. H. Clark, J. M. Youd, and J. M. B. Newman, Nutritive and non-nutritive blood flow: rest and exercise, *Acta Physiol. Scand.* **168**(4), 519-530 (2000).
6. J. H. Williams, and W. S. Barnes, The positive inotropic effect of epinephrine on skeletal muscle: a brief review, *Muscle & Nerve* **12**(12), 968-975 (1989).
7. R. J. Murphy, P. F. Gardiner, G. Rousseau, M. Bouvier, and L. Beliveau, Chronic beta-blockade increases skeletal muscle beta-adrenergic-receptor density and enhances contractile force, *J. Appl. Physiol.* **83**(2), 459-465 (1997).
8. J. A. Rall, Energetic aspects of skeletal muscle contraction: Implications of fiber types, *Exercise Sport Sci. Rev.* **13**, 33-74 (1985).
9. M. Schmitt, P. Meunier, A. Rochas, and J. Chatonnet, Catecholamines and oxygen uptake in dog skeletal muscle in situ, *Pflügers Arch.* **345**(2), 145-158 (1973).
10. C. D. Marsden, and J. C. Meadows, The effect of epinephrine on the contraction of human muscle, *J. Physiol.* **207**(2), 429-448 (1970).
11. E. A. Richter, N. B. Ruderman, and H. Galbo, Alpha and beta-adrenergic effects on metabolism in contracting, perfused muscle, *Acta Physiol. Scand.* **116**(3), 215-222 (1982).
12. B. Folkow, and H. D. Halicka, A comparison between "red" and "white" muscle with respect to blood supply, capillary surface area and oxygen uptake during rest and exercise, *Microvasc. Res.* **1**, 1-14 (1968).
13. E. Holmberg, and B. Waldeck, The effect of insulin on skeletal muscle contractions and its relation to the effect produced by beta-adrenoceptor stimulation, *Acta Physiol. Scand.* **109**(2), 225-229 (1980).
14. J. A. Flatman, and T. Clausen, Combined effects of epinephrine and insulin on active electrogenic Na^+-K^+ transport in rat soleus muscle, *Nature* **281**(5732), 580-581 (1979).
15. M. T. Crow, and M. J. Kushmerick, Phosphorylation of myosin light chains in mouse fast-twitch muscle associated with reduced actomyosin turnover rate, *Science* **217**(4562), 835-837 (1982).
16. M. T. Crow, and M. J. Kushmerick, Myosin light chain phosphorylation is associated with a decrease in the energy cost for contraction in fast twitch mouse muscle, *J. Biol. Chem.* **257**(5), 2121-2124 (1982).
17. J. K. Barclay, A delivery-independent blood flow effect on skeletal muscle fatigue, *J. Appl. Physiol.* **61**(3), 1084-1090 (1986).

ADAPTATION TO CHRONIC HYPOXIA DURING DIET-INDUCED KETOSIS

Michelle A. Puchowicz, Douglas S. Emancipator, Kui Xu, Danielle L. Magness, Obinna I. Ndubuizu, W. David Lust, and Joseph C. LaManna

Abstract: It is recognized that brain oxygen deprivation results in increased glycolysis and lactate accumulation. Moreover, glucose metabolism is altered during starvation or diet, resulting in increased plasma ketones (acetoacetate + ß-hydroxybutyrate; BHB). We investigated glucose and lactate adaptation to hypoxia in concurrence with diet-induced ketosis. Male Wistar rats were fed standard (STD), ketogenic (high fat; KG), or carbohydrate-rich (low fat; CHO) diets for 3 wks and then exposed to hypobaric (0.5 ATM) or normobaric atmosphere for 3 wks while on their diets. Lactate, ketones, and glucose concentrations were measured in plasma (mM) and brain tissue (mmol/g). Plasma and tissue ketone levels were elevated up to 12-fold in the KG fed groups compared with other groups (STD and CHO), with the hypoxic KG group reaching the highest levels (2.6 ± 1.3 mM and 0.3 ± 0.1 mmol/g; mean ± SD). Tissue lactate levels in the hypoxic ketotic rats (4.7 ± 1.3 mM) were comparable with normoxic STD (5.0 ± 0.7 mM) and significantly lower (ANOVA $P < .05$) than the hypoxic STD rats (6.1 ± 1.0 mM). These data indicate that adaptation to hypoxia did not interfere with ketosis, and that ketosis during hypoxia may lower lactate levels in brain, suggesting decreased glycolysis or increased glucose disposal.

1. INTRODUCTION

It is recognized that brain oxygen deprivation results in increased glycolysis and glucose transporter (GLUT-1), lactate accumulation, and decreased pH compared with normoxic conditions.[1, 2] The pH paradox is a concept explaining the observation that neuronal activity, as well as environmental changes (e.g. hypoxia and ischemia), stimulate glycolysis to a much greater extent than oxygen consumption.[3] The role of increased glycolysis during mild hypoxia is to compensate for the decrease in tissue CO_2

brought about by hyperventilation, producing a neutral pH_i. Glucose metabolism is also altered in response to prolonged fasting or starvation, resulting in increased ketone (ß-hydroxybutyrate + acetoacetate) plasma levels (ketosis).[4, 5] Ketones are alternate energy substrates to glucose and are a result of carbon derived from fatty acid ß-oxidation. McKenna et al. have reported that astrocytes from adult rat brain slices oxidize BHB (ß-hydroxybutyrate) and lactate faster than glucose, whereas synaptic terminals oxidize BHB and glucose faster than lactate.[6]

Ketogenic (high fat) diets have been successfully used to induce ketosis and have been used as an alternative treatment for controlling seizures in patients, particularly children, since the 1920's.[7] Based on the effectiveness of the ketogenic diet, and the well known fact that brain can switch from glucose as the principle energy substrate to ketones[4] without altering oxygen consumption[8, 9] or cerebral metabolic rate for glucose,[10] we have developed a rat model of ketosis by feeding a high fat diet to study glucose and ketone metabolism in brain. This study was designed to investigate whether ketones or lactate concentrations are altered in brain during conditions where lactate is known to accumulate, such as during hypoxia.

2. MATERIALS AND METHODS

2.1. Rat Diets

The ketogenic (KG) and high carbohydrate (CHO) diets were purchased from Research Diets (New Brunswick, NJ, U.S.A.) and standard (STD) diet, Teklab 8664, was provided by the CWRU animal facility. The KG diet protocol was chosen for its proven effectiveness for inducing moderate and stable ketosis in young adult Wistar rats and for its palatability.[10] The CHO diet is formulated as a control for the KG diet. All diets were formulated to meet the nutritional requirements for rats and fed as described in Table 1.

2.2. Animal Care and Hypoxic Exposure

Male Wistar rats (5 wks, 100-120 g) were fed STD diet one week prior to experiments during acclimatization to CWRU animal facilities. All rats were weighed, fasted for 24 hours, and re-weighed prior to starting on diets. Rats were then randomly assigned to one of three diet groups, KG, CHO, or STD, for three weeks. All rats were maintained on a 12:12 h light-dark cycle with food and water available *ad libitum*, and weighed three times a week. After 3 weeks of feeding the diets, a subgroup of rats from each diet group were kept in hypobaric chambers for a further 3 weeks at a constant pressure of 0.5 ATM (~10% O_2) while remaining on their assigned diets. The normoxic control rats of each diet groups were housed in the same room next to the hypobaric

Table 1. Nutritional Breakdown of Rat Diet.

	Fat (%)	Protein (%)	Carbohydrate (%)
Ketogenic (KG)	89.5	10.4	0.1
High carbohydrate (CHO)	11.5	10.4	78.1
Standard (STD)	27.5	20.0	52.6

chamber to ensure identical ambient conditions. Hypoxic exposure was continuous except for a maximum of 1 hour every other day to allow for husbandry and weighing.[11]

2.3. Surgical Procedures

On the final day of study, anesthesia was induced in hypoxic rats with a gas mixture of 2.5% halothane and 10% O_2 in N_2O. The rats were then tracheally intubated and ventilated with a mixture of 10% O_2/0.5% halothane/balance N_2O. Normoxic control rats underwent the same procedures except that the oxygen content in the gas mixture was ~21%. Tail artery catheters were placed for blood pressure monitoring, blood sampling, and drug administration. The rats were then paralyzed with Tracrium (2-5 mg/kg; Abbott Laboratories, IL) 2 minutes prior to funnel freezing procedure. The tidal volume (~1 ml/100 g body weight) and respiratory rate (~70 breaths/min) were set to yield an arterial $PaCO_2$ of ~24 mm Hg for hypoxic rats and ~40 mm Hg for normoxic control rats.[2, 12] Rats were frozen *in situ* about 30 mins after the funnel freezing procedure was completed.

2.4. *In situ* Fixation

Metabolic fixation in brain was accomplished by funnel freezing with liquid nitrogen.[13] Just before freezing, the skin of skull was incised and a plastic funnel was placed over the skull and sealed with vacuum grease around the base to prevent leakage of liquid nitrogen. Tracrium was administered in the tail vein to induce paralysis. Liquid nitrogen was then poured into the funnel and the funnel was maintained at least half full until 50% maBP or for 5 mins, at which point the rat was immersed in liquid nitrogen. Frozen rats were stored at -80°C until further processing.

2.5. Metabolic Assays Lactate, Glucose, Ketones (Acetoacetate and BHB)

The brains of frozen rats were removed in a glove box maintained at -30°C. Cortex samples (50 to 100 mg) were dissected from the right frontal lobe, weighed, and stored at -80°C prior to homogenization. Plasma (mM) and tissue (mmol/g wet wt) concentrations of lactate, glucose, and ketone (BHB and acetoacetate) were assayed by standard enzymatic assay on a Cobas Fara centrifuge analyzer (Roche). Briefly, an equal volume of 6% perchloric acid was added to plasma (~0.2 ml) and 0.5 ml of 6% perchloric acid was added to 0.1 g of tissue; the samples were vortexed and centrifuged. The supernatants were collected, neutralized with 2N KOH/0.5 N triethanolamine, and stored at -80°C for subsequent analysis.

2.6. Statistical Methods

Statistical analysis was performed using SPSS v10.0 for Windows. Group comparisons were made by 1-way and 2-way ANOVA using the Tukey HSD method or *t* test. Statistical significance was considered at the level of probability of less than 0.05. Auxiliary data analysis and graphical presentations were made using Microsoft Excel spreadsheets and SlideWrite 5.0 (32 edition).

Table 2. Physiological Parameters.

	Ketogenic (KG)		**Carbohydrate (CHO)**		**Standard (STD)**	
	Normoxia (n = 10)	Hypoxia (n = 14)	Normoxia (n = 10)	Hypoxia (n = 10)	Normoxia (n = 10)	Hypoxia (n = 10)
HCT (%)	44 ± 4	70 ± 7§	43 ± 2	68 ± 4§	45 ± 2	67 ± 4§
pH (units)	7.4 ± 0.05	7.4 ± 0.04	7.4 ± 0.04	7.4 ± 0.02	7.4 ± 0.05	7.4 ± 0.05
$PaCO_2$ (mm Hg)	38 ± 4	24 ± 2	41 ± 3	24 ± 2	40 ± 3	25 ± 3
PaO_2 (mm Hg)	98 ± 15	45 ± 5	94 ± 9	46 ± 5	98 ± 16	45 ± 5
mBP (mm Hg)	112 ± 8	101 ± 8	106 ± 8	104 ± 10	114 ± 14	105 ± 14

3. RESULTS

Table 2 shows the physiological parameters measured in each of the groups. Diet did not alter the hematocrit among the groups at any time point during the experimental period. Hematocrit levels were constant in the normoxic groups, but levels were significantly greater in the hypoxic groups (§, $P < .05$) compared with the normoxic groups for each of the diets. Other physiological parameters (pH, $PaCO_2$, PaO_2, maBP) were not significantly different between any of the groups (normoxic or hypoxic).

3.1. Body Weight

All rats in each of the groups gained weight over the experimental time of six weeks (Fig. 1). The mean body weights prior to fasting (122.7 ± 25; n = 64) changed among the groups following three weeks of feeding the diets just prior to hypoxic exposure. Hypoxic exposure resulted in a slower weight gain in all 3 groups; body weight was markedly altered upon initial exposure for the first two days, but then recovered to about 50% of the normoxic groups and was sustained for the remainder of the experimental period.

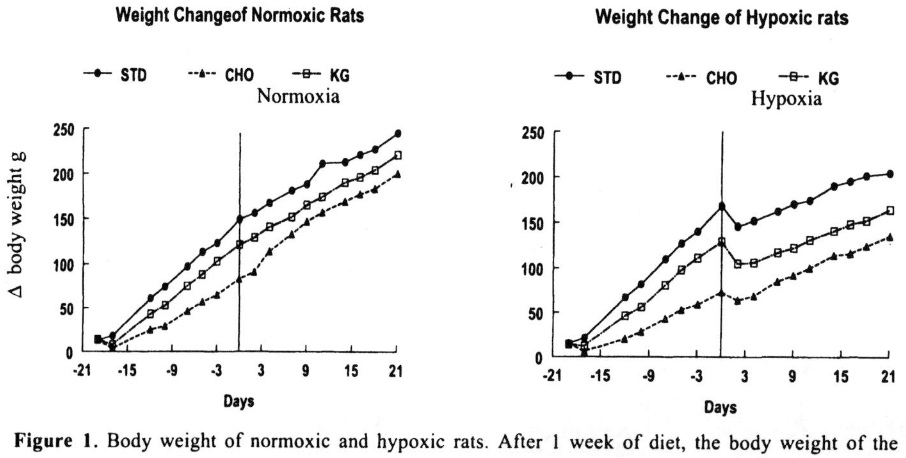

Figure 1. Body weight of normoxic and hypoxic rats. After 1 week of diet, the body weight of the CHO-normoxic rats was significantly lower ($P < .05$) compared with the STD-normoxic rats. There was a slight decrease in weight in all 3 diet groups during the beginning of hypoxic exposure (t = 0).

Table 3. Plasma Concentrations of Glucose, Lactate, and Ketones.

	Ketogenic (KG)		**Carbohydrate (CHO)**		**Standard (STD)**	
	Normoxia (n = 10)	Hypoxia (n = 14)	Normoxia (n = 10)	Hypoxia (n = 10)	Normoxia (n = 10)	Hypoxia (n = 10)
$Glucose_{Plasma}$ (mM)	9.7 ± 1.3	9.0 ± 2.4	11.0 ± 1.5	10.6 ± 2.3	11.0 ± 1.6	11.0 ± 2.1
$Lactate_{Plasma}$ (mM)	1.1 ± 0.1	2.2 ± 0.8**	1.8 ± 0.5	2.1 ± 0.4	1.7 ± 0.4	2.4 ± 0.7**
Ketones:						
BHB_{Plasma}	2.3 ± 1.3*	2.3 ± 1.3*	0.26 ± 0.2	0.3 ± 0.1	0.2 ± 0.1	0.4 ± 0.3
$AcAc_{Plasma}$	0.25 ± 0.2*	0.3 ± 0.3*	< 0.02	< 0.02	< 0.02	< 0.02

*$P < .05$; **$P < .05$

Though there was a significant difference in body weight between the KG- and CHO-fed groups compared with the STD-fed groups (normoxic and hypoxic), the mean difference varied by 11%, with the STD group being the higher limit and the CHO group being the lower limit. There was no difference in rate of weight gain in the groups by the latter end of the experimental period.

Table 3 shows the glucose, lactate, and ketone (BHB and acetoacetate) plasma concentrations in each of the groups. Glucose levels in both plasma and tissue remained unchanged in all groups (STD normoxic tissue glucose: 3.0 ± 1.0 mmol/g wet wt). There were no differences between the plasma lactate levels in any of the diet groups exposed to hypoxia, although the KG- and STD-hypoxic animals had significantly higher plasma lactate levels than their corresponding normoxic group (Table 3; **$P < .05$). In tissue, KG-hypoxic animals had significantly lower lactate levels than the STD-hypoxic group (Fig. 2a). Ketone plasma levels (BHB and acetoacetate) were greatest by 10-fold in the KG-diet fed animals compared with the other diet groups (Table 3; *$P < .05$). Tissue BHB levels were similar under both hypoxia and normoxia in all diet groups, but they were greatest in the KG-diet fed animals by about 4-fold (Fig. 2b).

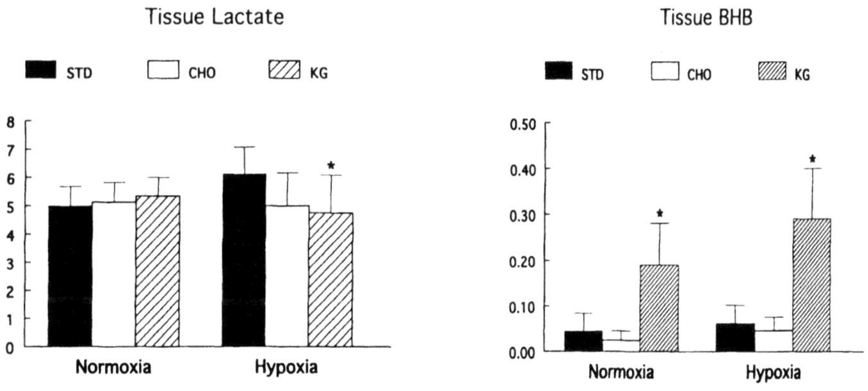

Figure 2. Lactate and BHB concentrations in tissue: (a) the tissue lactate level in the KG hypoxic group was lower than the STD hypoxic group; (b) BHB levels in the KG group were higher than in the other 2 diet groups under both normoxic and hypoxic conditions. Y-axis represents concentrations in μmol/g tissue. *$P < .05$.

4. DISCUSSION

Ketone influx into the brain is largely dependent on the blood concentration of ketones[14] and the activity of the monocarboxylic acid transporter system.[15] Recently, a study measuring levels of BHB and lactate in the brain of humans on short-term fast showed an increase in both metabolites.[16] The authors hypothesized that the increase in brain lactate likely results from ketones displacing lactate oxidation without altering glucose phosphorylation and glycolysis. However, it is unknown how ketosis may be altered in the brain when brain lactate levels are known to be elevated, such as during hypoxia.

The results of this study show that adaptation to hypoxia did not interfere with ketosis induced by feeding a high fat diet. This was demonstrated by the sustained high levels of ketones present in both plasma and brain tissue following three weeks of hypoxia while on the ketogenic diets. However, brain lactate levels were not significantly elevated in our animals following three weeks of hypoxia, as previously reported,[2, 12] suggesting adaptation to hypoxia. A possible explanation could be that younger aged rats adapt to hypoxia by three weeks of exposure. Interestingly, tissue lactate levels in the KG-hypoxic rats were significantly lower than the STD-hypoxic rats, suggesting that ketones altered glucose metabolism, possibly through inhibition of glycolysis or by increased lactate disposal. However, the pH paradox would predict that the inhibition of glycolysis during mild hypoxic exposure would result in lower blood and tissue pH. Although tissue pH was not measured in this study, blood pH remained unchanged in all the groups, and ketosis did not result in metabolic acidosis or interfere with adaptation to mild chronic hypoxia. Future studies measuring the response to acute hypoxic exposure in ketogenic rats would further address if ketosis alters lactate levels in brain.

5. ACKNOWLEDGMENTS

These studies were supported by the National Institutes of Neurological Disorders and Stroke NS46074, National Institute of General Medical Sciences GM066309, Medical Research Service of the Department of Veteran Affairs and the National Institutes of Health EY00300

6. REFERENCES

1. S. I. Harik, R. A. Behmand, and J. C, LaManna, Hypoxia increases glucose transport at blood-brain barrier in rats, *J. Appl. Physiol.* **77**, 896-901 (1994).
2. S. I. Harik, W. D. Lust, S. C. Jones, K. L. Lauro, S. Pundik, and J. C. LaManna, Brain glucose metabolism in hypobaric hypoxia, *J. Appl. Physiol.* **79**, 136-140 (1995).
3. J. C. LaManna, Hypoxia/ischemia and the pH paradox, *Adv. Exp. Med. Biol.* **388**, 283-292 (1996).
4. O. E. Owen, A. P. Morgan, H. G. Kemp, J. M. Sullivan, M. G. Herrera, and G. F. Cahill Jr., Brain metabolism during fasting, *J. Clin. Invest.* **46**, 1589-1595 (1967).
5. L. Sokoloff, Metabolism of ketone bodies by the brain, *Ann. Rev. Med.* **24**, 271-280 (1973).
6. M. C. McKenna, J. T. Tildon, J. H. Stevenson, R. Boatright, and S. Huang, Regulation of energy metabolism in synaptic terminals and cultured rat brain astrocytes: differences revealed using aminooxyacetate, *Develop. Neurosci.* **15**, 320-329 (1993).
7. J. M. Freeman, M. T. Kelly, and J. B. Freeman, *The Epilepsy Diet Treatment*, 2nd edition (Demos Vermande, New York, 1996).
8. G. F. J. Cahill, and O. E. Owen, Starvation and survival, *Trans. Am. Clin. Climat. Ass.* **79**, 13-18 (1967).

9. N. B. Ruderman, P. S. Ross, M. Berger, and M. N. Goodman, Regulation of glucose and ketone-body metabolism in brain of anaesthetized rats, *Biochem. J.* **138**, 1-10 (1974).
10. S. I. Harik, A. S. Al-Mudallal, J. C. LaManna, W. D. Lust, and B. E. Levin, Ketogenic diet and the brain, *Ann. NY Acad. Sci.* **835**, 218-224 (1997).
11. J. C. LaManna, K. A. McCracken, and K. P. Strohl, Changes in regional cerebral blood flow and sucrose space after 3-4 weeks of hypobaric hypoxia (0.5 ATM), *Adv. Exp. Med. Biol.* **248**, 471-477 (1989).
12. J. C. LaManna, L. M. Vendel, and R. M. Farrell, Brain adaptation to chronic hypobaric hypoxia in rats, *J. Appl. Physiol.* **72**, 2238-2243 (1992).
13. U. Ponten, R. A. Ratcheson, L. G. Salford, and B. K. Siesjo, Optimal freezing conditions for cerebral metabolites in rats, *J. Neurochem.* **21**, 1127-1138 (1973).
14. S. G. Hasselbalch, G. M. Knudsen, J. Jakobsen, L. P. Hageman, S. Holm, and O. B. Paulson, Blood-brain barrier permeability of glucose and ketone bodies during short-term starvation in humans, *Am. J. Physiol.* **268**, E1161-E1166 (1995).
15. D. Z. Gerhart, B. E. Enerson, O. Y. Zhdankina, R. L. Leino, and L. R. Drewes, Expression of the monocarboxylate transporter MCT2 by rat brain glia, *Glia* **22**, 272-281 (1998).
16. J. W. Pan, T. L. Rothman, K. L. Behar, D. T. Stein, and H. P. Hetherington, Human brain beta-hydroxybutyrate and lactate increase in fasting-induced ketosis, *J. Cereb. Blood Flow Metab.* **20**(10), 1502-1507

VARIED RESPONSE OF SPONTANEOUS TUMORS TO ANTIANGIOGENIC AGENTS

Bruce M. Fenton, Scott F. Paoni, Brian Grimwood, and Ivan Ding

Abstract: Since conventional therapies are directly dependent on the supply of either drugs or oxygen, a key question is whether antiangiogenic agents produce detrimental effects on tumor vascular function, thus compromising combination therapies. A second question is whether experimental results based on fast-growing, transplanted tumors mimic those in slowly developing spontaneous tumors, which may be more representative of response in human primary tumors. To investigate changes in tumor pathophysiology, three antiangiogenic agents were compared: a) endostatin, b) anti-VEGFR-2 (DC101), and c) celecoxib. Total blood vessels were identified using anti-CD31, perfused vessels using $DiOC_7$, and hypoxia by EF5 uptake. Although individual tumor growth rates varied substantially, DC101 produced the most striking inhibition. DC101 increased total and perfused vessel spacing as well as overall hypoxia, while endostatin increased total vessel spacing, and hypoxia and celecoxib had no marked effects. These results reinforce the idea that pathophysiological changes in spontaneous tumors are in general reflective of response in transplanted tumors. Furthermore, although DC101 inhibited growth in roughly half of the spontaneous tumors, the remaining tumors were unaffected. A key focus of future studies will be to investigate the underlying rationale for the widely varying antiangiogenic response among tumors that outwardly appear so similar.

1. INTRODUCTION

On the basis of successful preclinical data, numerous antiangiogenic agents are now in clinical trials, either alone or in combination with conventional therapies.[1] A key question that arises is whether antiangiogenic agents produce detrimental effects on tumor vascular function and oxygen delivery, since conventional therapies are usually directly dependent on the supply of either chemotherapeutic drugs or oxygen. A second question is whether experimental results based on transplanted tumors can be repeated in more slowly growing, spontaneous tumors that may be more representative of human primary tumors. Previous studies have indicated that substantial differences in vascularity may exist between transplanted and spontaneous tumors, which can then result in corresponding differences in susceptibility to antivascular therapies.[2] In order to investigate the effects of antiangiogenic strategies on tumor pathophysiology in spontaneous mammary tumors, three antiangiogenic agents were contrasted: a) endostatin, a naturally occurring inhibitor of endothelial proliferation and migration[3, 4]; b) DC101, an antibody to vascular endothelial growth factor receptor-2 (VEGFR-2)[5, 6]; and c) celecoxib, a COX-2 inhibitor that has also been shown to inhibit angiogenesis through inhibition of endothelial proliferation and induction of apoptosis.[7] Growth inhibition following these agents was related to corresponding changes in tumor blood vessel spacing and hypoxia.

2. METHODS

2.1. Animal Model

Spontaneous mammary carcinomas arose in retired C3H/Hej breeder mice (murine mammary tumor virus positive, The Jackson Laboratory) between 6-24 months of age. Tumors vary in time of appearance and site, but they are confined to the mammary line and demonstrate features of well-differentiated adenocarcinomas with a pseudoglandular architecture.[8] Tumor volumes were estimated by the formula: $\pi ab^2/6$, where a and b are the major and minor diameters of the tumor as measured using calipers. Guidelines for the humane treatment of animals were followed as approved by the University of Rochester Committee on Animal Resources.

2.2. Drug Treatments

Treatments were initiated upon discovery of spontaneous tumors, and mice were divided arbitrarily into 4 treatment groups: 1) controls, 2) anti-VEGFR-2 (DC101, ImClone Systems, Inc., New York, NY), administered *i.p.* at 45 mg/kg every 3 days, 3) recombinant murine endostatin (rmNYendo), given *i.p.* at 20 mg/kg once per day, or 4) celecoxib (kindly provided by Searle, Chicago, IL), mixed in the diet (600 PPM). Tumors were frozen and 9.0 μm frozen sections were cut for immunohistochemistry.

2.3. DiOC$_7$ Perfusion Marker and EF5 Hypoxic Marker

To visualize blood vessels open to flow, an *i.v.* stain, DiOC$_7$, was injected 1 min prior to tumor freezing, which preferentially stains cells immediately adjacent to the vessels.[9] Localized areas of tumor hypoxia were assessed by immunohistochemical identification of sites of 2-nitroimidazole metabolism.[9] A pentafluorinated derivative (EF5) of etanidazole was injected *i.v.* one hr before tumor freezing. Regions of high EF5 metabolism were visualized using a fluorochrome (Cy3, Amersham) conjugated to the ELK3-51 antibody. This antibody is extremely specific for the EF5 drug adducts that form when the drug is incorporated by hypoxic cells.[10] EF5 (made by NCI) and ELK3-51 were obtained from the University of Pennsylvania Imaging Service Center (C. Koch).

2.4. Immunohistochemistry and Image Analysis

Tumor sections were imaged using a 20× objective, digitized, background-corrected, and analyzed using Image-Pro software.[9] Color image montages from 16 adjacent microscope fields in each of four adjoining tumor regions were acquired and digitally combined under three staining conditions. First, images of the $DiOC_7$ were obtained immediately after cryostat sectioning. Following staining, the sections were returned to the same stage coordinates, and both EF5/Cy3 and anti-CD31 (Pharmingen, San Diego, CA) images were acquired. Vessels and hypoxia were quantified using automated image analysis techniques.[11] Briefly, distance map filters converted pixel intensities of the perfused vessel images to intensity levels proportional to the distances between tumor cells and the nearest vessels. This distance map was multiplied by an image of white grid points on a black background to obtain a spatial sampling of the distance map intensities, which are proportional to the distances to the nearest vessel.

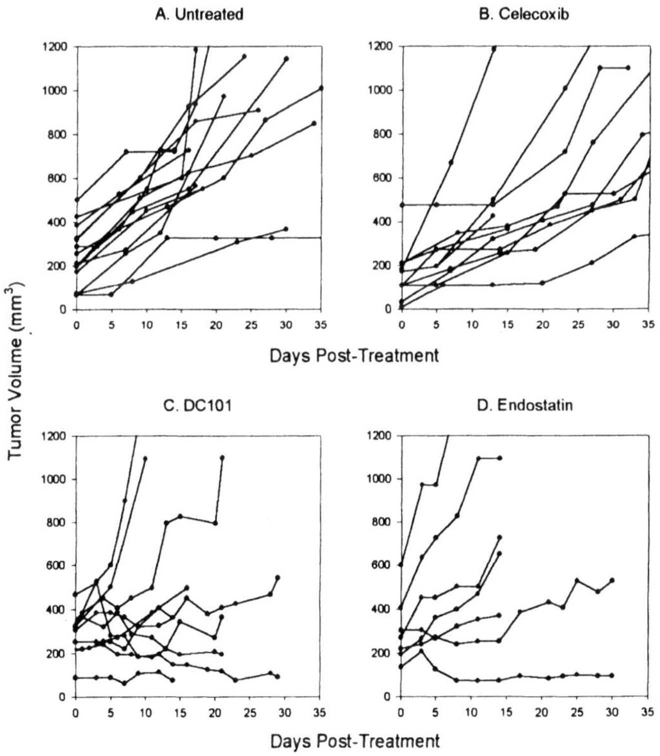

Figure 1. Effect of treatment on spontaneous tumor growth. Each line corresponds to a single tumor and treatment was initiated over a range of tumor volumes (50-600 mm^3).

3. RESULTS AND DISCUSSION

Limited studies have previously been published investigating the effects of antiangiogenic agents on spontaneous murine tumors.[6, 8, 12] Yokoyama et al.[12] reported that endostatin was useful in delaying tumor growth in a transgenic mammary model, and more recently Izumi et al.[6] showed that DC101 could also inhibit vessel and tumor growth in spontaneous tumors arising in retired breeder mice. However, no reports have looked at the effect of antiangiogenic strategies on tumor hypoxia in such models. The present work studied the response of these tumors to three different treatments, with the overall aims of determining: a) whether tumor growth could be inhibited in spontaneous tumors, and b) whether vascular function and oxygenation would subsequently be compromised.

Numerous studies have shown that tumor response to antiangiogenic agents can vary depending on stage of growth at onset of treatment.[13-15] In Rip1Tag2 pancreatic tumors, anti-VEGFR-2[14] and endostatin[15] were found to be much more effective when administered during the initiation stage of tumor development, rather than after tumors had become well established. Figure 1 demonstrates the effects of each of the current strategies on tumor growth inhibition. For untreated tumors, a linear fit was a close approximation to the data (12/14 tumors had correlation coefficients > 0.90), and slopes of the linear regressions were therefore used to characterize the growth curves. No spontaneous regressions were observed among these tumors. Also, neither the 14 untreated nor the 10 celecoxib-treated tumors, and only one of the 7 endostatin-treated tumors, had a slope of less than 5. In contrast, 4 of the 10 DC101-treated tumors had slopes of zero or less, indicating substantial growth inhibition (termed responders). In line with the previously described dependence on tumor stage,[14] response to endostatin, and to some extent DC101, also varied inversely with tumor volume at onset of treatment. For the endostatin (Figure 1D), the slope of the growth curve was, in fact, directly proportional to initial tumor volume ($r^2 = 0.78$). This relation could, however, also reflect inherent tumor aggressiveness, since treatment of faster-growing tumors would more likely be initiated at a somewhat larger volume. For the DC101, the effect of initial tumor volume was somewhat less pronounced, and some large tumors also responded well to treatment (see Figure 1C).

In transplanted tumor models, studies of the effects of antiangiogenic strategies on tumor oxygenation and blood flow have been somewhat contradictory. Following endostatin treatment, increased endothelial cell apoptosis and decreased angiogenesis suggest that this agent could result in a compromised vascular supply, which indeed has been described in some studies.[16] In contrast, acute administration of endostatin has also been reported to increase tumor oxygenation.[17] Surprisingly, extended endostatin treatment has also been shown to potentiate radiotherapy.[18] With DC101, previous work has almost invariably reported marked reductions in tumor growth and vascularization,[19] in some cases accompanied by an increase in tumor hypoxia.[5, 20] In some tumor models, this early antiangiogenic response has been shown to be followed by a second wave of vascular growth and improved tumor oxygenation,[20] possibly in response to hypoxia-induced upregulation of VEGF. DC101 has also been shown to improve radioresponse of tumors without a corresponding effect on radiobiological hypoxia fraction (as determined using paired survival curves),[21] suggesting that oxygen improvements may not be the critical parameter. For celecoxib, previous reports in our laboratory and others have

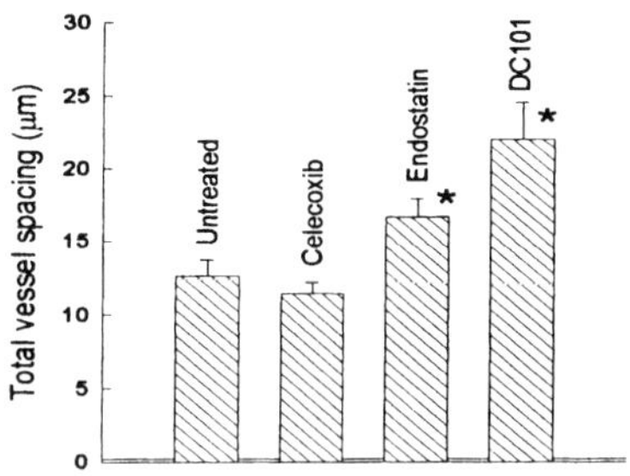

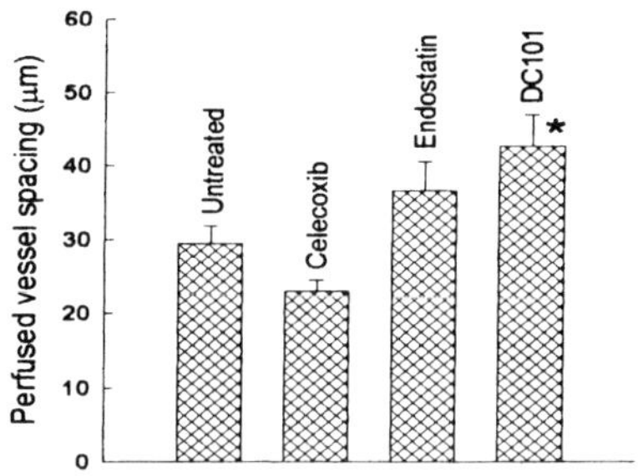

Figure 2. Effect of treatment on total (left) and perfused (right) blood vessel spacing. Plots were based on 16 untreated tumors, 7 celecoxib, 11 endostatin, and 10 DC101. Asterisks (*) denote significant differences from untreated tumors ($P < .05$).

generally shown tumor hypoxia to be unchanged in transplanted tumors.[22] Other studies, however, have also reported decreases in total vascularity following this agent.[23]

Figure 2 summarizes the effects of treatment on both total and perfused blood vessel spacing. Treatment with either endostatin or DC101 resulted in significant increases in total vessel spacing ($P = .02$ and $P < .001$, respectively), indicative of the commonly reported decrease in vascularity. However, only DC101 produced a corresponding decrease in perfused vessel spacing ($P = .007$), which is more indicative of vascular function. Although the number of DC101 tumors was somewhat limited, no significant differences were apparent between responders and non-responders in terms of vascular spacing or hypoxia. In contrast, celecoxib had no effect on either total or perfused vessel spacing.

Figure 3 illustrates changes in overall tumor hypoxia marker uptake (mean EF5/Cy3 intensity). While celecoxib treated tumors had levels of uptake that were equivalent to untreated tumors, overall hypoxia in both endostatin and DC101-treated tumors was significantly increased ($P = .005$ and $P = .0002$, respectively). The combination of a decrease in vascular functionality and a subsequent increase in tumor hypoxia would be

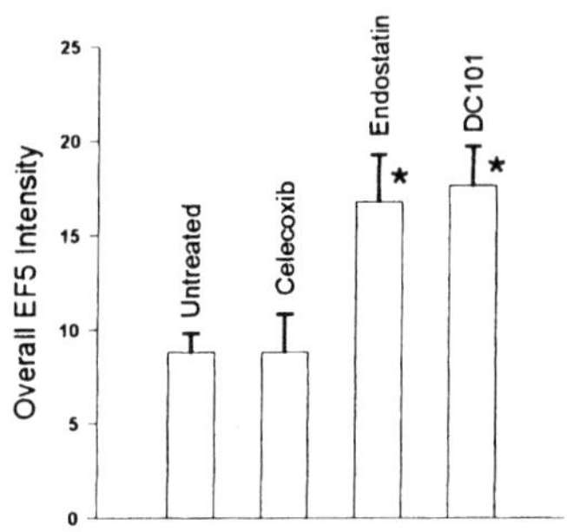

Figure 3. Overall hypoxia marker uptake, as measured by EF5/Cy3 intensities. Asterisks (*) denote significant differences from untreated tumors ($P < .05$).

expected to lead to a concomitant decrease in both tumor radioresponse and delivery of chemotherapeutic agents. However, this is in contradiction to previous reports that have instead found an enhancement in response when either endostatin or DC101 was combined with radiation.[18, 24] In general, however, previous work has not supported the idea that tumor oxygenation is the primary factor in potentiating the radiation response. Hanna et al.[18] suggested that the endothelial compartment was the more likely target, rather than an enhancement of tumor cell radiosensitivity. Our earlier studies in transplanted tumors, however, demonstrated a substantial reduction in overall tumor hypoxia following acute administration of endostatin,[3] again suggesting that the precise timing of radiation could be an important consideration.

4. CONCLUSIONS

In summary, the current experiments demonstrate that alternative antiangiogenic strategies can have varying effects on tumor pathophysiology. Overall, response of spontaneous tumors to the three agents was comparable to previous reports in transplanted tumor models. Additionally, the agents that most substantially inhibited tumor growth also tended to produce the most striking effects on tumor vasculature and hypoxia. Further studies of the temporal pathophysiological changes accompanying antiangiogenic treatments are clearly needed to better understand and to optimally schedule antiangiogenic agents with conventional therapy.

5. ACKNOWLEDGMENTS

Financial support was provided by NIH Grant CA52586.

REFERENCES

1. P. A. Burke, and S. J. DeNardo, Antiangiogenic agents and their promising potential in combined therapy, *Crit. Rev. Oncol. Hematol.* **39**, 155-171 (2001).
2. J. C. Lee, D. C. Kim, M. S. Gee, H. M. Saunders, C. M. Sehgal, M. D. Feldman, S. R. Ross, and W. M. Lee, Interleukin-12 inhibits angiogenesis and growth of transplanted but not in situ mouse mammary tumor virus-induced mammary carcinomas, *Cancer Res.* **62**, 747-755 (2002).
3. B. M. Fenton, S. F. Paoni, B. K. Beauchamp, B. Tran, L. Liang, B. Grimwood, and I. Ding, Evaluation of microregional variations in tumor hypoxia following the administration of endostatin, *Adv. Exp. Med. Biol.* **510**, 19-24 (2003).
4. T. Boehm, J. Folkman, T. Browder, and M. S. O'Reilly, Antiangiogenic therapy of experimental cancer does not induce acquired drug resistance, *Nature* **390**, 404-407 (1997).
5. C. J. Bruns, M. Shrader, M. T. Harbison, C. Portera, C. C. Solorzano, K. W. Jauch, D. J. Hicklin, R. Radinsky, and L. M. Ellis, Effect of the vascular endothelial growth factor receptor-2 antibody DC101 plus gemcitabine on growth, metastasis and angiogenesis of human pancreatic cancer growing orthotopically in nude mice, *Int. J. Cancer* **102**,101-108 (2002).
6. Y. Izumi, E. di Tomaso, A. Hooper, P. Huang, J. Huber, D. J. Hicklin, D. Fukumura, R. K. Jain, and H. D. Suit, Responses to antiangiogenesis treatment of spontaneous autochthonous tumors and their isografts, *Cancer Res.* **63**, 747-751 (2003).
7. K. M. Leahy, R. L. Ornberg, Y. Wang, B. S. Zweifel, A. T. Koki, and J. L. Masferrer, Cyclooxygenase-2 inhibition by celecoxib reduces proliferation and induces apoptosis in angiogenic endothelial cells in vivo, *Cancer Res.* **62**, 625-631 (2002).

8. P. K. Lala, N. Al-Mutter, and A. Orucevic, Effects of chronic indomethacin therapy on the development and progression of spontaneous mammary tumors in C3H/HeJ mice, *Int. J. Cancer* **73**, 371-380 (1997).
9. B. M. Fenton, S. F. Paoni, J. Lee, C. J. Koch, and E. M. Lord, Quantification of tumor vascular development and hypoxia by immunohistochemical staining and HbO_2 saturation measurements, *Br. J. Cancer* **79**, 464-471 (1999).
10. E. M. Lord, L. Harwell, and C. J. Koch, Detection of hypoxic cells by monoclonal antibody recognizing 2-nitroimidazole adducts, *Cancer Res.* **53**, 5721-5726 (1993).
11. B. M. Fenton, S. F. Paoni, B. K. Beauchamp, and I. Ding, Zonal image analysis of tumour vascular perfusion, hypoxia, and necrosis, *Br. J. Cancer* **86**, 1831-1836 (2002).
12. Y. Yokoyama, J. E. Green, V. P. Sukhatme, and S. Ramakrishnan, Effect of endostatin on spontaneous tumorigenesis of mammary adenocarcinomas in a transgenic mouse model, *Cancer Res.* **60**, 4362-4365 (2000).
13. K. Boggio, E. Di Carlo, S. Rovero, F. Cavallo, E. Quaglino, P. L. Lollini, P. Nanni, G. Nicoletti, S. Wolf, P. Musiani, and G. Forni, Ability of systemic interleukin-12 to hamper progressive stages of mammary carcinogenesis in HER2/neu transgenic mice, *Cancer Res.* **60**, 359-364 (2000).
14. G. Bergers, S. Song, N. Meyer-Morse, E. Bergsland, and D. Hanahan, Benefits of targeting both pericytes and endothelial cells in the tumor vasculature with kinase inhibitors, *J. Clin. Invest.* **111**, 1287-1295 (2003).
15. G. Bergers, K. Javaherian, K. M. Lo, J. Folkman, and D. Hanahan, Effects of angiogenesis inhibitors on multistage carcinogenesis in mice, *Science* **284**, 808-812 (1999).
16. D. R. Sorensen, T. A. Read, T. Porwol, B. R. Olsen, R. Timpl, T. Sasaki, P. O. Iversen, H. B. Benestad, B. K. Sim, and R. Bjerkvig, Endostatin reduces vascularization, blood flow, and growth in a rat gliosarcoma, *Neuro-oncol* **4**, 1-8 (2002).
17. B. M. Fenton, S. F. Paoni, B. G. Grimwood, and I. Ding, Disparate effects of endostatin on tumor vascular perfusion and hypoxia in two murine mammary carcinomas, *Int. J. Radiat. Oncol. Biol. Phys.* **57**(4), 1038-1046 (2003).
18. N. N. Hanna, S. Seetharam, H. J. Mauceri, M. A. Beckett, N. T. Jaskowiak, R. M. Salloum, D. Hari, M. Dhanabal, R. Ramchandran, R. Kalluri, V. P. Sukhatme, D. W. Kufe, and R. R. Weichselbaum, Antitumor interaction of short-course endostatin and ionizing radiation, *Cancer J.* **6**, 287-293 (2000).
19. M. Prewett, J. Huber, Y. W. Li, A. Santiago, W. O'Connor, K. King, J. Overholser, A. Hooper, B. Pytowski, L. Witte, P. Bohlen, and D. J. Hicklin, Antivascular endothelial growth factor receptor (fetal liver kinase 1) monoclonal antibody inhibits tumor angiogenesis and growth of several mouse and human tumors, *Cancer Res.* **59**, 5209-5218 (1999).
20. N. Hansen-Algenstaedt, B. R. Stoll, T. P. Padera, D. E. J. G. Dolmans, D. J. Hicklin, D. Fukumura, and R. K. Jain, Tumor oxygenation in hormone-dependent tumors during vascular endothelial growth factor receptor-2 blockade, hormone ablation, and chemotherapy, *Cancer Res.* **60**, 4556-4560 (2000).
21. S. V. Kozin, Y. Boucher, D. J. Hicklin, P. Bohlen, R. K. Jain, and H. D. Suit, Vascular endothelial growth factor receptor-2-blocking antibody potentiates radiation-induced long-term control of human tumor xenografts, *Cancer Res.* **61**, 39-44 (2001).
22. B. M. Fenton, B. K. Beauchamp, S. F. Paoni, P. Okunieff, and I. Ding, Characterization of the effects of antiangiogenic agents on tumor pathophysiology, *Am. J. Clin. Oncol.* **24**, 453-457 (2001).
23. J. L. Masferrer, K. M. Leahy, A. T. Koki, B. S. Zweifel, S. L. Settle, B. M. Woerner, D. A. Edwards, A. G. Flickinger, R. J. Moore, and K. Seibert, Antiangiogenic and antitumor activities of cyclooxygenase-2 inhibitors, *Cancer Res.* **60**, 1306-1311 (2000).
24. W. Shi, C. Teschendorf, N. Muzyczka, and D. W. Siemann, Gene therapy delivery of endostatin enhances the treatment efficacy of radiation, *Radiother. Oncol.* **66**, 1-9 (2003).

NON-INVASIVE MEASUREMENT OF TUMOR OXYGENATION USING EMBEDDED MICROPARTICULATE EPR SPIN PROBE

Govindasamy Ilangovan, Anna Bratasz, and Periannan Kuppusamy

Abstract: We have developed a novel procedure for *in situ* monitoring of oxygen concentration in growing tumors by electron paramagnetic resonance (EPR)-based oximetry using embedded paramagnetic particulates. The new approach uses spin probes that are permanently embedded or implanted in the tumor. A particular advantage of this procedure is that it is non-invasive, both in terms of implantation of the probe as well as readouts of oxygen. We implanted a mixture of RIF-1 tumor cells and microparticulates of lithium phthalocyanine (LiPc) in the upper hind leg of C3H mice to grow as solid tumor. This enabled repeated measurements of oxygen concentration from the implanted site (tumor) for more than two weeks during the progression of the tumor. The particulates that were embedded in the tumor were stable and non-toxic to tumor cells. There was no apparent inhibitory effect to cell proliferation or tumor growth rate. The measurements indicated that the pO_2 of the tumor decreased exponentially with tumor growth (size) and reached hypoxia (< 4 mm Hg). EPR imaging was used to identify the distribution of the particles in the tumor. The data showed a heterogeneous distribution of the probe particles within the tumor volume. Imaging of oxygen in the growing tumor demonstrated the development of significant hypoxia in the tumor within 4-6 days after inoculation. In summary, the EPR spectroscopy and imaging using embedded spin probe enabled accurate and repeated measurements of pO_2 under non-perturbing conditions in growing tumors.

1. INTRODUCTION

Tumor tissues are generally characterized by compromised vasculature and low oxygenation (hypoxia). There is strong evidence that some human tumors may not respond well to chemo- or radiation treatment if the tumor tissue is poorly oxygenated.[1]

Thus the level of oxygen concentration in tumors is not only a predictor of treatment outcome but also useful to select the appropriate therapeutic dose of regimens and to design strategies to increase tumor oxygenation to maximize cell killing with minimal damage to adjacent normal tissue.[2] Some of the recent developments to increase the sensitivity of tumors to treatment include additional agents for redox mediation, photosensitisers, oxygenation treatments like carbogen-breathing, and the addition of vasodilators to increase blood perfusion.[3-5] However, the methods currently available to measure oxygen levels in tissues under non-invasive conditions are rife with limitations. We describe here the use of electron paramagnetic resonance (EPR) to non-invasively measure pO_2 in an experimental tumor. The technique uses a particulate oxygen-sensing spin probe, namely the lithium phthalocyanine (LiPc), which is co-implanted along with tumor cells during transplantation. We demonstrate that oxygen concentration in the growing tumor can be measured accurately and repetitively over several days. The advantages of this method are: (i) the probe remains within the tumor and is not re-inserted every time, as is normally done in the case of polarographic electrodes or optical sensors; (ii) the probe is stable and non-toxic to the host tissue and remains localized within the tumor for weeks; (iii) the measurements are made from the same location; (iv) oxygen readings are obtained non-invasively; and (v) the measurements can be performed in less than a minute. Thus the measurements can be made from the same site of tumors enabling meaningful comparisons between data obtained on different days. The main aim of the present work was to demonstrate that the embedded LiPc probe can be used to study the changes in pO_2 as a function of tumor growth and it further has the potential of providing an oxygen concentration map in the solid tumor under non-invasive conditions.

2. EXPERIMENTAL DETAILS

2.1. The Probe, Calibration, and pO_2 Measurement

Lithium phthalocyanine microcrystalline powder (LiPc) was used as a probe for EPR oximetry. The probe was synthesized by electrochemical method.[6] The EPR line-width of the probe showed a linear relationship to pO_2 in the range 0 – 100 mm Hg with an oxygen sensitivity of 8.9 mG/mm Hg. Tissue pO_2 values were determined from a pre-constructed calibration curve obtained by equilibrating the LiPc in saline with known concentrations of oxygen.

2.2. EPR Measurements

The measurements were performed using an EPR imaging system consisting of an L-band EPR spectrometer, water-cooled gradient coils, a loop-gap surface resonator, and a personal computer-based data acquisition system. The EPR data acquisition and processing were performed using custom-developed data acquisition software. Image reconstructions were performed using a filtered-backprojection method.

2.3. Mice, Tumor Growth, and Transplantation

Female C3H mice (6 weeks old, 25 ± 3 g), supplied through Frederick Cancer Research Center Animal Production, Frederick, MD, were housed five per cage in climate-controlled rooms and allowed food and acidified water *ad libitum*. Experiments were conducted according to the principles outlined in the Guide for the Care and Use of Laboratory Animals prepared by the Institute of Laboratory Animal Resources, National Research Council. RIF-1 tumors were transplanted using the procedure described previously.[7] Briefly, a single cell suspension of 1×10^6 cells in 0.1 mL was injected subcutaneously in the right hind leg of mice. The animals were observed closely and the tumors became palpable approximately 5 days after injection. Tumors were allowed to grow to a size of about 8-10 mm in the greatest dimension. This cycle was continued with further transplantation throughout the work.

2.4. Co-Implantation, Tumor Volume Measurements, and Animal Preparation

A homogeneous suspension containing 1×10^6 RIF-1 cells and approximately 20 μg of LiPc microcrystals (particle size: 2 x 20-50 μm needles) in 60 μL saline was injected subcutaneously into the upper part of the right hind leg. In control mice, the suspension, without LiPc, was injected. The tumor size was measured using a vernier caliper and the volume was determined from the orthogonal dimensions (d_1, d_2, and d_3) and using the formula $\pi/6(d_1 \times d_2 \times d_3)$. The *in vivo* EPR measurements of co-implanted LiPc were made immediately after injection and continued daily for 2 weeks following the implantation. Histological studies revealed that there was no noticeable effect in the tumor tissue around the LiPc crystals. Isoflurane (1.5% mixed with room air) was used as inhalation anesthesia. The flow rate of the breathing gas mixture was maintained at 2 L/min. A thermister rectal probe was used to monitor body temperature. The body temperature was maintained at 37 ± 1°C using an infrared lamp.

3. RESULTS

3.1. Changes in pO_2 as a Function of Tumor Volume

The pO_2 values in the growing tumor of a set of 5 mice were obtained by measuring the EPR line-shape of the co-embedded LiPc probe. The measurements were performed from the day of injection of the probe to 14 days. The pO_2 value was high on the day of

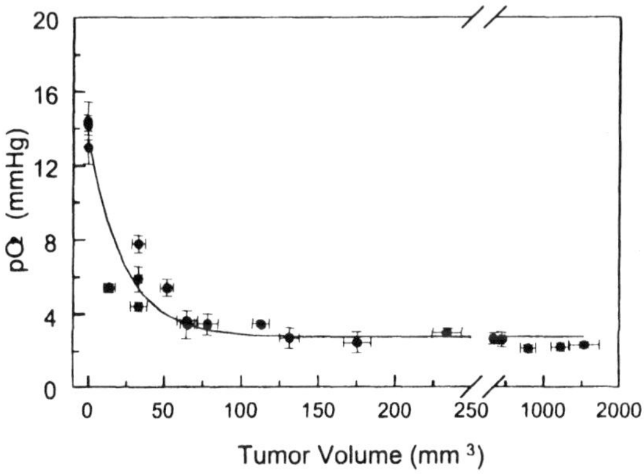

Figure 1. pO_2 as a function of tumor growth. The measurements were performed repeatedly in 5 mice using co-embedded LiPc in the growing tumor for 14 days. The error bars (SEMs) in x and y axes represent errors in the measurement of tumor size and pO_2 in each animal.

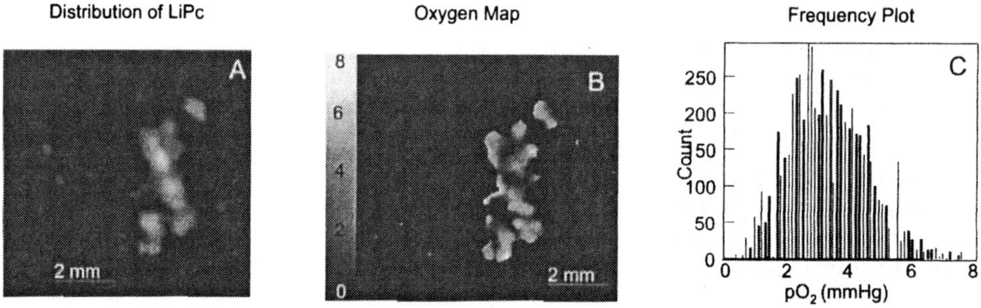

Figure 2. EPR images of LiPc and oxygen in the tumor. **A**: Distribution of the probe in a RIF-1 tumor. **B**: Distribution of oxygen concentration in the same tumor **C**: Frequency plot (number of occurrences of a particular pO_2 in the image) of oxygen distribution.

injection of the probe. However, the oxygen concentration started to fall off rapidly from day 2 onwards. On day 5, when palpable size of tumor was noticed, the pO_2 values were very low (~ 4 mm Hg). The mean pO_2 data plotted as a function of tumor volume showed that the pO_2 decreased exponentially during the first few days, when the tumor volume measured ~ 60 mm^3 (corresponding to 4 mm dia.) and remained constant thereafter (Figure 1).

3.2. Imaging of the Particulates and Oxygen in the Growing Tumor

Spatial EPR imaging was performed to monitor the location and redistribution of the injected particulates during tumor growth. While the particulates in the subcutaneous tissue were confined to a single spot of about 2 mm diameter on the day of injection, they were observed to redistribute within the tumor on subsequent days. Figure 2 shows the distribution of the probe in a typical tumor on day 12 (tumor volume ~500 mm^3) after inoculation. It was observed that probe particulates migrated in the tumor to as much as 6 mm along the vertical (longitudinal) direction of the leg. The exact location of the particulates with respect to the tumor was determined using T_2*-weighted MRI images (data not shown). As one can see from Figure 2A, the particles are not distributed homogeneously, but rather they are confined to a small region of about 2 x 2 x 6 mm^3. This suggests that the cell proliferation and solid tumor formation occurred around the core of the initial deposit and that the particles did not distribute themselves in the entire tumor volume.

Imaging of oxygen concentration in the tumor was performed by spectroscopic (spectral-spatial) imaging using the oxygen-dependent line-shape information in the image. Figure 2B shows the oxygen mapping obtained from the tumor shown in Figure 2A. Since line-shape data are obtained only from the regions where the particulates are present, oxygen concentration is not obtained in the regions outside the particulate image shown in Figure 2A. The distribution of oxygen concentration in the tumor shows that the tumor is significantly hypoxic and heterogeneous with respect to oxygenation. The heterogeneity in the pO_2 is further evidenced from the pO_2 histogram, which shows pO_2 ranging from 1 to 8 mm Hg with a median pO_2 of 4.2 ± 0.3 mm Hg. Overall the experiments demonstrated that non-invasive mapping of oxygen concentration in the tumor can be obtained with good precision and accuracy.

4. DISCUSSION

The present study demonstrates a novel procedure for *in situ* monitoring of oxygen concentration in growing tumors by EPR-based oximetry using embedded paramagnetic particulates. A particular advantage of this procedure is that it is non-invasive, both in terms of implantation of the probe, as well as readouts of oxygen. Most other methods of oxygen measurement require the infusion of an oxygen-sensor or insertion of a needle, catheter, or probe head in the desired location resulting in injury to the tissue. In addition, the invasive procedure needs to be repeated for oxygen readouts over a period of time extending from days to weeks. This will also result in two types of errors in the measurement. The measurements are often performed from a freshly injured site. Trauma associated with the injury, disruption of capillary blood vessels, and inflammation processes at the measurement site may lead to erroneous values of oxygen levels. The second type of error can occur in repeated measurements due to the inability of the procedures to place the probe exactly at the same location used previously. The present method, in addition to overcoming the above limitations of the other techniques, also enables accurate and precise determination of oxygen concentration in tissues.

We and others have previously demonstrated the unique capabilities of LiPc for measuring pO_2 in biological tissues.[8, 9] The probe is stable and sensitive to oxygen in tissues for weeks without apparent toxicity to the host tissue. It is the only known paramagnetic crystalline solid that exhibits the sharpest EPR spectrum with very high spin density. It takes only a few μg of the material to obtain a signal with reasonable intensity for line-width measurements. LiPc is especially useful for biological oximetry because, in the absence of oxygen, the line is very narrow (< 20 mG) and the line-width increases proportionately with the amount of oxygen (pO_2) that is in equilibrium with the particles.

We have previously shown that the pO_2 values measured by the implanted particle in the gastrocnemius muscle of normal (non-tumor) C3H mice remain unchanged at 16 ± 4 mm Hg, except for a small decrease during the first day of implantation of the particle attributed to tissue trauma induced by the fresh injury. We further measured the pO_2 values in RIF-1 tumor by implanting the probe in the tumor when the tumor was approximately 6 mm in diameter and found that the tumor was significantly hypoxic and it continued to be hypoxic thereafter.[8] The present study utilizing LiPc co-injected with the tumor cells enabled us to monitor the changes in the tissue oxygenation that occurred immediately after inoculation. It was observed that the pO_2 rapidly decreased from the normal tissue value of 16 ± 2 mm Hg to about 4 mm Hg within 5 days when the tumor volume was just ~ 60 mm^3 (corresponding to ~ 4 mm in diameter). Thus, it appears that tissue oxygenation falls rapidly in the very early stages of differentiation of the cells to form solid tumor.

The dependence of pO_2 on tumor size has been observed in several experimental tumors and spheroids.[10] However, the observation of an exponential decrease in pO_2 as a function of tumor size in the present study is a novel finding. It should also be noted that tumor growth volume is known to increase exponentially as a function of days after implantation of tumor cells.[11] This suggests that tumor hypoxia occurs very early in the tumor progression due to insufficient supply of oxygen to the exponentially increasing number of cells. The presently observed changes in pO_2 during *in vivo* tumor growth have some striking similarities to the results previously obtained in spheroids.[10] Using programmed insertion of polarographic electrodes into spheroids of different sizes, Sutherland et al.[10] observed that the central region in spheroids of 1-3 mm in diameter had almost zero oxygenation. It was further observed in the same spheroids that the tumor oxygenation decreased exponentially from the outer rim towards the center. Similar results were also reported by Kunz-Schughart et al.[12] In the present study we observed that the tumor cells tended to aggregate with the particulates in culture and, hence, it is reasonable to expect that the implanted tumor cells grow as aggregates *in vivo*. This may create a spheroid-like growth with the particulates entrapped in the center of the aggregates, and thus resulting in an exponential decrease in the tumor oxygenation during the first few days of the development of solid tumor.

The frequency plot of tumor oxygenation obtained in the present work shows a widespread heterogeneity in the pO_2 values. Similar distribution of pO_2 values was noticed with the use of polarographic electrodes. This frequency plot is different from the histogram normally obtained from the polarographic electrode,[13] where the polarographic electrode is physically inserted into different locations, unlike the present case where the values are obtained from the individual pixels in the image. Thus in a 125 x 125 image, as in the present case, the values are read at 125 x 125 points within the sampling area of 5 mm^2. Such a high-resolution pO_2 mapping is only possible with imaging techniques. The large scatter without any significant trend noticed in the relationship of tumor size versus median pO_2 using the polarographic electrode is also because of inaccuracy in measurement and values pooled from different animals. In the present approach, the same animals were used for pO_2 measurements throughout the period of interest so that the measurements were more consistent and significant, and so that the decay pattern in pO_2 was noticed with an increase in tumor size.

5. CONCLUSIONS

In the present study, we have demonstrated a novel procedure based on EPR spectroscopy and imaging coupled with an oximetry spin probe to follow changes in oxygen concentration during the growth of an experimental tumor. The probe was co-embedded into the tumor by means of mixing the tumor cells along with the probe particulates during implantation in mice. This enabled non-invasive measurements of oxygen in the growing tumor over a period of two weeks. EPR imaging measurements were performed in order to monitor the redistribution of probe and oxygen concentration within the tumor as a function of tumor development. It is demonstrated that this type of spin probe co-embedded during tumor transplantation produces more accurate and reliable pO_2 measurements. The pO_2 of the RIF-1 tumor was found to decrease exponentially in lower tumor volumes and reached a constant value in the hypoxic range of 3-4 mm Hg. The spatial mapping of the pO_2 in the tumor showed significant heterogeneity in the tumor oxygenation. Though the present technique may not be suitable in the clinical setting, it can be used for the study of experimental tumors, where accurate and non-invasive measurement of oxygenation is needed. However, oxygen measurements in the clinical setting could be performed by one-time implantation of the particulates (usually ~10 μg) in the tumor, a minimally invasive and advantageous procedure.

6. ACKNOWLEDGEMENTS

We acknowledge financial support from the NIH grants CA 78886, CA 102264, and EB 004031.

REFERENCES

1. H. B. Stone, J. M. Brown, T. L. Phillips, and R. M. Sutherland, Oxygen in human tumors: correlations between methods of measurement and response to therapy, Summary of a workshop held November 19-20, 1992, at the National Cancer Institute, Bethesda, Maryland, *Radiat. Res.* **136**, 422-434 (1993).
2. J. Overgaard, and M. R. Horsman, Modification of hypoxia-induced radioresistance in tumors by the use of oxygen and sensitizers, *Semin. Radiat. Oncol.* **6**, 10-21 (1996).
3. S. Y. Sun, W. Li, P. Yue, S. M. Lippman, W. K. Hong, and R. Lotan, Mediation of N-(4-hydoxyphenyl)retinamide-induced apoptosis in human cancer cells by different mechanisms, *Cancer Res.* **59**, 2493-2498 (1999).
4. B. Krammer, Vascular effects of photodynamic therapy, *Anticancer. Res.* **21**, 4271-4277 (2001).
5. M. E. Powell, D. R. Collingridge, M. I. Saunders, P. J. Hoskin, S. A. Hill, and D. J. Chaplin, Improvement in human tumour oxygenation with carbogen of varying carbon dioxide concentrations, *Radiother. Oncol.* **50**, 167-171 (1999).
6. G. Ilangovan, J. L. Zweier, and P. Kuppusamy, Electrochemical preparation and EPR studies of lithium phthalocyanine: Evaluation of the nucleation and growth mechanism and evidence for potential dependent phase formation, *J. Phys. Chem. B* **104**, 4047-4059 (2000).
7. P. Kuppusamy, H. Li, G. Ilangovan, A. J. Cardounel, J. L. Zweier, K. Yamada, M. C. Krishna, and J. B. Mitchell, Noninvasive imaging of tumor redox status and its modification by tissue glutathione levels, *Cancer Res.* **62**, 307-312 (2002).
8. G. Ilangovan, H. Li, J. L. Zweier, M. C. Krishna, J. B. Mitchell, and P. Kuppusamy, In vivo measurement of regional oxygenation and imaging of redox status in RIF-1 murine tumor: effect of carbogen-breathing, *Magn. Reson. Med.* **48**, 723-730 (2002).
9. K. J. Liu, P. Gast, M. Moussavi, S. W. Norby, N. Vahidi, T. Walczak, M. Wu, and H. M. Swartz, Lithium phthalocyanine: a probe for electron paramagnetic resonance oximetry in viable biological systems, *Proc. Natl. Acad. Sci. USA* **90**, 5438-5442 (1993).
10. R. M. Sutherland, B. Sordat, J. Bamat, H. Gabbert, B. Bourrat, and W. Mueller-Klieser, Oxygenation and differentiation in multicellular spheroids of human colon carcinoma, *Cancer Res.* **46**, 5320-5329 (1986).
11. S. M. Hahn, F. J. Sullivan, A. M. DeLuca, C. M. Krishna, N. Wersto, D. Venzon, A. Russo, and J. B. Mitchell, Evaluation of tempol radioprotection in a murine tumor model, *Free Radic. Biol. Med.* **22**, 1211-1216 (1997).
12. L. A. Kunz-Schughart, K. Groebe, and W. Mueller-Klieser, Threė-dimensional cell culture induces novel proliferative and metabolic alterations associated with oncogenic transformation, *Int. J. Cancer* **66**, 578-586 (1996).
13. M. F. Adam, M. J. Dorie, and J. M. Brown, Oxygen tension measurements of tumors growing in mice, *Int. J. Radiat. Oncol. Biol. Phys.* **45**, 171-180 (1999).

THE EFFECTS OF TUMOUR BLOOD FLOW AND OXYGENATION MODIFIERS ON SUBCUTANEOUS TUMOURS AS DETERMINED BY NIRS

Franklyn A. Howe, James P. Connelly, Simon P. Robinson, Roger Springett, and John R. Griffiths

Abstract: Modulation of tumour oxygenation may be used to increase or decrease tumour hypoxia in order to improve the effect of radiotherapy or bioreductive drugs, respectively. Magnetic resonance imaging (MRI) and near infrared spectroscopy (NIRS) are techniques sensitive to blood deoxyhemoglobin concentration (Hb) that can be used to investigate tumour hypoxia indirectly via blood oxygenation levels. In this study we have used NIRS to determine absolute Hb and changes in deoxyhemoglobin and oxyhemoglobin (HbO) in subcutaneous rodent tumours for challenges that alter blood flow and oxygenation, with the aim to better interpret our MRI data. Both carbogen [95% O_2 + 5% CO_2] and 100% O_2 breathing produced a similar and significant reduction in Hb and increase in HbO, but a negligible change in HbT (= Hb + HbO). In contrast, N_2 breathing to terminal anoxia and intravenous hydralazine produced a negligible increase in Hb, but large reductions in HbO and HbT. HbT is proportional to blood volume, so our data suggests large blood volume decreases occur with challenges likely to cause reduced arterial blood pressure. Hence MRI techniques that measure the R_2* relaxation rate, which varies linearly with total Hb, will underestimate the effects of hypotensive agents at increasing tumour hypoxia.

1. INTRODUCTION

The uncontrolled growth of tumour tissue results in the development of a structurally and functionally abnormal blood supply; consequently tumours frequently become hypoxic, leading to their well-known resistance to radiotherapy.[1] Modulation of tumour oxygenation so as to increase or decrease tumour hypoxia is required to improve the effect of radiotherapy or bioreductive drugs, respectively. Agents may also be used to increase tumour blood flow for improved drug delivery.[2, 3] Determination of baseline tumour oxygenation and subsequent changes induced by blood flow or oxygenation modifiers is therefore of relevance both experimentally and clinically for optimising cancer treatments. In order to monitor changes in tumour blood flow and oxygenation, we have been investigating magnetic resonance imaging (MRI) techniques in which image contrast is dependent on the R_2* relaxation rate,[4, 5] a parameter linearly dependent on the paramagnetic deoxyhemoglobin (Hb) concentration of tissue. However, interpreting R_2* changes in tumours in terms of blood oxygenation (with inference to tissue oxygenation changes) is not always straightforward, due to the heterogeneous and abnormal characteristics of their vasculature.[4] Near infrared spectroscopy (NIRS) can be used to investigate blood oxygenation via the NIR absorption characteristics of hemoglobin.[6, 7] Recent NIRS studies have investigated Hb and HbO changes in tumours following hypercapnic and hyperoxic challenges.[8, 9] NIRS is complementary to MRI[10, 11] and, in addition to providing information on tumour blood oxygenation non-invasively, it may aid in the interpretation of MRI data for assessing tumour oxygenation.

In this study we have used NIRS to determine blood oxygenation changes in subcutaneous rodent tumours using: a) 100% O_2 breathing, intended to improve blood oxygenation; b) carbogen [95% O_2 + 5% CO_2] breathing, intended to improve oxygenation and blood flow; c) Hydralazine, which is expected to increase tumour hypoxia by tumour blood flow reduction; and d) terminal 100% N_2 breathing to completely reduce blood oxygenation. Our aim is to better understand how MRI can be related to tumour blood oxygenation (and thence to tumour hypoxia) and used to assess changes induced by modifiers of blood oxygenation and flow.

2. METHODS

GH3 prolactinomas were grown subcutaneously to 1.2 – 2 cm dia. in female Wistar Furth rats with Hypnorm/Hypnovel used for anaesthesia, as described previously.[4, 5] Animals were supported prone with the tumour protruding at the side to rest on a

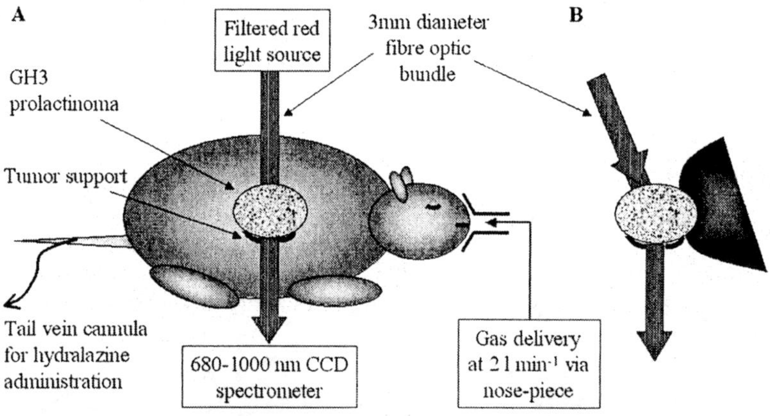

Figure 1. A) Representation of the animal cradle and NIRS system and optodes. B) Axial view demonstrating offset optode angle to minimise body wall light scattering.

platform containing the receive fibre optic cable of the NIRS system (see Fig. 1). Measurements were performed in a dark enclosure with heating to maintain animal core temperature at 36.5°C ± 1.5°C.

A custom NIRS system (constructed by R. Springett whilst at Dept. Medical Physics and Bioengineering, University College, London, UK) was operated in transmission mode (see Fig. 1) and comprised: tungsten light source (LOT Oriel Ltd, Surrey, UK); 3 mm diameter transmit and receive optodes (Schott Fibre Optics, Doncaster, UK); a Triax 320 spectrograph (JY Horriba Ltd, Stanmore, UK) with a cooled CCD system (Wright Instruments, Enfield, UK). Absolute deoxyhemoglobin Hb, and changes in deoxyhemoglobin (ΔHb), oxyhemoglobin (ΔHbO), and total hemoglobin (ΔHbT = ΔHbO + ΔHb) were determined quantitatively with custom software (written by R.S.) by analysis of the 2nd derivative spectrum using the water absorption as a reference.[6, 7] Statistical analyses were completed using paired or single *t* tests and the non-parametric Mann Witney.

Oxygen and carbogen breathing challenges were made on the same rats (n = 6), with carbogen (n = 4) or oxygen (n = 2) first, with 10 minutes air breathing between challenges. On another cohort (n = 3), 100% nitrogen breathing was used until post-mortem. On a separate cohort of rats (n = 7), hydralazine (5 mg kg^{-1}) was administered via tail vein catheter.

3. RESULTS

Average hemoglobin changes for each vascular challenge (measured at 30 minutes post-hydralazine and at 10 minutes post-challenge, otherwise) and initial absolute deoxy-hemoglobin concentrations (aHb) are summarised in Table 1. Initial aHb was variable between tumour groups, indicating tumour heterogeneity in vascular volume, but the values were found to be only just significant (Mann Witney, $P = .042$) between the oxygen/carbogen and the hydralazine cohort. The percentage change in vascular volume (V) can be estimated for the N_2 breathing results if we assume tumour blood becomes fully deoxygenated and the initial HbO = -ΔHbO. For constant hematocrit, HbT = k·V where k is a constant, hence ΔV/V = ΔHbT / HbT = ΔHbT / (aHb - ΔHbO). This gives –52% ± 1% change in vascular volume for terminal N_2 breathing. (A similar calculation for the hydralazine challenge gives –34% ± 8% vascular volume change, but the assumption that the tumour becomes fully deoxygenated is not necessarily valid).

Table 1. Absolute deoxyhemoglobin (aHb) and changes in deoxy- (ΔHb), oxy- (ΔHbO), and total (ΔHbT) hemoglobin concentration (mean ± s. d. in μmole l^{-1}).

	aHb (initial)	ΔHb	ΔHbO	ΔHbT
Oxygen (n = 6)	166 ± 75	-84 ± 25*$^{++}$	83 ± 25	-1 ± 10
Carbogen (n = 6)	166 ± 75	-89 ± 23*$^{++}$	85 ± 26^{++}	0 ± 6
Nitrogen (n = 3)	101 ± 13	23 ± 34	-142 ± 37^{+}	-119 ± 24^{+}
Hydralazine (n = 7)	97 ± 27	9 ± 15	-45 ± 19^{+++}	-48 ± 15^{+++}

Paired *t* test: aHb compared pre- to post-challenge: * $P < .001$.
Single *t* test: ΔHb, ΔHbO, and ΔHbT compared to zero baseline: $^{+}$ $P < .05$, $^{++}$ $P < .001$, $^{+++}$ $P < .0005$.

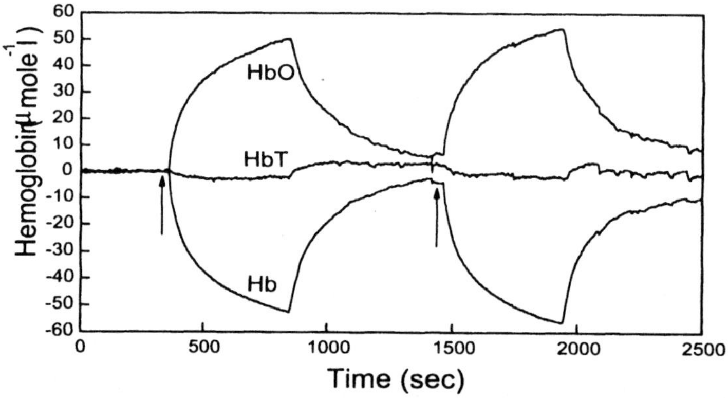

Figure 2. Arrows indicate start of the 10 minute episodes of carbogen breathing and subsequent 100% oxygen breathing, with 10 minutes air breathing in between. The change in HbO and Hb between the start and end of the carbogen and oxygen breathing episodes is similar. Results were similar for all tumours studied.

3.1. Carbogen and Oxygen Breathing

Both carbogen and oxygen breathing produced an increase in tumour HbO and a decrease in Hb with a subsequent return towards baseline on recommencing air breathing. There was no significant difference in the changes induced by 10 minutes of the host breathing carbogen, or oxygen with carbogen first (n = 2), nor was there any difference if oxygen breathing occurred first (n = 2). For both carbogen and oxygen challenges, HbT levels indicated a small but clear decrease for 2 tumours (e.g. Fig. 2), a small increase for one, and no change for the remaining three tumours.

3.2. Hydralazine

Hydralazine produced large reductions in HbO and HbT for all tumours, but a variable response in Hb. At 30 minutes post-hydralazine, Hb increased in 4 tumours (range 5 to 47 μmole l^{-1}), decreased in 1 tumour (-5 μmole l^{-1}), and was close to the pre-hydralazine level in 2 tumours. The example in Figure 3 shows an initial increase in Hb followed by a decrease to near its pre-hydralazine value.

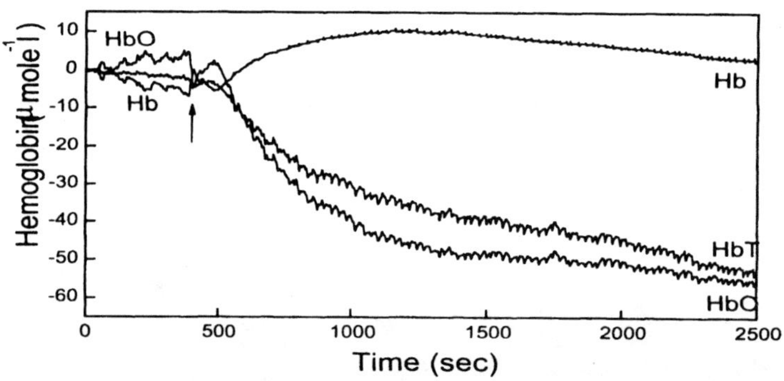

Figure 3. Changes in Hb (see text for variability of Hb changes between tumours), HbO, and HbT following hydralazine administration (arrow).

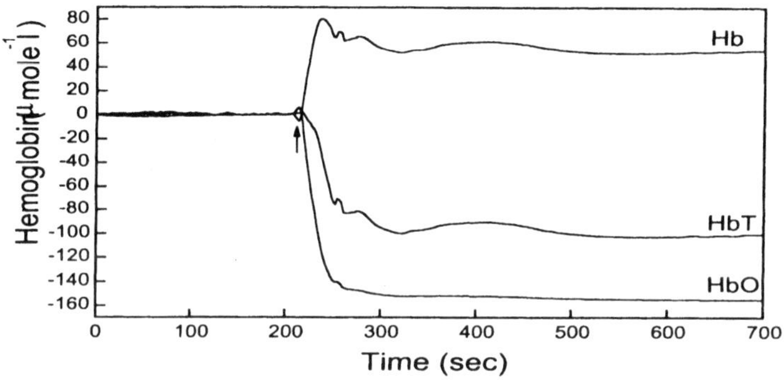

Figure 4. Changes in Hb (see text for variability of Hb changes between tumours), HbO, and HbT following terminal nitrogen breathing (arrow).

3.3. Terminal Nitrogen Breathing

With 100% N_2 breathing, all three tumours showed a rapid decrease in HbO and HbT that reached a minimum at approximately 10 minutes after commencement. Hb showed a rapid initial increase in all 3 tumours and a subsequent decrease in Hb, which for one tumour was reduced to below that measured during initial air breathing (-14 µmole l^{-1}) (Fig. 4).

4. DISCUSSION

The response of a tumour to blood flow and oxygenation modifiers depends on both the host's systemic response and the response of the individual tumour as determined by its particular vascular structure. In tissue with normal smooth muscle, hypercapnia can produce vasodilation due to elevated pH,[12] whereas hyperoxia may cause vasoconstriction from its interaction with $NO^{\cdot}$.[13] Tumours may have a combination of vessels, with or without smooth muscle, as well as major host vessels feeding the tumour, which may explain the varied response to hypercapnic and hyperoxic challenges: carbogen has been shown to be both more effective[4, 5, 14] and equally as effective[8, 9] as oxygen at increasing tumour blood flow and oxygenation. In this study we found that carbogen and oxygen breathing were equally effective at improving tumour blood oxygenation, and that when changes in HbT occurred, they were very small and similar for both gases. Since carbogen or oxygen breathing are not expected to alter systemic blood pressure,[15] we deduce that, in these current prolactinomas, the majority of vessels must lack smooth muscle, the increased tumour HbO being mainly a result of increased oxygenation of the arterial blood, with negligible effects due to vasoconstriction or vasodilation.

Hydralazine is known to produce vasodilation in host vessels and thereby increase tumour hypoxia by reducing tumour blood flow by the "steal" effect; an additional effect is reduced systemic blood pressure.[16] N_2 breathing is also expected to increase tumour hypoxia due to reduced arterial blood oxygen levels. We found that both intravenous hydralazine and N_2 breathing to terminal anoxia produced large reductions in HbO and HbT in all tumours, which would indicate increased tumour hypoxia, but Hb only increased significantly in some tumours. On average, a negligible Hb change occurred that was a consequence of the large drop in HbT and an indication that large reductions in tumour blood volume occur with hydralazine and N_2 breathing. Hence, although blood Hb concentration must necessarily rise, the tissue concentration of Hb does not increase proportionally. We believe that the HbT decrease results from vascular collapse due to the combination of high interstitial fluid pressure in tumours and reduced systemic blood pressure, either from the effects of hydralazine or from cardiac arrest with N_2 breathing. Similar effects are also seen by NIRS with some anaesthetics.[17]

In conclusion, although our hyperoxic challenges produced a decrease in tumour Hb concentration, the opposite was not always true for our hypoxic challenges. In tumours, with their abnormal vasculature, the relationship between tumour Hb concentration and hypoxia (or hyperoxia) may not be monotonic, particularly if systemic effects occur that lead to blood volume changes, such as those found with hypotensive agents, like hydralazine. Although R_2^* has been shown to correlate with Hb in the piglet brain with graded hypoxia,[11] in tumours R_2^* changes may underestimate the effectiveness of hypotensive agents designed to increase tumour hypoxia and may even suggest the contrary (i.e. decreased hypoxia), as has been observed with the decrease in R_2^* following N_2 breathing in some tumours.[4]

5. ACKNOWLEDGEMENTS

This work was supported by Cancer Research UK, grant C12/A1209.

REFERENCES

1. M. R. Horsman, Measurement of tumor oxygenation, *Int. J. Radiat. Oncol. Biol. Phys.* **42**, 701-704 (1998).
2. D. J. Chaplin, M. R. Horsman, and D. S. Aoki, Nicotinamide, Fluosol DA, and carbogen: A strategy to reoxygenate acutely and chronically hypoxic cells in vivo, *Br. J. Cancer* **63**, 109-112 (1991).
3. I. J. Stratford, G. E. Adams, J. C. Bremner, S. Cole, H. S. Edwards, N. Robertson, and P. J. Wood, Manipulation and exploitation of the tumor microenvironment for therapeutic benefit, *Int. J. Radiat. Biol.* **65**, 85-94 (1994).
4. F. A. Howe, S. P. Robinson, D. J. O. McIntyre, M. Stubbs, and J. R. Griffiths, Issues in flow and oxygenation dependent contrast (FLOOD) imaging of tumours, *NMR Biomed.* **14**, 497-506 (2001).
5. S. P. Robinson, D. R. Collingridge, F. A. Howe, L. M. Rodrigues, D. J. Chaplin, and J. R. Griffiths, Tumor response to hypercapnia and hyperoxia monitored by FLOOD magnetic resonance imaging, *NMR Biomed.* **12**, 98-106 (1999).
6. C. E. Cooper, C. E. Elwell, J. H. Meek, S. J. Matcher, J. S. Wyatt, M. Cope, and D. T. Delpy, The non-invasive measurement of absolute cerebral deoxyhemoglobin concentration and mean optical pathlength in the neonatal brain by second derivative near infrared spectroscopy, *Pediatr. Res.* **39**, 32-38 (1996).
7. R. Springett, J. Newman, M. Cope, and D. T. Delpy, Oxygen dependency and precision of cytochrome oxidase signal from full spectral NIRS of the piglet brain, *Am. J. Physiol. Heart. Circ. Physiol.* **279**, H2202-H2209 (2000).
8. E. L. Hull, D. L. Conover, and T. H. Foster, Carbogen-induced changes in rat mammary tumour oxygenation reported by near infrared spectroscopy, *Br. J. Cancer* **79**, 1709-1716 (1999).

9. Y. Gu, V. A. Bourke, J. G. Kim, A. Constantinescu, R. P. Mason, and H. Liu, Dynamic response of breast tumor oxygenation to hyperoxic respiratory challenge monitored with three oxygen-sensitive parameters, *App. Optics* **16**, 2960-2966 (2003).
10. J. F. Dunn, Y. Zaim-Wadghiri, B. W. Pogue, and I. Kida, BOLD MRI vs. NIR spectrophotometry: will the best technique come forward, *Adv. Exp. Med. Biol.* **44**, 103-113 (1998).
11. S. Punwani, R. J. Ordidge, C. E. Cooper, P. Amess, and M. Clemence, MRI measurements of cerebral deoxyhemoglobin concentration [dHb] – correlation with near infrared spectroscopy, *NMR Biomed.* **11**, 281-289 (1998).
12. C. Aalkjær, and L. Poston, Effects of pH on vascular tension: which are the important mechanisms? *J. Vasc. Res.* **33**, 347-359 (1996).
13. I. T. Demchenko, T. D. Oury, J. D. Crapo, and C. A. Piantidosi, Regulation of the brain's vascular response to oxygen, *Circ. Res.* **91**, 1031-1037 (2002).
14. J. A. Kruuv, W. R. Inch, and J. A. McCredie, Blood flow and oxygenation of tumors in mice, *Cancer* **20**, 51-59 (1967).
15. D. M. Brizel, S. Lin, J. L. Johnson, J. Brooks, M. W. Dewhirst, and C. A. Piantadosi, The mechanisms by which hyperbaric oxygen and carbogen improve tumour oxygenation, *Br. J. Cancer* **72**, 1120-1124 (1995).
16. G. M. Tozer, R. J. Maxwell, J. R. Griffiths, and P. Pham, Modification of the ^{31}P magnetic resonance spectra of a rat tumour using vasodilators and its relationship to hypotension, *Br. J. Cancer* **62**, 553-560 (1990).
17. R. G. Steen, D. A. Wilson, C. Bowser, J. P. Wehrle, S. S. Rajan, and J. D. Glickson, ^{31}P NMR spectroscopic and near infrared spectrophotometric studies of effects of anesthetics on in vivo RIF-1 tumors: relationship to tumor radiosensitivity, *NMR Biomed.* **2**, 87-92 (1989).

HYPERTHERMIA AND HYPERMETABOLISM IN FOCAL CEREBRAL ISCHEMIA

Edwin M. Nemoto, Charles Jungreis, Donald Larnard, Hiroto Kuwabara, Michael Horowitz, and Amin Kassam

Abstract: The reliable and reproducible creation of an animal model of focal cerebral ischemia is not easily accomplished. Using a transortibal approach, we showed that occlusion of the posterior cerebral artery (PCA), middle cerebral artery (MCA), and the contralateral anterior cerebral artery (ACA) created a large cortical and subcortical stroke in the non-human primate (NHP). Subsequently, we created the same stroke endovascularly in the NHP. Using the endovascular stroke model in the NHP, we measured brain temperature with thermocouples and cerebral blood flow (CBF) by stable xenon CT in one NHP, and $CMRO_2$ and CBF by positron emission tomography (PET) in another NHP.

Two female non-human primates (*M. mulatta*) weighing 7.0 and 8.0 kg, respectively, were studied under fentanyl-diazepam anesthesia with continuous monitoring of arterial blood pressure, rectal temperature, and end-tidal CO_2 with intermittent blood gas measurements. Using an endovascular approach, the PCA (P2), MCA (M1), and the ICA at the bifurcation and contralateral ACA produced a large hemispheric stroke. In the right ischemic hemisphere, temperatures increased by 2°C-3°C. PET measurement of CBF and $CMRO_2$ showed that $CMRO_2$ increased in the region of the ischemic stroke. We found that both hyperthermia and hypermetabolism occur in acute stroke.

1. INTRODUCTION

Creation of a reliable and reproducible animal model of focal cerebral ischemia (FCI) is not easily accomplished because of intra- and interspecies variability in collateral circulation. In the rat, FCI has been induced by direct surgical occlusion of the MCA via a craniotomy[1] or by insertion of a propylene suture of given size into the MCA followed by withdrawal.[2] In larger animals, such as the pig[3] or the nonhuman primate, FCI has been induced by a craniotomy via the transorbital approach[4, 5] that is highly invasive and especially problematic when changes in brain temperature are of interest. The craniotomy and loss of cerebrospinal fluid destroys the temperature dynamics and gradients within the brain.

In a series of studies in the non-human primate (NHP), we showed that transorbital occlusion of the posterior cerebral artery (PCA), middle cerebral artery (MCA), and the ipsilateral anterior cerebral artery (ACA) creates a large cortical and subcortical stroke in the NHP (Nemoto et al., unpublished). Subsequently, we created the same stroke by endovascular occlusion of the same vessels using coils in the NHP that also created a large MCA stroke.[6] Using the endovascular stroke model in the NHP, we measured brain temperature with thermocouples and cerebral blood flow (CBF) by stable xenon CT in one animal, and $CMRO_2$ and CBF by positron emission tomography (PET) in another animal.

2. METHODS AND MATERIALS

2.1. Temperature and Cerebral Blood Flow (CBF) by Stable Xenon CT

In a protocol approved by the Institutional Animal Care and Use Committee (IACUC), two non-human primates (*M. mulatta*) were anesthetized with ketamine 20 mg/kg and 1 mg atropine, i.m. Their tracheas were intubated with cuffed endotracheal tubes (4.5 mm OD) and their lungs mechanically ventilated on 1%-3% isoflurane/70% nitrous oxide/30% oxygen during insertion of a catheter into the femoral artery for continuous monitoring of arterial blood pressure and arterial blood samples for blood gas analyses. Thereafter, the animals were transported to the angiography suite where GDC coils (1 mm x 5 mm) were inserted into the posterior cerebral artery (PCA-P2), the ipsilateral anterior cerebral artery (ACA-A1) and into the middle cerebral artery (MCA-M1), and verified for complete occlusion by contrast injection (Fig. 1).

After endovascular occlusion for the MCA stroke, the NHP was immediately taken to the CT scanner where stable xenon CT CBF measurements were made. The NHP inhaled 28% medical-grade xenon gas in 40% oxygen (XeScan stable xenon in oxygen USP, Praxair Pharmaceutical Gases, Praxair, Inc., Danbury, CT) via facemask, for 4.3 minutes. Two scans were obtained at each of four CT levels before and six scans during Xe inhalation (General Electric Systems, Milwaukee, WI).

After the Xe/CT scan, the NHP was transported to the laboratory where 14 (7 in each hemisphere) thermocouple probes (1 mm x 25 mm) were inserted into the brain, which occurred at approximately 2 hours after the MCA stroke. Each temperature probe had 6 thermocouples at 5 mm intervals. The temperature measurements shown in Figure 2 are the temperatures obtained upon insertion of the probes into the brain. The time zero measurements were made two hours after MCA occlusion.

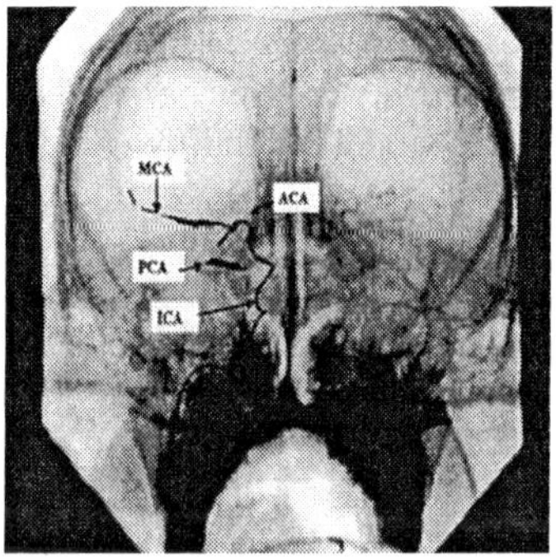

Figure 1. Angiogram showing coils inserted into the posterior cerebral artery (PCA-P2), at the bifurcation of the internal carotid artery (ICA) occluding also the anterior cerebral artery (ACA-A1), and in the middle cerebral artery (MCA-M1).

2.2. Temperature, Cerebral Blood Flow, and Oxygen Consumption by Positron Emission Tomography

In the second NHP, before induction of the stroke, a baseline anatomical MRI was done. Following the MRI, the NHP was taken to the laboratory where 6 temperature probes (1 mm x 25 mm) each with 6 thermocouples at 5 mm intervals were inserted into the hemispheres bilaterally. The NHP was then taken to the PET scanner for a baseline scan. Thereafter, the NHP was taken to the angiosuite where the same three-vessel occlusion was performed with coils, and immediately thereafter, the NHP was taken to the PET scanner for post-stroke scans. PET measurements of CBF, $CMRO_2$, and OEF were made for up to 6 hrs post-stroke.

2.2.1. MRI for PET Co-registration

MRI studies were performed on a 1.5-Tesla, whole-body Signa Scanner (General Electric Medical Systems, Milwaukee, WI). The NHP was positioned in a standard head coil and a brief scout T1-weighted image was obtained from which axial 3 mm T1 weighted images and FLAIRS were prescribed. T2 and proton density images were acquired to exclude unexpected pathology. The total time for the MRI scan was about 30 minutes. The MRI data were transferred to the PET facility over the electronic network and coregistered with the PET data on a SPARC station. The co-registered MRI was used as an individualized anatomic map for the selection of regions of interest (ROI) and for each of the planes.

2.2.2. Positron Emission Tomography

2.2.2a. PET Scanner System. The PET studies were performed at the UPMC PET facility on a Siemens//CTI HR+, high-resolution tomograph. The HR+ has an axial field-of-view of 15.2 cm covering the whole brain and acquires 63 transaxial slices 2.4 mm thick. The measured in-plane spatial resolution is 4.5 mm full-width at half-maximum, and the axial resolution is 3.5 mm. Studies were acquired in two-dimensional (2D) mode.

2.2.2b. CBF Measurements. Continuous arterial blood sampling (6 ml/min) was initiated at the time of injection of 50 mCi ^{15}O-water in 5-7 cc saline as an intravenous bolus. Arterial blood was withdrawn by a Master flex peristalsis pump (model #7550-90) through a Siemens Activity Counter. A 180 second dynamic PET scan (10 x 6 seconds, then 10 x 12 seconds) began upon injection. Blood withdrawal continued for 30 seconds beyond the end of the PET scan, for a total withdrawal time of about 210 seconds and a total blood volume of about 20 mls, which was returned to the animal after data acquisition was completed. Blood activity data was automatically accumulated in a Sun SPARC station for later processing.

2.2.2c. Calculation of CBF. The ^{15}O-water PET data were analyzed using a 2-compartment (1-tissue compartment) model approach described by Ohta et al.[7] This involved 3 parameters that provided a measure of the clearance of ^{15}O-water from blood to brain (K_1, ml/min/ml), brain efflux (K_2, min^{-1}), and arterial input function timing delays.[8] CBF was calculated as $K_1 * 100$ (ml/100g/min) assuming a tissue density of one.

2.2.2d. $CMRO_2$ Measurement. Each $CMRO_2$ measurement began with inhalation of 200 cc of medical air containing 50 mCi of ^{15}O-oxygen gas from a Victoreen Gas Delivery System. Upon breathing the tracer, a 180 second dynamic PET scan (10 x 6 seconds, then 10 x 12 seconds) with continuous arterial blood sampling was begun, as previously described for the CBF measurements. An arterial blood sample was obtained at the end of each scan and analyzed for arterial oxygen content, blood gases, and hematocrit.

2.2.2e. Calculation of $CMRO_2$. Calculation of $CMRO_2$ was performed using a two-compartment model approach described by Mintun et al.[9] and modified by Ohta et al.[10] $CMRO_2$ (ml/100g/min) was calculated as follows: $CMRO_2 = K_1^O \cdot C_aO_2 \times 100$, where K_1^0 is the rate of uptake of O_2 gas from blood to brain and C_aO_2 is the arterial oxygen content (ml O_2/ml blood).

2.2.2f. Calculation of OEF. Oxygen extraction fraction (OEF) was calculated as the K_1^o/K_1^wratio for individual regions. This ratio relates oxygen gas uptake to CBF, which relates to the rate of oxygen delivery. As previously defined, $CMRO_2$ is the value of $K1^0$ multiplied by the arterial oxygen content.

3. RESULTS

3.1. Temperature and Cerebral Blood Flow Study

A large cortical and subcortical stroke was observed with zero blood flow in much of the MCA territory clearly showing the effectiveness of the technique in inducing a large stroke (Fig. 3). Temperatures in the ischemic hemisphere in the frontal and medial parietal cortex approached 38°C to 40°C (Fig. 2) despite the exposed cranium, which caused temperatures in the normal hemisphere to range between 27°C and 29°C. Thus, the temperature gradient between normal and ischemic hemispheres was as high as 10°C.

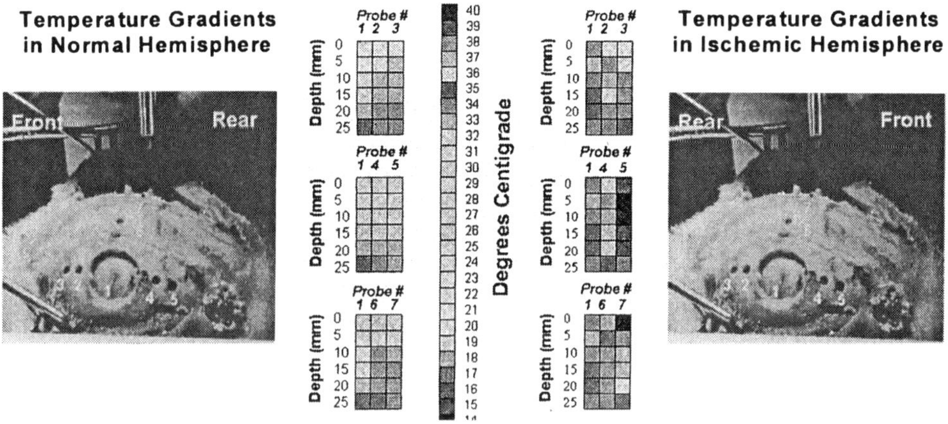

Figure 2. Temperature profiles 2h after focal cerebral ischemia with probe locations numbered and temperatures in the normal left hemisphere (left panel) and the ischemic right hemisphere (right panel). Temperatures in the normal hemisphere range between 28°C-30°C in the upper layers (0-10 mm) due to craniectomy and exposure. Despite this exposure, temperatures in the ischemic hemisphere ranged as high as 40°C in probes located in the frontal (#5) and medial (#7) cortex.

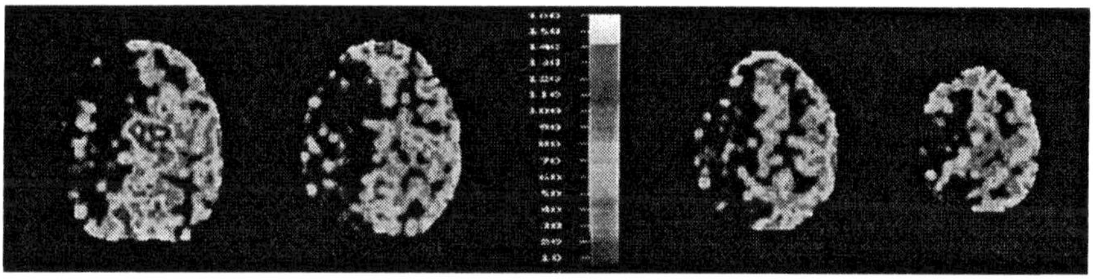

Figure 3. Stable xenon cerebral blood flow maps at four axial levels in the NHP at 60 min after endovascular MCA occlusion with a color scale. Left hemisphere image is right hemisphere and right hemisphere image is left hemisphere. Note large regions of cerebral hypoperfusion with CBF values less than 20 ml/100 g/min.

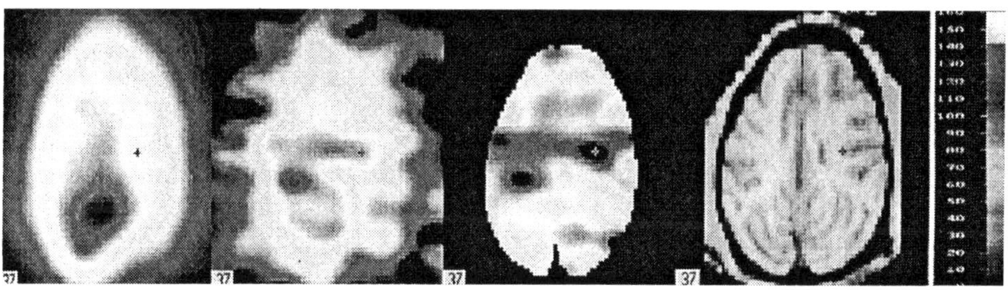

Figure 4. Positron emission tomography (PET) images of cerebral blood flow (CBF), cerebral metabolic rate for oxygen ($CMRO_2$), and oxygen extraction fraction (OEF) with a magnetic resonance image (MRI). Color scale on the right is applicable for all PET images. Right image is the right hemisphere and left image is left hemisphere.

3.2. Cerebral Blood Flow and Oxygen Consumption by Positron Emission Tomography

Endovascular occlusion in the second NHP produced temperature increases of up to 2°C, which was somewhat lower compared with the previous study. The explanation is probably that the MCA was punctured and a subarachnoid hemorrhage was created, as revealed by inspection of the brain after the study. Nevertheless, an ischemic pattern was observed, as shown in Figure 4. Also illustrated in Figure 4 are the changes in $CMRO_2$, OEF, and the MRI image. A mark (+) in the figure provides a point of reference between figures, which illustrates the fact that in the region of hypoperfusion, one can see that $CMRO_2$ and OEF are both elevated with low CBF. Also notable is the fact that the increased $CMRO_2$ and OEF are seen not only in the ischemic region but also in the normal areas of the brain.

4. DISCUSSION

An increase of 2°C-3°C in brain temperature was observed within the first two hours during ischemia. The increase in brain temperature was associated with an increase in $CMRO_2$ in parallel with an increase in OEF, despite a reduction in CBF. The increase in $CMRO_2$ observed by PET occurred not only in the ischemic hemisphere, but also in the normal hemisphere where there was no increase in temperature, but rather a decrease. This apparent paradox could be explained by a stress-induced, catecholamine-mediated increase in circulating catecholamines which impact on brain metabolism to increase $CMRO_2$ in both the ischemic and normal hemisphere. However, in the ischemic hemisphere, the low CBF allows the build up of the heat of metabolism whereas in the normal, non-ischemic hemisphere the accompanying increase in blood flow would allow the clearance of the heat generated by metabolism.

The detrimental effects of even a one degree centigrade rise in temperature on the severity of ischemic[11] or traumatic brain injury[12] has been well documented. The correlation between fever and severity of ischemic injury and outcome after stroke is also well-recognized.[13, 14] However, not well appreciated is the fact that brain temperature itself increases as a result of ischemia. Busto et al.[11] reported a decrease in brain temperature with ischemia in a rat, whose small, and thin calvarium is prone to heat loss. On the other hand, Minamisawa et al.[15] completely enclosed the rat in a humidified temperature controlled box and showed an increase in brain temperature with global brain ischemia.

The question may be asked as to why a *decrease* in brain temperature rather than an *increase* has been observed in most of the experimental studies reported previously. We believe that many of the studies have been done in mice and rats whose thin calvarium allows rapid heat loss through the skull especially when the skin has been reflected or removed for insertion of temperature electrodes in the skull. In these small animals, the flow of heat is from the cerebrovasculature to the brain and out through the skull. In large animal models, occlusion of the middle cerebral artery requires either a pteriotomy and a subtemporal approach or a transorbital approach to the MCA, both of which completely destroy the thermal dynamics of the brain as well as cause a loss of cerebrospinal fluid[4, 5] resulting in a route of heat loss from blood to tissue and out through the skull. This is clearly different in the circumstance where the thermal dynamics of the brain and the skull are intact and the cerebral circulation dissipates the heat of brain metabolism. The directional flow of heat is then from the brain parenchyma and to the blood perfusing the brain, at least as long as brain metabolism continues.

The impact of the increase in brain temperature as a direct result of the reduction in perfusion and as a secondary result of the catecholamine-mediated stress response remains to be determined. It is clear, however, that cerebral hyperthermia and hypermetabolism in the face of hypoperfusion is detrimental to the ischemic or traumatized brain.

REFERENCES

1. A. Tamura, D. I. Graham, J. McCulloch, and G. M. Teasdale, Focal cerebral ischemia in the rat. I. Description of technique and early neuropathological consequences following middle cerebral artery occlusion, *J. Cereb. Blood Flow Metab.* **1**, 53-60 (1981).
2. K. Minematsu, L. Li, M. Fisher, C. H. Sotak, M. A. Davis, and M. S. Fiandaca, Diffusion-weighted magnetic resonance imaging: rapid and quantitative detection of focal brain ischemia, *Neurol.* **42**(1), 235-240 (1992).
3. M. Sakoh, L. Ostergaard, L. Rohl, D. F. Smith, C. Z. Simonsen, J. C. Sorensen, P. V. Poulsen, C. Gyldensted, S. Sakaki, and A. Gjedde, Relationship between residual cerebral blood flow and oxygen metabolism as predictive of ischemic tissue viability: sequential multitracer positron emission tomography scanning of middle cerebral artery occlusion during the critical first 6 hours after stroke in pigs, *J. Neurosurg.* **93**(4), 647-657 (2000).
4. J. G. Frazee, X. Luo, G. Luan, D. S. Hinton, D. A. Hovda, M. S. Shiroishi, and L. T. Barcliff, Retrograde transvenous neuroperfusion: a back door treatment for stroke, *Stroke* **29**(9), 1912-1916 (1998).
5. J. Huang, J. Mocco, T. F. Choudhri, A. Poisik, S. J. Popilskis, R. Emerson, R. L. DelaPaz, A. G. Khandji, D. J. Pinsky, and E. S. Connolly Jr., A modified transorbital baboon model of reperfused stroke, *Stroke* **31**(12), 3054-3063 (2000).
6. M. Horowitz, A. Kassam, E. M. Nemoto, C. A. Jungreis, G. R. Rao, and J. Arimoto, An endovascular primate model for the production of a middle cerebral artery ischemic infarction, *Intervent. Radiol.* **7**, 223-228 (2001).
7. S. Ohta, E. Meyer, H. Fujita, D. C. Reutens, A. Evans, and A. Gjedde, Cerebral [^{15}O]water clearance in humans determined by PET: I. Theory and normal values, *J. Cereb. Blood Flow Metab.* **16**, 765-780 (1966).
8. H. Iida, I. Kanno, S. Miura, M. Murakami, K. Takahashi, and K. Uemura, Error analysis of a quantitative cerebral blood flow measurement using $H_2{}^{15}O$autoradiography and positron emission tomography with respect to the dispersion of the input function, *J. Cereb. Blood Flow Metab.* **6**, 536-545 (1986).
9. M. A. Mintun, M. E. Raichle, W. R. Martin, and P. Herscovitch, Brain oxygen utilization measured with O-15 radiotracers and positron emission tomography, *J. Nucl. Med.* **25**, 177-187 (1983).
10. S. Ohta, E. Meyer, and C. J. Thompson, Oxygen consumption of the living human brain measured after a single inhalation of positron emitting oxygen, *J. Cereb. Blood Flow Metab.* **12**, 179-192 (1992).
11. R. Busto, W. D. Dietrich, M. Y. T. Globus, I. Valdes, P. Schienberg, and M. D. Ginsberg, Small differences in intraischemic brain temperature critically determine the extent of ischemic neuronal injury, *J. Cereb. Blood Flow Metab.* **7**, 729-738 (1987).
12. R. F. Albrecht 2nd, C. T. Wass, and W. L. Lanier, Occurrence of potentially detrimental temperature alterations in hospitalized patients at risk for brain injury, *Mayo Clin. Proc.* **73**(7), 629-635 (1988).
13. J. Castillo, A. Davalos, J. Marrugat, and M. Noya, Timing for fever-related brain damage in acute ischemic stroke, *Stroke* **29**, 2455-2460 (1998).
14. M. D. Ginsburg, and R. Busto, Combating hyperthermia in acute stroke: A significant clinical concern, *Stroke* **29**, 529-534 (1998).
15. H. Minamisawa, M. L. Smith, and B. K. Siesjo, The effect of mild hyperthermia and hypothermia on brain damage following 5, 10, and 15 minutes of forebrain ischemia, *Ann. Neurol.* **28**, 26-33 (1990).

VARIABILITY OF CEREBRAL HEMOGLOBIN CONCENTRATION IN VERY PRETERM INFANTS DURING THE FIRST 6 HOURS OF LIFE

Kurt von Siebenthal, Matthias Keel, Jean-Claude Fauchère, Vera Dietz, Daniel Haensse, Ursula Wolf, Urs Helfenstein, Oskar Bänziger, Hans U. Bucher, and Martin Wolf

Abstract: Cerebral hemoglobin concentration (cHbc), a major determinant of oxygen transport capacity in the brain, shows a considerable variability due to physiological and methodological factors. In order to determine the (relative) contribution of these factors, the cHbc variability within the first 6 hours of life was studied in 28 very preterm infants using near infrared spectrophotometry (NIRS). Mean cHbc values were 46.4 ± 14.1 μmol/l (2.75 ± 0.84 ml/100 g). Is the variability in cHbc related to the methodology of cHbc measurements or to physiological variables? A statistical model of stepwise regression (backward selection) with 13 independent variables and with cHbc as a dependent variable showed that, from the total variability of ± 14.1 μmol/l, only 3.7 μmol/l (26%) were of methodological origin, while the major portion, 9.3 μmol/l (66%) were related to four physiological variables: birth weight, gestational age, blood glucose and transcutaneous carbon dioxide tension. The remaining 1.1 μmol/l (7.8%) were unexplained.

We conclude that NIRS, which allows continuous monitoring of cerebral oxygenation and metabolism even in the first hours of postnatal life, is a valid technique to measure cHbc in very preterm infants. The major portion of the large variability of early cHbc registrations can be attributed to physiological factors.

1. INTRODUCTION

The mortality of very low birth weight infants has decreased over the last few decades. Cerebral morbidity and long-term sequelae, however, have remained constant.[1]

Neonatal research needs to be focused on the etiology of cerebral morbidity in order to understand the pathogenesis of neurodevelopmental impairment. Several methods have been used for this purpose. Analysis of perinatal risk factors showed only a weak correlation with development in later years.[2] Neuroimaging[3] and long-term measurements of neurophysiological variables appear to be better predictive tools, but they are expensive and demand high levels of expertise.[4, 5] An alternative method for exploring early cerebral dysfunction is near infrared spectrophotometry (NIRS). This technique is non-invasive and can be used at the bedside for long-term monitoring of even very tiny preterm infants.

One of the parameters measured by NIRS is cerebral hemoglobin concentration (cHbc in μM), which is the major indicator of oxygen transport capacity in the brain. cHbc varies widely within and between subjects. This variability is of instrumental or physiological origin. If the instrumental part can be reduced, the physiological variability prevails, making cHbc a promising candidate for assessing cerebral perfusion. Previously, we were able to reduce the instrumental variability by setting quality standards.[6]

The aim of this study was to determine the relative contribution of physiological and methodological factors to the cHbc variability within the first 6 hours of postnatal life.

2. METHODOLOGY

2.1. Near Infrared Spectrophotometry (NIRS)

The method of NIRS has been described in detail elsewhere.[7-10] In the neonatal brain, NIRS measures concentration changes in cerebral oxyhemoglobin (O_2Hb in μM), deoxyhemoglobin (HHb in μM), and their sum, total hemoglobin (tHb). By inducing slow changes in arterial oxygen saturation (SaO_2 in %), it is possible to determine cHbc.[6] Cerebral blood volume (CBV) in ml/100 g can be derived if the hemoglobin concentration in the peripheral blood is known.[11] The assessment of cHbc requires a slow oxygenation change over a period of at least 1.2 minutes by altering the inspired oxygen fraction. The concomitant change in the SaO_2 is measured by pulse oxymetry, while the concentration changes of cerebral O_2Hb, HHb, and tHb are measured by NIRS. A valid evaluation requires a change in SaO_2 of at least 4%, while remaining within the normal range of 85% to 95%. The slope between the changes in O_2Hb and SaO_2 is determined by a linear regression. It represents the amount of additional O_2Hb per % change in SaO_2. The absolute cHbc corresponds to the amount of O_2Hb for an extrapolated 100% change in SaO_2, i.e. $cHbc = 100 * dO_2Hb/dSaO_2$ (μM).

The neonatal sensor of the Critikon Cerebral Oxygenation Monitor 2001 was placed fronto-parietally. It emitted light at 774, 815, 867, 904 nm. A flexible fibre bundle conducted the light to the infant. Transmitted light pulses emerging from the tissue were detected by a silicon photodiode converting the light into electrical signals. Attenuation changes were converted to concentration changes of the chromophores.[12] The interoptode distance was 35 mm and the differential pathlength factor was 4.4.[13]

2.2. Measurement Procedure

NIRS measurements began within the first 6 hours of life and continued for 2 hours. Transcutaneous pO_2 ($tcpO_2$, Hellige Oxymonitor) and pCO_2 ($tcpCO_2$, Hellige Kapnomonitor), SaO_2 (Nellcor 200 or Radiometer Oximeter), heart rate, and mean arterial pressure recorded from an umbilical artery catheter (Hellige Vicom-SM) were recorded with the NIRS data. These variables were kept within normal ranges during the study.

A cranial ultrasound (Acuson 128XP, 7 MHz transducer) was, whenever possible, performed within the first 24 hours of life (1 infant with a germinal matrix hemorrhage was examined on day 3). Intracranial hemorrhages were graded,[14] and the most severe degree during the first 3 days of postnatal life was included in the statistical analysis.

This study was approved by the local ethical committee of our institution and parental informed consent was obtained from one or both parents.

2.3. Statistical Analysis

A statistical model of stepwise regression (backward selection) was performed with the following 13 independent variables: 1) gestational age, 2) birth weight, 3) head circumference, 4) sex, 5) umbilical arterial pH at birth, 6) hemoglobin concentration in peripheral blood, 7) blood glucose, 8) initial inspired oxygen fraction, 9) SaO_2, 10) $tcpCO_2$, medication [11) analgesia, 12) sedation] and 13) cranial ultrasound. cHbc was the dependent variable. The number of independent variables had to be restricted to 13 for statistical reasons. The main inclusion criterion for categorical variables (11, 12, and 13) was that at least 3 infants differed from the main level. The stepwise regression procedure is a method to identify a combination of independent variables that have a significant influence on the dependent variable cHbc. In the backward selection, initially all independent variables are entered into the model and then the least important variables are removed in steps until all remaining variables are significant. The procedure tests whether the accuracy of the model improves if any of the removed variables were to be entered again.[15]

2.4. Test/Retest Variability (TRV)

In an earlier study performed with the same equipment, we described the methodology in detail.[6] Based on 218 slow oxygenation changes in 39 NIRS sessions, we demonstrated a TRV of 17.3%. The prerequisite for the calculation of a valid TRV was that cHbc remained constant during all of the slow oxygenation changes in one session. This was fulfilled if tHb did not vary by more than 1.5 μM. Other factors influencing the cHbc measurements, such as the duration of the oxygenation change, the changes in SaO_2, tHb, mean arterial blood pressure, and $tcpCO_2$, and tHb versus oxygen index, were studied and standards for valid measurements were set.

2.5. Patients

Twenty-eight very preterm infants (15 female, 13 male) admitted to our neonatal intensive care unit were studied within the first 6 hours of life. Infants with severe birth asphyxia or with malformations were excluded. Mean ± standard deviation of gestational age was 28 ± 2.2 weeks, birth weight 1125 ± 412 g, head circumference of 25.9 ± 2.6 cm, and umbilical arterial pH at birth 7.26 ± 0.08. 19 infants were mechanically ventilated (9 conventionally, 10 with high frequency oscillation at 15 Hz) and 9 infants required nasal CPAP. All infants received supplemental oxygen with a mean inspired oxygen fraction of 0.42 ± 0.15%. The following parameters were determined at the time of measurement: hemoglobin 16.6 ± 2.7 g/100 ml, blood glucose 4.9 ± 2.2 mM, SaO_2 91.2% ± 2.3 %, and $tcpCO_2$ 5.6 ± 1.1 kPa. Twenty infants received pethidine as analgesic and 5 received diazepam for sedation. In 4 infants, an intracranial hemorrhage of grade 1 or 2 was diagnosed during their first 3 days of post-natal life.[14]

3. RESULTS

3.1. Cerebral Hemoglobin Concentration (cHbc) and Blood Volume (CBV)

Mean cHbc values were 46.4 ± 14.1 μM (corresponding to 2.75 ± 0.84 ml/100 g by using the same assumptions as in the literature[11]).

3.2. Test/Retest Variability (TRV)

One hundred and thirty one (n = 131) slow oxygenation changes (average 4.7 per infant) were induced in 28 infants. Due to the repeated measurements, the influence of the TRV on the results is reduced by a factor of $2.2 = \sqrt{4.7}$, which is equal to 7.8% of the mean cHbc of 46.4 μM. Thus, methodological factors contribute 3.7 μM (26%) to the total variability of 14.1 μM (Table 1).

3.3. Physiological Variables

Out of 13 variables, only birth weight, gestational age, pCO_2, and blood glucose contributed significantly to the cHbc variability during the first 6 hours of life (Table 2). The r^2 value of the model was 0.66; in other words, 66% of the cHbc variability can be explained by these variables, accounting for 9.3 μM of the cHbc variability of 14.1 μM. Only 1.1 μM/l (7.8%) of the variability remains unexplained in our model (Table 1).

Table 1. Factors contributing to cHbc variability.

	Variability of cHbc		
Influencing factors	Proportion of mean of 46.4 μM	Proportion of SD of 14.1 μM	cHbc variability
NIRS methodology: TRV of 17.3%	7.8%	26%	3.7 μM
Physiological variables: birth weight, gestational age, blood glucose, and pCO_2.	20%	66%	9.3 μM
Unknown	2.3%	7.8%	1.1 μM

Table 2. Physiological variables influencing cHbc.

	Estimate	*P* value
Birth weight	-0.06	< .0001
Gestational age	8.6	.0004
pCO_2	4.7	.0095
Blood glucose concentration	-2.2	.0335

4. DISCUSSION

We found a mean cHbc of 46.4 ± 14.1 μM and CBV of 2.75 ± 0.84 ml/100 g. These results are comparable with those from other studies examining brain circulation. Using NIRS in transmission mode, we found CBV values of 2.22 ± 0.44 ml/100 g (37.46 ± 6.75 μM) in 12 preterm and term infants.[11] One study including only preterm infants demonstrated slightly higher CBV values of 3.7 ± 1.1 ml/100 g.[16] Other studies performed with positron emission tomography and single photon emission computer tomography and with NIRS in adults reported higher CBV values.[10, 17, 18] In preterm infants, however, the variability of cHbc or CBV when measured with NIRS appears to be larger when compared with adult values. Some studies have suggested that immature cerebral autoregulation in preterm infants could be responsible for this increased variability.[19, 20] The main question is whether this variability of cHbc in very preterm infants can be explained by instrumental errors of NIRS or by physiological variables.

In a first step, we reduced the variability due to methodological errors (e.g. movement artifacts or other everyday interferences) that affect the steady state conditions by modifying the previous method,[11] especially with regard to the quality standards. Without applying those quality standards, the TRV was higher than 40%. With the quality standards,[6] the TRV was reduced to 17.3% by rejecting 50% of the measurements. Despite the strict standards in this study, we found a large inter-individual variability of 30.4% of the mean cHbc (14.1 μM of 46.4 μM). Due to repeated measurements (4.7 per infant), the influence of the TRV was further reduced from 17.3% to 7.8%, corresponding to 3.7 μM (7.8% of 46.4 μM). We therefore conclude that the TRV, which represents the methodological errors, contributes 3.7 μM to the total variability of 14.1 μM, i.e. 26%.

Physiological variables, however, have a much greater influence on cHbc than the instrumental errors. Birth weight, gestational age, blood glucose, and $tcpCO_2$ were found to contribute significantly to the cHbc variability. The r^2 value of 0.66 indicates that 66% (9.3 μM of 14.1 μM) of the cHbc variability is explained by these variables (Table 1). Birth weight showed a highly significant negative correlation with an effect of –0.06, which means that for each 100 g increase in birth weight, the cHbc decreased by 6 μM. On the other hand, an increase in gestational age of one week increased cHbc by 8.6 μM. For a preterm infant gaining weight at a rate of 140 g per week, the decrease in cHbc of 8.4 μM would be leveled out by the concomitant increase of cHbc of 8.6 μM due to gestational age. Infants who are small for their gestational age showed high cHbc values. Furthermore, there was a positive correlation between $tcpCO_2$ and cHbc. For every kPa increase in $tcpCO_2$, cHbc increased by 4.7 μM. pCO_2 is one of the most important factors influencing cerebral autoregulation and its increase leads to an increase in cHbc.[21]

As known from cerebral blood flow measurements,[22] blood glucose values correlate negatively with cHbc. A decrease in blood glucose was associated with increased cHbc values (per 1 mM fall in blood glucose, there was a 2.2. μM increase in cHbc). With the *P* value just being significant ($P = .03$), it was surprising to find such a strong effect in the multivariate analysis. This seems to indicate that blood glucose levels and perhaps other metabolic factors could strongly influence cerebral hemodynamics.

5. CONCLUSION

We conclude that, in preterm infants, cHbc evaluation using NIRS in reflectance mode is a valid method. The cHbc values are in a similar range as those obtained using NIRS in transmission mode or other methods. The cHbc values have a higher variability in preterm infants than in adults. The major portion (66%) of this variability can be explained by four physiological variables: birth weight, gestational age, blood glucose, and tcpCO_2 (Table 1). Only 26% of the variability can be attributed to the technical errors; a small percentage (7.8%) of the variability remains unexplained.

NIRS is a non-invasive technique that disturbs neither infants nor the neonatal staff, and it allows continuous monitoring of cerebral oxygenation, circulation, and metabolism. There is hope that such registrations will contribute to a better understanding of cerebral dysfunction in very tiny preterm infants and thereby lead to new insights in their management, which may ultimately improve their later psychomotoric development.

6. ACKNOWLEDGMENTS

This work was funded in part by a Johnson & Johnson International Academic Research Support Award and by a grant of the Swiss National Research Council (No. 32-25812.86). The authors thank Dr. S. Kundu for her helpful suggestions.

REFERENCES

1. B. Hagberg, G. Hagberg, I. Olow, and L. van Wendt, The changing panorama of cerebral palsy in Sweden, VII. Prevalence and origin in the birth year period 1987-90, *Acta Paediatrica* **85**, 954-960 (1996).
2. R. Largo, D. Pfister, L. Molinari, S. Kundu, A. Lipp, and G. Duc, Significance of prenatal, perinatal and postnatal factors in the development of AGA preterm infants at five and seven years, *Dev. Med. Child Neurol.* **31**, 440-456 (1989).
3. M. Graham, M. Levene, J. Trounce, and N. Rutter, Prediction of cerebral palsy in very low birth weight infants: prognostic ultrasound study, *Lancet* **2**(8559), 593-596 (1987).
4. D. Wertheim, E. Mercuri, J. Faundez, M. Rutherford, D. Acolet, and L. Dubowitz, Prognostic value of continuous electroencephalographic recording in full term infants with hypoxic ischaemic encephalopathy, *Arch. Dis. Child* **71**, F97-F102 (1994).
5. L. Hellström-Westas, I. Rosen, and N. Svenningsen, Predictive value of early continuous amplitude integrated EEG recordings on outcome after severe birth asphyxia in full term infants, *Arch. Dis. Child* **72**, F32-F38 (1995).
6. M. Wolf, H. U. Bucher, V. Dietz, M. Keel, K. von Siebenthal, and G. Duc, How to evaluate slow oxygenation changes to estimate absolute cerebral haemoglobin concentration by Near Infrared Spectrophotometry, *Adv. Exp. Med. Biol.* **411**, 495-501 (1996).
7. J. E. Brazy, Near-infrared spectroscopy, *Clin. Perinatol.* **18**, 519-534 (1991).
8. K. von Siebenthal, G. Bernert, and P. Casaer, Near-infrared spectroscopy in newborn infants, *Brain Dev.* **14**, 135-143 (1992).
9. H. U. Bucher, A. D. Edwards, A. E. Lipp, and G. Duc, Comparison between near infrared spectroscopy and ^{133}Xenon clearance for estimation of cerebral blood flow in critically ill preterm infants, *Pediatr. Res.* **33**, 56-60 (1993).
10. C. E. Elwell, *A Practical Users Guide to Near Infrared Spectroscopy*, (Hamamatsu Photonics, Japan, 1995).
11. J. S. Wyatt, M. Cope, D. T. Delpy, C. E. Richardson, A. D. Edwards, S. Wray, and E. O. Reynolds, Quantitation of cerebral blood volume in human infants by near-infrared spectroscopy, *J. Appl. Physiol.* **68**, 1086-1091 (1990).
12. S. Matcher, C. Elwell, C. Cooper, M. Cope, and D. Delpy, Performance comparison of several published tissue near-infrared spectroscopy algorithms, *Anal. Biochem.* **227**, 54-68 (1995).

13. J. S. Wyatt, M. Cope, D. T. Delpy, P. van der Zee, S. Arridge, A. D. Edwards, and E. O. Reynolds, Measurement of optical path length for cerebral near-infrared spectroscopy in newborn infants, *Dev. Neurosci.* **12**, 140-144 (1990).
14. L. Papile, J. Burstein, and H. Koffler, Incidence and evolution of subependymal and intraventricular haemorrhage: A study of infants with birth weights less than 1500 g, *J. Pediatr.* **92**, 529-534 (1978).
15. A. J. Miller, *Subset Selection in Regression* (2nd edition), (Chapman & Hall/CRC Press, London & New York, 2002).
16. N. C. Brun, and G. Greisen, Cerebrovascular responses to carbon dioxide as detected by near-infrared spectrophotometry: comparison of three different measures, *Pediatr. Res.* **36**, 20-24 (1994).
17. R. Grubb, M. Raichle, C. Higgins, and J. Eichling, Measurement of regional blood volume by emission tomography, *Ann. Neurol.* **4**, 322-328 (1978).
18. F. Sakai, K. Nakazawa, Y. Tazaki, K. Ishii, H. Hino, H. Igarashi, and T. Kanda, Regional cerebral blood flow and blood volume and hematocrit measured in normal human volunteers by single-photon emission computed tomography, *J. Cereb. Blood Flow. Metab.* **5**, 207-213 (1985).
19. V. Ramaekers, P. Casaer, H. Daniels, and G. Marchal, Upper limits of brain flow autoregulation in stable infants of various conceptional age, *Early Human. Dev.* **24**, 249-258 (1990).
20. J. S. Wyatt, D. A. Edwards, M. Cope, D. T. Delpy, D. C. McCormick, A. Potter, and E. O. Reynolds, Response of cerebral blood volume to changes in arterial carbon dioxide tension in preterm and term infants, *Pediatr. Res.* **29**, 553-557 (1991).
21. V. Dietz, M. Wolf, M. Keel, K. von Siebenthal, O. Baenziger, and H. U. Bucher, CO_2 reactivity of cerebral haemoglobin concentration in healthy newborns measured by near infrared spectrophotometry, *Biol. Neonate* **75,** 85-90 (1999).
22. O. Pryds, G. Greisen, and B. Friis-Hansen, Compensatory increase of CBF in preterm infants during hypoglycaemia, *Acta Paediatr. Scand.* **77**, 632-637 (1988).

COMPUTATIONAL STUDY ON USE OF SINGLE-POINT ANALYSIS METHOD FOR QUANTITATING LOCAL CEREBRAL BLOOD FLOW IN MICE

Michelle A. Puchowicz, Krishnan Radhakrishnan, Kui Xu, Danielle L. Magness, and Joseph C. LaManna

Abstract: The benefits of a mouse model are efficiency and availability of transgenics/ knockouts. Quantitation of cerebral blood in small animals is difficult because the cannulation procedure may introduce errors. The [^{14}C]-iodoantipyrine autoradiography (IAP) method requires both the tissue concentration and the time course of arterial concentration of the [^{14}C] radioactive tracer. A single point-analysis technique was evaluated for measuring blood flow in mice (30 g ± 0.3 g; n = 11) by using computational models of sensitivity analysis, which quantitates relationships between the predictions of a model and its parameters. Using [^{14}C]-IAP in conjunction with mathematical algorithms and assumed arterial concentration-versus-time profiles, cortical blood flow was deduced from single-point measurements of the arterial tracer concentration. The data showed the arterial concentration profile that produced the most realistic blood flows (1.6 ± 0.4; mean ± SD, ml/g/min) was a profile with a ramp time of 30 sec followed by a constant value over the remaining time period of 30 sec. Sensitivity analysis showed that the total experimental time period was a more important parameter than the lag period and the ramp period. Thus, it appears that the accuracy of the assumption of linearly increasing arterial concentration depends on the experimental time period and the final arterial [^{14}C]-iodoantipyrine concentration.

1. INTRODUCTION

The recent development of mutant mice make the mouse a useful model for studying neurological diseases by combining local cerebral blood flow measurements[1, 2] and molecular mechanisms associated with brain disease and injury. The iodoantipyrine (IAP) method for measuring local cerebral blood flow requires quantification of both local tissue and arterial concentrations of [^{14}C]-IAP. Use of this method has been well developed in the rat,[3-6] but it is not widely practiced in smaller animals, such as the mouse, because of the technical difficulties associated with the small size.[1] Erroneous blood flow values will result if corrections for catheter lag time, dead space washout, and blood loss are not taken into account.

A recent methodological study on the quantitative measurement of local cerebral blood flow demonstrates use of a modified version of the [^{14}C]-IAP technique in anesthetized mice.[2] We explored use of this method for quantitating cortical blood flow in mice and found that the assumption of a linear arterial tracer concentration-versus-time profile over a 60 sec period resulted in unrealistic blood flow values that were 3-fold greater than previously reported.[2-4] However, use of an arterial tracer concentration profile that included a ramp up to a time of 30 sec followed by a constant value over the remaining 30 sec produced blood flows that were similar to previously reported values. In addition, sensitivity analysis was used to study effects of uncertainties associated with three models of assumed arterial tracer concentration-versus-time profiles deduced from single point measurements of arterial [^{14}C] tracer concentration.

2. METHODS

2.1. Determination of Cortical Blood Flow in Mice

Cortical blood flow was determined using [^{14}C]-iodoantipyrine (IAP) autoradiography in conjunction with a mathematical algorithm. Adult C57Bl male mice (26.5 ± 2.6 g; n = 11) were anesthetized with halothane and 2.5 μCi of [^{14}C]-IAP was administered by intraperitoneal injection. The mice were immersed in liquid nitrogen 60 sec after injection.[2, 4] The brains were then removed and sectioned in a cryotome (–20°C). Slides containing brain sections were placed on a Hyperfilm β-max autographic film with [^{14}C] standards for 45 days. The images were digitized using the BIOQUANT image analysis system and the tissue radioactivity concentration calculated. Blood from the frozen heart was sampled to obtain the final arterial blood [^{14}C] radioactivity.

2.2. Quantitation of Blood Flow

Cortical blood flows were calculated according to the algorithm described by Sakurada et al.[3] In combination with autoradiography, a convolution integral (Eq. 1) and an assumed arterial [^{14}C] tracer concentration-versus-time profile was used to determine blood flow rates in each of the mice from the final [^{14}C] tracer concentration in the blood sampled from the frozen heart:

$$C^*(T) = \lambda\kappa\int_0^T C_a^*(t)e^{-\kappa(T-t)}dt \tag{1}$$

In this equation, $C^*(T)$ is local tissue radioactivity (dpm/g) at time T (end of the experimental period), $C_a^*(t)$ is arterial blood radioactivity (dpm/ml) at any time t, λ is tissue:blood partition coefficient (0.8 for IAP), and $K = (f\,m)/(\lambda V)$, where f is blood flow rate (ml/min), m is diffusion coefficient of the tracer across the blood-brain barrier, and V is tissue mass (g).[2] Three specific arterial [^{14}C] tracer concentration-versus-time profiles were used to evaluate the calculated blood flows in the mice: (a) ramp (linear rise) over the time interval [t_1, t_3], (b) step function (constant value) over the time interval [t_1, t_3], and (c) a ramp over the time interval [t_1, t_2], followed by a constant value over the time interval [t_2, t_3] (Fig. 1).

Each of the three models comprised a short time delay (t_1) of 3.3 sec.[2] This time period was followed by a temporal profile of the arterial [^{14}C] tracer concentration from zero over the experimental time period (t_3) of 60 sec. The value of t_2 (30 sec) was estimated on the basis of the arterial tracer concentration-versus-time obtained by Maeda et al.[2] An expression for the tissue [^{14}C] tracer concentration at the sampling time (60 sec) was then derived from the convolution integral and the blood flow deduced by an iterative method. By assuming instantaneous achievement of steady state, the second profile (Fig. 1b) produces a lower bound to the blood flow and thus serves as an accuracy check of the blood flow computed with the more realistic model.

2.3. Evaluation of Concentration-Versus-Time Profiles

In addition to the blood flows, sensitivity analysis was used to assess the effects of uncertainties in the three time points (t_1, t_2, and t_3) on the computed blood flows produced by the three profiles (Fig. 1). Sensitivity analysis provides information on the effects of

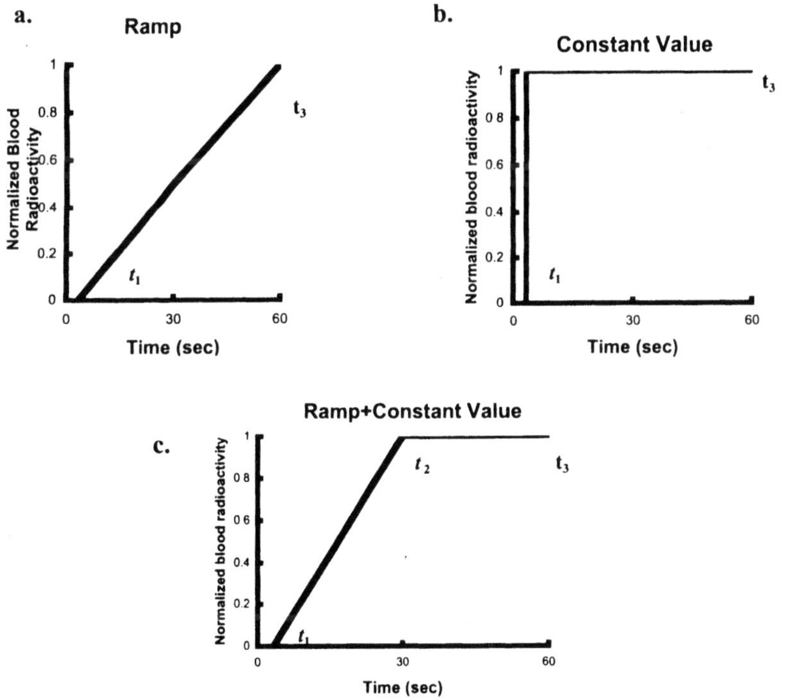

Figure 1. Three assumed concentration-versus-time profiles over a 60 sec period with 3.3 sec lag time: **a**. ramp (linear rise) over the time interval [t_1, t_3], **b**. step function (constant value) over the time interval [t_1, t_3], and **c**. ramp over the time interval [t_1, t_2], followed by a constant value over the time interval [t_2, t_3].

uncertainties or changes in the input parameters of the problem on the solution.[7-10] By using sensitivity analysis to assess the effects of the uncertainties in the three time points, we were able to examine the relative importance of the selected three time values to the computed blood flow. As discussed previously, nominal values for the time points for the three profiles were either selected on the basis of previously reported values or estimated.

3. RESULTS

The results of this study showed that determining blood flow in mice using the modified [^{14}C]-IAP method and an assumed linear profile of the arterial [^{14}C] tracer over the 60 sec period (Fig. 1a) produced blood flow values 3-fold greater than expected (Table 1). Therefore we examined the use of two other profiles (Fig. 1b-c) to identify the most appropriate profile for our conditions and found that a linear ramp time up to 30 sec followed by a constant value over the remaining 30 sec, produced blood flows similar to previously reported values in mice.[1, 2] The constant value profile (Fig. 1b) produced lower blood flow values as expected. Also, the arterial [^{14}C] tracer concentration was three times greater than previously reported using a modified IAP method despite the similarities in both the initial dose (2.5 mCi/30 g mouse) and specific activity of the tracers (40-60 mCi/mmol).[2]

Sensitivity analysis revealed that t_3 resulted in the greatest percent change in blood flow due to a 1% change in uncertainty compared to the other time points, t_1 or t_2 (Table 1). The ramp + constant value profile yielded the greatest error in percent change in blood flow in t_3 compared to the other profiles. For all three profiles, the percent change in blood flow was similar due to a 1% change in uncertainty in t_1.

4. DISCUSSION

The measurement of local cerebral blood flow using [^{14}C]-IAP with an intraperitoneal tracer and a single blood sample from the heart at the end of the experiment can be relatively accurate, since this method does not require cannulation of blood vessels or frequent blood sampling. A study where this method was used to quantitate local cerebral blood flow in anesthetized mice revealed that the final blood concentration of [^{14}C] tracer sampled from the heart (454 ± 242 nCi/ml) correlated to that

Table 1. Blood flow and normalized sensitivity coefficients.

Profile	Blood flow (ml/min/g)	Normalized Sensitivity Coefficients		
		<S(t_1)>	<S(t_2)>	<S(t_3)>
Ramp	3.48 ± 1.49	0.006 ± 0.00	--	-1.005 ± 0.00
Constant Value	1.16 ± 0.34	0.006 ± 0.00	--	-1.005 ± 0.00
Ramp + Constant Value	1.56 ± 0.46	0.003 ± 0.00	0.42 ± 0.02	-1.35 ± 0.02

sampled from the artery.[2] The authors concluded that the modified [^{14}C]-IAP method can be used to measure blood flow in mice. This method assumes a linear rise from [^{14}C] tracer entering the arterial blood following 30 to 120 sec from the time of the intraperitoneal injection to the value measured in the blood sample from the frozen heart.

The results of the present study using the modified [^{14}C]-IAP method indicated that the computed blood flow is a function of the arterial concentration profile, thus illustrating the importance of an accurate input profile. Errors in the input profile will result in erroneous blood flow values. Physiological measurements of the arterial profile would provide the most accurate input data. However, measurements in a small animal, such as the mouse, are subject to constraints in the surgical procedures required for collecting serial blood samples. It would therefore be useful to establish the minimal number of serial samples and sampling frequency needed in order to quantitate blood flow with adequate accuracy, as substantiated by sensitivity analysis. Data from our preliminary experiments in mice indicate that the arterial concentration is given by a lag time followed by a ramp and then a constant value, as assumed in Figure 1. The preliminary data also indicate a lag time greater than the assumed value (Fig. 1). This difference in the assumed lag time[2] of 3.3 sec and what we measured (25 sec) could be attributed to the invasive nature of the surgical procedure required to obtain arterial blood samples over time as well as the extended exposure to anesthesia.

For arterial tracer concentration-versus-time profiles that are not a constant linear rise over the experimental time period, the assumed profile needs to be modified. We found that the ramp + constant value profile resulted in more realistic blood flow values. An explanation for requiring the ramp + constant profile is that the uptake of the tracer from the intraperitoneal cavity to the venous system is saturated, resulting in a non-linear profile. This process would be analogous to a "leaky" capacitor that results in slow discharge or uptake from one compartment to another. In addition, the final blood concentration of [^{14}C] tracer in our studies was three times greater than that reported by Maeda et al.[2] However, the uptake profile from the intraperitoneal cavity into the venous system should not be affected by the initial amount given unless the saturation time point shifts to the left with increasing [^{14}C] tracer concentration. Thus, the discrepancies in the blood flow values computed with the linear profile may be unsuitable for higher [^{14}C] tracer blood concentrations or when uptake of the tracer from the intraperitoneal cavity to the venous system is saturated and does not follow a linear profile. These differences among studies of computed blood flows when using the modified IAP method remain to be resolved, perhaps by using other comparative methods, such as those using isopropyl-[^{14}C]-iodoamphetamine (IPIA).[11, 12]

Sensitivity analysis revealed that the total time period is a more important parameter than the earlier time periods in that a 1% uncertainty in the former time has a larger effect on the computed blood flow; that is, t_3 is more important than t_2 or t_1. Thus, sampling at an earlier time, when the arterial tracer concentration is more likely to be a ramp (linear range) would reduce error introduced when using a blood flow model that assumes a linear arterial tracer concentration profile. Thus, the ideal sampling time would most likely be at the point where the ramp meets the constant value.

In conclusion, the modified [^{14}C]-IAP method can be used in a mouse model with careful consideration given to the assumed arterial tracer concentration profile. Blood flow methods that require the tracer to be taken up by the venous system and assume the shape of an arterial profile would be more sensitive to errors. Future experiments will include comparing arterial profiles when sampling from artery versus heart to determine if a shift in the saturation time point occurs with increasing tracer concentration as well as using other blood flow indicator analogs, such as IAPA.

5. ACKNOWLEDGMENTS

These studies were supported by the National Institutes of Neurological Disorders and Stroke NS46074, National Institute of General Medical Sciences GM066309, Medical Research Service of the Department of Veteran Affairs and the National Institutes of Health EY00300.

REFERENCES

1. T. M. Jay, G. Lucignani, A. M. Crane, J. Jehle, and L. Sokoloff, Measurement of local cerebral blood flow with [^{14}C]iodoantipyrine in the mouse, *J. Cereb. Blood Flow Metab.* **8**, 121-129 (1988).
2. K. Maeda, G. Mies, L. Olah, and K. A. Hossmann, Quantitative measurement of local cerebral blood flow in the anesthetized mouse using intraperitoneal [^{14}C]iodoantipyrine injection and final arterial heart blood sampling, *J. Cereb. Blood Flow Metab.* **20**, 10-14 (2000).
3. O. Sakurada, C. Kennedy, J. Jehle, J. D. Brown, G. L. Carbin, and L. Sokoloff, Measurement of local cerebral blood flow with iodo [^{14}C] antipyrine, *Am. J. Physiol.* **234**, H59-H66 (1978).
4. H. K. Richards, A. H. J. Lovick, and J. D. Pickard, A modification of the method for the measurement of cerebral blood flow using [^{14}C]iodoantipyrine in small animals, *J. Cereb. Blood Flow Metab.* **7**, 124-126 (1987).
5. H. N. Ferimer, K. L. Kutina, and J. C. LaManna, Methyl isobutyl amiloride delays normalization of brain intracellular pH after cardiac arrest in rats, *Crit. Care Med.* **23**(6), 1106-1111 (1995).
6. K. L. Lauro, H. Kabert, and J. C. LaManna, Methyl isobutyl amiloride alters regional brain reperfusion after resuscitation from cardiac arrest in rats, *Brain Res.* **831**(1-2), 64-71 (1999).
7. K. Radhakrishnan, Combustion Kinetics and Sensitivity Analysis Computations, in: *Numerical Approaches to Combustion Modeling*, edited by E. S. Oran and J. P. Boris (American Institute of Aeronautics and Astronautics, Washington, DC, 1991) pp. 83-128.
8. K. Radhakrishnan, *LSENS–A General Chemical Kinetics and Sensitivity Analysis Code for Homogeneous Gas-Phase Reactions*, NASA RP-1328, Washington, DC. (1994).
9. K. Radhakrishnan, J. C. LaManna, and M. E. Cabrera, A quantitative study of oxygen as a metabolic regulator, *Appl. Cardiopulm. Pathophysiol.* **9**, 363-367 (2000).
10. K. Radhakrishnan, LSENS: Multipurpose kinetics and sensitivity analysis code for homogeneous gas-phase reactions, *AIAA J*, **41**, 848-855 (2003).
11. M. Ringel, R. M. Bryan, and R. C. Vannucci, Regional cerebral blood flow during hypoxia-ischemia in the immature rat: comparison of iodoantipyrine and iodoamphetamine as radioactive tracers, *Brain Res. Dev. Brain Res.* **59**, 231-235 (1991).
12. R. M. Byran, and R. B. Duckrow, Radial columns in autoradiographs from tracer methods for measuring cerebral cortical blood flow, *Am. J. Physiol.* **269**, H583-H589 (1995).

EFFECTS OF TEMPERATURE ON OXYGEN TRANSPORT IN SHEETS AND SPHERES OF RESPIRING TISSUES

Michael McCabe, and David J. Maguire

Abstract: The effect of temperature upon the oxygen partial pressure profiles (and hence upon flux) of oxygen through respiring tissues of differing architecture is examined. We have considered the two situations of respiring sheets of tissue and of respiring spheres.

Sheets of respiring tissue can model to some extent the behaviour of skin (which abandons its own temperature stasis in response to its obligations in the control of overall body temperature).

The oxygen profiles of spheres of respiring tissues subject to temperature shifts is investigated since it is a model for solid tumour oxygen kinetics where a spherical tumour, inadequately supplied with a capillary network, is being treated by one or another form of hyperthermia during cancer therapy.

1. INTRODUCTION

It is over 80 years since Krogh and Erlanger[1] first applied the laws governing diffusion to explain the oxygen profiles within the zones of tissues surrounding capillaries. Their work was extended by Warburg[2] who was able to use their approach to calculate the maximum permissible thickness for a slice of tissue to be fully oxygenated by diffusion from the cut surfaces. Since then we have examined the possibility of relating overall flux of oxygen as an explicit function of temperature.[3] It is this problem which we continue to address here.

Krogh and Warburg both used definitions of diffusion coefficient and respiration rate that are non-standard. Their diffusion coefficient, known as the Krogh diffusion coefficient, is expressed as ml of oxygen (reduced to STP) passing cm^2 of tissue surface per minute in response to an oxygen gradient of 1 atmosphere pressure per cm of tissue. Their respiration rate is expressed as ml of oxygen consumed (reduced to STP) per ml of tissue, per minute.

As their equations stand, there is no overt reference to temperature. However, it is known that both diffusion and the kinetics of respiration are functions of temperature.

2. THE EFFECT OF TEMPERATURE ON RESPIRATION

The rate relationship for a complex system such as respiration has been discussed by Borgmann,[4] who deduced an equation of the form

$$a = CT \exp(-u/RT) \tag{1}$$

where a is a respiration rate; C is an empirical constant for any given tissue; T is the absolute temperature; R is the gas constant in cal. $mol^{-1}T^{-1}$, and u is a complex temperature coefficient, which is in turn a function of the energies of activation of all of the enzyme systems taking part in the overall respiration. A number of authors have shown that this form of equation is appropriate to describe the respiration of isolated cell suspensions (at pO_2 levels where diffusion profiles are not limiting the respiration, see Farr & Furhman,[5] Furhman,[6] and Longmuir[7]). These authors have convincingly demonstrated that the respiration of real tissue systems does approximate to Arrhenius-type kinetics. These plots allow temperature coefficients to be measured. The Arrhenius coefficients for respiring cell suspensions, measured over temperature ranges of 273 to 300 degrees Kelvin, fall within the range 8,500 to 16,000 cal per mole, depending somewhat upon the actual tissue under investigation.

3. THE EFFECT OF TEMPERATURE UPON DIFFUSION

MacDougall and McCabe[8] have shown that the Krogh and Fick diffusion coefficients are related by an equation with the following form:

$$D(\text{Krogh}) = 60\alpha\, D(\text{Fick}) \tag{2}$$

where α is the Bunsen absorption coefficient of oxygen gas within the tissue, expressed as volumes of gas (reduced to STP) dissolved in unit volume of the (non-respiring) tissue under one atmosphere pressure of oxygen. The value of α cannot always be approximated to that for water since it has been shown that there is frequently considerable oxygen gas exclusion within some tissues.[9, 10] Despite the problem of exclusion of oxygen from certain (connective) tissues, it remains generally true that the relationship between oxygen gas solubility and temperature for tissues is approximated by a general thermodynamic relationship given by Glasstone.[11]

The Fick diffusion coefficient can also be expressed as a function of temperature.[12] This latter relationship requires an examination of the value of the viscosity of the tissue. In fact, the required viscosity value is the microscopic viscosity of the tissue (see the discussion by Ogston et al.[13]). This microscopic viscosity of tissue is presumed, as a first approximation, to be close to that for water. There are numerous equations which connect water viscosity with temperature, including those with an Arrhenius (exponential) form. When all of these substitutions are made into the original Krogh equation, an equation is obtained which relates the Krogh diffusion coefficient to temperature, as well as to the Bunsen absorption coefficient for oxygen, α; the differential heat of solution of oxygen into the tissue, δH; the hydrated radius of the oxygen molecule, r_o; and B, an experimentally derived constant from the (Arrhenius type) relationship between temperature and water viscosity.[14] R is the gas constant.

$$D(\text{Krogh}) = A\,T\,\exp\{(-\delta H + B)/RT\} \quad (3)$$

where $A = 10\alpha R/\pi r_o$. The validity of this equation has been tested by McCabe et al.[3] and found to be consistent with experimental data obtained by Ackerman and Gainer.[15]

This relationship can be inserted into the Warburg equation. For sheets of respiring tissue, the substituted Warburg equation takes the form

$$H = B^{*}C^{1/2}\exp(f/2RT)$$

$$B^{*} = 80\alpha_0 R/\pi r_0; \qquad f = \mu - (\delta H + B) \quad (4)$$

where H is the limiting thickness for the slice which is just oxygenated to its centre, and μ is a complex temperature coefficient which is a function of the energies of activation of the enzymes within the respiratory complex (see McCabe et al.[3]).

For spheres of respiring tissue, the equation takes a slightly different form as follows:

$$R_L = 0.866\,B^{*}C^{1/2}\exp(f/2RT) \quad (5)$$

where the limiting radius (R_L) represents the radius of tissue which is just oxygenated to its core by diffusion from its surface.

Investigation of these two functions has shown that although the functions are somewhat dependent upon the values of the constants involved, they are nevertheless dominated by the exponential component. Thus whether the limiting thickness or the limiting radius of the oxygenated tissue will increase or decrease with temperature depends upon the sign of the function f. This function has been evaluated over the range of temperatures from 0°C to 50°C (see McCabe et al.[3]).

4. RESULTS & DISCUSSION

4.1. For Sheets of Respiring Skin

Human skin commonly functions over a wide range of temperatures. This is because whole body thermostasis is partially accomplished by abandoning temperature homeostasis within the skin. Skin is supplied with oxygen by diffusion from the subdermal capillaries of the *rete subpapillare*, which forms the capillary network at the boundary between the papillary and reticular layers of the skin. By inserting an appropriate experimental value for the activation energy of the respiratory complex of skin cells (about 8,500 cal.), it can be shown that an adequate supply of oxygen at 30°C becomes inadequate at 40°C, since by then only about 80% of the respiring sheet would still be oxygenated. However, we can observe that temperature elevation of the skin produces a consequent vasodilation of the subdermal capillaries. This in turn generates an increased blood flow, which together with a local temperature-induced shift of the oxygen-haemoglobin binding isotherm in the blood flowing through the capillaries, provides the necessary increased oxygen partial pressure gradient to completely oxygenate the whole of the tissue.

This strategy for skin oxygenation is seen to be an economical one since it avoids the need to supply an overabundance of oxygen to skin at lower temperatures in order to accommodate the demand at higher temperatures. Thus oxygen homeostasis for skin is maintained over a considerable range of temperatures by the expedient of controlling the blood flow within the *rete subpapillare*, presumably through local mechanisms responsive to both temperature and the extent of oxygenation of the skin. The sheet model may also be a model for cell culture and tissue engineering studies, where temperature changes during short periods (as in microscopic examination of material in petri dishes) may lead to transient oxygen deprivation or over-supply.

4.2. For Spheres of Respiring Tissue

Diffusion of oxygen from the outer surface into the core of a sphere of respiring tissue is important since it can be a model for the oxygenation of tumours in their early stages, either preceding capillary infiltration, or where angiogenesis is inadequate. In studying tumour oxygen kinetics there does exist the possibility of induced variations of temperature, since local or even whole body hyperthermia has been applied as a part of the approach to tumour therapy.

The insertion of experimental values appropriate for tumour cell respiration into the equation describing the limiting respiring radius for a spherical tissue also shows that, as the temperature for respiration is raised, the extent of oxygenation will diminish. This is important since there is a generally protective effect of hypoxia or anoxia in minimising induced damage, certainly by radiation, but also possibly by other agents. The tendency to increased anoxia for the tumour can be diminished or overcome by increasing the oxygenation of the surrounding capillarised tissues.

The spherical model is also a model for the earliest stages of foetal development, although there is little likelihood of DT in the protected environment of the foetus. However, temperature variations may occur during *in vitro* fertilization, when early embryos are removed from incubators for the purpose of examination.

4.3. Consequences for Re-warming of Cooled Tissues

Significant temperature changes can occur within respiring tissues when the body is cooled, either by design as in some cardiac surgery, or by accident, due to immersion in freezing water. As the body or organ temperature is lowered, there is a reduction in oxygen demand, accompanied by a diminution in diffusion coefficient and an increased oxygen binding to blood. The substituted Warburg equation shows that the diminished respiration is dominant and outweighs the reduction in oxygen supply due to diminished diffusion. This results in an unwanted excess of free oxygen within the tissue. However, the diminished temperature also shifts the oxygen-haemoglobin binding isotherm of whole blood to the left, and this left shift results in a reduction of free oxygen difference between arterial and venous blood, thus diminishing the available oxygen for diffusion. Re-warming of previously cooled tissues is thought to represent a time of maximum risk for oxygen-induced damage to tissues, and the substituted equations which we have generated should be of use in predicting the likely consequence on tissue oxygenation for any sequence of changes of organ perfusion during re-warming, which should allow the risk of damage to be minimised.

REFERENCES

1. A. Krogh, The number and distribution of capillaries in muscles with calculations of the oxygen pressure head necessary for supplying the tissue, *J. Physiol.* **52**, 409-415 (1919).
2. O. H. Warburg, Versuche an uberlebendem Carcinomgewebe [Investigations into surviving carcinoma tissue], *Biochem. Z.* **142**, 317-334 (1923).
3. M. McCabe, K. Adam, and D. Maguire, The effect of temperature upon the combined diffusional and kinetic parameters of tissue respiration, *J. Theor. Biol.* **78**(1), 51-59 (1979).
4. U. Borgmann, A theoretical description of biological rates, *J. Theor. Biol.* **45**, 171-182 (1974).
5. D. A. Farr, and F. A. Fuhrman, Role of diffusion of oxygen in the respiration of tissues at different temperatures, *J. Appl. Physiol.* **20**, 637-646 (1965).
6. F. A. Fuhrman, Oxygen consumption of mammalian tissues at reduced temperatures, *Natl. Acad. Sci.* **451**, 50-51 (1956).
7. I. S. Longmuir, Respiration rate of rat-liver cells at low oxygen concentrations. *Biochem. J.* **65**(2), 378-382 (1957).
8. J. D. B. MacDougall, and M. McCabe, Diffusion coefficient of oxygen through tissues, *Nature* **215**(106), 1173-1174 (1967).
9. K. Greven, The oxygen diffusion coefficient of the liver, renal cortex and cerebral cortex under various conditions, *Pflugers Arch. Gesamte. Physiol. Menschen. Tiere.* **271**, 14-22 (1960).
10. R. Zander, Experimental Research on the O2 supply of the eye, *Albrecht Von Graefes Arch. Klin. Exp. Ophthalmol.* **195**(4), 215-230 (1975).
11. S. Glasstone, First and second laws of thermodynamics, in: *Textbook of Physical Chemistry, 2nd Edition* (Macmillan & Co., London, 1951), pp. 695-712.
12. S. Glasstone, Surface phenomena, in: *Textbook of Physical Chemistry, 2nd Edition* (Macmillan & Co., London, 1951), pp. 1257-1272.
13. A. G. Ogston, B. N. Preston, and J. D. Wells, On the transport of compact particles through solutions of chain polymers, *Proc. R. Soc. (London)* **333**, 297-316 (1973).
14. R. Battino, and H. L. Clever, The Solubility of Gases in Liquids, *Chem. Rev.* **66**, 395-463 (1966).
15. A. Akgerman, and J. L. Gainer, Predicting gas-liquid diffusivities, *J. Chem. Eng. Data* **17**, 372-377 (1972).

MODELING OF THE RESPONSE OF p_tO_2 IN RAT BRAIN TO CHANGES IN PHYSIOLOGICAL PARAMETERS

Oleg Y. Grinberg, Huagang Hou, Marcie A. Roche, Jennifer Merlis, Stalina A. Grinberg, Nadeem Khan, Harold M. Swartz, and Jeff F. Dunn

Abstract: It is known that oxygen tension in tissue (p_tO_2) will change in response to an alteration of physiological parameters including: pCO_2 in arterial blood, blood flow, capillary density, oxygen carrying capacity, and p50 of hemoglobin. We have used modeling to compute the change of p_tO_2 in response to changes of each physiological parameter and related these changes to experimental data.

The oxygen distribution in a Krogh cylinder was computed assuming a linear decrease of hemoglobin saturation from the arterial to the venous end of the capillary. Parameters of the model were used to compute the baseline cerebral p_tO_2 expressed as the mean value of the p_tO_2 over the whole cylinder. These parameters were adjusted to derive p_tO_2 values close to those measured at the relevant experimental conditions. Then each desired parameter was varied to calculate the change in p_tO_2 related to this parameter.

Effects of different factors on cerebral p_tO_2 were modeled and compared with experimental values obtained with various experimental interventions including: changing CBF, modifying p50 with the allosteric modifier RSR13, modification of capillary density, and hemoglobin content. An acceptable agreement of the computed and the experimental changes of the cerebral p_tO_2 was obtained for these experimental conditions.

1. INTRODUCTION

Oxygen tension in tissue (p_tO_2) is an important parameter for many physiological and pathophysiological processes, especially in the brain. It is known that p_tO_2 will change in response to an alteration of several physiological parameters including: pCO_2 in arterial blood,[1] blood flow,[1, 2] capillary density,[3] oxygen carrying capacity,[3] and the p50 of hemoglobin.[4] As the role of oxygen in pathophysiology becomes more apparent, there is a growing need to be able to predict the effect of any physiological manipulation on the p_tO_2. The interpretation of experimental data on p_tO_2 would be aided by appropriate, simple modeling of the factors that influence cerebral p_tO_2. Models that attempt to predict the absolute value of p_tO_2 are difficult to use due to the requirement for appropriate input data. In this paper, we allow the initial input data to vary, in order to determine the "baseline" value for p_tO_2. Then we use the model to predict the impact on p_tO_2 of variations in any single input parameter. This allows us to predict the deviation in oxygen tension in tissue (Δp_tO_2) caused by variation of these parameters. We hypothesize that the Δp_tO_2 could be predicted quantitatively better than the absolute value of p_tO_2, especially if the variation of the parameter is small.

2. METHODS

2.1. Computing Procedure

Physiological parameters were used first to compute the baseline cerebral p_tO_2 expressed as the mean value of the p_tO_2 over the Krogh cylinder.[5, 6] These parameters were taken from the accepted ranges (or close to those) for the rat brain and were adjusted to derive p_tO_2 values close to those measured at the relevant experimental conditions. Then one desired parameter was changed to evaluate the change (response) in cerebral p_tO_2. Effects of different parameters on cerebral p_tO_2 were modeled and compared with experimental data where changes were induced in: CBF with different anesthetics, CBF with hyperventilation, the p50 using the allosteric modifier RSR-13, and acclimation to chronic hypoxia. The cerebral p_tO_2 was computed using Mathcad.

2.2. Oxygen Distribution in a Krogh Capillary

Assuming that oxygen consumption of the tissue throughout the length of the cylinder, as well as radially, and linear velocity of the blood through the tissue are constant, Kety[7] came to the conclusion that the oxygen concentration of the blood must fall linearly with distance as the blood traverses the capillary from arterial to the venous end. Using this assumption, oxygen saturation at point x_i of a capillary is presented as follows:

$$S(x_i) = S_a - (CMRO_2 * x_i)/(CBF * C) \tag{1}$$

where S_a is the oxygen saturation of blood at the arterial end of the capillary, $CMRO_2$ is the cerebral metabolic rate of oxygen (oxygen consumption per unit volume) of normal rat brain, CBF is cerebral blood flow, C (C = 0.0091 moles of oxygen per liter of blood) is a carrying capacity of blood (calculated based on accepted data that content of hemoglobin in rat blood is 15 g/dl and one gram of hemoglobin binds 1.36 ml of oxygen), and x_i is a dimensionless distance from the arterial end of the capillary, $0<x_i<1$, $i = 1,\ldots,N$ (in this calculation N = 40). The pO_2 in blood plasma $p_pO_2(x_i)$ in a capillary at point x_i can be expressed by the Hill equation:

$$p_pO_2(x_i) = \left(S(x_i)/[1 - S(x_i)]\right)^{1/n} p50 \tag{2}$$

Appropriate values for rat blood were used (n = 2.6, and p50 = 38 mm Hg).[8]

2.3. Tissue p_tO_2 Distribution in a Krogh Cylinder

The oxygen distribution in a Krogh cylinder was computed assuming negligible oxygen diffusion along the capillary. p_tO_2 in any point (x_i, r_j) of a Krogh cylinder is presented by substituting Eq.[2] into the Krogh equation:

$$p_tO_2(r_j,x_i) = p_pO_2(x_i) - CMRO_2 * \{R^2 \ln(r_j/r0)^2 - (r_j^2 - r_0^2)\}/4\alpha D \tag{3}$$

where $r_0 \geq r_j \geq R$, r_0 is the capillary radius, j = 1,…,N, α is the oxygen solubility, D is the oxygen diffusion coefficient, R is the radius of the Krogh cylinder, $R = 1/\sqrt{3.5 * L_v}$, and L_v is length of capillaries per unit volume. Following the analysis of the p_tO_2 measurements by electrodes carried out by Toma-Dasu et al,[9] we assumed that the electron paramagnetic resonance data on p_tO_2 reflects the mean value of p_tO_2 over the surface of the implanted material. From this, we argue that the EPR measured p_tO_2 can be presented as an average over the Krogh cylinder:

$$< p_tO_2 > = \frac{1}{N^2} \sum_i \sum_j p_tO_2(r_j, x_i) \tag{4}$$

where i and j represent arbitrary distance units along the capillary and away from the capillary, respectively, and the sum represents the p_tO_2 over the volume.

Note that the size of the Krogh cylinder in the rat brain is significantly smaller than the surface of the implanted material (about 1-2 mm^2). To model the measured p_tO_2, one may assume that a local p_tO_2 value (p_tO_2 in a small local area of the surface of the implanted material) randomly reflects the p_tO_2 value of adjacent Krogh cylinders. These values should be considered as a boundary condition of oxygen diffusion inside the volume of the implanted material. To simplify the desired solution, a mean p_tO_2 value could be presented by averaging these many local p_tO_2 values. In the absence of such information, the measured p_tO_2 values in this modeling are approximated by averaging p_tO_2 values over one Krogh cylinder.

2.4. Microvascular Morphometrics from Rats

Quantification of cerebrovascular morphology for this study was undertaken by Dunn et al.[10] using an infusion of dextran-conjugated fluoroescein isothiocyanate, rapid freezing, fixation, sectioning, and confocal microscopy. Quantitative measurements, corrected for slice thickness (volume = section thickness X area), were used to determine capillary radius (r_0), length volume (L_v), and intercapillary distance in a region of interest in the cortex. Capillary radius (r_0 = 2.2 μm) was averaged from measurements perpendicular to the longitudinal axis of the vessels. Length per volume of the capillaries was calculated from the total length of capillaries, $L_v = 805 \pm 27$ mm/mm^3. Intercapillary distance was measured from the average distance between microvessels, $2R = 40 \pm 1.8$ μm.[10] These parameters are consistent with previous measurements in rat brain by other groups.[11-13]

2.5. Parameters for Modeling

There are several other parameters that are required for modeling. The oxygen diffusion coefficient in water (D) is $2.3*10^{-5}$ cm^2/s at 38°C.[14] There are a range of estimates of oxygen diffusion coefficient D in tissue. For instance, Dutta and Popel[15] give an estimate of the intracellular oxygen diffusion coefficient in muscles as $8.5*10^{-5}$ - $2.2*10^{-5}$ cm^2/s. A value of $1*10^{-5}$ cm^2/s was used in this study. A typical value for the solubility of oxygen in blood plasma is $\alpha = 1.34*10^{-9}$ M/ml/mm Hg.[16] We used this value despite the fact that, on average, it should be higher in brain due to the presence of lipid membranes. Cerebral blood flow (CBF) has been measured over a range of 0.44-1.28 ml/g/min.[17, 18] Values for the cerebral metabolic rate ($CMRO_2$) of oxygen range from 26-70 $nM/s/cm^3$.[19-21] The initial values for modeling p_tO_2 were largely within these ranges. However, as it was mentioned above, this simplified model is intended to describe the change in p_tO_2, and not the absolute value itself. For instance, we found that using a value for $CMRO_2$ of 70-110 $nM/s/cm^3$ resulted in a good approximation of the initial values for p_tO_2.

3. RESULTS AND DISCUSSIONS

3.1. The Effect of Anesthesia on Cerebral p_tO_2

The effect of anesthetics on p_tO_2 in rat brain measured by EPR oximetry was reported by Hou et al.,[22] who measured p_tO_2 in rat brain at different levels of inspired oxygen. Figure 1 shows the experimental data and the predicted changes in p_tO_2 based on the model. As a first approximation, we modeled the effect of anesthesia on p_tO_2 in rat brain by the variation of CBF levels. Using CBF = 0.5ml/g/min, $CMRO_2$ was adjusted first to derive the p_tO_2 value close to that measured with $FiO_2 = 0.33$ (the left-most point). In order to obtain the correspondence between the p_tO_2 and P_aO_2 for the curves shown at Figure 1, CBF was varied and shown in the figure legend. An equal correspondence could have been obtained if one varied $CMRO_2$. The main point is that, once a correspondence has been achieved, it is now relatively simple to determine the influence of the single variable, in this case the P_aO_2 on p_tO_2.

3.2. The Effect of RSR13

A synthetic allosteric modifier of hemoglobin, RSR13, may be used to manipulate p_tO_2 for therapeutic purposes. Kunert et al.[23] observed a right-shift of p50 from 38 mm Hg to 58 mm Hg in rats 1 hour after the infusion of 200 mg/kg of RSR13. Although a modeling study indicated that Hb saturation would decline with infusion of RSR13 in a spontaneously breathing animal,[24] this was not observed to occur in a study where animals were ventilated.[4] Measurements in this study, with a ΔP_aO_2 value of 36.8 mm Hg

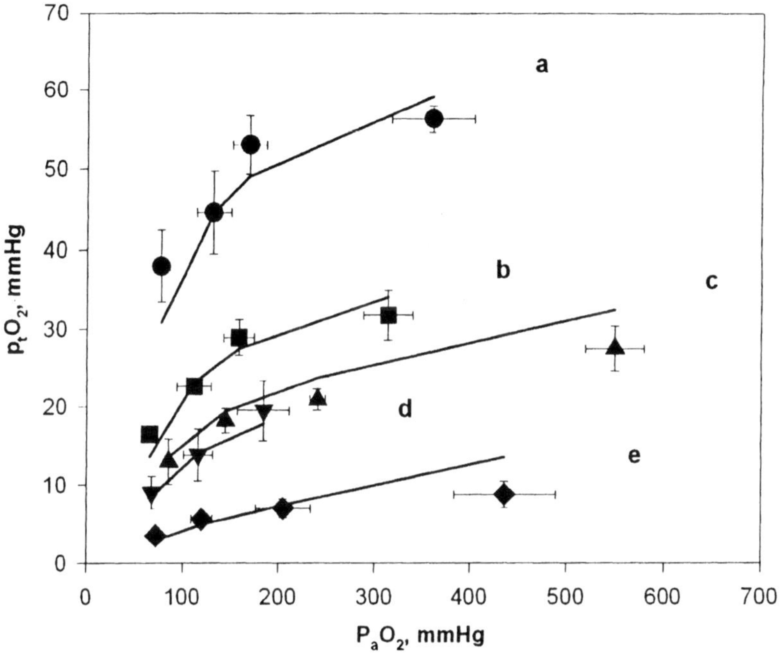

Figure 1. Effect of P_aO_2 on cerebral p_tO_2 with different anesthetics. The points are experimental; the lines are the result of modeling: a) isoflurane (CBF = 1.45 ml/ml/s), b) halothane (CBF = 0.7 ml/ml/s), c) α-chloralose/urethane (CBF = 0.5 ml/ml/s), d) pentobarbital (CBF = 0.4 ml/ml/s), e) ketamine/xylazine (CBF = 0.1 ml/ml/s).

and earlier measured a right-shift of p50,[23] suggesting that, in ventilated animals, a right-shift of the oxygen saturation curve due to the administration of RSR13 changes mainly the pO_2 in arterial blood, and Hb saturation changes insignificantly.

Three input parameters were adjusted to derive p_tO_2 values close to those measured in rat brain (P_aO_2 = 70 mm Hg, $CMRO_2$ = 80 nM/s/cm^3, and CBF = 0.8 ml/g/min). Using these data, the model predicted a Δp_tO_2 of 22.3 mm Hg, while the measured Δp_tO_2 was 11 mm Hg. The observed change in Δp_tO_2 was about half of the predicted value, which suggests that RSR13 influences tissue p_tO_2 through additional mechanisms to that of a change in p50.

3.3. Cerebral p_tO_2 in Acclimated Rats

It has been shown that acclimation to hypoxia (1/2 atm) significantly increases cerebral p_tO_2.[3] We modeled the impact of varying individual and multiple physiological parameters known to change with acclimation to low oxygen. The Hct increased from 41% to 70% (Dunn, *personal communication*). The L_v changed from 805 to 1265 mm/mm^3, and the capillary radius changed from 2.16 to 2.69 μm.[10] The P_aO_2 was previously measured at 98 mm Hg pre-acclimation and 103 mm Hg post-acclimation

Table 1. Measured and predicted changes in p_tO_2 in brain with acclimation to hypoxia.

Measured p_tO_2, mm Hg			Model predicted change with acclimation, predicted Δp_tO_2, mm Hg			
Pre-acclimation p_tO_2, mm Hg	Post-acclimation p_tO_2, mm Hg	Change with acclimation, Δp_tO_2, mm Hg	Increase only L_v	Increase only r_0	Increase only Hct	Change all parameters
26	62	36	9	3	13.5	24

The measured values are taken from Dunn et al.[3]

while breathing normoxic gas.[25] Assuming that chronic hypoxia does not change the oxygen affinity of blood,[25] and assuming a constant $CMRO_2$ and CBF, we computed changes in p_tO_2 with acclimation (Table 1).

Assuming the modeling reflects correctly the p_tO_2 measurements, this indicates that the desired Δp_tO_2 could not be computed using any one parameter. The fact that satisfactory correspondence between the computed and measured values of Δp_tO_2 in chronic rats could not be achieved even with simultaneous changes of all three parameters (hematocrit, length, and radius of capillaries) probably indicates that there also is an increase in CBF. This might be expected if animals that were adapted to low oxygen were acutely exposed to high oxygen (such as these were), resulting in an increase in $PaCO_2$ and a stimulation of CBF.[26]

3.4. Hyperventilation

Hyperventilation has been used for many years in the management of head injury. It has been shown[1] that hyperventilation results in a decrease of p_tO_2, an increase of P_aO_2, and a significant increase in oxygen saturation of the arterial blood. We investigated if this simple modeling could help to better understand the effect of hyperventilation on brain tissue p_tO_2. In the model, we incorporated the widely accepted dual effect of pCO_2 on p_tO_2: both blood acidification and an effect on CBF.

3.4.1. Modeling of Effect of Blood Acidification by CO_2

To describe blood acidification by CO_2, the modification of the Bohr formula ($BC = \Delta \log pO_2/\Delta pH$) in the form of $\Delta pO_2/pO_2 = BC*\Delta pH$, and Siggaard-Andersen and Engel nomogram in a form $\Delta pH = -0.63*\Delta pCO_2/pCO_2$ can be used. By combining these two equations and taking into account that the Bohr coefficient for Sprague Dawley rats at 50% saturation of blood (SO_2 = 50%) equals BC = -0.643,[8] one can derive $\Delta pO_2/pO_2 = 0.4*\Delta pCO_2/pCO_2$. This formula can be used as a first approximation to describe the effect of CO_2 on p50:

$$\Delta p50/p50 = 0.4*\Delta pCO_2/pCO_2 \quad (5)$$

where ΔpCO_2 is the change in arterial carbon dioxide. This formula allows us to predict that the p50 = 31 mm Hg under hyperventilation, when the arterial P_aCO_2 = 20 mm Hg.

3.4.2. Modeling of the Effect of pCO_2 on CBF

It has been shown[27] that the variation of pCO_2 in the range of 20 to 80 mm Hg causes reproducible changes in CBF. Within this physiological range, whole-brain CBF in anesthetized animals changes from 0.1 to 0.2 $ml \cdot g^{-1} \cdot min^{-1}$ for each 10 mm Hg change in pCO_2 of arterial blood. The variation of the coefficient from 0.1 to 0.2 $ml \cdot g^{-1} \cdot min^{-1}$ reflects the nonlinearity of CBF response to a change in pCO_2. To simplify the modeling, linear approximation of CBF response to a change in pCO_2 and the mean value of this coefficient were used in this study:

$$\Delta CBF = 0.015 * \Delta pCO_2 \tag{6}$$

Equation (6) predicts $\Delta CBF = -0.25$ $ml \cdot g^{-1} \cdot min^{-1}$ under hyperventilation with experimental arterial $P_aCO_2 = 20$ mm Hg. Using these data, the computed change $\Delta p_tO_2 = -10.2$ mm Hg fits surprisingly well to the experimental value of -13 mm Hg when using the values of p50 = 38, CBF = 0.8, $CMRO_2 = 85$, $r_0 = 2.2$, $P_aCO_2 = 37$, and $P_aO_2 = 136$, despite raw linear approximation of CBF response to a change in pCO_2.

4. CONCLUSIONS

This simple model, based on a Krogh model, allows one to predict some changes in p_tO_2 in rat brain while single parameters are varied. An acceptable agreement of the modeled and experimental changes of the cerebral p_tO_2 was obtained for several different experimental conditions. However, the model did not provide full quantitative agreement with some experimental results.

1. CBF, as adjustable parameter, models the effect of anesthetics on cerebral p_tO_2 and predicts a wide range of values of CBF for the different anesthetics.
2. The model does not support the hypothesis that the effect of RSR13 on cerebral p_tO_2 is due only to a shift in p50.
3. Even taking into account all parameters known to change with chronic hypoxia, the model underestimated the measured change in cerebral p_tO_2 in rats acclimated to hypoxia (1/2 atm). This indicates that some other variable, such as CBF, is likely to change when hypoxia-adapted animals are exposed to normoxia.
4. The model acceptably describes the effect of hyperventilation on brain tissue p_tO_2.

This paper indicates that the use of a "relative change" model to predict changes in p_tO_2 may allow one to understand better the influence of individual physiological parameters on p_tO_2.

5. ACKNOWLEDGEMENTS

This work was supported by NIH (NIBIB) grants P01 EB002180, and RO1 EB002085, and used the facilities of the EPR Center for the Study of Viable Systems supported by NIH (NIBIB) Grant P41 EB002032. RSR13 was provided by Allos Therapeutics, Inc., Westminster, CO.

REFERENCES

1. C. I. Nwaigwe, M. A. Roche, O. Grinberg, and J. F. Dunn, Effect of hyperventilation on brain tissue oxygenation and cerebrovenous pO_2 in rats, *Br. Res.* **868**, 150-156 (2000).
2. H. Lei, O. Grinberg, C. I. Nwaigwe, H. G. Hou, H. Williams, H. M. Swartz, and J. F. Dunn, The effects of ketamine-xylazine anesthesia on cerebral blood flow and oxygenation observed using nuclear magnetic resonance perfusion imaging and electron paramagnetic resonance oximetry, *Br. Res.* **913**(2), 174-179 (2001).
3. J. F. Dunn, O. Grinberg, M. Roche, C. I. Nwaigwe, H. G. Hou, and H. M. Swartz, Non-invasive assessment of cerebral oxygenation during acclimation to hypobaric hypoxia, *J. Cereb. Blood Flow Metab.* **20**, 1632-1635 (2000).
4. O. Y. Grinberg, M. Miyake, H. Hou, R. P. Steffen, and H. M. Swartz, The dose-dependent effect of RSR13, a synthetic allosteric modifier of hemoglobin, on physiological parameters and brain tissue oxygenation in rats, *Adv. Exp. Med. Biol.* **530**, 287-296 (2003).
5. A. Krogh, The number and distribution of capillaries in muscle with calculations of the oxygen pressure head necessary for supplying the tissue, *J. Physiol.* (London) **52**, 409 (1919).
6. A. Krogh, *The Anatomy and Physiology of Capillaries*, Yale University Press, New Haven (1922).
7. S. S. Kety, Determination of tissue oxygen tension, *Fed. Proc.* **16**, 666-670 (1957).
8. C-F Cartheuser, Standard and pH-affected hemoglobin-O_2 binding curves of Sprague-Dawley rats under normal and shifted P_{50} conditions, *Comp. Biochem. Physiol.* **106A** (4), 775-782 (1993).
9. I. Toma-Dasu, A. Waites, A. Dasu, and J. Denekamp, Theoretical simulation of oxygen tension measurement in tissue using a microelectrode: I. The response function of the electrode, *Physiol. Meas.* **22**, 713-725 (2001).
10. J. F. Dunn, M. A. Roche, R. Springett, M. Abajian, J. Merlis, C. P. Daghlian, S. Y. Lu, and M. Makki, Monitoring angiogenesis in brain using steady-state quantification of dR2 with MION infusion, *Magn. Reson. Med.* **51**(1), 55-61 (2004).
11. T. Bar, The vascular system of the cerebral cortex, *Adv. Anat. Embryol. Cell Biol.* **59**(I-VI), 1-62 (1980).
12. Y. P. Ma, A. Koo, H. C. Kwan, and K. K. Cheng, On-line measurement of the dynamic velocity of erythrocytes in the cerebral microvessels in the rat, *Microvascular Res.* **8**(1), 1-13 (1974).
13. D. A. Tata, and B. J. Anderson, A new method for the investigation of capillary structure, *J. Neurosci. Methods* **113**(2), 199-206 (2002).
14. C. E. Stenis, and C. J. D. Fell, Diffusivity of oxygen in water, *Can. J. Chem. Engineer* **49**(6), 885 (1971).
15. A. Dutta, and A. S. Popel, A theoretical analysis of intracellular oxygen diffusion, *J. Theoret. Biol.* **176** (4), 433-445 (1995).
16. A. S. Popel, Theory of oxygen-transport to tissue, *Crit. Rev. Biomed. Engineer.* **17**(3), 257-321 (1989).
17. Y. Magata, T. Temma, T. Iida, M. Ogawa, T. Mukai, T. Morimoto, J. Konishi, and H. Saji, Development of injectable O-15 Oxygen and estimation of rat OEF, *J. Cereb. Blood Flow Metabol.* **23**(6), 671-676 (2003).
18. K. Norberg, B. Scatton, and J. Korf, Amphetamine-induced increase in rat cerebral blood flow - apparent lack of catecholamine involvement, *Br. Res.* **149**(1), 165-174 (1978).
19. V. L. Baughman, W. E. Hoffman, D. J. Miletich, and R. F. Albrecht, Cerebrovascular and cerebral metabolic effects of N_2O in unrestrained rats, *Anesthesiol.* **73**, 269-272 (1990).
20. W. J. Thoman, D. Gravenstein, van der Aa, and S. Lampotang, Autoregulation in a simulator-based educational model of intracranial physiology, *J. Clin. Monitor. Comput.* **15**(7-8), 481-491 (1999).
21. Z. H. Zhu, Y. Zhang, R. X. Tian, H. Lei, N. Y. Zhang, X. L. Zhang, H. Merkle, K. Ugurbil, and W. Chen, Development of O-17 NMR approach for fast imaging of cerebral metabolic rate of oxygen in rat brain at high field, *Proc. Nat. Acad. Sci. USA* **99**(20), 13194-13199 (2002).
22. H. Hou, O. Y. Grinberg, S. Taie, S. Leichtweis, M. Miyake, S. Grinberg, H. Xie, M. Csete, and H. Swartz, Electron paramagnetic resonance assessment of brain tissue oxygen tension in anesthetized rats, *Anesth. Analg.* **96**(5),1467-1472 (2003).
23. M. P. Kunert, J. F. Liard, and D. J. Abraham, RSR-13, an allosteric effector of hemoglobin, increases systemic and iliac vascular resistance in rats, *Am. J. Physiol.* **271**, H602-613 (1996).
24. B. D. Kavanagh, T. W. Secomb, R. Hsu, P-S Lin, J. Venitz, and M. W. Dewhirst, A theoretical model for the effects of reduced hemoglobin-oxygen affinity on tumor oxygenation, *Int. J. Radiat. Oncol. Biol. Phys.* **53**, 172-179 (2002).
25. J. C. LaManna, Rat brain adaptation to chronic hypobaric hypoxia, *Adv. Exp. Med. Biol.* **317**, 107-114 (1992).
26. J. B. Jensen, B. Sperling, J. W. Severinghaus, and N. A. Lassen, Augmented hypoxic cerebral vasodilation in men during 5 days at 3,810 m altitude, *J. Appl. Phys.* **80**(4), 1214-1218 (1996).
27. J. E. Brian Jr., Carbon dioxide and cerebral circulation, *Anesthesiol.* **88**, 1365-1386 (1998).

BLACK MAGIC AND EPR OXIMETRY:

From Lab to Initial Clinical Trials

Nadeem Khan, Huagang Hou, Patrick Hein, Richard J. Comi, Jay C. Buckey, Oleg Grinberg, Ildar Salikhov, Shi Y. Lu, Hermine Wallach, and Harold M. Swartz

Abstract: EPR oximetry is a technique that can make repeated non-invasive measurements of the pO_2 in tissues. To extend the application of EPR oximetry to humans, India ink is the probe of choice because appropriate India inks have EPR signals whose line widths are sensitive to changes in oxygen concentrations, and, most importantly, India ink already has been used extensively in humans as a marker in the skin, lymphatics, various organs during surgery, tumors, and for decoration as tattoos.

We have developed an India ink that has good sensitivity to oxygen, high stability in tissues, good signal intensity, and minimal toxicity. In this article we describe the various properties of this India ink, results obtained from our animal experiments, and our first preliminary clinical results, which are part of the first systematic clinical use of EPR oximetry. The clinical results indicate that it is possible to do repeated measurements over several months and probably years after the injection of the ink, indicating that long-term follow-up studies are feasible. We are very encouraged with these results and are confident that EPR oximetry using India ink will be a non-invasive, fast, and reliable technique for pO_2 measurements in clinical studies.

1. INTRODUCTION

It would be very useful to have a method to directly and repetitively measure the partial pressure of oxygen (pO_2) in tissues. This would be especially desirable for planning and evaluating therapy for tumors and vascular insufficiencies. A variety of techniques are available for measuring pO_2 in tissues; however, none of these techniques has been shown to have the properties needed for optimal experimental and clinical use (i.e. sensitivity, accuracy, ease, and ability to make measurements repeatedly). EPR oximetry is one of the most promising techniques for accomplishing this goal. It is based on the fact that molecular oxygen can interact with paramagnetic materials such as nitroxides, lithium phthalocyanine, coals such as fusinite, chars, and India ink, affecting their EPR spectra in a reproducible manner that is proportionate to the amount of oxygen. In the studies described here we use India ink, measuring the oxygen-induced change in the line width of the EPR spectrum, which is calibrated against different known pO_2 values.[1, 2]

In order to extend the application of EPR oximetry to humans, India ink is the probe of choice because appropriate India inks have EPR signals whose line widths are sensitive to changes in oxygen concentrations[3, 4]; most importantly, India ink already has been used extensively in humans (as a marker in the skin, lymphatic, mucosal tissue, and tumors, and for decoration as tattoos).[5-8]

To facilitate the application of EPR oximetry in human subjects, we have developed an India ink that has good sensitivity to oxygen, high stability in tissues, good signal intensity, and minimal toxicity. In this article we describe the various properties of this India ink, results obtained from our animal experiments, and our first preliminary clinical results, which are part of the first systematic clinical use of EPR oximetry.

2. MATERIALS AND METHODS

2.1. India Ink

We identified an effective India ink, Higgins black magic waterproof ink (No. 4465), through a search of local commercial sources of inks. This ink has an excellent signal to noise ratio. To increase the signal intensity, the ink was concentrated to 20% of its original volume by heating at approximately 90°C-100°C for six to seven hours. We also studied this ink after dialysis to remove soluble components that might be involved in potential toxicity. Prior to injection, the ink was autoclaved for one hour at 121°C. The calibration of the ink was done against 0%, 1%, 2%, and 5% perfused oxygen concentrations at 37°C on a 1.2 GHz spectrometer. The change in line width followed a second order polynomial relation with changes in the oxygen concentration. The coefficients of the fit were determined and used to convert the line widths obtained into pO_2 values.

2.2. Animal Preparation and Experimental Protocol

Nine male Sprague-Dawley rats, 200-250 g, (Charles River Laboratories, Wilmington, MA) were used in this study. The experimental techniques and protocol were approved by the Dartmouth College Animal Care and Use Program. The rats were anesthetized using 2.5%-3.0% isoflurane, and approximately 15-20 μl of the India ink was injected intramuscularly or subcutaneously in rats, in the left or right hind limbs, respectively (day 0). The intramuscular injection of the ink was approximately 3 mm deep from the surface of the skin. These injections were done using a 1 ml syringe fitted with a 23 gauge needle. The tissue pO_2 was measured after ink injections at days 1, 4, 7, 14, 21, and 28. After pO_2 measurements at each time point, one rat was sacrificed and the tissue was prepared for histological assessment. The rats were maintained with 1.25%-1.5% of isoflurane with 26% oxygen through a sealed nose cone at a rate of 2 liters/minute throughout the experiment. The animal temperature was maintained by using a heated water blanket and warm air forced through the gap of the magnet. After baseline pO_2 measurements (10-15 minutes), the leg muscle was temporarily compressed using a rubber band for not more than 5-7 minutes, to restrict the blood supply temporarily. During the compression, the tissue pO_2 was measured again.

2.3. EPR Measurements

In vivo EPR measurements were carried out on a 1.2 GHz (L-band) EPR spectrometer. The rats were placed in the magnet, and the extended loop resonator was placed directly above the site of ink injection. Typical settings for the spectrometer were: incident microwave power, 80 mW; magnetic field center, 400 gauss; scan range, 120 gauss; and scan time 12-15 seconds. Modulation amplitude was set at less than one-third of the EPR line width. Usually 6-9 spectra were averaged to achieve a better signal to noise ratio. For *in vitro* experiments, we used a 9.2 GHz (X-band) Bruker EPR spectrometer fitted with a variable temperature controller.

3. RESULTS

We compared the *in vitro* signal to noise ratio of the Higgins black magic ink with the ink that was used in our laboratory for preliminary *in vivo* experiments (obtained from the radiation oncology department of the Dartmouth-Hitchcock Medical Center). These comparisons were made with the L-band EPR spectrometer under similar experimental conditions (Table 1). The change in the line width of the EPR signal of the ink was studied at different incident microwave powers of the L-band spectrometer and different perfused oxygen concentrations (Figure 1).

Table 1. *In vitro* signal to noise ratio of the India ink measured with the L-band EPR spectrometer under similar experimental conditions.

India ink (20% of original volume, concentrated by heating)	Perfused oxygen concentration		
	1% O_2	2% O_2	5% O_2
Higgins Black ink 46730 (Old ink)	38	41	37
Higgins Black Magic 4465 (New ink)	128	149	150

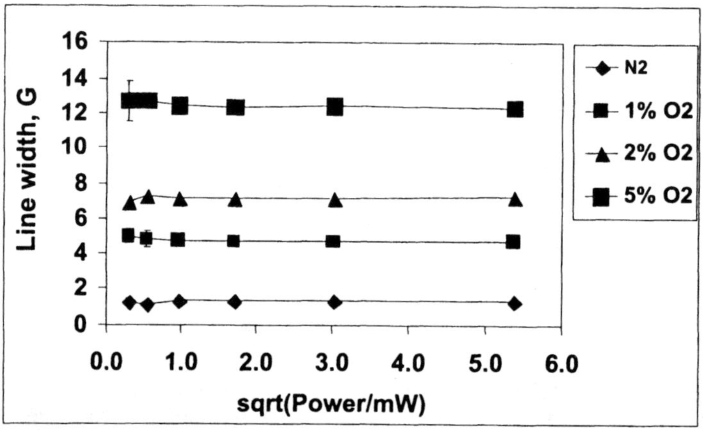

Figure 1. Change in the line width of Higgins black magic ink with different microwave powers and at different perfused oxygen concentrations.

We measured the effect of various experimental conditions (i.e., pH, a mild broadening agent, a mild oxidizing agent, and temperature) on the EPR line width of this ink (Table 2). The only significant effect was a modest decrease in line width with increasing temperature. We measured the stability and sensitivity of this ink in rodents. The pO_2 measured in the skeletal muscle and subcutaneous tissue of rats over several time points before and after compression after ink injection is shown in Figure 2.

The non-dialyzed, concentrated black magic ink was then used in our first clinical experiments. The measurements in volunteers were carried out in the first specifically developed clinical L-band EPR spectrometer at the Dartmouth-Hitchcock Medical Center. We injected the ink in the feet of healthy volunteers at the sites of greatest risk for patients with diabetic peripheral vascular diseases (under the first metatarsal head) and also at the site where transcutaneous oxygen measurements often are made (first interosseous dorsal space) (Figure 3a). The typical EPR spectrum obtained from these injections (7-10 μl of ink) is shown in Figure 3b. The tissue pO_2 has been measured over several weeks (Figure 4).

Table 2. Effect of various experimental conditions on the apparent pO_2 (EPR line width) of black magic ink measured at 9.2 GHz.

Experimental Conditions	LW (G)	pO_2, mm Hg
pH 4, 37°C	9.49	36.12
pH 7, 37°C	9.41	35.63
pH 10, 37°C	9.6	36.86
$K_3Fe(CN)_6$, 5mM, 37°C	9.65	37.11
Ascorbic acid, 1mM, 37°C	9.53	36.37
Temperature 27°C	11.48	49.57
32°C	10.66	43.79
42°C	9.17	34.17
47°C	8.32	29.03

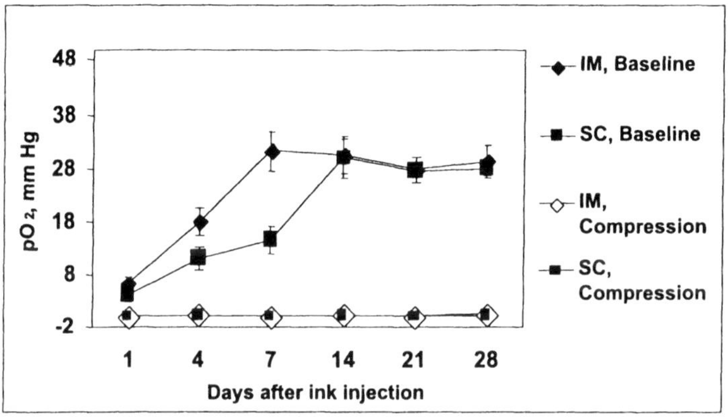

Figure 2. The pO_2 in skeletal muscle (IM) and subcutaneous tissue (SC) in rats injected with dialyzed concentrated black magic ink, Mean ± SD, (n = 7-9).

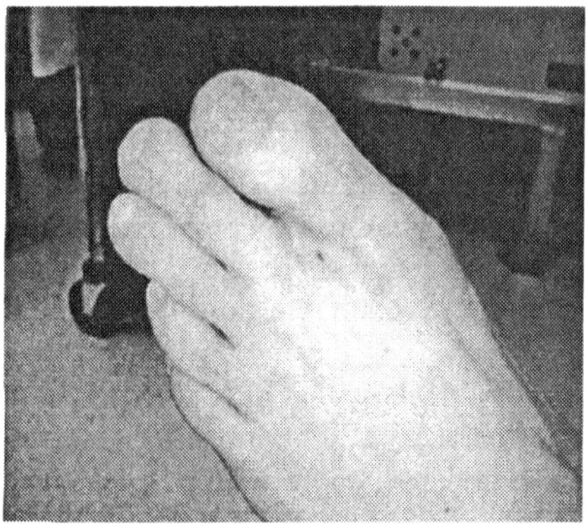

3a

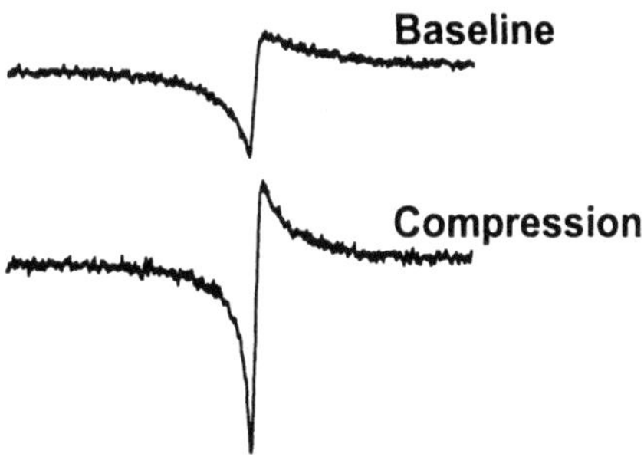

3b

Figure 3. a: The cosmetic result of non-dialyzed, concentrated Higgins black magic ink injection in the first interosseous dorsal space (between the first and second toes). **b:** shows the EPR signals obtained from only 7-10 μl of concentrated ink before (upper tracing) and after (lower tracing) temporary muscle compression.

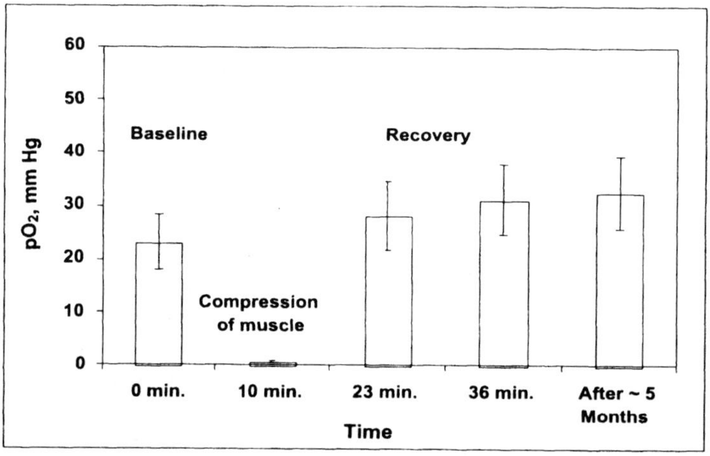

Figure 4. pO_2 measured by EPR using concentrated Higgins black ink (~5 μl) under the plantar first metatarsal head of a healthy volunteer. Measurements of baseline tissue pO_2, pO_2 after muscle compression, and pO_2 after compression was released, i.e. recovery, were completed on Dec. 11, 2002. Tissue pO_2 was measured again on April 30, 2003. The EPR measurements for baseline and recovery were done for a period of ~15 minutes and for ~ 5 minutes during muscle compression (Mean ± SD).

4. DISCUSSION

This ink was reduced to 20% of its original volume to achieve better signal intensity with minimal volume. Results indicate that this ink has an approximately three-to-four times better signal to noise ratio as compared with the original India ink (Table 1). The line width is sensitive to changes in oxygen concentrations, and no power saturation was observed at 1.2 GHz (Figure 1). The line width of the EPR signals was not significantly affected by changes in pH, mild broadening, mild oxidizing, or reducing agents; however, significant changes were observed with change in temperature (Table 2). The *in vivo* results in animals indicate that a stable pO_2 was observed by day 14 (Figure 2). At all time points, the line width of the EPR signal observed after muscle compression was similar to that obtained with 0% perfused oxygen in the calibration. These results indicate that there is no change in the calibration of this ink, and the suitable time for the EPR measurements is 7-14 days after injection of the ink.

We carried out our first preliminary clinical experiments using EPR oximetry and this ink. Only 7-10 μl of the concentrated ink was necessary to obtain an EPR signal, which is shown in Figure 3B. The signal to noise ratio observed was nearly 18-20 in single EPR scans with a scan time of 15 seconds. The results shown in Figure 4 indicate that it is possible to do repeated measurements over several months and perhaps years after the ink injection, indicating that long-term follow-up studies are feasible. We are very encouraged with these results and are confident that the EPR technique using India ink will be a non-invasive, fast, and reliable technique for pO_2 measurements in clinical studies. At present, we plan to carry out a systematic study of these injected sites over longer periods and also to extend the study to more volunteers to obtain more data with the aim of developing EPR oximetry as a valuable tool for clinical use.

5. ACKNOWLEDGMENTS

This work was supported by PO1 EB002180, "Measurement of pO_2 in Tissues *In Vivo* and *In Vitro*," and used the facilities of P41 EB002032, "EPR Center for the Study of Viable Systems."

REFERENCES

1. H. M. Swartz, and R. B. Clarkson, The measurement of oxygen *in vivo* using EPR techniques, *Phys. Med. Biol.* **43**(7), 1957-1975 (1998).
2. J. F. Dunn, and H. M. Swartz, *In vivo* electron paramagnetic resonance oximetry with particulate materials, *Methods* **30**(2), 159-166 (2003).
3. H. M. Swartz, K. J. Liu, F. Goda, and T. Walczak, India ink: a potential clinically applicable EPR oximetry probe, *Magn. Reson. Med.* **31**(2), 229-232 (1994).
4. H. M. Swartz, and T. Walczak, Developing *in vivo* EPR oximetry for clinical use, *Adv. Exp. Med. Biol.* **454**, 243-252 (1998).
5. R. Nizam, N. Siddiqi, S. K. Landas, D. S. Kaplan, and P. G. Holtzapple, Colonic tattooing with India ink: benefits, risks, and alternatives, *Am. J. Gastroenterol.* **91**(9), 1804-1808 (1996).
6. K. K. Ellis, and M. B. Fennerty, Marking and identifying colon lesions. Tattoos, clips, and radiology in imaging the colon, *Gastrointest. Endosc. Clin. N. Am.* **7**(3), 401-411 (1997).
7. N. Price, M. R. Gottfried, E. Clary, D. C. Lawson, J. Baillie, K. Mergener, C. Westcott, S. Eubanks, and T. N. Pappas, Safety and efficacy of India ink and indocyanine green as colonic tattooing agents, *Gastrointest. Endosc.* **51**(4.1), 438-442 (2000).
8. B. A. Shatz, L. B. Weinstock, P. E. Swanson, and E. P. Thyssen, Long-term safety of India ink tattoos in the colon, *Gastrointest. Endosc.* **45**(2),153-156 (1997).

AXIAL OXYGEN DIFFUSION IN THE KROGH MODEL:

Modifications to account for myocardial oxygen tension in isolated perfused rat hearts measured by EPR oximetry

Oleg Grinberg, Boris Novozhilov, Stalina Grinberg, Bruce Friedman, and Harold M. Swartz

Abstract: The cylindrical steady-state model developed by Krogh with Erlang has served as the basis of understanding oxygen supply in living tissue for over eighty years. Due to its simplicity and agreement with some observations, it has been extensively used and successfully extended to new fields, especially for situations such as drug diffusion, water transport, and ice formation in tissues. However, the applicability of the model to make even a qualitative prediction of the oxygen level of specific volumes of the tissue is still controversial. We recently have developed an approximate analytical solution of a steady-state diffusion equation for a Krogh cylinder, including oxygen concentration in the capillary. This model was used to explain our previous experimental data on myocardial pO_2 in isolated perfused rat hearts measured by EPR oximetry. An acceptable agreement with the experimental data was obtained by assuming that a known limitation of the existing EPR methods—a tendency to over-weight low pO_2 values—had resulted in an under-estimate of the pO_2. These results are consistent with recent results of others, which stress the importance of taking into account the details of what is measured by various methods.

1. INTRODUCTION

The cylindrical steady-state model developed by Krogh with Erlang has served as the basis of understanding oxygen supply in living tissue for over eighty years. The use of a circulation unit with one axial capillary permitted a description of the oxygen transport in tissue using simple analytical functions. The simplicity of the Krogh-Erlang model, which was achieved by ignoring axial oxygen diffusion of oxygen and blood as an oxygen carrier, still attracts researchers even though more complex and potentially more accurate extensions of the concepts have been developed. These include more sophisticated mathematical models for oxygen transport in tissue (see reviews by Popel[1] and Hellums et al.[2]), studies that take into account the vascular structure,[3] and considerations of regulation of oxygen delivery.[4, 5]

1.1. Incorporation of Axial Oxygen Diffusion

One of the earliest and still widely used approaches is the Kety assumption[6] that assumes a linear decrease of oxygen concentration in the capillary from the arterial to the venous end. Several groups have now calculated the effect of axial oxygen diffusion along the capillary by solving the diffusion equation using digital methods. Fletcher and Schubert[7] have developed a model that takes into account axial diffusion and wall permeability effects in perfused capillary-tissue structures. The venous boundary conditions and the characteristic "permeability coefficient" of the vessel wall were incorporated into the model. Lagerlund and Low[8] included axial diffusion in blood and in surrounding tissue. In this simulation, it was assumed that the oxygen consumption of nerve tissue obeys Michaelis-Menten kinetics rather than zero-order kinetics, as had been assumed in the Krogh model. Analytical solutions for the radially averaged, axially distributed modified Krogh model were developed by Schubert and Zhang.[9] In this model they averaged the tissue oxygen radially, using a mass-transfer coefficient to maintain radial transport, and adding axial diffusion in the capillary and tissue.

1.2. Applications to Processes Where Hemoglobin is Not Involved in Oxygen Diffusion

Using numerical methods and an advanced Krogh model, Gabet[10] simulated injection of a drug into a capillary and its diffusion in tissues, and obtained results that were close to the experimental data. Millard and Gorman[11] used a similar approach to describe substrate concentration in tissue. Using the Krogh approach, Rubinsky and Pegg[12] developed a mathematical-model for the freezing process in biological tissue. Bischof et al.[13]; Devireddy and Bischof[14] and Devireddy et al.[15] have used the Krogh cylinder model to simulate water transport and ice formation in both isolated hepatocytes and whole tissue slices.

These studies show that the Krogh model still has great applicability to make a qualitative prediction under different experimental circumstances. These results also demonstrate that the use of digital methods makes analysis more complex. We recently have developed an approximate analytical solution of a steady-state diffusion equation for a Krogh cylinder including oxygen concentration in the capillary. This solution was used to reanalyze our previous experimental data on myocardial pO_2 in isolated perfused rat hearts measured by EPR oximetry[16] without an oxygen carrier.

2. DIFFUSION EQUATION AND BOUNDARY CONDITIONS

We considered a diffusion equation in a Krogh cylinder using cylindrical coordinates:

$$\frac{1}{r}\frac{\partial}{\partial r}\left(r\frac{\partial U}{\partial r}\right)+\frac{\partial^2 U}{\partial z^2}-\frac{Q}{D}=0 \tag{1}$$

boundary conditions in tissue $r_0 \le r \le R$, $0 \le z \le L$:

$$z=0,\ \frac{\partial U}{\partial z}=0;\ z=L,\ \frac{\partial U}{\partial z}=0; \tag{2a}$$

$$r=R\ ,\ \frac{\partial U}{\partial r}=0;\ r=r_0\ ,\ U_c=U\big|_{r_0} \tag{2b}$$

and boundary conditions in the capillary $0 < r < r_0$, $0 \le z \le L$:

$$\pi r_0^2 V\frac{dU_c}{dz}=2\pi r_0 D\frac{\partial U}{\partial r}\bigg|_{r_0}\ ,\ U_c(0)=U_0 \tag{3}$$

where $U=U(r,z)$, $Uc=U(r,z)$ are oxygen concentrations in tissue and capillary, respectively, Q is zero order oxygen consumption, D is oxygen diffusion coefficient, R and L are radius and length of Krogh cylinder respectively, r_0 is radius of capillary, V is speed of media in the capillary, and U_0 is oxygen concentration in capillary at the arterial end of capillary.

To solve the problem the following dimensionless variables and parameters were introduced:

$$\varphi=\frac{U}{U_0},\ \varphi_c=\frac{U_c}{U_0},\ \xi=\frac{r}{R},\eta=\frac{z}{R} \tag{4a}$$

$$a=\frac{r_0}{R},\ l=\frac{L}{R} \tag{4b}$$

$$u=\frac{r_0 V}{2D},\ q=\frac{QR^2}{DU_0} \tag{4c}$$

Using these variables and parameters, one can write the dimensionless diffusion equation and boundary conditions: diffusion equation in tissue

$$a \le \xi \le 1,\ 0 \le \eta \le l$$

$$\frac{1}{\xi}\frac{\partial}{\partial \xi}\left(\xi\frac{\partial \varphi}{\partial \xi}\right)+\frac{\partial^2 \varphi}{\partial \eta^2}-q=0 \tag{5}$$

with boundary conditions:

$$\eta=0,\ \frac{\partial \varphi}{\partial \eta}=0;\ \eta=l,\ \frac{\partial \varphi}{\partial \eta}=0;\ \xi=1,\ \frac{\partial \varphi}{\partial \xi}=0 \tag{6}$$

and in capillary:

$$0 \le \eta \le l$$

$$u\frac{d\varphi_c}{d\eta}=\left.\frac{\partial \varphi}{\partial \xi}\right|_a \tag{7}$$

$$\eta=0,\ \varphi_c=1 \tag{8}$$

3. RESULTS AND DISCUSSION

3.1. The Solutions of the Diffusion Equation in a Krogh Cylinder

The solutions of the Eqs. (4)-(8) are expressed as follows:

$$\varphi=A_0-\frac{q}{4}\left(2\ln\xi-\xi^2\right)+\sum_{n=1}^{\infty}A_nF_n(\xi)\cos\gamma_n\eta \tag{9}$$

where:

$$F_n(\xi)=K_1(\gamma_n)I_0(\gamma_n\xi)+I_1(\gamma_n)K_0(\gamma_n\xi) \tag{10a}$$

$$\gamma_n=\frac{n\pi}{l},\ n=\ 1,\ 2,\ \ldots \tag{10b}$$

and:

$$\varphi_c=1-\frac{q\left(1-a^2\right)}{2au}\eta+\frac{1}{u}\sum_{n=1}^{\infty}A_nM_n\sin\gamma_n\eta, \tag{11}$$

where:

$$M_n=K_1(\gamma_n)I_1(\gamma_na)-I_1(\gamma_n)K_1(\gamma_na) \tag{11a}$$

and A_n should be found from the following set of linear equations:

$$A_0 = 1 + \frac{q}{4}\left[2\ln a - a^2 - \frac{l(1-a^2)}{au}\right] + \frac{1}{\pi u}\sum_{k=1}^{\infty}\frac{1-(-1)^k}{k}A_k M_k \tag{12}$$

$$A_n F_n = \frac{1-(-1)^n}{n^2}\frac{ql(1-a^2)}{\pi^2 au} + \frac{2}{\pi u}\sum_{k=1}^{\infty}k\frac{1-(-1)^{k+n}}{k^2-n^2}A_k M_k, \quad n = 1,2.. \tag{12a}$$

3.2. Calculation Procedure and Results of Computing

Using Eqs. (9)-(12), the discrete oxygen distribution in a Krogh cylinder was digitally derived (40x40 points) using numerical values of parameters for the rat heart that were chosen based on measurements of others and our experimental data. Linear Eqs. (12)-(12a) were solved using MathCAD software. This software also was used for all other calculations. Three modes of approximation (n = 3) were used for calculations in this modeling.

The discrete oxygen distribution in the plane perpendicular to the capillary at any point z_i of the capillary can be presented using the classical Krogh formula:

$$pO_2(z_i,r_j) = pO_2^{cap}(z_i) - \frac{Q}{4\alpha D}\left(2R^2\ln\left(\frac{r_j}{r_0}\right) - (r_j^2 - r_0^2)\right) \tag{13}$$

where $pO_2^{cap}(z_i)$ is the oxygen concentration in the capillary at the point z_i, $\alpha = 1.32*10^{-9}$ moles/ml/mmHg is oxygen solubility, $D = 1*10^{-5}$ cm^2/sec is oxygen diffusion coefficient, $r_0 = 2.2$ μm is capillary radius, Q is oxygen consumption of 1 cm^3 of tissue, $R = \sqrt{1/(2\sqrt{3}D_c)}$ is radius of Krogh cylinder, $D_c = 2500$ mm^{-2} is capillary density, r_j is radial coordinate, and $r_0 \geq r_j \geq R$.

Assuming that both the oxygen consumption of the tissue throughout the length of the cylinder as well as radially, and the linear velocity of the blood through the tissue are constant, Kety came to the conclusion that the oxygen concentration of the blood must fall linearly with distance as the blood traverses the capillary from arterial to the venous end. Blood is not involved in oxygen transport in isolated rat hearts perfused with crystalloid solutions. Therefore, using the Kety assumption, the oxygen concentration in the capillary $\alpha*pO_2^{cap}(z_i)$ can be expressed as follows:

$$\alpha*pO_2^{cap}(z_i)=\alpha*pO_2^{in}-(Q*z_i)/(F*L) \tag{14}$$

where pO_2^{in} is influent oxygen, F is flow rate per 1 cm^3 of tissue, and L=0.02 cm is capillary length.

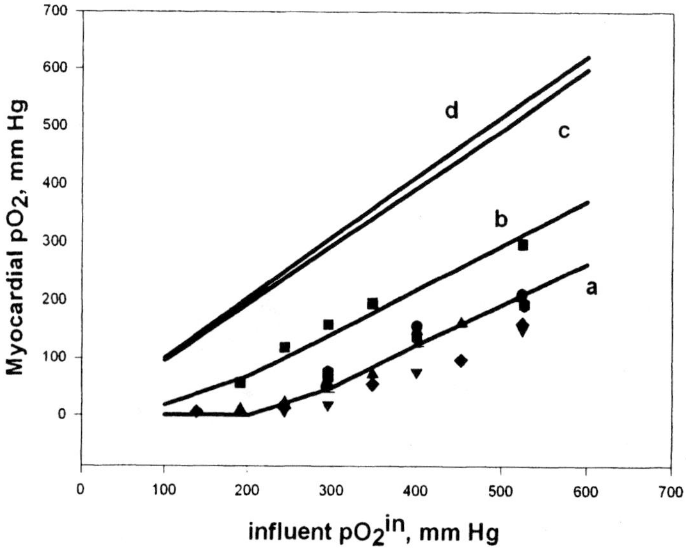

Figure 1. Modeling of myocardial pO_2 at varying influent oxygen levels using the Krogh model with (*a, b*), and without axial oxygen diffusion (*c, d*). pO_2 was derived using the LW of averaged EPR spectrum (*a, c*) and the mean pO_2 over Krogh cylinder (*b, d*). Experimental data from several different isolated hearts are represented by different symbols.[16]

We found that the oxygen distribution derived using Eqs. (9)-(12) was significantly different from the distribution given by the classical Krogh model [Eqs. (13)-(14)]. In order to compare these two models with experimental data,[16] the mean pO_2 value $<pO_2(z_i, r_j)>$ over Krogh cylinder was calculated using our analytical solution [Eqs. (9)-(12)] and the classical Krogh model with the Kety assumption [Eqs. (13)-(14)]:

$$< pO_2(z_i, r_j) >= \frac{1}{N^2} \sum_i \sum_j pO_2(z_i, r_j) \tag{15}$$

In those experiments, a constant flow experimental setup was used. When the effect of influent oxygen on myocardial pO_2 was investigated, the effect of pO_2^{in} on oxygen consumption was observed to be: [$Q=10^{-9}*(45.9+0.148*pO_2^{in})$ moles/sec/g]. In the set of experiments where the effect of flow rate (F) on myocardial pO_2 was investigated, the effects of F on influent oxygen [$pO_2^{in}=313.3+469.9*F$] and on oxygen consumption [$Q = 10^{-9}*(87.5+125*F)$ moles/sec/g] were found (flow rate was varied from 10 to 20 ml/min).

Figures 1 and 2 show the results of these calculations. One can see that the solutions with axial oxygen diffusion give a better fit to the experimental results: the mean pO_2 computed using measured oxygen consumption and flow rate is closer to the measured myocardial pO_2 in isolated perfused rat hearts. Of course, it is possible to improve the agreement with the classical Krogh model, but this requires increasing the oxygen consumption in the calculations to unexpected levels.

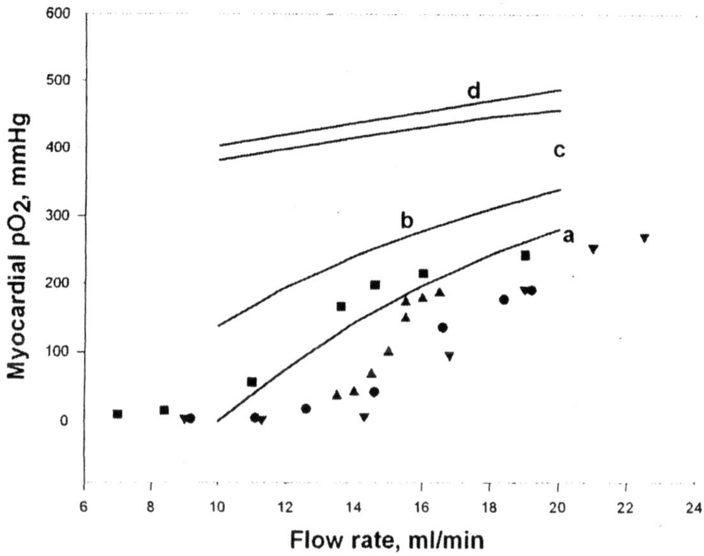

Figure 2. Modeling of myocardial pO_2 at varying flow rates using the Krogh model with (a, b) and without (d, c) axial oxygen diffusion. pO_2 was derived using the mean pO_2 over the Krogh cylinder (b, d) and LW of averaged EPR spectrum (a, c). Experimental data in isolated hearts are indicated by different symbols.

In addition, Figures 1 and 2 also show that both calculations give higher pO_2 levels than were observed experimentally. We could get a result that was consistent with expected physiology and the model by assuming that a known limitation of the existing EPR methods—a tendency to over-weight low pO_2 values—had resulted in an under-estimate of the pO_2. The pO_2 at each point of the Krogh cylinder was simulated as individual EPR lines, and the experimentally observed EPR line was computed by averaging these 40x40 spectra.

$$LW_{i,j} = a + b * pO_2(x_i, r_j) \tag{16}$$

$$RPRLine(h) = \frac{1}{N^2}\sum_i\sum_j \frac{LW_{i,j} * h}{[3*(LW_{i,j})^2 + h^2]^2} \tag{17}$$

Then $pO_2(EPR) = c+d*LW$ is derived using the LW of the EPRline(h), where a, b, c, and d are calibration coefficients derived by calibration of LiPc crystals *in vitro* (N=40). This assumption is equivalent to the assumption that small LiPc crystals are located at each point of the Krogh cylinder. The mean pO_2 value then was calculated using this averaged spectrum and the LiPc calibration. Using this approach, an acceptable agreement with the experimental data was obtained: both the dependences of pO_2 versus flow rate and pO_2 versus influent oxygen were successfully described, Figures 1 and 2. We are developing methods of spectral analysis that will enable us to correct for the tendency to over-weight low values in a heterogeneous environment; in the future we should be able to utilize fully the improved model.

4. CONCLUSIONS

The inclusion of axial oxygen diffusion in the Krogh model and the mean pO_2 value calculated using the LW of the averaged EPR spectrum lead to better agreement between the computed and the experimental pO_2 in the isolated rat heart for these experimental conditions. These results are consistent with recent results of others that stress the importance of taking into account the details of what is measured by various methods, so that they can be compared accurately with other results and evaluated properly by models.[17]

5. ACKNOWLEDGEMENTS

This work was supported by PO1 EB002180, "Measurement of pO_2 in Tissues in Vivo and in Vitro," and used the facilities of P41 EB002032, "EPR Center for the Study of Viable Systems."

REFERENCES

1. A. S. Popel, Theory of oxygen transport to tissue, *Crit. Rev. Biomed. Eng.* **17**(3), 257-321 (1989).
2. J. D. Hellums, P. K. Nair, N. S. Hugang, and N. Ohshima, Simulation of intraluminal gas transport process in microcirculation, *Ann. Biomed. Eng.* **24**(1), 1-24 (1996).
3. M. Sharan, and A. Popel, A compartmental model for oxygen transport in brain microcirculation in the presence of blood substitutes, *J. Theor. Biol.* **216**(4), 479-500 (2002).
4. F. Hyder, R. G. Shulman, and D. L. Rothman, A model for the regulation of cerebral oxygen delivery, *J. Appl. Physiol.* **85**(2), 554-564 (1998).
5. C. A. Lodi, A. T. Minassian, L. Beydon, and M. Ursino, Modeling cerebral outoregulation and CO_2 reactivity in patients with severe head injury, *Am. J. Physiol.* **274** (5 Pt. 2), H1729-H1741 (1998).
6. S. S. Kety, Determination of tissue oxygen tension, *Fed. Proc.* **16**, 666-670 (1957).
7. J. E. Fletcher, and R.W. Schubert, Axial diffusion and wall permeability effects in perfused capillary-tissue structures, *Biosystems* **20**(2), 153-174 (1987).
8. T. D. Lagerlund, and P. A. Low, Axial diffusion and Michaelis-Menten kinetics in oxygen delivery in rat peripheral-nerve, *Am. J. Physiol.* **260** (2 Pt. 2), R430-R440 (1991).
9. R. W. Schubert, and X. Zhang, The equivalent Krogh cylinder and axial oxygen transport, Oxygen Transport to Tissue XVIII, *Adv. Exper. Med. Biol.* **411**, 191-202 (1997).
10. L. Gabet, Capillary net and injection modeling, *Int. J. BioMed. Comp.* **31**(1), 25-36 (1992).
11. C. A. Millard, and A. D. Gorman, A model for substrate concentrations in tissue surrounding single capillaries, *Math. Comp. Model.* **25**(11), 1-7 (1997).
12. B. Rubinsky, and D. E. Pegg, A mathematical-model for the freezing process in biological tissue, *Cryobiol.* **25**(6), 546 (1988).
13. J. C. Bischof, C. M. Ryan, R. G. Tompkins, M. L.Yarmush, and M. Toner, Ice formation in isolated human hepatocytes and human liver tissue, *ASAIO J.* **43**(4), 271-278 (1997).
14. R. V. Devireddy, and J. C. Bischof, Measurement of water transport during freezing in mammalian liver tissue: Part II - The use of differential scanning calorimetry, *J. Biomech. Eng. (Transactions of the ASME)* **120**(5), 559-569 (1998).
15. R. V. Devireddy, J. E. Coad, and J. C. Bischof, Microscopic and calorimetric assessment of freezing processes in uterine fibroid tumor tissue, *Cryobiol.* **42**(4), 225-243 (2001).
16. B. J. Friedman, O. Y. Grinberg, K. A. Isaacs, T. M. Walczak, and H. M. Swartz, Myocardial oxygen-tension and relative capillary density in isolated-perfused rat hearts, *J. Mol. Cell. Cardiol.* **27**(12), 2551-2558 (1995).
17. I. Tomas-Das, A. Waites, A. Das, and J. Denekamp, Theoretical simulation of oxygen tension measurement in tissues using a microelectrode: I. The response function of the electrode, *Physiol. Measur.* **22**(4),713-725 (2001).

DIFFERENTIATING HEMODYNAMIC COMPROMISE BY THE OEF RESPONSE TO ACETAZOLAMIDE IN OCCLUSIVE VASCULAR DISEASE

Edwin M. Nemoto, Howard Yonas, Hiroto Kuwabara, Ronda Pindzola, Donald Sashin, Yuefang Chang, Tudor Jovin, James Gebel, Maxim D. Hammer, and Lawrence Wechsler

Abstract: Identification of increased stroke risk in a population of symptomatic patients with occlusive vascular disease (OVD) is presently accomplished by measurement of oxygen extraction fraction (OEF) or cerebrovascular reserve (CVR). However, many regions identified by compromised CVR are not identified by OEF. Our aim was to determine whether the response of OEF to acetazolamide, namely, oxygen extraction fraction response (OEFR) would identify those hemispheres in hemodynamic compromise with normal OEF. Nine patients symptomatic with transient ischemic attacks and strokes, and with occlusive vascular disease were studied. Anatomical MRI scans and T_2-weighted images were used to identify and grade subcortical white matter infarcts. PET cerebral blood flow (CBF) and OEF were measured after acetazolamide. The relationship between CVR and oxygen extraction fraction response (OEFR) showed that positive OEFR occurred after acetazolamide despite normal baseline OEF values. The two hemispheres with positive OEFR were also associated with severe (> 3 cm) subcortical white matter infarcts. We found that the OEFR was highly correlated with CVR and identified hemispheres that were hemodynamically compromised despite normal baseline OEF.

1. INTRODUCTION

Identification of cerebral hemodynamic compromise in symptomatic patients suffering strokes or transient ischemic attacks (TIA) is important in determining whether they may benefit from flow promoting therapies, such as bypass surgery. Presently, two primary imaging methods are being used to identify hemodynamic compromise in these patients. These methods are: 1) Positron emission tomography (PET) measurement of resting cerebral oxygen extraction fraction (OEF); and 2) Measurement of cerebrovascular reserve (CVR) after a vasodilatory challenge with acetazolamide (ACZ). Whether these methods are equivalent and identify the same hemispheres in hemodynamic compromise remains to be rigorously evaluated.

In an earlier study,[1] we reported that the two measures of hemodynamic compromise, namely the measurement of cerebrovascular reserve (CVR) by stable Xe/CT cerebral blood flow (CBF) and OEF by PET, do not identify the same hemispheres in hemodynamic compromise. Hemispheres were identified in hemodynamic compromise by CVR but not by OEF. That is to say, CVR was low or impaired whereas OEF was normal or even low. The objective of this study was to determine whether those regions with compromised CVR and normal or low OEF are still under hemodynamic stress and can be differentiated by the response to a vasodilatory challenge with ACZ.

2. METHODS AND MATERIALS

2.1. Subjects

This study followed a protocol approved by the Institutional Review Board. Nine patients, ranging in age from 20 to 77 years (6 males, 3 females), were recruited into the study with informed consent. All but one patient had an occlusion of the carotid artery (diagnosed by angiography or CT angiography): the exception being one patient with an occlusion of the middle cerebral artery (MCA). Patients were symptomatic with either recurrent transient ischemic events (n = 9) or a fixed neurologic deficit (n = 4). The time from the last ischemic event and the PET study was 2 months or longer.

Following informed consent, patients were screened for the anatomical MRI and the PET study. After the MRI scan, an arterial catheter was inserted for arterial blood sampling along with a peripheral venous catheter for drug and ^{15}O-water injection for the PET studies. The paradigm for the PET studies was: ^{15}O-gas| ^{15}O-water| acetazolamide 15 mg/kg, i.v.| ^{15}O-water| ^{15}O-gas. CBF measurements were made just before and beginning 15 min after ACZ administration (15 mg/Kg i.v.) to ensure appropriate timing for measurement of CVR. The CBF and OEF measurements were made between 30 and 45 min after ACZ administration, within the time of maximal vasodilatory effects.

2.2. MRI for PET Co-Registration

MRI studies were performed on a 1.5-Tesla, whole-body Signa Scanner (General Electric Medical Systems, Milwaukee, WI). The subjects were positioned in a standard head coil and a brief scout T1-weighted image was obtained from which axial 3 mm T1 weighted images and FLAIRS were prescribed. T2 and proton density images were acquired to exclude unexpected pathology. The total time for the MRI scan was about 30 minutes. The MRI data were transferred to the PET Facility over the electronic network and co-registered with the PET data on a SPARC station using software routinely used for this purpose.[2] The co-registered MRI were used as an individualized anatomic map for the selection of regions of interest (ROI).

2.3. MRI Analysis for Subcortical Infarcts

One researcher (HY), blind to the OEF and CVR values of the patients, graded the subcortical infarcts on the T2-weighted images. Subcortical infarcts were graded: 0 = none; light = non-confluent and < 1 cm diam; moderate = non-confluent and 1 to 3 cm diam; severe = confluent and > 3 cm diam. However, in the analysis of the association with CVR and OEF values, they were all combined in one group of lesions versus no lesions because the number within any one subcategory was too small for statistical analysis.

2.4. PET Measurements of OEF

2.4.1. PET Scanner System

The PET studies were performed at the UPMC PET facility on a Siemens/CTI HR+ high-resolution tomograph. The HR+ has an axial field-of-view of 15.2 cm covering the whole brain and acquires 63 transaxial slices 2.4 mm thick. The measured in-plane spatial resolution is 4.5 mm full-width at half-maximum, and the axial resolution is 3.5 mm. Studies were acquired in two-dimensional (2D) mode (septa extended).

2.4.2. CBF and Cerebral Metabolic Rate for Oxygen ($CMRO_2$) Scanning Sessions

A 20-gauge radial artery catheter was inserted for collection of arterial blood samples and blood gas analyses. The patient was placed onto the scanner bed and fitted with a thermoplastic facemask for immobilization. Non-invasive blood pressure was measured with a blood pressure cuff around the forearm contralateral to the radial artery catheter. A 10 min transmission scan with a ^{68}Ge/Ga 68 ring-source was obtained for attenuation correction.

2.4.3. CBF Measurements

Continuous arterial blood sampling (6 ml/min) was initiated at the time of injection of 50 mCi ^{15}O-water in 5-7 cc saline as an intravenous bolus. Arterial blood was withdrawn by a Master flex peristalsis pump (model #7550-90) through a Siemens Liquid Activity Counter. A 180 second dynamic PET scan (10 x 6 seconds, then 10 x 12 seconds) began upon injection. Blood withdrawal continued for 30 seconds beyond the end of the PET scan for a total withdrawal time of about 210 seconds and a total of blood volume of about 20 mls. Blood activity data was automatically accumulated in a Sun SPARC station for later processing.

The ^{15}O-water PET data were analyzed using a 2-compartment (1-tissue compartment) model approach described by Ohta et al.[3] This involved 3 parameters that provided a measure of the clearance of ^{15}O-water from blood to brain (K_1, ml/min/ml), brain efflux (K_2, min^{-1}), and arterial input function timing delays.[4] CBF was calculated as K_1*100 (ml/100g/min) assuming a tissue density of one.

2.4.4. $CMRO_2$ Measurements

The patient inhaled 200 cc of medical air containing 50 mCi of ^{15}O-oxygen gas from a Victoreen Gas Delivery System. Upon breathing the tracer, a 180 second dynamic PET scan (10 x 6 seconds, then 10 x 12 seconds) with continuous arterial blood sampling was begun as previously described for the CBF measurements. An arterial blood sample was obtained at the end of each scan and analyzed for arterial oxygen content, blood gases, and Hematocrit.

A two-compartment model approach was used for $CMRO_2$ calculation as described by Mintun et al.[5] and modified by Ohta et al.[6] $CMRO_2$ (unit: ml/100g/min) was calculated as follows: $CMRO_2 = K_1^O \cdot C_aO_2 \times 100$,where K_1^0 is the rate of uptake of O_2 gas from blood to brain and C_aO_2 is the arterial oxygen content (ml O_2/ml blood).

2.4.5. Calculation of OEF

Oxygen extraction fraction (OEF) was calculated as the K_1^o / K_1^w ratio for individual regions. This ratio relates oxygen gas uptake to CBF and thereby the rate of oxygen delivery. As previously defined, $CMRO_2$ is the value of K_1^0 multiplied by the arterial oxygen content.

2.4.6. Regions of Interest (ROI)

2.4.6a. Level Analysis. MRI images were co-registered with PET images and the middle cerebral artery (MCA) regions of interest (ROI) drawn at 4 axial levels, two at the level of the basal ganglia, one at the level of the cerebral ventricles, and on the roof of the ventricles.

2.4.6b. Hemispheric Analysis. Four successive planes, each 2.5 mm thick and thickness of 10 mm, were used for each of 4 levels which were averaged for hemispheric averages.

2.4.6c. Data Analysis. Statistical analyses were done using the Prophet (AB Tech Corporation) statistical program using one-way ANOVA and the *t* test with Bonferroni correction. Regression analyses were completed using Microsoft Excel. All data are expressed as mean ± SD.

3. RESULTS

The relationship between CVR and OEF before ACZ, by 4-level analysis did not show any meaningful correlation between the two variables (Fig. 1A). There were 3 levels that had OEF values greater than 50%. After ACZ, a significant (P = .0001), negative linear correlation (R = -0.57) relationship was observed (Fig. 1B), described by the equation:

$$\text{CVR } (\%) = -9.143 \text{ (OEF\%)} + 100.87 \quad (1)$$

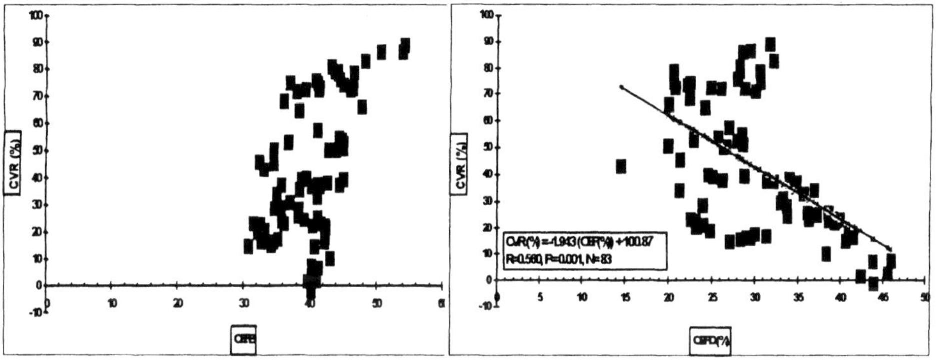

Figure 1. A 4-level analysis of the correlation between cerebrovascular reserve (CVR%) and oxygen extraction fraction (OEF) before (OEFB, panel A) and after (OEFD, panel B) acetazolamide 15 mg/kg, i.v., in 9 patients with occlusive vascular disease.

A plot of the relationship between the OEF response (OEFR) and OEF at baseline (OEFB) shows that the brain regions associated with the worst subcortical white matter injury (Fig. 2) showed an increase in OEF, i.e. OEFR, in response to an ACZ challenge.

4. DISCUSSION

In an earlier study, we reported that although there was a significant correlation between CVR and OEF, suggesting concordance between these two measures of hemodynamic compromise, we also showed that a substantial number of patients (~40%) were identified as compromised by CVR but with normal OEF.[1] Subsequently, we found that these brain regions showing compromised CVR and normal or low OEF were

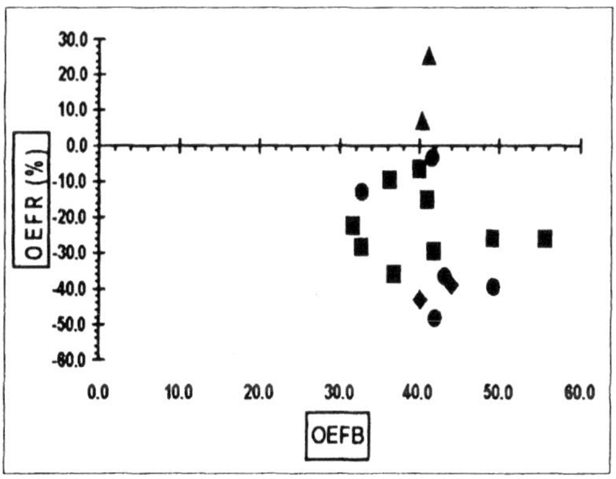

Figure 2. Correlation between the response of oxygen extraction fraction (OEFR) in response to an acetazolamide 15 mg/kg, i.v., and baseline oxygen extraction fraction (OEFB) in the middle cerebral artery territory of each hemisphere in 9 symptomatic stroke patients with occluded carotids. Subcortical white matter injury: > 3+ cm = ▲, 3 cm = ◆, < 1 cm = ●, none = ■.

significantly associated with subcortical infarctions,[7] suggesting that these brain regions may have suffered ischemic injury or deafferentation leading to a primary reduction in cerebral metabolic rate for oxygen ($CMRO_2$) and a coupled fall in OEF. Whether these regions with low or compromised CVR and normal or low OEF are at risk for stroke is a question that requires an answer. The results obtained in this study suggest that these regions are at greater risk for stroke because the brain regions with the worst subcortical injury were also the hemispheres that increased their OEF in response to the acetazolamide challenge. That is to say, OEF increased in response to ACZ, indicating that compromised flow persisted in those regions despite the fact that their normal baseline OEF values were well within the range of normal as shown in Figure 2.

Previous studies in which the relationship between CVR and OEF have been examined suggest that CVR and OEF do not identify the same patients or hemispheres in hemodynamic compromise, despite an overall correlation between the two variables.[8-10] These discrepancies were left unexplained. Nariai et al.[9] showed that of 14 brain regions with compromise by CVR, OEF was elevated in 8 and normal in 6. In 19 patients, 8 showed no correlation between CVR and OEF.[8] In a correlation between CVR and OEF, 9 of 64 regions were identified as compromised by CVR but not by OEF.[10] These discrepancies were likely presumed to be a failure in CVR rather than OEF. Yamauchi et al.,[11] in a 5-year follow-up study, reported that of a total of 11 strokes, 5 occurred in 7 patients with high OEF, and 6 occurred in 33 patients with normal OEF or 54% (6/11) of the strokes occurred in subjects with normal OEF. Whether CVR was also impaired in these patients is unknown. These observations suggest that while OEF may be an independent predictor of hemodynamic compromise, its reliability as a single value predictor of hemodynamic compromise may be questionable. It also suggests that we have an incomplete understanding of the relationships between OEF and CVR and their relative ability in detecting hemodynamic compromise.

Our observations are in accordance with the observations by the investigators cited above, who chose not to discuss the discordance between CVR and OEF in identifying hemodynamic compromise. The reason for the discordance between CVR and OEF or the false negatives of OEF in identifying hemodynamic compromise may be explained by a primary reduction in tissue oxygen demand due to ischemic injury and a coupled reduction in OEF. Our data are suggestive and therefore require confirmation in a larger study.

REFERENCES

1. E. M. Nemoto, H. Yonas, H. Kuwabara, R. Pindzola, D. Sashin, C. C. Meltzer, J. C. Price, and Y. F. Change, Detection of stage II compromised cerebrovascular reserve by Xenon-CT cerebral blood flow with acetazolamide and oxygen extraction fraction by positron emission tomography, in: *Brain Imaging Using PET*, edited by M. Senda, Y. Kimura, and P. Herscovitch (Academic Press, New York, 2002), pp. 259-267.
2. R. P. Woods, S. R. Cherry, and J. C. Mazziotta, Rapid automated algorithm for aligning and reslicing PET images, *J. Comp. Assist. Tomog.* **16**, 620-633 (1992).
3. S. Ohta, E. Meyer, H. Fujita, D. C. Reutens, A. Evans, and A. Gjedde, Cerebral [^{15}O] water clearance in humans determined by PET: I. Theory and normal values, *J. Cereb. Blood Flow Metab.* **16**, 765-780 (1966).
4. H. Iida, I. Kanno, S. Miura, M. Murakami, K. Takahashi, and K. Uemura, Error analysis of a quantitative cerebral blood flow measurement using $H_2{}^{15}O$ autoradiography and positron emission tomography with respect to the dispersion of the input function, *J. Cereb. Blood Flow Metab.* **6**, 536-545 (1986).

5. M. A. Mintun, M. E. Raichle, W. R. Martin, and P. Herscovitch, Brain oxygen utilization measured with O-15 radiotracers and positron emission tomography, *J. Nucl. Med.* **25**, 177-187 (1983).
6. S. Ohta, E. Meyer, and C. J. Thompson, Oxygen consumption of the living human brain measured after a single inhalation of positron emitting oxygen, *J. Cereb. Blood Flow Metab.* **12**, 179-192 (1992).
7. E. M. Nemoto, H. Yonas, H. Kuwabara, R. R. Pindzola, D. Sashin, C. C. Meltzer, J. C. Price, Y. Chang, and D. W. Johnson, Identification of hemodynamic compromise by CVR and OEF in occlusive vascular disease, *J. Cereb. Blood Flow Metab.* (In press, October 2004).
8. I. Kanno, K. Uemura, S. Higano, M. Murakami, H. Iida, S. Miura, F. Shishido, A. Inugami, and I. Sayama, Oxygen extraction fraction at maximally vasodilated tissue in the ischemic brain estimated from the regional CO_2 responsiveness measured by positron emission tomography, *J. Cereb. Blood Flow Metab.* **8**(2), 227-235 (1988).
9. T. Nariai, R. Suzuki, K. Hirakawa, T. Maehara, K. Ishii, and M. Senda, Vascular reserve in chronic cerebral ischemia measured by the acetazolamide challenge test: comparison with positron emission tomography, *Am. J. Neuroradiol.* **16**(3), 563-570 (1995).
10. M. Imaizumi, K. Kitagawa, K. Hashikawa, N. Oku, T. Teratani, M. Takasawa, T. Yoshikawa, P. Rishu, T. Ohtsuki, M. Hori, M. Matsumoto, and T. Nishimura, Detection of misery perfusion with split-dose ^{123}I-iodoamphetamine single-photon emission computed tomography in patients with carotid occlusive diseases, *Stroke* **33**(9), 2217-2223 (2002).
11. H. Yamauchi, H. Fukuyama, Y. Nagahama, H. Nabatame, M. Ueno, S. Nishizawa, J. Konishi, and H. Shio, Significance of increased oxygen extraction fraction in five-year prognosis of major cerebral arterial occlusive diseases, *J. Nucl. Med.* **40**, 1992-1998 (1999).

THE ANOMALOUS EINSTEIN-STOKES BEHAVIOUR OF OXYGEN AND OTHER LOW MOLECULAR WEIGHT DIFFUSANTS

Michael McCabe, David J. Maguire, and Nicholas A. Lintell

Abstract: Almost a century ago, Einstein and Sutherland independently derived equations that describe the relationship between diffusion of solutes and the molecular parameters of those solutes. In that time it has been recognized that, although the equations adequately describe the diffusion of large and medium-sized molecules, there is deviation from this relationship for small molecules. Many authors have attempted to redefine the equations for diffusion, with varying degrees of success, but generally have not attempted to consider the fundamental events that may be occurring at the molecular level during the diffusion of small molecules. In this presentation, we attempt to provide such an explanation, particularly with respect to the diffusion of oxygen through water. We consider the possibility of a random rotational model that complements the (slower) translational process of traditional diffusion and thereby provides accelerated diffusion of small molecules. It is hoped that our description of this model may provide a basis for the development of mathematical modelling of the process.

1. INTRODUCTION

We have investigated the extent to which low molecular weight substances, including oxygen, depart from the Einstein–Stokes equation relating diffusion coefficient through water to the molar volume of the diffusant. Our results indicate that caffeine, with a molecular weight of 194, is close to obeying the Einstein-Stokes relationship, while smaller molecular weight substances which we have investigated, namely oxygen, nitrogen, and tritiated water, all show anomalously high values of diffusion in water. The diffusion of these small and light molecules was better accounted for by an empirical equation of the form:

$$D = KT/(n\pi\eta R) \tag{1}$$

where n is less than the number 6 predicted by the Einstein-Stokes equation. An alternative, but also empirical, relationship has also been tested for fit as follows:

$$d = c\eta^{-\alpha} \tag{2}$$

where c is an empirical constant, and α is 2/3. The experimental data obtained previously show an excellent fit to a relationship of this form. The results suggest that there may be a microviscosity effect (i.e. the apparent viscosity of the solvent in the immediate vicinity of the diffusant is diminished from that of the bulk solvent). The values for self diffusion of water which we have obtained accord with a model depicting water as having a hard spherical core of diameter 2·75 Angstrom which takes part in hydrogen bonding at 4 locations, proposed by Evans.[1] Our results lead us to suspect that the apparent reduced water viscosity in the immediate neighbourhood of the diffusants is a consequence of transient hydrogen bond formation between the diffusant (oxygen, nitrogen, and water) with already diffusing water. While translational diffusion of the background water will occur equally in all directions and would therefore not be expected to enhance the transfer of the diffusant, the rotational diffusion will always promote an overall forward movement of the diffusing species, and it is this which we see as the probable cause of the observed low microviscosity in the immediate region of the diffusants.

2. THE EINSTEIN-STOKES RELATIONSHIP

In the century since Einstein[2, 3] and Sutherland[4] independently published papers dealing with the relationship between the diffusion of solutes and their molecular parameters, a great deal has been measured and said regarding the departure from the Stokes-Einstein relationship for a wide variety of diffusants, by virtue of their degree of hydration and/or their departure from spherical. Both of these causes for deviation from the Einstein-Stokes relationship have the consequence of retarding the speed of diffusion to less than that predicted for a molecule of equivalent mass, but with dimensions of a sphere and without a hydrated envelope. In fact, diminished diffusion is frequently taken as an indication of the extent of departure from a perfect sphere or the extent of hydration existing in the solution.

When the relationship was first described it was recognised that the law was restricted to the case where the dimensions of diffusant molecules were larger than those of solvent molecules. It is clear that oxygen will not necessarily be accommodated within the Einstein-Stokes relationship since its molecular size is of the same general magnitude as the solvent water molecules. There have been numerous attempts to extend the Einstein-Stokes law to the situation where the diffusant is of the same order of magnitude as the solvent molecules,[5-7] or is considerably smaller.[8] We have decided to re-investigate oxygen diffusion, along with some other similarly sized molecules, to see the extent of

Table 1. Values for diffusion coefficients through water, measured in Fick units as $cm^2 sec^{-1}$ and corrected where necessary to equivalent values at 25°C.

Diffusant	mol. wt.	D (Fick)	Method Used	Authors
Tritiated water	20	$2.02 \cdot 10^{-5}$	through gels	McCabe et al. (1975)[9]
Nitrogen	28	$2.53 \cdot 10^{-5}$	through gels	McCabe et al. (1975)[9]
Oxygen	32	$2.05 \cdot 10^{-5}$	through gels	McCabe et al. (1975)[9]
Oxygen	32	$2.2 \cdot 10^{-5}$	polarographic	McCabe et al. (1992)[10]
Caffeine	194	$6.79 \cdot 10^{-6}$	Gouy diffusiometer	McCabe (1972)[11]
Caffeine	194	$6.8 \cdot 10^{-6}$	through gels	McCabe (1972)[11]
Cytochrome c	15.600	$1.03 \cdot 10^{-6}$	analytical c'fuge	Setlow et al. (1962)[12]
Ribonuclease	12.900	$1.38 \cdot 10^{-6}$	analytical c'fuge	Setlow et al. (1962)[12]

departure from the Stokes-Einstein law, and of course to examine the extent of conformity to the predictions of other models and equations.

The diffusion coefficient of oxygen in water at 25°C has already been reported by us. Additionally we have measured the diffusion of nitrogen and the self-diffusion of tritiated water through water. Additionally we have previously reported the variation of oxygen diffusion through aqueous solutions of varying sucrose concentration (and hence varying viscosity). We have also measured and previously reported the diffusion of the relatively low molecular weight substance caffeine through water, also at 25°C. The values obtained are summarised in Table 1, along with values for other low molecular weight, and some larger diffusants, which we have found in the literature.

The Einstein-Stokes relationship can be expressed as:

$$D = RT/[6\pi\eta N(3MV/4\pi N)^{1/3}] \tag{3}$$

where MV is the molar volume in cm^3 (i.e. the gram-molecular weight divided by the density), N is the Avogadro number, R is the Gas constant, T is the absolute temperature, and η is the viscosity of the solvent.

By substituting numerical values for the constants in the above equation, we have, for a spherical molecule in water:

$$D^{25}(\text{sphere}) = 33.06 \times 10^{-6}/(MV)^{1/3} \tag{4}$$

where D^{25}(sphere) is the diffusion coefficient at 25°C calculated from the assumption that the Einstein-Stokes relationship holds and that the molecules are spherical and not hydrated.

Thus a plot of D^{25}(sphere) versus $1/(MV)^{1/3}$ will give a straight line passing through the origin. The measured value of D^{25} for each of the substances investigated is then compared with that calculated. The ratio of D(observed)/D(calculated) will provide information on the extent of departure from the Einstein-Stokes relationship. The results of such an analysis is shown in Figure 1.

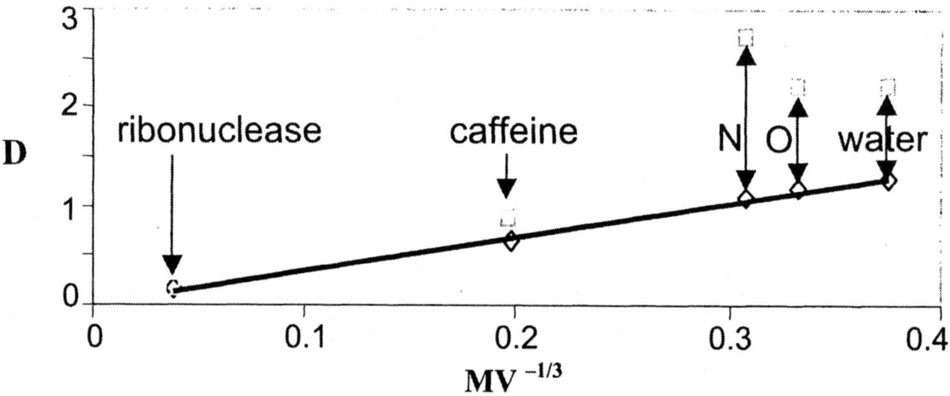

Figure 1. Showing the relationship between the molecular volume of an equivalent sphere and the consequent diffusion coefficient predicted by the Einstein-Stokes relationship. For comparison, measured diffusion coefficients are added to the graph at the position equivalent to their mass.

It can be seen that, for peptides and protein molecules, the theoretical value calculated from measurements of partial specific volume and the measured values taken from the literature are in good agreement. For caffeine, with a molecular weight of 194, but with molecular dimensions only marginally larger than water, it is apparent that the measured diffusion coefficient is 19% greater than that predicted for an equivalent unhydrated perfect sphere. From this molecular size downwards, the deviations become considerably more significant. Thus for nitrogen the calculated value, taking the volume of one mole of liquid nitrogen as 34 ml (the volume of one mole of the liquid at its boiling point) is $0{\cdot}94 \times 10^{-5}$ cm^2sec^{-1}, while the measured value is $2{\cdot}5 \times 10^{-5}$ cm^2sec^{-1}. This is the largest enhancement we have found and amounts to an increase of 266%. Similar calculations show that oxygen diffusion is also very significantly augmented at 191%, while tritiated water is augmented in a rather similar amount by 188%. Interestingly, while all of the molecules have a significant capacity to take part in hydrogen bonding with the water, nevertheless their diffusion is enhanced. At first sight this may seem paradoxical since the tendency to form larger but unstable and short lived complexes with the solvent water might be thought to effectively increase the drag and therefore diminish the measured diffusion coefficients.

3. THE SELF-DIFFUSION OF WATER

Self-diffusion of water has been measured by several authors.[9, 13] Attempts to account for the measured values have been proposed by Berendsen et al,[14] who have described an empirical model for water (the SPC model) in the liquid state. Values for the self-diffusion of water calculated from this model are even higher than the measured and accepted values. More recently, Yu et al.[15] have proposed a model for water structure which predicts a value of $2{\cdot}6 \times 10^{-5}$ cm^2sec^{-1}, which is reasonably close to experimental values of 2 to $2.2 \times 10^{-5} cm^2sec^{-1}$. Evans[1] has also proposed a model for water as having a hard spherical core with a diameter of 2·75Angstrom, with 4 hydrogen bonding locations in accordance with the lone pairs and the two hydrogens. The model gave predicted self-diffusion values of $2{\cdot}4 \times 10^{-5}$ cm^2sec^{-1}.

4. DIFFUSION OF OXYGEN THROUGH WATER

There are several experimental measures for oxygen diffusion through water or dilute gels[9, 16] or through aqueous solutions of varying viscosity.[10] While a simple Einstein-Stokes relationship predicts a diffusion coefficient for oxygen of about $1 \times 10^{-5} cm^2 sec^{-1}$, the experimental values are all approximately twice that value. Castillo et al.,[17] Pollock et al.[18] and Kowert et al.[19] have discussed the situation where the mass and dimensions of the diffusing molecule are comparable with those of the solvent. They have proposed an empirical relationship of the form:

$$D = \text{Constant} \times \eta^{-\alpha} \quad (5)$$

where η is the effective viscosity around the diffusant, and α has a value of 2/3.

Additionally, Murarka et al.[20] have investigated low molecular weight diffusion and related it to the self-diffusion of the solvent. They have suggested that there is a weak power law dependence of solute diffusion on the mass of the solvent particles.

The fit of our experimental results for oxygen diffusing through water and through sucrose solutions to an empirical relationship as above has also been tested. It was observed that there is a good fit for the D versus $1/\eta^{2/3}$ plot, suggesting one possibility that there is a microviscosity around the diffusant which differs from that of the bulk solvent. Thus effectively the diffusant "slips" through the water.

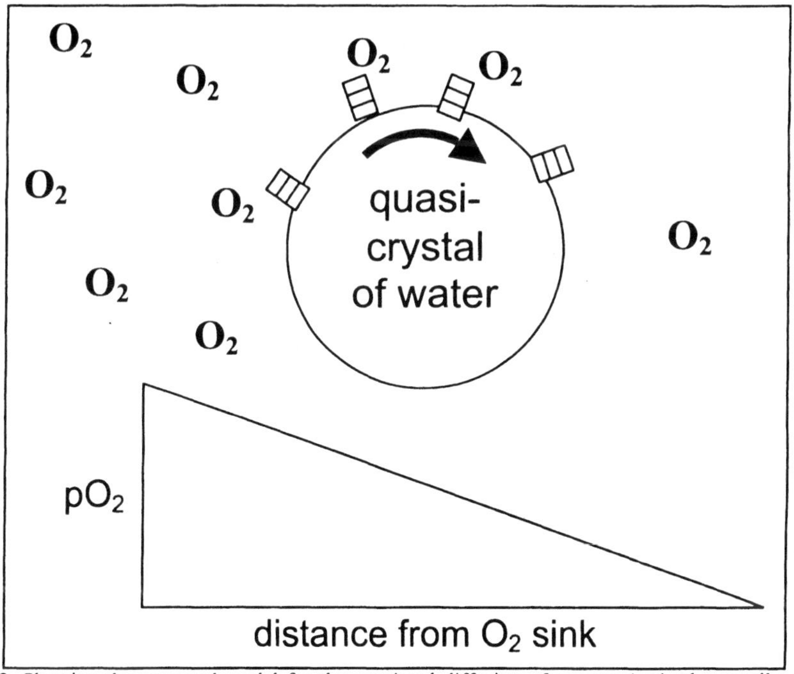

Figure 2. Showing the proposed model for the rotational diffusion of oxygen (and other small molecules capable of hydrogen-bonding to water) through water. Hydrogen bonds are depicted as lined rectangles and intentionally spaced in front of and behind the crystalline water structure in order to emphasize the spherical nature of the latter. The distributions of hydrogen bonds in this diagram is not intended to reflect structural reality, but instead to illustrate the time-course for one such bond involved in the proposed mechanism.

Using the graphical approach of Murarka et al.,[20] we have also tested for the possibility of a relationship between the (translational) self-diffusion coefficient of water and the observed enhancement of the comparable sized diffusants measured. However, the graph of ln(1/D) versus $\ln(m_1/m_2)$ shows only a weak dependence on such solute diffusion.

The results suggest to us the possibility of hydrogen bonding at any of the four positions, to adjacent water molecules, which subsequently may rotate before the hydrogen bonds break and subsequently reform. We envisage the process diffusion for oxygen as a kind of Grotthus phenomenon, whereby the oxygen (and other hydrogen bonded diffusants) are handed from water molecule to water molecule by rapid breaking and reformation of hydrogen bonds to adjacent clusters of water molecules which are, of course, taking part in self diffusion as well as rotational diffusion. Such a scheme is illustrated in Figure 2. It may be that the enhancement due to the rotational component of the water diffusion is the significant promoter of translational diffusion for oxygen.

REFERENCES

1. G. T. Evans, Force correlation and the diffusion coefficient of water, *J. Chem. Phys.* **117**, 11284-11291 (2002).
2. A. Einstein, Uber die von der molekular-kineticshen theorie der warme geforderte bewegung von in ruhenden flussigkeiten suspendierten teilchen, *Ann. Physik.* **17**, 549-560 (1905).
3. A. Einstein, Elementare theorie der brownschen bewegung, *Zeitschrift für Elektochemie* **XIV**, 235-239 (1908).
4. W. Sutherland, A dynamical theory of diffusion for non-electrolytes and the molecular mass of albumin, *Philos. Mag.* **9**, 781-785 (1905).
5. R. Mills, Self-diffusion in normal and heavy water in the range 1-45 degrees, *J. Phys. Chem.* **77**, 685-688.
6. A. K. Harrison, and R. Zwanzig, Transport on a dynamically disordered lattice, *Phys. Rev. A*, **32**(2), 1072-1075 (1985).
7. G. Phillies, Translational diffusion coefficient of macroparticles in solvents of high viscosity, *J. Phys. Chem.* **85**, 2838-2843 (1981).
8. L. A. Bunimovitch, and G. S. Ya, Statistical properties of the Lorentz gas with a periodic configuration of scatterers, *Comm. Math. Phys.* **78**, 479--497 (1980).
9. M. McCabe, and T. C. Laurent, Diffusion of oxygen, nitrogen and water in hyluronate solutions, *Biochem. Biophys. Acta.* **399**, 131-138 (1975).
10. M. McCabe, and D. J. Maguire, The measurement of the diffusion coefficient of oxygen thorough small volumes of viscous solution: implications for the flux of oxygen through tissues, *Adv. Exp. Med. Biol.* **316,** 467-473 (1992).
11. M. McCabe, The diffusion coefficient of caffeine through agar gels containing a hyaluronic acid-protein complex. A model system for the study of the permeability of connective tissues, *Biochem. J.* **127**, 249-253 (1972).
12. R. B. Setlow, and E. C. Pollard, *Molecular Physics* (Addison-Wesley Publishing Co., Massachusetts, 1962).
13. K. Krynicki, C. D. Green, and D. W. Sawyer, Pressure and temperature-dependence of self-diffusion in water, *Faraday Discuss.* **66**, 199-208 (1978).
14. H. J. C. Berendsen, J. P. M. Postma, W. F. van Gunsteren, and J. Hermans, Interaction models of water in relation to protein hydration. In *Intermolecular Forces: Proceedings of the Fourteenth Jerusalem Symposium on Quantum Chemistry and Biochemistry*. B. Pullman, editor, (D. Reidel Publishing Company, Dordrecht, The Netherlands) pp. 331-342 (1981).
15. Y. X. Yu, and G. H. Gao, Study on self-diffusion in water, alcohols and hydrogen fluoride by the statistical associating fluid theory, *Fluid Phase Equilib.* **179**, 165-179 (2001).
16. J. Jordan, E. Ackerman, and R. L. Berger, Polarographic Diffusion Coefficients of Oxygen Defined by Activity Gradients in Viscous Media *J. Am. Chem. Soc.* **78**(13), 2979-2983 (1956).
17. R. Castillo, C. Garza, and S. Ramos, Brownian Motion at the Molecular Level in Liquid Solutions of C60, *J. Phys. Chem.* **98**, 4188-4190 (1994).

18. G. L. Pollack, R. P. Kennan, J. F. Himm, and D. R. Stump, Diffusion of xenon in liquid alkanes: Temperature dependence measurements with a new method. Stokes–Einstein and hard sphere theories, *J. Chem. Phys.* **92**, 625-630 (1990).
19. B. A. Kowert, K. T. Sobush, N. C. Daug, G. L. Seck, C. F. Fukua, and C. L. Mapes, Diffusion of dioxygen in 1-alkenes and biphenyl in perfluoro-n-alkanes, *Chem. Phys. Lett.* **353**, 95 (2002).
20. R. K. Murarka, S. Bhattacharyya, and B. Bagchi, Diffusion of small light particles in a solvent of large massive molecules, *J. Chem. Phys.* **117**, 10730-10738 (2002).

GUT DYSOXIA:

Comparison of sites to detect regional gut dysoxia

Elaine M. Fisher, and Joseph C. LaManna

Abstract: Dysoxia, a state in which O_2 supply is inadequate to meet tissue metabolic needs, is often first detected in regional organs such as the gut. An increase in PCO_2 is believed to reflect the development of gut dysoxia. The stomach is a well-documented clinical site for detecting gut PCO_2; however, measurement issues make this a less than ideal monitoring site. Other sites along the GI tract may be equally sensitive to detect changes in PCO_2. Rectal CO_2 measurement may have the advantage of being less invasive, low risk, and continuous without the limitations associated with gastric monitoring. In this study, we compared PCO_2 at two sites (gastric, rectum) at baseline and during a dysoxic challenge, cardiac arrest. We obtained similar values of PCO_2 at both sites.

Ten male Wistar rats were anesthetized with 1%-2% Isoflurane/50% nitrous oxide/balanced O_2 and the tail artery and right atrium were cannulated. Severinghaus-type active tip PCO_2 electrodes (Microelectrode Inc, Bedford, NH) were calibrated and one electrode was surgically inserted into the stomach (G-PCO_2) and a second electrode was placed in the rectum (R-PCO_2). Animals were stabilized following surgery. Cardiac arrest was induced by administering a rapid injection of norcuron (0.1-0.2 mg/kg) and potassium chloride solution (0.5 M/L; 0.12 mL/100 gm of body weight). Animals were monitored for 15 minutes post-arrest. Data were collected at one minute intervals using the software Data Collect. All data are reported as mean ± SD.

Baseline G-PCO_2 was 64 ± 17 torr, not significantly different from R-PCO_2, 58 ± 7 torr. After 15 minutes of cardiac arrest, G-PCO_2 rose to 114 ± 42 torr, again not significantly different from R-PCO_2, which reached 112 ± 35 torr. Monitoring PCO_2 in the rectum is less invasive and might provide similar information when compared with gastric monitoring at baseline and during a dysoxic challenge.

1. INTRODUCTION

Tissue PCO_2 is a clinically recognized biochemical marker of perfusion failure resulting from a variety of conditions (hemorrhagic shock, sepsis, trauma) that arise when tissue oxygen requirements can no longer be sustained and anaerobic metabolism is initiated.[1, 2] The term dysoxia represents the point at which the O_2 supply and demand become mismatched and tissue damage begins.[3] Damage progresses hierarchically from cell to organ, culminating in damage to whole body function. Dysoxia is often detected first in the regional organs (i.e., gut, kidney, muscle) because of the initiation of compensatory mechanisms that shunt blood away from these non-vital organs toward the vital organs (i.e., brain, heart, and lungs). Dysoxia in the gut has been clinically linked to the development of the potentially lethal complications of shock and multiple organ failure (i.e., lung, kidney, heart, etc.). Early detection and treatment of sub-clinical signs of gut dysoxia may reverse or limit dysoxia, thereby preventing the development of complications and/or death.

The stomach is a well-documented site for measuring gastric PCO_2 at the bedside via gastric tonometry[1, 2, 4]; however, gastric PCO_2 monitoring is confounded by measurement issues such as the poor diffusibility of CO_2 out of the stomach following elevations related to ischemic events.[5] The back diffusion of CO_2 generated from the titration of gastric acid by pancreatic HCO_3^- in the duodenum and stomach brings into question whether changes in gastric CO_2 reflect tissue dysoxia or the neutralization of gastric acid (e.g., HCl) by pancreatic HCO_3^-.[6] Additionally, coating of the tonometer balloon by gastric contents such as tube feeding may alter the diffusion of CO_2 into the balloon, hence providing the practitioner with inaccurate values on which to make clinical decisions. Other sites along the gut (small intestine, sigmoid colon, rectum), when compared with gastric monitoring, have been equally or more sensitive for detecting changes in PCO_2, but the insertion of clinical monitoring devices into the small intestine or colon would be impractical. Limited research using indirect indicators of gut perfusion have been reported for monitoring tissue PCO_2 of the rectum,[7] a potential monitoring site that has the advantage of being assessable to monitoring via minimally invasive techniques and having less interference than measurement of gastric PCO_2.

The purpose of this study was to compare tissue PCO_2 measured in the rectum and stomach during a dysoxic challenge, cardiac arrest. We hypothesized that changes in gastric and rectal PCO_2 would be of similar value and magnitude at baseline and arrest.

2. METHODS

2.1. Surgical Preparation

This study was approved by the institutional Animal Care and Use Committee. Ten male Wistar rats (350 to 450 grams) were fasted overnight but continued to have free access to water. The rats were anesthetized with 1%-2% Isoflurane/50% nitrous oxide/balanced O_2 via nose cone. Anesthesia was titrated to maintain the mean arterial pressure (MAP) between 80-110 torr. The depth of anesthesia was evaluated periodically via foot compression. Tail artery cannulation was performed (PE 50, Intramedic) and the catheter was connected to a transducer (P-300, Grass Instruments) to monitor blood pressure (BP) (Astro-Med Inc., West Warwick, RI). The BP transducer was leveled to the mid-chest area. The right external jugular vein was cannulated and advanced into the right atrium. Catheters were intermittently flushed with 0.9% saline/heparin 5 IU/mL. The body temperature was maintained at 37.0°C using a TR-100 temperature controller with animal heating pad (Fine Science Tools, Foster City, CA). Severinghaus-type active tip PCO_2 electrodes (Microelectrode Inc, Bedford, NH) were calibrated according to manufacturer's instructions and connected to a meter (Thermo-Orion, Beverly, MA). One electrode was surgically inserted into the gastric lumen and a second electrode was placed in the rectum.

2.2. Experimental Protocol

Animals were stabilized following surgery (30-60 minutes) and the baseline measurement of gastric and rectal PCO_2 was the average of 15 minute values prior to cardiac arrest. An arterial blood sample was drawn at baseline to identify the adequacy of ventilation and evaluate parameters of systemic oxygenation and electrolyte balance. Cardiac arrest was induced by administering a rapid injection of norcuron (0.1-0.2 mg/kg) immediately followed by a potassium chloride bolus (0.5M/L; 0.12 mL/100 gm of body weight). Animals were monitored for 15 minutes post-arrest.

Baseline blood gas tensions and pH were determined using a blood gas analyzer. Oxygen saturation, content, and hemoglobin concentration were measured via Co-oximetery. Electrolytes (Na^+, Cl^-, K^+, Ca^{++}), glucose, and lactate were determined via the ABL-625; EML-550 unit (Radiometer, Westlake OH). Linearity of each PCO_2 electrode was tested by comparing the digital display on the Orion CO2 meter with the PCO_2 values of four known solutions ($PCO_2 \approx$ 45-50 torr; 75-80 torr; 100-110 torr; 170-180 torr) pre- and post-protocol. Gastric and rectal PCO_2 (G-PCO_2; R-PCO_2) were collected at one minute intervals using the software Data Collect (Thermo-Orion, Beverly, MA).

2.3. Data Analysis

Statistical analysis was performed using the SAS based program, JMP IN (Version 4). Pearson product moment correlation coefficients were calculated to examine the relationship between G- PCO_2 and R-PCO_2. We compared means by using the Student *t* test. To compare mean changes between baseline and arrest, we used a 2-way ANOVA. Results are presented as mean ± STD. We considered $P < .05$ to be significant.

Table 1. Mean (± SD) of systemic variables at baseline (n = 10)

Variable	Mean		SD	Variable	Mean		SD
pH	7.38	±	0.01	Lactate, mmol/L	2.9	±	1
PCO_2, torr	39.1	±	10	Glucose, mmol/L	9.6	±	1.3
PO_2, torr	145	±	24	Na^+, mmol/L	138	±	1.4
HCO_3^-, mmol/L	22.5	±	4.1	K^+, mmol/L	4	±	0.5
O_2 Sat, %	95.4	±	13	Cl^-, mmol/L	103	±	1.5
SBE, mmol/L	-1.9	±	4.2	Ca^{++}, mg/dL	4.6	±	0.4
p50, mm Hg	36.7	±	30	tHb, g/dL	13	±	5.5

PCO_2, partial pressure of carbon dioxide; PO_2, partial pressure of oxygen; HCO_3^- bicarbonate; O_2 Sat, oxygen saturation; SBE, standard base excess; Na^+, sodium; K^+, potassium; Cl^-, chloride; Ca^{++}, corrected calcium; tHb, total hemoglobin.

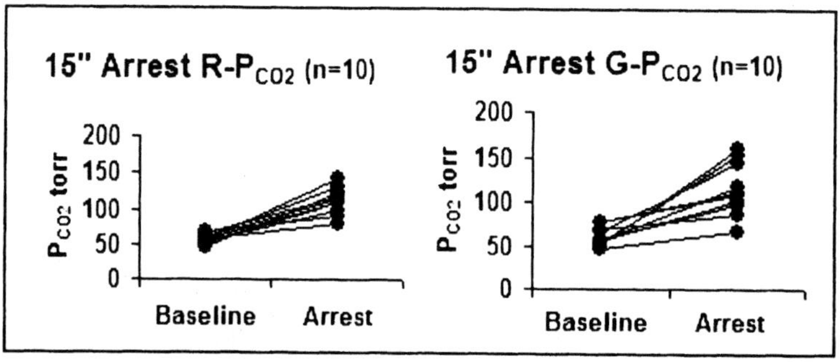

Figure 1. The relationship of R-PCO_2 and G-PCO_2 at baseline and arrest in each of ten rats.

3. RESULTS

Ten male Wistar rats weighing 345-454 grams were used in this study. Indicators of baseline ventilatory status, systemic oxygenation, and electrolytes are reported in Table 1. MAP at baseline ranged from 85-115 torr.

3.1. Relationship Between R-PCO_2 and G-PCO_2

The mean R-PCO_2 and G-PCO_2 at baseline and during 15 minutes arrest in each of 10 rats is displayed in Figure 1. The model was tested to ensure that rectal and gastric PCO_2 changed significantly between baseline and arrest (P = .002). The group means at baseline and arrest for R-PCO_2 (58.3 ± 6.7, 112.1 + 34.8 torr) and G-PCO_2 (63.7 ± 16.5, 114.2 + 41.5 torr) are displayed in Figure 2. The average difference between values of gastric and rectal PCO_2 at baseline and during 15 minutes arrest were 12.8 ± 22 torr, 0.74 ± 31.1 torr, respectively. Similar values of PCO_2 were obtained at the rectal and gastric site when paired comparisons were made (P = .28). A strong correlation was obtained between the measures of R-PCO_2 and G-PCO_2 throughout the experiment (R = 0.86). Moderate correlations were observed when comparisons were made at baseline and for the arrest phase (0.58, 0.6, respectively).

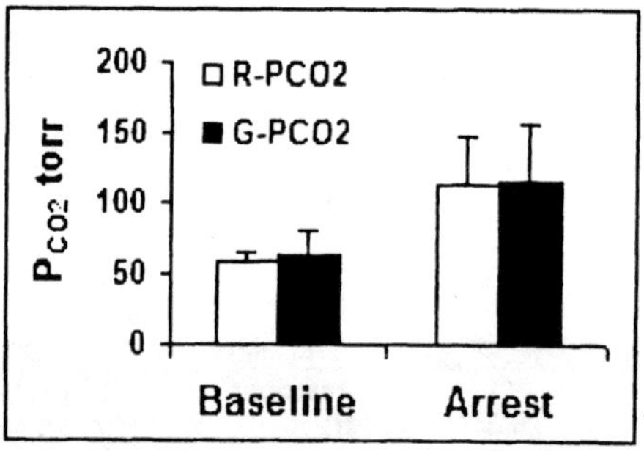

Figure 2. The average R-PCO_2 and G-PCO_2 at baseline and arrest.

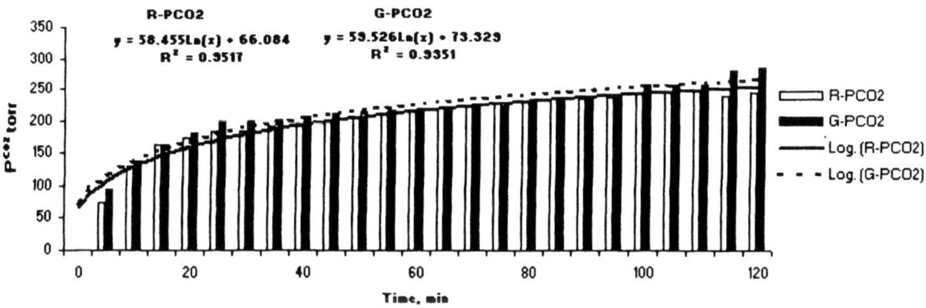

Figure 3. Five-minute average of R-P_{CO_2} and G-P_{CO_2} during prolonged arrest (n = 4).

3.2. Equilibration of R-P_{CO_2} and G-P_{CO_2} During Prolonged Arrest

Variations in P_{CO_2} at the rectal and gastric sites post-arrest may have resulted in a failure of P_{CO_2} to reach equilibrium at the sites. We tracked the values of R-P_{CO_2} and G-P_{CO_2} in four additional rats for a total arrest time of 90-120 minutes (Figure 3). Values at both sites rapidly increased followed by a slower rate of increase over time. The change in P_{CO_2} with time was logarithmic and asymptoting at approximately 250 torr. The curves between the two lines were similar.

3.3. Electrode Calibration Pre- and Post-Protocol

We evaluated electrode drift by comparing the regression coefficients, slopes, and y-intercepts pre- and post-data acquisition. Regression coefficients ranged from 0.94 to 0.98 with an average R^2 of 0.99 ± .02. Paired comparisons of regression coefficients pre- and post-experiment for each rat did not significantly differ ($P < .001$). The mean slope ranged between 0.92-0.96. The mean Y-intercept tended to vary to a greater degree than the slope and ranged from 0.5 to 4.9.

4. DISCUSSION

Similar values of tissue-P_{CO_2} were obtained from the rectal and gastric site at baseline and during 15 minutes of cardiac arrest. The rise in tissue P_{CO_2} during perfusion failure has been described in numerous laboratory and clinical trials. In this study, monitoring at both sites detected rapid changes in P_{CO_2} and reached a similar equilibrium during prolonged arrest; thus, the choice of a monitoring site should be the site that provides not only ease of accessibility, but also permits monitoring with the least measurement interference.

Identical P_{CO_2} values at the gastric and rectal sites were not found, as evidenced by the variation in the means, standard deviations, and correlations. We attribute these differences to measurement interference rather than changes in the relative amounts of P_{CO_2} along the gut. Gastric contents were present in the stomach despite 18 hours of fasting, and blood leaking into the stomach at the time of surgical insertion may have coated the diffusion surface of the electrode and interfered with measurement. At the rectal site, the only measurement interference we encountered occurred when stool coated the electrode tip. In this and the earlier clinical study involving open-heart surgery patients,[8] when the diffusion surface of the rectal electrode became coated with fecal material, an immediate and sudden rise in PCO_2 was detected. The electrode was easily removed, rinsed, replaced, and data collection was resumed. Thus, monitoring the rectal site was associated with less interference that could readily be remedied.

Differences between pre- and post-calibrations of the electrodes revealed that the slope was fairly constant (≈ 1), while the y-intercept values had greater variability. Electrode drift and changes in the diffusion surface likely altered the reliability of the electrodes.

4.1. Clinical Application

Monitoring conditions along the gut affect the ability of current clinical devices to accurately detect changes in tissue PCO_2. Tonometric measurement at the gastric site has been limited by the presence of gastric contents that may alter diffusion, the poor diffusibility of CO_2 out of the stomach following elevation, and the influence of gastric juice pH on gastric CO_2 production. In addition to site conditions, gastric feeding may alter measurement. An early treatment goal for all critically ill patients is to begin gastric feedings as soon as possible, ideally within 12 hours of admission, to maintain the integrity of gut and prevent bacterial translocation and possible sepsis. Feeding solutions are thick and capable of coating the balloon diffusion surface. The effect of prolonged dwelling time of these instruments in the stomach and changes in diffusion have not been reported.

Using the rectal site is a logical alternative because minimally invasive devices can be used and there is limited measurement interference at this site. Stool blocking the diffusion surface was the only limitation we identified. Continuous gastric feeding has been associated with the development of frequent bouts of diarrhea. However, current feeding solutions using a 2-calorie/mL base allow for continuous gastric feedings at lower rates and have reduced the incidence of diarrhea for critically ill patients. Of patients who do have diarrhea, infectious sources are frequently identifiable. In the clinical setting, the ability to differentiate whether changes in a measurement device result from a 'true' physiological change versus measurement error is critical to the treatment decision. The ease with which a measurement device could be removed, inspected, cleaned, and reinserted are an asset to using rectal PCO_2 monitoring. Further work using anaesthetized animals in a hemorrhage shock or hypercapnic model is necessary to validate use of the rectal site. It is our continued goal to develop clinically relevant technologies to monitor rectal PCO_2 in patients experiencing hemodynamic imbalances.

5. ACKNOWLEDGMENTS

We thank Doug Mitchell for assistance during data collection, Dr. Richard Steiner for statistical support, and Charlene Calabrese for her technical expertise and editing this manuscript. This work was supported by a Faculty Research Grant, The University of Akron, Akron, OH; Delta Omega Chapter, Sigma Theta Tau, Akron, OH.

REFERENCES

1. R. Schlichtig, and S. Bowles, Distinguishing between aerobic and anaerobic appearance of dissolved CO_2 in intestine during low flow, *J. Appl. Physiol.* **76**, 2443-2451 (1994).
2. R. Rozenfeld, M. Dishart, T. Tonnessen, and R. Schlichtig, Methods for detecting local intestinal ischemic anaerobic metabolic acidosis by P_{CO_2}, *J. Appl. Physiol.* **81**, 1834-1842 (1996).
3. R. Connett, C. Hoing, T. Gayeski, and G. Brooks, Defining hypoxia: A systems view of V_{O2}, glycolysis, energetics, and intracellular PO_2, *J. Appl. Physiol.* **68**, 833-842 (1990).
4. G. Gutierrez, F. Palizas, G. Doglio, A. Wainsztein, J. Gallesio, J. Pacin, and A. Dubin, Gastric intramucosal pH as a therapeutic index of tissue oxygenation in critically ill patients, *Lancet* **339**, 195-199 (1992).
5. M. Stevens, R. Thirlby, and M. Feldman, Mechanism for high P_{CO_2} in gastric juice: Roles of bicarbonate secretion and CO_2 diffusion, *Am. J. Physiol.* **253**, G527-G530 (1987).
6. R. Fiddian-Green, G. Pittenger, and W. Whitehouse, Back diffusion of CO_2 and its influence on the intramural pH in gastric mucosa, *J. Surg. Res.* **33**, 39-48 (1982).
7. A. Chendrasekhar, P. Srikumar, J. Fagerli, L. Barringer, J. Dulaney, and G. Timberlake, Rectal pH measurement in tracking cardiac performance in a hemorrhagic shock model, *J. Trauma Injury Infect. Crit. Care* **40**, 963-967 (1996).
8. E. Fisher, M. Kerr, L. Hoffman, R. Steiner, R. Baranak, and R. Schlichtig, A comparison of gastric and rectal CO2 in cardiac surgery patients, *submitted* (2004).

IMAGING OXYGEN PRESSURE IN THE RETINA OF THE MOUSE EYE

David F. Wilson, Sergei A. Vinogradov, Pavel Grosul, Akiko Kuroki, and Jean Bennett

Abstract: The phosphorescence lifetime imaging system previously used to image oxygen in the retina of the cat eye[1] was modified to allow imaging of phosphorescence lifetimes in the much smaller mouse eye. Following the lead of Shonat and coworkers,[2] a frequency domain approach was used in which the excitation light source was modulated in a 50% on:50% off square wave while the gate of the intensified CCD camera was similarly modulated but delayed with respect to the excitation. These were analyzed by fitting the intensity at each pixel to a sinusoid. The phase of the phosphorescence relative to the excitation was determined and from the phase shift and frequency, the phosphorescence lifetime was calculated. The Stern-Volmer relationship was then used to calculate the oxygen pressure at each pixel of the image array. High resolution maps of phosphorescence lifetime and oxygen pressure in the retina of the mouse eye have been attained. The retinal veins draining into the optic head appear as large, highly phosphorescent vessels against a lower phosphorescence background with a network of smaller vessels. The oxygen pressure in the retinal veins is typically from 20 to 30 mm Hg while the background has somewhat higher oxygen pressures. Experiments are underway to resolve the oxygen in the choroid from that in the retina. The arteries on the retinal surface can be observed, but their small diameter, relatively high oxygen pressures (> 90 mm Hg), and surrounding tissue with much lower oxygen pressures, makes accurate determination of the oxygen pressure a challenge.

1. INTRODUCTION

There are many diseases of the eye for which oxygen delivery to the retina is insufficient, and this is either a causative or a contributory factor in the resulting loss of vision. Oxygen measurements in the retina have been difficult, and when this was accomplished, it was generally through the insertion of small oxygen electrodes.[1-4] This approach is invasive and generally results in point measurements, although oxygen profiles through the retina have been reported.[2-4] Oxygen dependent quenching of phosphorescence, which provides a minimally invasive method for making the measurements,[5, 6] has been used for measurements in the retina,[7-10] but this type of measurement was initially limited to relatively large eyes, such as in cats and pigs. Shonat and Kight[10] have recently reported obtaining good phosphorescence lifetime images and oxygen pressure maps in rodents, a technical advance that will allow its use in a wide range of models for eye pathology. In the present paper, we report that, with a modification of our phosphorescence lifetime imaging system using an approach similar to that described by Shonat and Kight,[10, 11] we were able to obtain high resolution oxygen maps in the mouse eye.

2. METHODS

2.1. Instrumentation

A Xybion (now ITT Night Vision) ISG 750 intensified CCD camera was used to image the emission of a phosphor that had been injected via the tail vein. A frequency domain approach was used for determining the phosphorescence lifetime. The retina of the eye was imaged through long working distance (18 mm) microscope optics and excitation was effected on axis illumination by a modulated LED light source. The excitation light was reflected into the optical axis using a dicroic mirror and the emission imaged through a long pass Shott glass filter. In order to calculate the phosphorescence lifetime, both the excitation light and the camera intensifier were operated at the same frequency with a square wave (on or off) having a 50% duty cycle. A set of phosphorescence intensity images were collected with the intensifier operation delayed by different phase angles from 0 to 180 degrees relative to the excitation light. These were typically each 20 or 30 degrees, resulting in sets of either 7 or 10 images.

Once the images had been collected, the intensities at each pixel of the image array was fitted to a sinusoid. The phosphorescence lifetime was then calculated from the phase shift relative to the excitation light and the frequency. The oxygen pressure was then calculated using the Stern-Volmer relationship and the calibration coefficients for the phosphor, as shown in the following equation:

$$T^o/T = 1 + k_Q * T^o * PO_2 \qquad (1)$$

where T^o and T are the phosphorescence lifetimes at oxygen pressures (pO_2) of zero and the experimental value, respectively, and k_Q is a second order rate constant related to the frequency of collision of excited state phosphor molecules with molecular oxygen and the probability that energy transfer will occur in each collision. Equation (1) is used to calculate the oxygen pressure at each pixel of the image array from the phosphorescence lifetime map.

2.2. Phosphor for Oxygen Measurement

In the present study, we used Oxyphor G2 (2nd generation glutamate dendrimer of Pd-tetrabenzoporphyrin) as described by Vinogradov and coworkers[11-14] and calibrated by Dunphy et al.[15] This phosphor has strong absorption bands at 444 nm (Soret band) and at 635 nm (alpha band), while the phosphorescence emission is near 790 nm. We chose to excite the phosphorescence at the Soret band (455 nm) in order to limit the depth of penetration of the excitation light into the tissue. The Oxyphor G2 was injected i.v. as a solution (1.6 mg/ml; 0.15 ml/mouse) in unbuffered physiological saline at pH 7.5. The values of T^o and k_Q at pH 7.4 and 38° are 255 μsec and 280 mm Hg/sec, respectively.[15]

2.3. Animal Experiments

All procedures were carried out by trained personnel who followed procedures approved by the local IACUC committee. The mice were anesthetized by i.p. injection of 0.2 ml of a solution containing ketamine (25 mg/ml) and xylazine (25 mg/ml) dissolved in phosphate buffered saline. After they were fully anesthetized, the phosphor was injected in the tail vein and a drop of 1% Mydriacyl® (Alcori) was placed in each eye to dilate the pupil. As soon as the pupil was fully dilated, a drop of Goniosol® (CIBAVision Ophthalmics) was placed on the eye followed by a piece of thin plastic sheet or glass coverslip. Care was taken that no pressure was applied to the eye. The retina was imaged through the flat optical surface provided by the glass or plastic cover. The animals were anesthetized throughout the experimental procedure.

3. RESULTS AND DISCUSSION

3.1. Phosphorescence Intensity Images

A set of images of phosphorescence intensity measured with different phase delays relative to the excitation light and at a frequency of 1030 Hz are shown in Figure 1. The phase delay is given in the upper left corner of each image and ranged from 0 to 180 degrees. The measured intensity increased slightly from zero with increasing delay as expected for phosphorescence, and then decreased as the phase delay was further increased. The images are of the region of the optical nerve head where the blood enters through arteries and exits through the veins. The result is a pattern with the vessels appearing to radiate from a central avascular region that is 100 to 150 μm in diameter. The veins are most easily seen due to their large size and low oxygen pressures, both of which increase the phosphorescence intensity. The image taken with a 20 degree phase delay has been labeled and the different structural elements identified. The point of entry of the optic nerve is clearly outlined by the veins (V) and arteries (A) radiating from it. There is one identifiable retinal artery (A) in the image. The artery appears very similar to the veins in the phosphorescence intensity images and becomes apparent only when the phase shift, lifetime, and oxygen pressure maps are calculated.

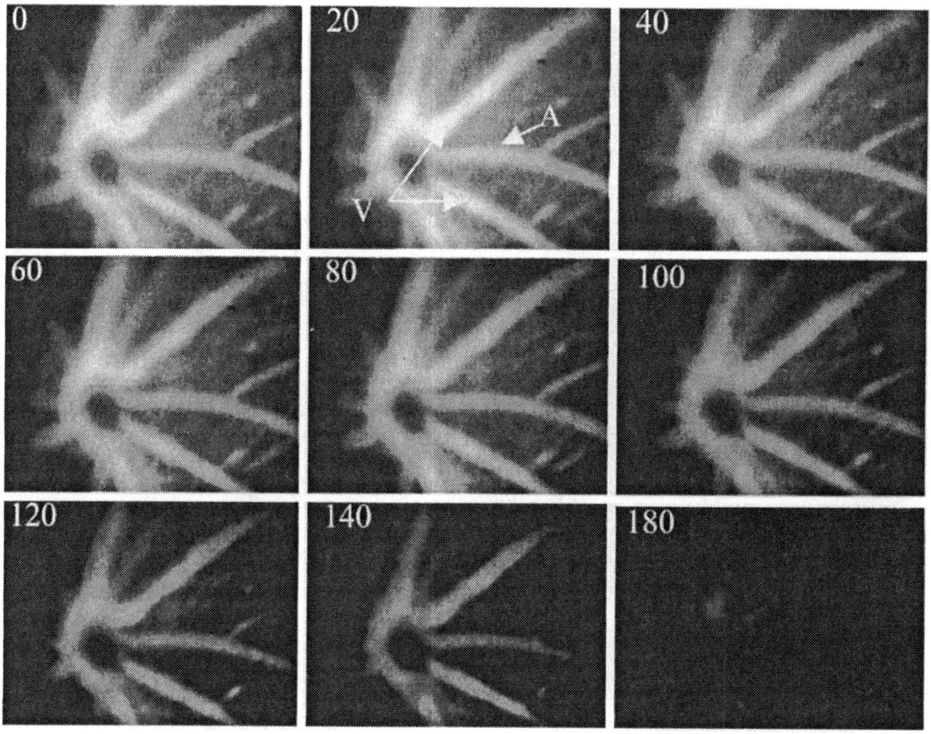

Figure 1. Images of the phosphorescence intensity from the mouse eye measured at 1,000 Hz and the dependence on the phase delay between the excitation light pulse and the detector gating. The image is of a region 1.1 mm wide and 0.95 mm high. The phase delay for each measurement is given in the upper left corner of the image.

3.2. Phase Shift, Lifetime, and Oxygen Pressure Maps

The intensity of the phosphorescence at each pixel of the image set is fitted to a sinusoid of the frequency used to collect the data. The phase shift calculated in the fitting procedure (least squares) is used to construct a map (image) of the phase shift at each pixel of the image set (not shown). When the frequency of the measurement is properly selected, the phase shift in the region of interest should be near 28 degrees, the value used for calibration of the phosphor. As can be seen, in the retinal image the phase shifts can be quite different for the different regions of the image, ranging from about 10 degrees to greater than 40 degrees. As the phase shift deviates from that used for calibration, there is increasing potential for error in the calculated oxygen pressure. For maximal accuracy, the frequency should be selected for which the phase shift in the region of interest is between about 20 and 35 degrees. Once the phase shift has been obtained, the phosphorescence lifetime can be calculated from the frequency and phase shift. A map can then be constructed showing the lifetime at each pixel (Figure 2). The phosphorescence lifetime map in Figure 2 shows the values range from about 20 μsec to 100 μsec. The spatial resolution is quite good, and it is possible to detect small vessels in

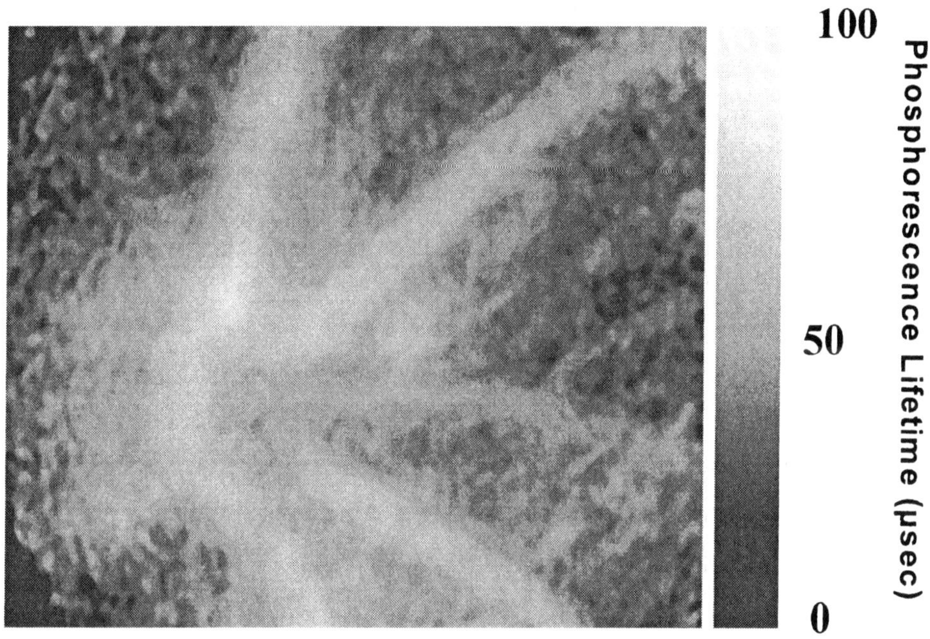

Figure 2. The phosphorescence lifetime map calculated from the image set shown in Figure 1. The lifetime scale in gray levels is shown on the right.

the retina even using the current optics. These small vessels in the retina (diameter about 10 microns) are at the limits of resolution of the current optics, but higher resolution would be possible if this were desired. It should, in particular, be possible to use pseudoconfocal techniques to remove the out of focus contribution to the phosphorescence images. This should significantly sharpen the images and improve the resolution.

Maps of the oxygen pressure have been calculated from the maps of phosphorescence lifetime (Figure 3). These indicate that, for the mouse being examined in these experiments, the oxygen pressures in the veins are between 20 and 30 mm Hg, whereas those from regions other than the veins are much higher. It is difficult to obtain accurate oxygen pressures in these regions for two reasons: 1. When using excitation light of 455 nm, it is expected that some of the excitation will pass through the retina and into the choroid. The blood in the very dense vasculature of the choroid has much higher oxygen pressure than does the blood in the retinal vessels and the relative contributions of the two signals could not be determined; 2. The phosphorescence intensities are much lower than from the veins and this substantially decreases the accuracy of the fit to a sinusoid. In imaging a system with a wide range of oxygen pressures (phosphorescence lifetimes) such as are present in tissue, resolution of the oxygen pressures in the different regions may require imaging at more than one frequency. Although the phase shift for the veins is near 28 degrees at frequencies between 600 and 1000 Hz, for the artery this requires a measurement frequency of 2 to 4 kHz. When measured at a phase shift near 28 degrees, the arterial oxygen pressure was between 75 and 80 mm Hg.

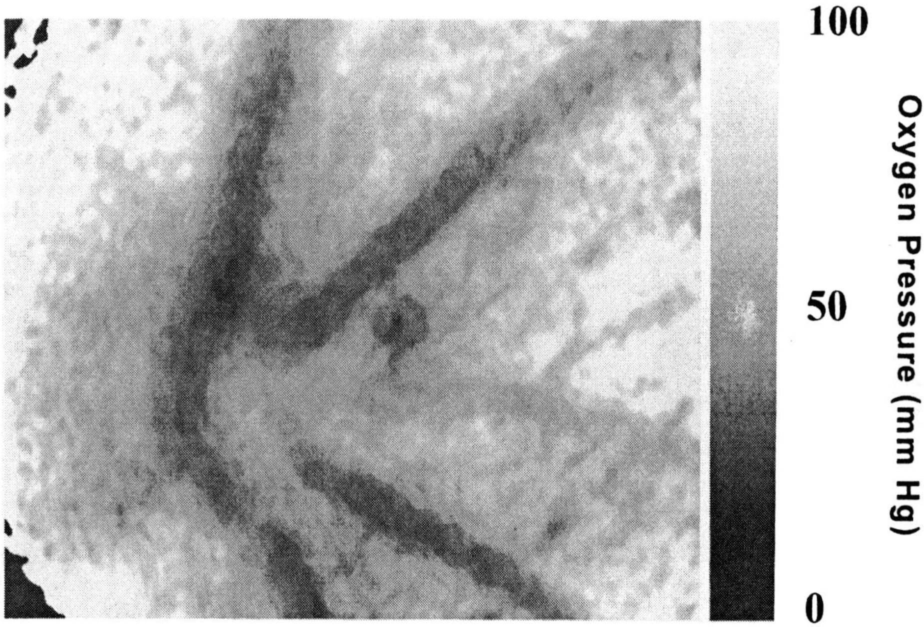

Figure 3. The oxygen pressure map calculated from the image set in Figure 1. The oxygen pressure scale in gray levels is presented on the right side of the map.

The frequency domain imaging system used in this study has significant advantages over the previous time domain instrument. This is due to: 1. the greater stability of the LED light sources compared with the flash lamps used in the latter, decreasing the variation in intensity between images; and 2. the flash lamps had a significant "tail" that required there be a significant delay after the flash before the camera was turned on, resulting in decreased efficiency in collecting the signal.

4. CONCLUSIONS

Phosphorescence lifetime imaging can provide high resolution maps of the oxygen distribution in the retina and choroid of the mouse. This technology provides an important new method for research and diagnosis in diseases of the eye.

5. ACKNOWLEDGEMENT

Supported in part by grants NS-31465 and HD041484.

REFERENCES

1. R. A. Linsenmeier, R. D. Braun, M. A. McRipley, L. B. Padnick, J. Ahmed, D. L. Hatchell, D. S. McLeod, and G. A. Lutty, Retinal hypoxia in long-term diabetic cats, *Invest. Ophthalmol. Vis. Sci.* **39**, 1647-1657 (1998).
2. D. Y. Yu, V. A. Alder, S. J. Cringle, E. N. Su, and M. Burns, Intraretinal oxygen distribution in urethan-induced retinopathy in rats, *Am. J. Physiol.* **274**, H2009-2017 (1998).
3. D. Y. Yu, S. J. Cringle, V. Alder, and E. N. Su, Intraretinal oxygen distribution in the rat with graded systemic hyperoxia and hypercapnia, *Invest. Ophthalmol. Vis. Sci.* **40**, 2082-2087 (1999).
4. D. Y. Yu, S. J. Cringle, E-N Su, and P. Ku, Intraretinal oxygen levels before and after photoreceptor loss in the RCS rat, *Invest. Ophthalmol. Vis. Sci.* **41**, 3999-4006 (2000).
5. J. M. Vanderkooi, G. Maniara, T. J. Green, and D. F. Wilson, An optical method for measurement of dioxygen concentration based on quenching of phosphorescence, *J. Biol. Chem.* **262**, 5476-5487 (1987).
6. D. F. Wilson, W. L. Rumsey, T. J. Green, and J. M. Vanderkooi, The oxygen dependence of mitochondrial oxidative phosphorylation measured by a new optical method for measuring oxygen, *J. Biol. Chem.* **263**, 2712-2718 (1988).
7. R. D. Shonat, D. F. Wilson, C. E. Riva, and S. D. Cranstoun, Effect of acute increases in intraocular pressure on intravascular optic nerve head oxygen tension in cats, *Invest. Ophthalmol. Vis. Sci.* **33**, 3174-3180, (1992).
8. R. D. Shonat, D. F. Wilson, C. E. Riva, and M. Pawlowski, Oxygen distribution in the retinal and choroidal vessels of the cat as measured by a new phosphorescence imaging method, *Appl. Opt.* **33**, 3711-3718, (1992).
9. S. Blumenröder, A. J. Augustin, and F. H. J. Koch, The influence of intraoccular pressure and systemic oxygen tension on the intravascular pO_2 of the pig retina as measured with phosphorescence quenching, *Surv. Ophthalmol.* **42**(1), S118-S126 (1997).
10. R. D. Shonat, and A. C. Kight, Frequency domain imaging of oxygen tension in the mouse retina, *Adv. Exp. Med. Biol.* **510**, 243-247 (2003).
12. S. A. Vinogradov, and D. F. Wilson, "Dendritic" porphyrins: New protected phosphors for oxygen measurements in vivo, *Adv. Exp. Med. Biol.* **428**, 657-662 (1997).
11. S. A. Vinogradov, and D. F. Wilson, Metallotetrabenzoporphyrins. New phosphorescent probes for oxygen measurements, *J. Chem. Soc. Perkin Trans. II*, 103-111 (1994).
13. S. A. Vinogradov, L. W. Lo, and D. F. Wilson, Dendritic polyglutamic porphyrins: Probing porphyrin protection by oxygen-dependent quenching of phosphorescence, *Chem. Eur. J.* **5**, 1338-1347 (1999).
14. S. A. Vinogradov, E. Kim, and D. F. Wilson, Pd tetrabenzoporphyrin-dendrimers – near infrared phosphors for oxygen measurements by phosphorescence quenching in Biomedical Nanotechnology Architectures and Applications, *Proc SPIE*, **4626**, 193-200 (2002).
15. I. Dunphy, S. A. Vinogradov, and D. F. Wilson, Oxyphor R2 and G2: Phosphors for measuring oxygen by oxygen dependent quenching of phosphorescence, *Anal. Biochem.* **310**, 191-198 (2002).

FLUORESCENCE-MEDIATED DETECTION OF A HETEROGENEITY IN A HIGHLY SCATTERING MEDIA

Hanzhu Jin and Kyung A. Kang

Abstract: Tumors usually become localized absorbers at near infrared (NIR) wavelengths due to the increase in hemoglobin amount around the tumor, which is caused by angiogenesis. When a tumor is small and/or deeply seated, the contrast by the hemoglobin only, however, may not be strong. For such situation, contrast agents may be helpful, because they are preferentially accumulated in the tumor due to the unorganized tumor vasculature.

In this study, indocyanine green (ICG) was used as a contrast enhancer. ICG is safe, absorbs NIR, and also generates fluorescence. A breast tissue-like model, embedded with a tumor model (1.2 x 0.7 x 0.5 cm) with/without ICG at a 1 cm depth, was constructed and the surface was scanned by a NIR time-resolved spectroscopy instrument. Enhanced contrast by ICG was confirmed in both absorption and fluorescence. For absorption, transmittance contrast was approximately two times higher than reflectance. In reflectance, the contrast by fluorescence was approximately four times higher than absorption. This study result shows that the information on both the absorption and fluorescence by ICG can be effectively used in detecting a tumor. A study of the ICG effect on deeper absorber detection is in progress.

1. INTRODUCTION

Near infrared (NIR) has been very successful in detecting breast tumors, even some of those not detected by X-ray. The usual mechanism of detecting tumors by NIR is the high photon absorption by the hemoglobin accumulated around a tumor, which is usually caused by tumor angiogenesis. When a tumor is, however, very small and deeply seated, the optical contrast of the tumor can be weak.[1]

Due to the unorganized vasculature of tumors, when a contrast agent is injected into an organ with a tumor, the agent tends to accumulate in the tumor. Therefore, optical contrast agents, such as dyes with high light absorption or fluorophores, can be effectively used to enhance the sensitivity of NIR spectroscopy. The absorbing agent can enhance absorption, or, if the dye is fluorescent, then the tumor becomes a light source at a wavelength different from the input light.

Indocyanine green (ICG), which has been approved by the United States Food and Drug Administration for human use, can absorb NIR and also generate NIR fluorescence. Previous results by Honar and Kang[2] showed that the source/detector configuration in NIR measurements affected the NIR detectability for an absorber in highly scattering media: the transmittance source/detector configuration had better performance for localizing an absorber in two dimensions, while the reflectance measurement provided depth-related information. A preliminary study using ICG as a NIR contrast agent has been conducted, measuring the absorption and fluorescence of an ICG-containing heterogeneity by both reflectance and transmittance. The obtained TRS spectra were converted to the frequency domain in order to observe the frequency response.[2, 3] The changes in magnitude ratio (ΔMR: the light intensity) and the phase shift ($\Delta\Phi$: mean photon path length) due to the presence of an ICG-containing absorber were analyzed at a modulation frequency of 100 MHz.

2. METHODS

2.1. Model Construction

2.1.1. Breast Model

Breast models were constructed using 2.2% (w/v) of agar (Sigma Chemical, St. Louis, MO), 22.6% (w/v) of dried milk (Kroger Co., Cincinnati, OH), and 0.00046% (v/v) of India ink (Design Higgins®; Sanford Co.; Bellwood, IL) in 1476 ml of water to make 1750 ml of the total mixture volume. This composition was selected to simulate the average optical properties of the human female breast: the absorption coefficient (μ_a) 0.03 cm^{-1} and the reduced scattering coefficient (μ_s') 6 cm^{-1}.[4] The mixture was then poured into a plastic container with the dimension of 24 x 14 x 5 cm. The thickness of 5 cm was selected to simulate the average thickness of slightly compressed breasts between two mammogram plates.[5]

2.1.2. ICG-Containing Tumor Model

ICG strongly absorbs light at around 780 nm and emits fluorescence at about 830 nm.[6] ICG (Sigma Chemical; St. Louis, MO) was added (25 μg/ml) to the same ingredient used for the breast model and the mixture was injected into a Vitamin E capsule (1.5 x 1.0 x 1.0 cm; National Vitamin Company, LLC.; Las Vegas, NV). Due to the thickness of the capsule shell, the actual size of the ICG-containing tumor model (inside of the Vitamin E capsule) was 1.2 x 0.7 x 0.5 cm. The surface of the Vitamin E capsule was coated with a clear acrylic top coat (Sally Hansen®; Del Laboratories, Inc.; Farmingdale,

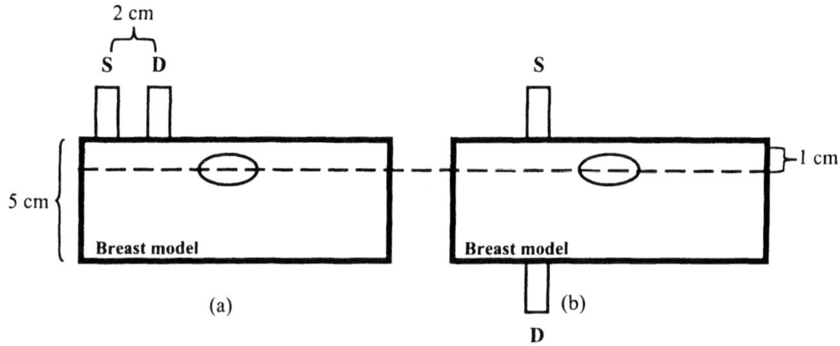

Figure 1. A schematic diagram of the breast model with an ICG-containing tumor model and the source/detector configuration for the measurement. Source-detector configurations: (a) Reflectance and (b) Transmittance; S: source, D: detector.

NY) to prevent it from being dissolved inside the breast model. Once the capsule was solidified, it was suspended 1 cm deep in the breast model, using a thin wire (Figure 1).

2.2. Instrument

The NIR-TRS instrument used in this study has a 780 nm laser diode with a pulse repetition rate of 5 MHz, and a single photon counting, multichannel plate photomultiplier tube (MCP-PMT; Hamamatsu photonics; Tokyo, Japan). This system was specially built by our research group with the financial support of the National Scientific Foundation. The photons released from a source optical fiber (250 μm of core diameter; OZ Optics; Ontario, Canada) travel through the system of interest, and those arrived at the detector fiber (3 mm of diameter; Thermal Orieal; Stratford, CT) are measured by the MCP-PMT. The photon energy is converted by the amplified electronic energy from the MCP-PMT and then the signal is sent to a constant fraction discrimination instrument (CFD; Oxford Instrument, Inc.; Oak Ridge, TN). The CFD sends this signal to a time-to-amplitude converter (TAC; Oxford Instruments, Inc.; Oak Ridge, TN), which converts the photon arrival time to the amplitude. For fluorescence measurement, a long pass filter (B+W 093 IR 830 nm; Schneider Optics, Inc.; New York, NY) was placed between the detector fiber and the MCP-PMT to allow passing of the fluorescence generated by the ICG, while minimizing the excitation light. The transmission rates of the filter are 59% at the 830 nm light and 1.6% at 780 nm. By overlapping two filters, the transmission of the excitation light was reduced 0.026%, providing more effective filtering.[7]

2.3. Measurement and Data Reduction

Figure 1 shows a schematic diagram of measurement configurations. After solidification of the breast model, the surface of the model was scanned in reflectance and transmittance with (for fluorescence image)/without (for absorption image) the filters. The measurements were performed on an area of 5 x 5 cm near to the absorber at an interval of 1 cm. For reflectance measurement of the source and detector separation

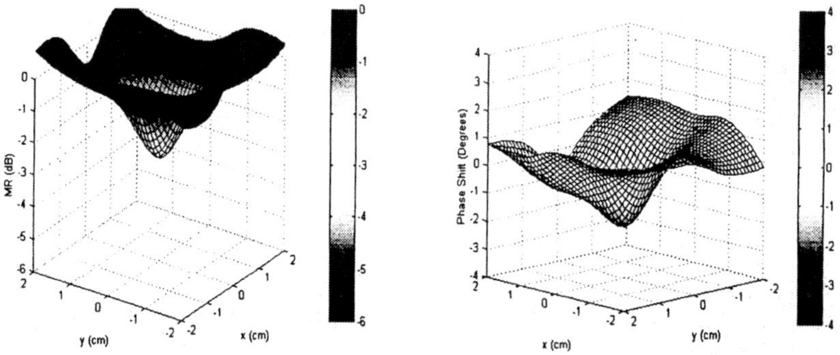

Figure 2. Changes in (a) the light intensity (ΔMR) and (b) the phase by the presence of an absorber when the measurements were taken in reflectance at the S-D separation of 2 cm. The modulation frequency: 100 MHz.

(S-D separation), 2 cm was used because the previous study results[2] showed that 2 cm was an effective distance for an embedded absorber located 1 cm deep. Obtained TRS spectra were converted to the frequency domain using a computer code developed for Fourier transformation.[8]

3. RESULTS AND DISCUSSION

3.1. Absorption by the ICG-Containing Tumor Model

Figure 2 (a) shows ΔMR values at a modulation frequency of 100 MHz obtained from the area (5 x 5 cm) around the tumor location when the measurement was taken in reflectance at 2 cm S-D separation. Due to the high absorption of ICG at 780 nm, as the source and detector unit (S-D unit) got closer to the ICG-containing tumor model, ΔMR decreased. The maximum decrease of 3 dB was obtained near the location of the tumor model. The phase changes (ΔΦ) for the reflectance measurement (Figure 2b) showed that the ICG-containing tumor model intersects the longer photon path, resulting in decreased ΔΦ values. The maximum Φ change was -2° at the tumor model location (Figure 2b).

For the transmittance (Figure 3a), the ΔMR decrease by the ICG-containing tumor was much greater than that for the reflectance, verifying the result by Honar and Kang[2] that an absorbing heterogeneity can be located better by transmittance. The maximum decrease of MR was about 5 dB. The phase or the mean photon path length in transmittance measurement was not significantly affected by the presence of the absorber (Figure 3b). This insensitivity of phase in transmittance also has been reported by Honar and Kang.[2]

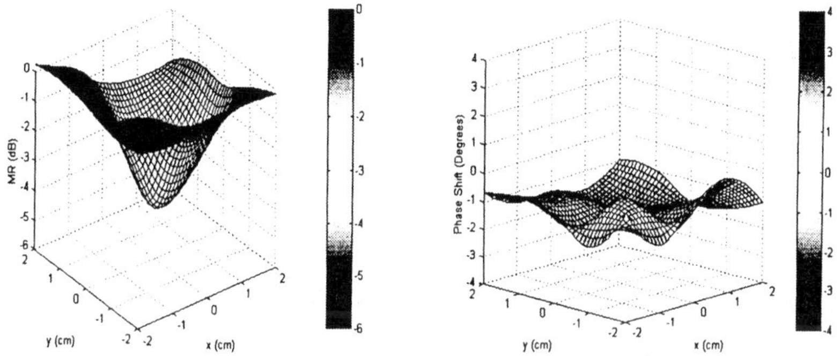

Figure 3. Changes in (a) the light intensity (ΔMR) and (b) the phase by the presence of an ICG-containing absorber when the measurements were taken in transmittance. The modulation frequency: 100 MHz.

3.2. Fluorescence by the ICG-Containing Tumor Model

As previously stated, ICG can generate fluorescence at 830 nm wavelength when excited at 780 nm light. Reflectance and transmittance measurements were performed for the same model used for the absorption measurement with two overlapped long pass filters. Figure 4 (a) shows ΔMR values obtained from the reflectance measurement at a S-D separation of 2 cm. For the fluorescence, the trend was found to be opposite to the absorption image. As the S-D unit got closer to the tumor model, the intensity of the fluorescence increased and the maximum increase was 21 dB. The phase increased as the S-D unit approached the absorber. The maximum increase of ΔΦ was about 34°, which is much more sensitive than the absorption image (Figure 4b). The path length of fluorescence and Φ in a highly scattering media for fluorescence is still under investigation.

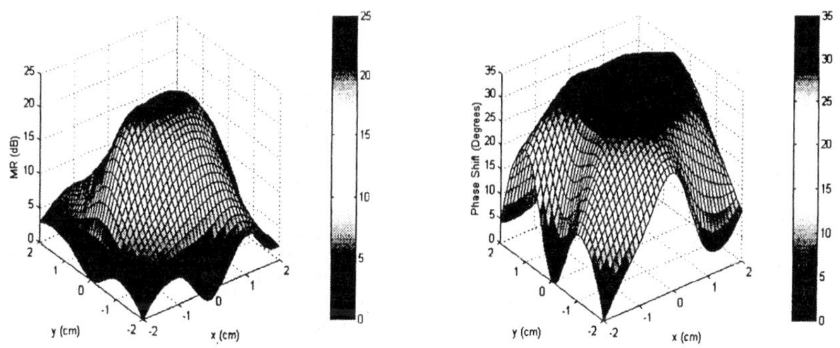

Figure 4. Changes in (a) the fluorescent light intensity (ΔMR) and (b) the phase measured by the presence of a fluorescent object when the measurements were taken in reflectance at the S-D separation of 2 cm. The modulation frequency: 100 MHz.

In case of the transmittance for fluorescent imaging, the fluorescence signals were very small, probably because most of the generated fluorescence was absorbed by the model (distance between the Vitamin E capsule and detector probe was at least 4 cm). In summary, for a shallow fluorescent object, reflectance provided better contrast than transmittance.

4. CONCLUSIONS

A preliminary study on the fluorescent absorber localization by NIR-TRS was performed for a shallow (1 cm depth) fluorescent object (1.2 x 0.7 x 0.5 cm) in a scattering media with the optical properties similar to the breast tissue. Enhanced contrast was confirmed by both absorption and fluorescence of ICG. For absorption, transmittance gave a better contrast than reflectance. For fluorescence, reflectance provided a better contrast than transmittance. For this particular shallow absorber depth, the fluorescence contrast was significantly higher than that of absorption. Both the light intensity and phase shift were decreased as the S-D unit got closer to the absorber for the absorption, and the opposite trend was found for the fluorescence image. By combining the absorption and fluorescence of ICG, the heterogeneity position can be clearly located.

Future studies will include the relationship between the detectability and system/operation parameters such as size of the tumor, the concentration of ICG, depth of the tumor, and the S-D separation. It is also suggested that fluorophores with a higher quantum yield and a more effective fluorescent filtering system would provide better detectability.

REFERENCES

1. M. A. Franceschini, K. T. Moesta, S. Fantini, G. Gaida, E. Gratton, H. Jess, W. W. Mantulin, M. Seeber, P. M. Schlag, and M. Kaschke, Frequency-domain techniques enhance optical mammography: Initial clinical results, *Proc. Natl. Acad. Sci. USA* **94**, 6468-6473 (1997).
2. A. L. Honar, and K. A. Kang, Effect of the source and detector configuration on the detectability of breast cancer, *Comp. Biochem. Physiol. A* **132**, 9-15 (2002).
3. K. A. Kang, D. F. Bruley, J. M. London, and B. Chance, Highly scattering optical system identification via frequency response analysis of NIR-TRS spectra, *Anal. Biomed.* **22**, 240-252 (1994).
4. T. Durduran, R. Choe, J. P. Culver, L. Zubkov, M. J. Holboke, J. Giammarco, B. Chance, and A. G. Yodh, Bulk optical properties of healthy female breast tissue, *Phys. Med. Biol.* **47**, 2847-2861 (2002).
5. R. Cubeddu, A. Pifferi, P. Taroni, A. Torricelli, and G. Valentini, Time-resolved transmittance imaging with a diffusion model, *Proc. SPIE* **2626**, 328-333 (1995).
6. E. Kywana, and E. M. Sevick-Muraca, Fluorescence lifetime spectroscopy in multiply scattering media with dyes exhibiting multiexponential decay kinetics, *Biophys. J.* **83**(2), 1165-1176 (2002).
7. Y. Chen, C. P. Mu, X. Intes, D. Blessington, and B. Chance, Near-infrared phase cancellation instrument for fast and accurate localization of fluorescent heterogeneity, *Rev. Sci. Instrum.* **74**(7), 3466-3473 (2003).
8. D. F. Bruley, Pulse reduction code written for process identification (personal communication), 1974.

DEVELOPING STRATEGIES FOR THREE-DIMENSIONAL IMAGING OF OXYGEN TENSION IN THE RODENT RETINA

Ross D. Shonat and Adam S. Norige

Abstract: Insufficient oxygen delivery and retinal hypoxia have been implicated as causal in the development of many devastating diseases of the eye. While the two-dimensional imaging of retinal oxygen tension (PO_2) has now been applied in a variety of different animal models, it is fundamentally a luminescence-based system lacking depth discrimination. However, mammalian retinal tissue is nourished by two distinct vascular beds, the retinal and the choroidal vasculatures, and they are exceedingly difficult to separate using traditional two-dimensional imaging strategies. Numerous studies have demonstrated that retinal and choroidal PO_2 differ substantially. Therefore, the single PO_2 value currently returned through data analysis cannot accurately represent the separate contributions of the choroidal and retinal vasculatures to the state of retinal oxygenation. Such a separation would significantly advance our understanding of oxygen delivery dynamics in these two very distinct vasculatures. In this study, we investigate new strategies for generating separate retinal and choroidal PO_2 maps in the rodent retina using our existing phosphorescence-based lifetime imaging system.

1. INTRODUCTION

The survival of the retina is dependent on an adequate supply of oxygen and nutrients to fuel the metabolic processes that are vital for primary visual function. In many retinal diseases, including diabetic retinopathy, age-related macular degeneration, and retinopathy of prematurity, developing retinal hypoxia has been implicated in the disease process. However, causal links relating retinal hypoxia to disease progression have not yet been definitively established. While the two-dimensional imaging of retinal oxygen tension (PO_2), based on phosphorescence quenching, has now been applied in a variety of different animal models,[1-3] it is fundamentally a luminescence-based system lacking depth discrimination. Because the mammalian retina is nourished by two distinct vascular beds, the retinal vasculature and the choroidal vasculature, it is exceedingly difficult to separately image them using traditional two-dimensional imaging strategies. Numerous studies have demonstrated conclusively that retinal PO_2 is substantially lower than choroidal PO_2, with choroidal PO_2 being quite close to systemic arterial levels.[4] However, without an ability to reject out-of-focus light, *in vivo* phosphorescence is collected from both of these vascular beds during image acquisition. As a consequence, the single PO_2 value currently returned through data analysis does not accurately represent the separate contribution of the choroidal and retinal vasculatures to the state of retinal oxygenation. Such a separation would significantly advance our understanding of oxygen delivery dynamics in these two very distinct vasculatures, and increase our knowledge of the disease state. In this study, we investigate new strategies for generating separate retinal and choroidal PO_2 maps in the rodent retina using our existing phosphorescence-based lifetime imaging system.[3]

2. TWO-DIMENSIONAL IMAGING OF RETINAL OXYGEN TENSION

2.1. Theory and Instrumentation

Methodology for the two-dimensional imaging of retinal PO_2 in the mouse and rat has been described in detail elsewhere.[3] Briefly, an oxygen-sensitive, palladium (Pd)-porphyrin derivative is injected into the systemic vasculature of a rodent animal model and allowed to distribute throughout the cardiovascular system, including the retinal and choroidal vessels of the eye. This vascular oxygen probe is then excited by sinusoidally-modulated light at an appropriate wavelength, and the resulting phosphorescence emission is collected by an intensified CCD camera. If the sensitivity of the intensifier is modulated at the same frequency as the excitation light and a number of phosphorescence intensity images (approximately 10-15) are collected at different intensifier phase delays (see Figure 1), then the resulting stack of intensity images can be used to determine the phase delay of the phosphorescence emission at each pixel, according to Eq. (1):

$$I(\theta_D) = k[Pd]\{1 + 1/2\, m_D\, m \cos(\theta - \theta_D)\} \tag{1}$$

Figure 1. Representative phosphorescence intensity images in a rat retina at three different intensifier phase delays [$\theta_D = 9^\circ$ (left), 100° (center), and 160° (right)].

Figure 2. Representative phosphorescence intensity image (left) and corresponding two-dimensional maps of θ (0-40°), τ (0-100 μs), PO_2 (0-140 mm Hg), and R^2 (0.96-1.0) (second left to right) in a mouse retina. The four maps to the right have been converted to a grey-scale format from their original color representations.

where θ_D is the known phase delay of the intensifier and θ is the unknown phosphorescence phase.

There are three unknowns in this equation, necessitating at least three intensity images to calculate θ using the method of least-squares. θ is related to the apparent phase lifetime, τ, by Eq. (2):

$$\tan\theta = \omega\,\tau \tag{2}$$

and τ is converted to PO_2 using the Stern-Volmer equation, Eq. (3):

$$\tau_o / \tau = 1 + k_Q\,\tau_o\,PO_2 \tag{3}$$

where τ_o (μs) is the lifetime in the absence of O_2, and k_Q ($mmHg^{-1}/\mu s$) is the bimolecular rate, or quenching, constant. Representative maps of phase delay (θ_D), lifetime (τ), PO_2, and coefficient of determination (R^2) are shown in Figure 2.

2.2. Problems with Two-Dimensional Retinal PO_2 Maps

The retina of mammals is supplied with oxygen and nutrients from two separate and distinct vasculatures: the retinal vasculature, distributed within the inner layers of the retina, and the choroidal vasculature, situated behind the retina. The light-absorbing photoreceptors, which have an extremely high oxygen demand, are located within an avascular retinal layer located between these two vasculatures. PO_2 values in the choroidal vasculature are very high (> 60 mm Hg),[4] a physiologic state hypothesized as necessary for maintaining an adequate diffusive flux of oxygen to the highly oxidative photoreceptors.[5] By contrast, PO_2 values in the retinal circulation are similar to the values observed in most other vascular beds (15-30 mm Hg). Thus, a measure of PO_2 in the retina is complicated by the presence of two distinct vasculatures with different oxygen dynamics. Because the current phosphorescence imaging system does not reject out-of-focus light, phosphorescence is detected from both of these vascular beds during image acquisition. This is clearly evident in the capillary regions of the oxygen map generated in Figure 2. Capillary PO_2 is shown near 100 mm Hg, consistent with choroidal PO_2, and is even higher than measured retinal arterial PO_2. Choroidal phosphorescence dominates in these capillary regions between the larger retinal vessels. Unfortunately, two (or more) distinct phase shifts, representing retinal and choroidal PO_2, cannot be reliably extracted from the limited number of phase-sensitive phosphorescence images currently collected, and alternate strategies for separating the retinal and choroidal contribution are required.

3. STRATEGIES FOR THREE-DIMENSIONAL IMAGING OF PO_2

To discriminate between the retinal and choroidal vasculatures and permit a separate study of the retinal and choroidal contributions to the oxygenation state of the retina, both hardware and software strategies currently are being investigated.

3.1. Hardware Imaging Strategies

Optical sectioning strategies for separating the retinal and choroidal phosphorescence on the imaging CCD detector include a standard laser-based confocal imaging strategy and a new *pseudo*-confocal strategy based on a grating imaging system.[6, 7] The optical sectioning characteristics of a standard fluorescence microscope can be improved dramatically by the introduction and manipulation of a Ronchi grating mask in the illumination optics of the microscope. Unlike a confocal laser scanning microscope, this "grating image system" retains the parallel data capture characteristics of a CCD camera, making it considerably faster than the more costly confocal systems. More importantly, a grating image system is hardware compatible with the frequency-domain PO_2 imaging system currently in use. A customized grating imaging system (OptiGrid, Ronchi grating = 5 LP/mm, Triptar Lens Co., Rochester, NY) was recently incorporated into the frequency-domain PO_2 imaging system, and *in vivo* tests were performed in the mouse retina. Unfortunately, the phosphorescence "stripes" necessary for optical sectioning were not always visible with the Ronchi grating in place, particularly in the capillary regions of the mouse retina. This is most likely due to the poor optical transmission characteristics of the rodent eye. Therefore, it was determined that a robust grating imaging system is likely not feasible for *in vivo* rodent eye studies.

Another hardware strategy for separating the retinal and choroidal vasculature involves the use of both blue and green excitation light. Because blue light is more highly

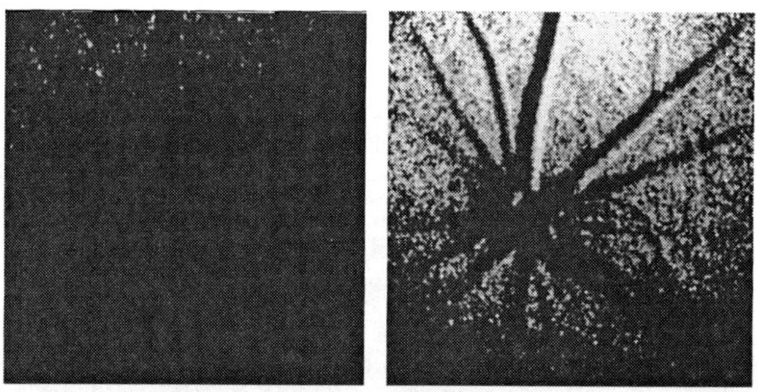

Figure 3. PO_2 maps (0-150 mm Hg full scale) of the rat retina taken first with blue (left, λ = 412 nm) and then green (right, λ = 532 nm) excitation light. These PO_2 maps have been converted to a grey-scale format from their original color representations. To preserve the visual contrast, both PO_2 maps are shown using the same grey-scale.

absorbed by retinal tissue and does not penetrate as efficiently to the choroid, excitation of the probe with blue light should reduce the choroidal influence. In preliminary tests, the capillary PO_2 in maps produced with blue (λ = 412 nm) excitation light, as seen in Figure 3 (left), are lower than that seen with green (λ = 532 nm) excitation light (Figure 3, right) when taken one right after the other. These differences can likely be exploited to separate the retinal and choroidal PO_2 contributions.

3.2. Software Image Processing Strategies

Because the calculated phase shift (θ) necessary for generating a measure of PO_2 is dependent on the sinusoidal excitation frequency, ω (Eq. 2), image processing strategies that involve the use of two or more excitation frequencies are also being investigated for separating retinal and choroidal PO_2 contributions. For example, residual *in phase* fluorescence or light scattering signal can be removed mathematically by collecting phosphorescence image stacks at two distinct sinusoidal excitation frequencies. It is hypothesized that a multi-frequency analysis can also be used to separate retinal and choroidal PO_2 contributions.

4. CONCLUSIONS

Ocular diseases that affect the retina, including diabetic retinopathy and numerous proliferative retinopathies, are devastating diseases leading to major vision loss in humans. While the etiologies of these diseases are different, abnormalities in the delivery and consumption of oxygen are thought to be significant in all of them. Despite this common factor, quantitative studies that adequately correlate disease progression with changes in retinal and choroidal PO_2 are lacking. An ability to generate both retinal and choroidal PO_2 maps would substantially advance our understanding of retinal dynamics. Ultimately, an ability to independently generate both retinal and choroidal PO_2 maps would permit studies investigating the separate contributions of the retinal and choroidal vasculatures in retinal disease.

5. ACKNOWLEDGMENTS

The authors acknowledge the technical assistance of Kevin Hawkins. This research was supported in part by a research grant from the Whitaker Foundation, Rosslyn, VA.

REFERENCES

1. R. D. Shonat, D. F. Wilson, C. E. Riva, and S. D. Cranstoun, Effect of acute increases in intraocular pressure on intravascular optic nerve head oxygen tension in cats, *Invest. Ophthalmol. Vis. Sci.* **33**, 3174-3180 (1992).
2. R. D. Shonat, D. F. Wilson, C. E. Riva, and M. Mawlowski, Oxygen distribution in the retinal and choroidal vessels of the cat as measured by a new phosphorescence imaging method, *Appl. Optics* **31**, 3711-3718 (1992).
3. R. D. Shonat, and A. C. Kight, Oxygen tension imaging in the mouse retina, *Ann. Biom. Engin.* **31**, 1084-1096 (2003).

4. N. D. Wangsa-Wirawan, and R. A. Linsenmeier, Retinal oxygen: Fundamentals and clinical aspects, *Arch. Ophthalmol.* **121**, 547-557 (2003).
5. R. A. Linsenmeier, and L. Padnick-Silver, Metabolic dependence of photoreceptors on the choroid in the normal and detached retina, *Invest. Ophthalmol. Vis. Sci.* **41**, 3117-3123 (2000).
6. F. Lanni, and T. Wilson, Grating image systems for optical sectioning fluorescence microscopy of cells, tissues, and small organisms, in: *Imaging Neurons: A Laboratory Manual*, edited by R. Yuste, F. Lanni, and A. Konnerth (Cold Spring Harbor Laboratory Press, Cold Spring Harbor, NY, 2000), pp. 8.1-8.9.
7. M. A. A. Neil, R. Juskaitis, and T. Wilson, Method of obtaining optical sectioning by using structured light in a conventional microscope, *Optics Lett.* **22**, 1905-1907 (1997).

CEREBRAL PtO_2, ACUTE HYPOXIA, AND VOLATILE ANESTHETICS IN THE RAT BRAIN

Huagang Hou, Oleg Y. Grinberg, Stalina A. Grinberg, Nadeem Khan, Jeff F. Dunn, and Harold M. Swartz

Abstract: We describe our results on the effect in rats of two commonly used, volatile anesthetics on cerebral tissue PO_2 (PtO_2) and other physiological parameters at FiO_2 levels ranging from 0.35 to 0.1. The study was performed in 12 rats that had lithium phthalocyanine (LiPc) crystals implanted in the left cerebral cortex. FiO_2 was maintained at 0.35 during surgical manipulation and baseline EPR measurements, after which time, each animal was exposed to varying levels of FiO_2 (0.26, 0.21, 0.15, and 0.10) for 30 minutes at each level. No significant difference in PtO_2 was observed between the isoflurane and halothane groups at any FiO_2 level, and the cerebral arterial PO_2 (PaO_2) also was similar for both groups. However, the cerebral PtO_2 under both isoflurane and halothane anesthesia was lower during hypoxia ($FiO_2 \leq 0.15$) than under normoxia ($FiO_2 = 0.21$) and there was a significant difference in mean arterial blood pressure (MABP) between isoflurane and halothane groups under both mild and severe hypoxia. The pH and cerebral arterial PCO_2 ($PaCO_2$) were similar for the halothane and isoflurane groups during normoxia ($FiO_2 = 0.21$) and mild hypoxia ($FiO_2 = 0{,}15$), but following severe hypoxia ($FiO_2 = 0.10$), both parameters were lower in the halothane anesthetized animals. These results confirm that cerebral PO_2 cannot be inferred directly from measurements of other parameters, indicating that methodology incorporating continuous direct measurement of brain oxygen will lead to a better understanding of cerebral oxygenation under anesthesia and hypoxia.

1. INTRODUCTION

While overall the deleterious effects of hypoxia on various aspects of brain function are well recognized, presumably due to decreased cerebral PtO_2, our understanding of the role of oxygen in the pathophysiology of brain remains empirical and incomplete. This is due, in part, to a lack of quantitative data on the relationship between cerebral PtO_2 and the deleterious effects that are observed. It has been shown that the brain can maintain an energy balance down to a PtO_2 of 8.8 mmHg when under ketamine/xylazine anesthesia.[1] However, under clinical conditions, isoflurane and halothane are more widely used. Since anesthesia has varying effects on cerebral blood flow (CBF) and cerebral metabolic rate of oxygen ($CMRO_2$),[2-4] it is likely that volatile anesthetics will have different effects on brain PtO_2 during hypoxia. The aim of this study was to investigate brain PtO_2 by EPR oximetry during hypoxia, using two different volatile anesthetics, isoflurane and halothane. The effects of various anesthetics under a range of FiO_2 (0.21-1.0) on brain oxygenation have been measured previously using EPR oximetry.[2, 5]

2. MATERIALS AND METHODS

2.1. Paramagnetic LiPc Crystals

Oxygen sensitive lithium phthalocyanine (LiPc) crystals were synthesized in our laboratory. The line width of LiPc is a linear function of PO_2 and is independent of local metabolic processes, the presence of other paramagnetic species, and pH.[6] The LiPc crystals equilibrate with local tissue PtO_2 in less than 30 sec, and the response of the line width to changes in PtO_2 is stable for at least 30 days in the brain. The spectra reflect the average partial pressure of oxygen on the surface of the crystals.[7] The high density of unpaired spins, combined with a narrow intrinsic line width of LiPc, allows measurements of PtO_2 in the brain using 1-4 crystals with a total diameter of ~200 μm.

2.2. Animal Preparation and Anesthesia

Twelve male Sprague-Dawley rats, 250-350 g, (Charles River Laboratories, Wilmington, MA) were used. One week prior to the PtO_2 measurements, rats were anesthetized with ketamine/xylazine (80/8 mg/kg, i.m.), and LiPc crystals were placed via a spinal needle directly into the brain at a depth of 2.0 mm from the surface of the skull, through 1.0 mm drilled holes located 3.0 mm from the midline and 1.0 mm in front of bregma.

Rats with LiPc crystals in their brain were randomly assigned to either isoflurane or halothane experimental groups and accordingly were anesthetized with 3.0% isoflurane or 2.5% halothane in 26% oxygen/74% nitrogen. After an adequate level of anesthesia was achieved, endotracheal intubation was performed using laryngoscopy and an "over-the-needle" 14 gauge catheter. Femoral arterial and venous catheters (PE 50 tubing) filled with heparinized saline were inserted and all wound sites were infiltrated with 2.0% lidocaine. The arterial catheter was connected to a pressure transducer for continuous blood pressure monitoring and periodic blood gas measurements, while the venous catheter was used for administration of drugs and fluids. Total surgical preparation time was 30-40 min (measured from the start of induction). Following surgery, the concentration of the inspired anesthetic agent (measured with a Datex Capnomac II Anesthetic Agent Analyzer) was reduced to either 0.7 MAC isoflurane (0.9%-1.0%) or 0.7 MAC halothane (0.7%-0.8%) in 26% oxygen/74% nitrogen. Next, we paralyzed the rats with pancuronium (0.2 mg/kg/h), then ventilated with a small animal ventilator (Model 683, Harvard Apparatus Inc, South Natick, MA). These doses/concentrations were chosen to produce comparable acute levels of anesthesia based on the literature. Rectal temperature was controlled at 37.0°C ± 0.5°C via a heated pad.

2.3. EPR Measurements

Spectra of LiPc were obtained using an EPR spectrometer constructed in our laboratory with a low frequency (1.2 GHz, "L - band") microwave bridge.[8] Each rat was placed in the magnet and the head positioned so that the brain was directly under the extended loop resonator, which was adjusted to obtain the maximum signal from the LiPc in the cerebral cortex. Typical settings for the spectrometer were: incident microwave power, 10 mW; magnetic field center, 425 gauss; scan range, 1 gauss; modulation frequency, 27 kHz. Modulation amplitude was set at less than one-third of the EPR line width. Scan time was 2 min, and 3-5 scans were usually averaged to achieve a better signal to noise ratio. The line widths were converted to PO_2 using a calibration curve determined for each batch of crystals.

2.4. Experimental Protocol

All six animals in each group underwent the same protocol. The only difference between the two groups was the anesthetic. FiO_2 was maintained at 0.35 during vascular and airway access, and then animals were allowed to stabilize for 10-15 min before beginning data collection. Then the rats were exposed for 30 minutes each to FiO_2 of 0.26, 0.21, 0.15, and 0.10. Fluid balance was maintained with 1.0 ml/hr of saline (i.v.). Blood pressure (mean arterial blood pressure, MABP) was continuously monitored by a pressure transducer (Biopac Systems, MA). Blood pressure was not controlled pharmacologically in order to avoid drugs with direct effects on cerebral vasculature. Arterial blood gas analysis was performed using a Ciba-Corning 288 blood gas analyzer (Ciba-Corning Diagnostic Corp, MA) on 0.1 ml of blood collected from the femoral artery. Samples were collected 10 min before commencing a change of FiO_2 and at the termination of the experiment. At the end of the experiment, the rats were euthanized. Gross and microscopic examination (H & E staining) of the tissue around the implanted LiPc confirmed that crystals were in the cerebral cortex and that there was no significant inflammatory infiltrate or necrosis around the LiPc. The experimental techniques and protocol were approved by Dartmouth College Animal Care and Use Program.

2.5. Statistical Analysis

Data analysis was done by means of Student t test. $P < .05$ was considered statistically significant. All data are expressed as means ± SE. A paired t test was used to compare the variables between control and hypoxia within the same treatment group. Comparisons between two anesthetics in the normoxic and the hypoxic state at the same time point were made using the Student t test for unpaired samples.

Table 1. Mean blood pressure (mm Hg), heart rate (beats/minute), rectal temperature (°C), blood gas (mm Hg), pH, and cerebral PtO_2 (mm Hg) during exposure to each anesthetic agent at different FiO_2 levels.

	FiO_2					
Variables	0.35	0.26	0.21	0.15	0.10	0.35
			Isoflurane			
MABP	105.6 ± 5	96.3 ± 5	98.0 ± 4	$82.2 \pm 3^{+}$	$67.4 \pm 11^{+}$	106.5 ± 3
HR	360.9 ± 18	348.6 ± 14	375.7 ± 29	378.9 ± 23	351.3 ± 22	391.5 ± 21
Temp	36.8 ± 0.2	37.1 ± 0.2	37.3 ± 0.2	37.2 ± 0.2	36.7 ± 0.3	37.2 ± 0.3
PaO_2	$142.6 \pm 7^{++}$	$111.3 \pm 4^{++}$	84.0 ± 2	$45.9 \pm 3^{++}$	$37.7 \pm 4^{++}$	$154.5 \pm 6^{++}$
$PaCO_2$	39.2 ± 1	38.7 ± 1	37.9 ± 1	40.3 ± 2	37.2 ± 1	37.8 ± 1
pH	$7.42 + 0.01$	7.42 ± 0.01	7.43 ± 0.01	7.41 ± 0.03	7.37 ± 0.02	7.41 ± 0.01
PtO_2	$28.9 \pm 2.8^{++}$	$22.3 \pm 3^{++}$	14.4 ± 2.5	$6.8 \pm 1^{++}$	$2.8 \pm 0.3^{++}$	$27.3 \pm 4^{++}$
			Halothane			
MABP	$93.8 \pm 3^{*}$	$84.8 \pm 2^{*}$	$86.6 \pm 7^{*}$	$70.1 \pm 5^{*++}$	$38.7 \pm 2^{**++}$	$95.6 \pm 5^{*+}$
HR	331.2 ± 12	344.0 ± 9	348.7 ± 12	374.4 ± 10	342.9 ± 20	393.2 ± 17
Temp	36.6 ± 0.2	37.0 ± 0.2	37.3 ± 0.3	37.4 ± 0.4	37.2 ± 0.4	37.7 ± 0.3
PaO_2	$132.9 \pm 4^{++}$	$106.1 \pm 5^{++}$	76.1 ± 3	$46.4 \pm 4^{++}$	$36.7 \pm 2^{++}$	$134.6 \pm 12^{+}$
$PaCO_2$	40.4 ± 1	40.8 ± 2	40.2 ± 2	41.1 ± 2	$34.4 \pm 0.7^{*+}$	38.4 ± 1
pH	7.42 ± 0.01	7.41 ± 0.01	7.43 ± 0.01	7.41 ± 0.02	$7.26 \pm 0.04^{*++}$	7.42 ± 0.02
PtO_2	$28.0 \pm 2^{++}$	$21.6 \pm 2^{++}$	13.7 ± 3	$8.6 \pm 2^{++}$	$4.7 \pm 0.8^{++}$	$27.6 \pm 4^{++}$

Values are given as means ± SE. * = $P < .05$; ** = $P < .01$, significant differences as compared with isoflurane (unpaired *t* test). $^{+}$ = $P < .05$; $^{++}$ = $P < .01$, significant differences as compared with 0.21 FiO_2 in same group (paired *t* test). N = 6 in each group.

3. RESULTS

Table 1 summarizes the data. The heart rate and rectal temperature did not change with decreasing FiO_2. The MABP did not vary significantly from baseline ($FiO_2 = 0.21$) in either group at the higher FiO_2 levels of 0.35 or 0.26; however, the MABP did vary significantly during hypoxia, at FiO_2 levels of 0.15 and 0.10. There was significant difference in MABP between the two anesthetics at all FiO_2 levels, which became more pronounced under hypoxia. These differences occurred even though they are well in the normal cerebral regulatory ranges at the FiO_2 level of 0.10 in the isoflurane group.

The $PaCO_2$ and pH did not change significantly from baseline in the isoflurane group, but did in the halothane group when the FiO_2 level was 0.10. At a FiO_2 level of 0.10, the $PaCO_2$ and pH in the halothane group were significantly lower than in the isoflurane group. With both anesthetics, the PaO_2 decreased significantly with decreasing FiO_2; at the same FiO_2, there were no significant differences in PaO_2 between anesthetic groups. Similarly, the PtO_2 decreased significantly with decreasing FiO_2 for both anesthetics. Again, there was no statistically significant difference in PtO_2 between the groups at any FiO_2.

4. DISCUSSION

The absolute values obtained by EPR in this study suggested that cerebral PtO_2 in rats anesthetized by isoflurane or halothane at FiO_2 levels of 0.35 and 0.26 are in agreement with data from prior studies of brain PtO_2. Brain PtO_2 at FiO_2 levels of 0.21 are lower than in some previous reports[9, 10] (14.4 ± 2.5 mm Hg and 13.7 ± 2.6 mm Hg, respectively) despite maintenance of normal $PaCO_2$, arterial pH, and rectal temperature. It may be that the level of FiO_2, PaO_2, and concentration of anesthetics were different between the studies. The low values of cerebral PtO_2 at normoxic FiO_2 levels of 0.21 in the two groups indicate that the usual practice of using room air as the breathing gas for rodents can result in cerebral PtO_2 values that may affect the physiology. This occurred even though these animals were carefully ventilated so that the $PaCO_2$ and pH remained within the normal range and the body temperature was kept at the physiological level. It therefore appears highly desirable to use a higher FiO_2 in all studies with rodents involving anesthesia.

Brain tissue oxygen tension falls almost immediately when the partial pressure of inspired oxygen is reduced even moderately. In moderate and severe hypoxia, brain blood volume and blood flow are expected to increase,[11, 12] but these compensatory mechanisms were not sufficient to avoid decreasing the cerebral PtO_2. Our absolute cerebral PtO_2 values during isoflurane or halothane anesthesia at FiO_2 levels of 0.10 appear low. Whether these absolute values led to damage, however, cannot be assessed from the present study since the animals were terminated at the conclusion of the measurements and neither neurologic nor histological assessments were made. A previous study using ketamine/xylazine anesthesia showed that energy impairment began when the PtO_2 fell below 8.8 mm Hg.[1] In the present study, after FiO_2 was switched from 0.10 to 0.35, cerebral PtO_2 and other physiologic parameters returned to normal. Further outcome studies will be required to assess the functional effect of these PtO_2 values on neuronal integrity.

Cerebral PtO_2 is expected to be determined by the local balance of metabolic rate and delivery of O_2. Lebrun-Grandie et al.[13] measured O_2 uptake and blood flow in regions of the normal human brain and found, in general, that in a given region, blood flow was proportional to O_2 consumption. There was, however, complex and considerable variation in O_2 consumption and blood flow when animals and humans were anesthetized by different anesthetics and in the hypoxia state. Because CBF and $CMRO_2$ were not measured, underlying mechanisms that contributed to PtO_2 are not delineated directly by our data, but some inferences may be made. MABP in rats maintained with isoflurane was significantly higher than that of rats treated with halothane, but all values from isoflurane group were well within the limits of cerebral autoregulation (80-160 mmHg in rats[14]) even at a FiO_2 level of 0.10, while at the same level, values from the halothane group were lower than the limits of cerebral autoregulation. Thus, the differences in MABP evident during this study would not be expected to alter CBF in the isoflurane group. CBF may have changed in the halothane group, but it was not directly measured in this study. Although we did observe higher average brain PtO_2 in the halothane group, this difference failed to reach statistical significance. Perhaps this is a result of the small sample of animals in which the data were available (n = 6 each group), which decreased our statistical power.

Recognition of the differences in cerebral PtO_2 during isoflurane and halothane anesthesia in the hypoxic state may be of value in understanding the effects of anesthetics in experimental pathological states, as well as in evaluating the effects of anesthetic on intracranial pressure (ICP). For example, increases in CBF may be associated with parallel increases in cerebral blood volume and ICP.[15] Todd and Drummond[16] reported that increases in ICP were identical for both isoflurane and halothane, despite greater increases in CBF during halothane anesthesia. Hypoxia also increases CBF and blood volume.[11, 12] It is unknown if the effects of hypoxia and inhalation anesthetics on cerebral PtO_2 and ICP have a cooperative action; therefore, understanding the differences in cerebral PtO_2 and ICP and the relationship between them during isoflurane and halothane anesthesia may have applications in experimental pathological states, such as stroke models, in which the regional effects of anesthetic agents may alter the location of a focal infarction.[17]

In summary, the general pattern of observed changes in cerebral PtO_2 suggests that the response to hypoxia under halothane does not differ greatly from isoflurane. However, the magnitude of induced MABP, $PaCO_2$, and pH changes varied significantly with the different agents. This demonstrates the value of making direct measurements of PtO_2. The ability to obtain such data with EPR oximetry indicates the usefulness of this method.

5. ACKNOWLEDGMENTS

This work was supported by a NIH (NIBIB) grant P01 EB002180, RO1 EB002085, and used the facilities of the EPR Center for the Study of Viable Systems supported by NIH (NIBIB) Grant P41 EB002032.

6. REFERENCES

1. E. L. Rolett, A. Azzawi, K. J. Liu, M. N. Yongbi, H. M. Swartz, and J. F. Dunn, Critical oxygen tension in rat brain: a combined ^{31}P-NMR and EPR oximetry study, *Am. J. Physiol. Regulatory Integrative Comp. Physiol.* **279**(1), R9-R16 (2000).
2. H. Lei, O. Y. Grinberg, C. I. Nwaigwe, H. G. Hou, H. Williams, H. M. Swartz, and J. F. Dunn, The effects of ketamine/xylazine anesthesia on cerebral blood flow and oxygenation observed using nuclear magnetic resonance perfusion imaging and electron paramagnetic resonance oximetry, *Brain Res.* **913**(2), 174-179 (2001).
3. W. E. Hoffman, D.J. Miletich, and R.F. Albrecht, The effects of midazolam on cerebral blood flow and oxygen consumption and its interaction with nitrous oxide, *Anesth. Analg.* **65**(7), 729-733 (1986).
4. R. A. Berkowitz, W. E. Hoffman, F. Cunningham, and T. McDonald, Changes in cerebral blood flow velocity in children during sevoflurane and halothane anesthesia, *J. Neurosurg. Anesthesiol.* **8**(3), 194-198 (1996).
5. H. G. Hou, O. Y. Grinberg, S. Taie, S. Leichtweis, M. Miyake, S. Grinberg, H. Xie, M. Csete, and H. M. Swartz, Electron paramagnetic resonance assessment of brain tissue oxygen tension in anesthetized rats, *Anesth. Analg.* **96**(5), 1467-1472 (2003).
6. K. J. Liu, P. Gast, M. Moussavi, S. W. Norby, N. Vahidi, T. Walczak, M. Wu, and H. M. Swartz, Lithium phthalocyanine: A probe for electron paramagnetic resonance oximetry in viable biological systems. *Proc. Natl. Acad. Sci. USA* **90**(12), 5438-5442 (1993).
7. H. M. Swartz, and R. B. Clarkson, The measurement of oxygen in vivo using EPR techniques, *Phys. Med. Biol.* **43**(7), 1957-1975 (1998).
8. M. J. Nilges, T. Walczak, and H. M. Swartz, 1 GHz in vivo ESR spectrometer operating with a surface probe, *Phys. Med.* **5**, 195-201 (1989).
9. A. I. Mass, W. Fleckenstein, D. A. De Jong, and H. van Santbrink, Monitoring cerebral oxygenation: experimental studies and preliminary clinical results of continuous monitoring of cerebrospinal fluid and brain tissue oxygen tension, *Acta Neurochir. Suppl. (Wien)* **59**, 50-57 (1993).

10. W. A. van den Brink, I. K. Haitsma, C. J. Avezaat, A. B. Houtsmuller, J. M. Kros, and A. I. Maas, Brain parenchyma/pO2 catheter interface: a histopathological study in the rat, *J. Neurotrauma* **15**(10), 813-824 (1998).
11. B. A. McKinley, W. P. Morris, C. L. Parmley, and B. D. Butler, Brain parenchyma PO2, PCO2, and pH during and after hypoxic, ischemic brain insult in dogs, *Crit. Care Med.* **24**(11), 1858-1868 (1996).
12. N. Dahlgren, Local cerebral blood flow in spontaneously breathing rats subjected to graded isobaric hypoxia, *Acta Anaesthesiol. Scand.* **34**(6), 463-467 (1990).
13. P. Lebrun-Grandie, J. C. Baron, F. Soussaline, C. Loch'h, J. Sastre, and M. G. Bousser, Coupling between regional blood flow and oxygen utilization in the normal human brain. A study with positron tomography and oxygen 15, *Arch. Neurol.* **40**(4), 230-236 (1983).
14. M. J. Hernandez, R. W. Brennan, and G. S. Bowman, Cerebral blood flow autoregulation in the rat, *Stroke* **9**(2), 150-155 (1978).
15. J. B. Madsen, G. E. Cold, E. S. Hansen, and B. Bardrum, The effect of isoflurane on cerebral blood flow and metabolism in humans during craniotomy for small supratentorial cerebral tumors, *Anesthesiol.* **66**(3), 332-336 (1987).
16. M. M. Todd, and J. C. Drummond, A comparison of the cerebrovascular and metabolic effects of halothane and isoflurane in the cat, *Anesthesiol.* **60**(4), 276-282 (1984).
17. D. G. Nehls, M. M. Todd, R. F. Spetzler, J. C. Drummond, R. A. Thompson, and P. C. Johnson, A comparison of the cerebral protective effects of isoflurane and barbiturates during temporary focal ischemia in primates, *Anesthesiol.* **66**(4), 453-464 (1987).

RATE OF CHANGE IN CEREBRAL OXYGENATION AND BLOOD PRESSURE IN RESPONSE TO PASSIVE CHANGES IN POSTURE:

A comparison between pure autonomic failure patients and controls

Ilias Tachtsidis, Clare E. Elwell, Terence S. Leung, Katharine Bleasdale-Barr, Katharine Hunt, Nathan Toms, Martin Smith, Christopher J. Mathias, and David T. Delpy

Abstract: The cardiovascular and cerebrovascular responses to head-up postural change are compromised in pure autonomic failure (PAF) patients because of sympathetic denervation. The aim of this study was to characterize the rate of change of systemic mean blood pressure (MBP) and cerebral haemodynamics in response to passive posture changes. Nine PAF patients and 9 age-matched controls took part in this study. MBP and oxy- (O_2Hb), deoxy-haemoglobin (HHb), and tissue oxygenation index (TOI) on the forehead were continuously monitored non-invasively using the Portapres® and near-infrared spectroscopy (NIRS), respectively. From visual inspection of the haemoglobin difference signal ($Hb_{diff} = O_2Hb\text{-}HHb$), seven distinct phases were marked (1: supine, 2: start passive tilt, 3: head up to 60° degrees, 4: end of tilt, 5: tilt reversal, 6: return to supine, 7: rest); the same time points were used for all of the other signals. For each phase, the slope was calculated using a linear regression algorithm. Significant differences were found between PAF patients and controls in the Hb_{diff} slope magnitudes for phases 3 ($P < .05$) and 5 ($P = .01$), and the duration of phase 2 ($P < .05$). MBP slope magnitudes showed significant differences for phases 2 ($P < .01$) and 5 ($P < .01$). These differences in the rate of change suggest differences in blood vessel resistance related to sympathetic activation.

1. INTRODUCTION

On assumption of the upright posture, venous pressure increases in the feet and approximately 500 ml of blood pools in the lower extremities and splanchnic circulation.[1] This reduces venous return to the heart, which lowers the cardiac stroke volume, and further reduces the central arterial pressure. Healthy individuals adjust their cardiovascular system to this gravitational displacement of blood by increasing systemic vascular resistance through autonomic reflex activity.[2] The baroreceptor-mediated sympathetic mechanism immediately restores blood pressure.

In normal subjects, there is no fall in blood pressure during head-up tilt, unlike patients with pure autonomic failure (PAF), in whom blood pressure falls promptly and remains low, with a blood pressure overshoot on return to the horizontal position. Postural (orthostatic) hypotension (a fall of over 20 mm Hg in systolic blood pressure) is a cardinal manifestation of sympathetic vasoconstriction failure; therefore, the cardiovascular and cerebrovascular responses to head-up postural change are severely compromised.[3] Postural hypotension often results in cerebral hypoperfusion symptoms (such as syncope, dizziness, and visual disturbances); suboccipital/paracervical 'coat hanger' neck pain, lower back pain, and calf pain due to muscle hypoperfusion can also occur frequently.[4]

The aim of this study was to characterize the rate of change of systemic blood pressure, cerebral oxygenation, and haemodynamics in response to passive changes in posture in patients with PAF and healthy age-matched controls. We provide a template that fully describes the pattern of changes seen in the systemic and cerebral haemodynamics during a passive tilt protocol. In 1997, Colier et al.[5] identified a similar pattern of the rate of change in cerebral haemodynamics in a group of healthy and hypovolemic volunteers during passive tilt. We extend this work by more fully characterizing the patterns seen in a PAF patient group compared with healthy controls.

2. METHODS

2.1. Participants

Nine PAF patients, mean age 61 ± 7 years, and 9 age-matched healthy volunteers, mean age 62 ± 7 years, took part in this study. The local ethics committee of the National Hospital for Neurology and Neurosurgery (London, UK) approved the protocol for the study, and all subjects gave informed consent for participation. Pure autonomic failure is one clinical classification of primary chronic autonomic failure, and was formerly known as 'idiopathic orthostatic hypotension'.[6] The PAF patients studied did not have any associated neurological disorders or any signs of central nervous system lesions.

2.2. Measurements

A continuous wave near-infrared spectrometer (NIRS) with a sampling rate of 6 Hz (NIRO 300, Hamamatsu Photonics KK) was used to measure absolute cerebral tissue oxygenation index (TOI) over the frontal cortex using the spatially resolved reflectance spectroscopy technique,[7] together with changes in oxy-haemoglobin ($\Delta[O_2Hb]$) and deoxy-haemoglobin (Δ[HHb]) by utilizing the modified Beer-Lambert law.[8] The probe was placed on the forehead (taking care to avoid the midline sinuses) and was shielded from ambient light by using an elastic bandage and a black cloth; the studies took place in a darkened room. An optode spacing of 5 cm was used, and optical filters were also used where necessary to optimise signal:noise ratio. The differential pathlength factor (DPF) was calculated for each subject using the age dependence equation[9]:

$$DPF_{780} = 5.13 + 0.07 A_y^{0.81} \quad (1)$$

where DPF_{780} is the DPF at 780 nm, and A_y is the age of the subject in years.

A Portapres® system (TNO Institute of Applied Physics, Biomedical Instrumentation) was used to continuously and non-invasively measure blood pressure from the finger. The mean blood pressure (MBP) data was collected to a PC via a serial link at 100 Hz sampling rate, and was later resampled at 6 Hz.

2.3. Protocol

Each subject was strapped on an electrical moving tilt table and underwent three postural changes. After an initial 10 min resting period in the supine position, the subject was tilted upright to an angle of 60° for another 10 min, and then returned to the supine position for an additional 10 min. Head-up tilt was interrupted if severe symptoms of orthostatic intolerance occurred, whereupon the patient was immediately returned to the supine position. The tilting movement process was performed via an electrical motor and lasted 6 sec for both the up and down movements.

2.4. Signal Analysis

The NIRS and blood pressure data were low pass filtered (cut-off frequency 0.2 Hz) to remove respiratory noise; additionally, a Savitsky-Golay filter (a 10 sec moving 2nd order polynomial) was used to further smooth the data. From visual inspection of the haemoglobin difference ($\Delta[Hb_{diff}]$) signal (derived by subtracting $\Delta[O_2Hb]$ from Δ[HHb]), total haemoglobin (Δ[HbT]) signal (derived from adding $\Delta[O_2Hb]$ to Δ[HHb]), TOI signal, and MBP signal, distinct phases were recognized and marked. For each phase, the slope was calculated using a linear regression algorithm. Figure 1 shows an illustration of the pattern seen in the $\Delta[Hb_{diff}]$ signal.

Not all of the subjects clearly showed all the phases, so data from missing phases has been excluded from the group analysis (see Tables 1 and 2). Direct comparison of the magnitude of the slopes between the PAF patients and the controls was then performed for each signal using the Student *t* test.

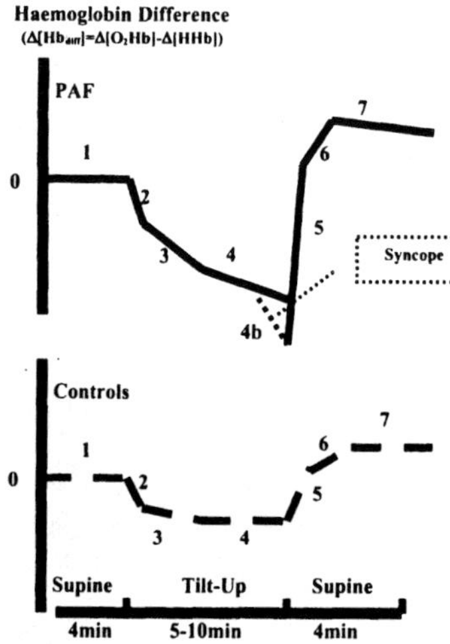

Figure 1. A schematic representation of the pattern seen in the signals. During phase 1, the subject lies supine resting, and when tilt begins, there is a passive circulatory response, recognised as phase 2. This leads to phase 3, during which any active vascular control mechanisms kick in; the head-up posture finishes with phase 4, except in a symptomatic patient who exhibits a further fall (phase 4b). Tilt reversal starts with a fast passive response, shown as phase 5, that leads to phase 6, which demonstrates the hyperaemia effect, and finally phase 7, a recovery to baseline which marks the end of the experiment. For clarity only, the $\Delta[Hb_{diff}]$ is displayed.

Table 1. Number of subjects in which each phase was clearly detected.

Number of subjects (Controls/PAF) for each phase.	Phase 1	Phase 2	Phase 3	Phase 4	Phase 5	Phase 6	Phase 7
Hb_{diff}	9/9	6/9	9/9	9/9	9/9	6/9	9/9
HbT	9/9	8/9	8/9	9/9	9/9	8/7	9/9
TOI	9/8	6/8	8/8	9/7	9/7	4/5	9/7
MBP	9/9	9/9	3/8	9/9	9/9	2/9	9/9

Table 2. The number of phases seen in every subject for each signal.

Number of phases seen in each subject.

Controls	Hb_{diff}	HbT	TOI	MBP	Patients	Hb_{diff}	HbT	TOI	MBP
Control 1	7	7	6	7	PAF 1	7	7	7	7
Control 2	7	6	6	5	PAF 2	8	8	7	7
Control 3	7	7	7	5	PAF 3	7	7	7	7
Control 4	7	7	7	5	PAF 4	7	7	7	7
Control 5	7	7	7	6	PAF 5	7	5	3‡	7
Control 6	5	7	5	6	PAF 6	7	7	N/A†	7
Control 7	7	7	7	6	PAF 7	8	8	7	6
Control 8	5	7	5	5	PAF 8	7	7	7	7
Control 9	5	5	5	5	PAF 9	7	6	7	7

‡ Patient hyperventilated when head-up; this affected the TOI measurements; therefore data after this was excluded.

† The TOI measurement was contaminated by artefact; therefore the TOI data from this patient was excluded.

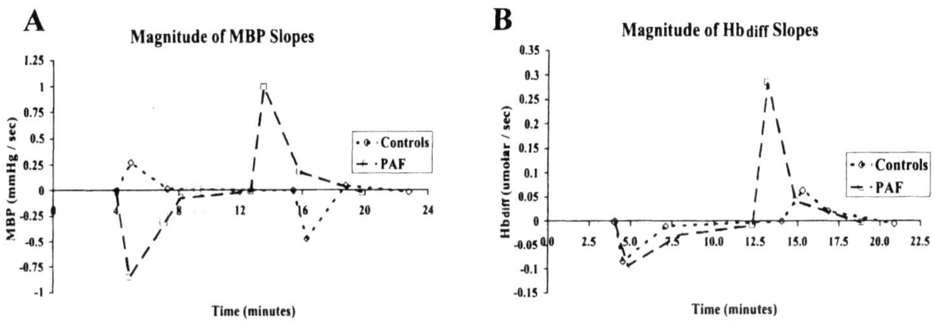

Figure 2. PAF patients versus healthy controls: A. Magnitude of MBP slopes with mean time; B. Magnitude of $\Delta[Hb_{diff}]$ slopes with mean time.

3. RESULTS

3.1. MBP

Using an unpaired Student *t* test, significant differences were found in the magnitude of slopes between PAF patients and healthy controls for phase 2 ($P < .01$) and phase 5 ($P < .01$) (see Fig. 2A).

3.2. NIRS Signals

Using an unpaired Student *t* test, significant differences were found in the $\Delta[Hb_{diff}]$ slope magnitudes between PAF patients and healthy controls for phase 3 ($P < .05$) and phase 5 ($P = .01$) (see Fig. 2B). Also, there was significant difference in the duration of phase 2 ($P < .05$), as PAF patients demonstrated a longer phase period (n = 9, period = 47.7 ± 16.6 sec) than healthy controls (n = 6, period = 29.9 ± 14.6 sec). No significant differences were found in the magnitude of slopes and the periods for the $\Delta[HbT]$ and the TOI.

4. DISCUSSION

This study provides a comprehensive template for the cerebral oxygenation, cerebral haemodynamics, and blood pressure changes with time during head-up tilt in both PAF patients and healthy age-matched controls. We demonstrated that there were significant differences in the rate of change of $\Delta[Hb_{diff}]$ and blood pressure between PAF patients and healthy controls. The rates of change in $\Delta[HbT]$ and TOI did not demonstrate significant differences, presumably due to large individual subject variability and small group numbers.

In the healthy controls during head-up tilt, the cardiovascular responses were consistent with sympathoneural activation in response to gravitational stress, leading to a rise in MBP; this was seen as a rapid positive phase 2 (n = 9, slope = 0.28 ± 0.14 mm Hg/sec). These results are in agreement with other studies in normal humans, reporting activation of compensatory mechanisms that maintain blood pressure when upright. In the PAF patients, head-up tilt caused a fall in MBP, seen as a large negative phase 2 (n = 9, slope = -0.85 ± 0.4 mm Hg/sec), leading to substantial postural hypotension. In the healthy controls following tilt reversal, MBP returned promptly, although not entirely, to pre-tilt levels, exhibiting a very high negative phase 5 (n = 9, slope = -0.46 ± 0.46 mm Hg/sec). In the PAF group, blood pressure promptly rose, settling slightly higher than the pre-tilt levels and exhibiting a larger absolute slope magnitude of phase 5 when compared with controls (n = 9, slope = 1 ± 0.46 mm Hg/sec).

The change in arterial blood pressure in healthy volunteers is a consequence of continuous changes in the activity of the nerves to the systemic resistance blood vessels; this involves a set of complex adjustments to maintain arterial blood pressure so as to ensure adequate perfusion of vital organs, including the brain. In this study, we used an NIRS system to interrogate the brain and provide information on cerebral haemodynamics and oxygenation. $\Delta[Hb_{diff}]$ can be used as an indicator of oxygenation by looking for any mismatch between oxygen demand and oxygen supply. Following the orthostatic hypotension, PAF patients exhibited symptoms of cerebral hypoperfusion, which often leads to syncope. $\Delta[Hb_{diff}]$ signal fell quite rapidly immediately following the head-up tilt, and continued to fall after the transition, as demonstrated in the magnitude of phase 3 (n = 9, slope = -0.03 ± 0.01 μmol/sec); this was not seen in the healthy controls, where the rate of change reached a steady state (n = 9, slope = -0.009 ± 0.02 μmol/sec) during phase 3, which indicates a completed circulation compensation. On tilt reversal, recognised as phase 5, there was a return of peripherally pooled blood and fluid into the central vascular compartment and head. The recovery phase 5 also demonstrated the inability of the PAF patients to control their vasculature from the passive changes, exhibiting a large positive slope that led to hyperaemia (n = 9, slope = 0.29 ± 0.2 μmol/sec); once again, this was not observed in the healthy controls (n = 9, slope = 0.07 ± 0.05 μmol/sec).

The mechanism that controls peripheral resistance, and hence the rate of change in blood flow and oxygenation, is the sympathetic system, which is absent in PAF patients who lack the ability to regulate their systemic and cerebral vasculature. Further evidence to support this was found when investigating the periods of each $\Delta[Hb_{diff}]$ phase. There were significant differences during phase 2, which represents the passive circulatory response immediately after the start of the tilt; PAF patients demonstrated a longer phase period (n = 9, period = 47.7 ± 16.6 sec) than healthy controls (n = 6, period = 29.9 ± 14.6 sec). During the initial stage of tilt, the effective peripheral vascular resistance of the lower body is low and blood flow to the lower limbs increases. The sympathetic system, which clearly responds in the healthy controls, regulates blood pressure and limits the pooling of blood to the lower extremities.

The most striking difference in the NIRS measurements between the PAF patients and healthy controls was found in $\Delta[Hb_{diff}]$; $\Delta[HbT]$ and TOI did not show any significant differences. $\Delta[HbT]$ can be used as an indicator of blood volume and TOI, which is the ratio of $\Delta[O_2Hb]$ to $\Delta[HbT]$, indicates the balance between oxygen delivery and tissue oxygen consumption. $\Delta[HbT]$ and TOI rate of change did not reach significance for either phase 3 ($P = .054$ and $P = .056$, respectively) or phase 5 ($P = .1$ and $P = .099$, respectively). This may be due to small group numbers and the large inter-subject variability of these measurements. In a recent study, Al-Rawi et al.[10] measured haemoglobin concentrations and TOI during carotid surgery, and showed that the sensitivity of TOI to intracranial changes was 87.5%, with a specificity of 100%. This raises confidence in assuming that TOI reflects changes in cerebral oxygenation; however, TOI is also affected by the arterial:venous volume ratio, which may change significantly during these tilt studies.

In this study, we have provided a template for investigating the rate of change of the systemic and cerebral effects of posture change in PAF patients and healthy age-matched controls. Given the wide range of conditions and symptoms related to autonomic failure, it is suggested that this template will be of use in describing the neurophysiology of these groups with studies on sufficient number of patients.

5. ACKNOWLEDGMENTS

The authors would like to thank all the volunteers who participated in this study, and Hamamatsu Photonics KK for providing the NIRO 300 spectrometer. This work was supported by the EPSRC/MRC, grant No GR/N14248/01 (IT), and the Wellcome Trust grant No GR/062558 (TSL).

6. REFERENCES

1. M. Lye, and T. Walley, Haemodynamic responses in young and elderly healthy subjects during ambient and warm head-up tilt, *Clin. Sci.* **94**, 493-498 (1998).
2. C. J. Mathias, Orthostatic hypotension: causes mechanisms and influencing factors, *Neurology* **45**(S5), S6-S11 (1995).
3. C. J. Mathias, and R. Bannister, in: *Autonomic Failure. A textbook of clinical disorders of the autonomic nervous system*, edited by C. J. Mathias, and R. Bannister (Oxford Medical Publications, 2002), pp. 169-195.
4. C. J. Mathias, R. Mallipeddi, and K. Bleasdale-Barr, Symptoms associated with orthostatic hypotension in pure autonomic failure and multiple system atrophy, *J. Neurol.* **246**, 893-898 (1999).
5. W. N. J. M. Colier, R. A. Binkhorst, M. T. E. Hopman, and B. Oeseburg, Cerebral and circulatory haemodynamics before vasovagal syncope induced by orthostatic stress, *Clin. Physiol.* **17**, 83-94 (1997).
6. R. Bannister, and C. J. Mathias, in: *Autonomic Failure. A textbook of clinical disorders of the autonomic nervous system*, edited by C. J. Mathias, and R. Bannister (Oxford Medical Publications, 2002), pp. 307-316.
7. S. Suzuki, S. Takasaki, T. Ozaki, and Y. Kobayashi, A tissue oxygenation monitor using NIR spatially resolved spectroscopy, *Proc. SPIE* **3597**, 582-592 (1999).
8. D. T. Delpy, and M. Cope, Quantification in tissue near-infrared spectroscopy, in: *Philosophical Transactions of the Royal Society of London Series B-Biological Sciences*, **352**(1354), 649-659 (1997).
9. A. Duncan, J. H. Meek, M. Clemence, C. E. Elwell, P. Fallon, L. Tyszczuk, M. Cope, and D. T. Delpy, Measurement of cranial optical path length as a function of age using phase resolved near infrared spectroscopy, *Pediatr. Res.* **39**(5), 889-894 (1996).
10. P. G. Al-Rawi, P. Smielewski, and P. J. Kirkpatrick, Evaluation of a near-infrared spectrometer (NIRO 300) for the detection of intracranial oxygenation changes in the adult head, *Stroke* **32**, 2492-2500 (2001).

VALIDATION OF THE CAS NEONATAL NIRS SYSTEM BY MONITORING VV-ECMO PATIENTS:

Preliminary results

Paul B. Benni, Bo Chen, Francine D. Dykes, Scott F. Wagoner, Micheal Heard, April J. Tanner, Teresa L. Young, Khodayar Rais-Bahrami, Oswaldo Rivera, and Billie L. Short

Abstract: The CAS neonatal NIRS system determines absolute regional brain tissue oxygen saturation (SnO_2) and brain true venous oxygen saturation ($SnvO_2$) non-invasively. Since NIRS-interrogated tissue contains both arterial and venous blood from arterioles, venules, and capillaries, SnO_2 is a mixed oxygen saturation parameter, having values between arterial oxygen saturation (SaO_2) and cerebral venous oxygen saturation (SvO_2). To determine a reference for SnO_2, the relative contribution of SvO_2 to SaO_2 drawn from a brain venous site vs. systemic SaO_2 is approximately 70:30 (SvO_2:SaO_2). If the relationship of the relative average contribution of SvO_2 and SaO_2 is known and does not change to a large degree, then NIRS true venous oxygen saturation, $SnvO_2$, can be determined non-invasively using SnO_2 along with SaO_2 from a pulse oximeter.

1. METHODS

1.1. NIRS Methodology

The CAS NIRS system is a continuous wave (CW), three wavelength (775, 800, 850 nm), low power laser system that uses a specially designed sensor for neonates. The sensor's light source to detector separation distance is 25 mm, to allow for sufficient brain tissue interrogation of neonatal subjects. The sensor is usually attached to the subject at a hairless site on the forehead to monitor forebrain regional oxygenation. The

CAS NIRS algorithm[1] is based on an expanded version of the Modified Beer-Lambert Law to determine SnO_2:

$$A_\lambda = -\log(I/I_o)_\lambda = \alpha_\lambda * C * d * B + G_\lambda + F_\lambda + N_\lambda \quad (1)$$

A is the optical attenuation in tissue at wavelength λ (units: optical density OD); I_o is the incident light intensity (W/cm^2); *I* is the detected light intensity (W/cm^2); α_λ is the wavelength-dependent absorption coefficient of the chromophore ($OD * cm^{-1} * \mu M^{-1}$); *C* is the concentration of chromophore (μM); *L* is the light source to detector distance (cm); *B* is the light scattering differential pathlength factor (unit-less); G_λ is the light scattering loss (OD); F_λ is the "fixed" background absorption (OD); and N_λ accounts for any instrumentation related errors. $E_\lambda = G_\lambda + F_\lambda + N_\lambda$ for non-hemoglobin losses (OD).

The differential wavelength form of Eq. (1) minimizes the effects of E_λ, as shown in Eq. (2). The three-wavelength form of Eq. (2) to solve for Hb (deoxyhemoglobin) and HbO_2 (oxyhemoglobin) are shown in Eqs. (3)-(4). Then $SnO_2 = HbO_2 / (HbO_2 + Hb)$ as shown in Eq. (5), where Ψ_{Hb} and Ψ_{HbO2} are empirical NIRS calibration coefficients which are set once by experimental data or by optical phantoms. NIRS tissue SnO_2 is then related to the calibration reference, $SmvO_2$, determined from the weighted venous and arterial oxygen saturations from venous (Kv) and arterial (Ka) compartment contributions shown in Eq. (6). NIRS true venous $SnvO_2$ is determined from Eq. (7):

$$A_{\lambda 1} - A_{\lambda 2} = \Delta A_{\lambda 12} = (\Delta\alpha_{Hb\lambda 12} * Hb + \Delta\alpha_{HbO2\lambda 12} * HbO_2)* d * B + \Delta E_{\lambda 12} \quad (2)$$

$$\begin{vmatrix} \Delta A_{\lambda 12} \\ \Delta A_{\lambda 13} \end{vmatrix} [\Delta\alpha']^{-1}(d*B)^{-1} - \begin{vmatrix} \Delta E_{\lambda 12} \\ \Delta E_{\lambda 13} \end{vmatrix} [\Delta\alpha']^{-1}(d*B)^{-1} = \begin{vmatrix} Hb \\ HbO_2 \end{vmatrix} \quad (3)$$

$$\begin{vmatrix} A_{Hb} \\ A_{HbO2} \end{vmatrix} (d*B)^{-1} - \begin{vmatrix} \Psi_{Hb} \\ \Psi_{HbO2} \end{vmatrix} (d*B)^{-1} = \begin{vmatrix} Hb \\ HbO_2 \end{vmatrix} \quad (4)$$

$$SnO_2\% = (A_{HbO2} - \Psi_{HbO2})/(A_{HbO2} - \Psi_{HbO2} + A_{Hb} - \Psi_{Hb})*100\% \quad (5)$$

$$SnO_2 \approx SmvO_2 = [Ka * SaO_2 + Kv * SvO_2], \text{ where } Ka + Kv = 1 \quad (6)$$

$$\text{NIRS true venous } SnvO_2 = [SnO_2 - Ka * SaO_2] / Kv \quad (7)$$

1.2. Clinical Setting

Extracorporeal membrane oxygenation (ECMO) is a procedure to treat infants and children with life-threatening cardiorespiratory failure that is unresponsive to conventional therapy.[2] Venous jugular bulb or cephalad catheters are sometimes used during veno-venous ECMO to increase cerebral venous drainage and provide a means to monitor internal jugular oxygen saturation ($SjvO_2$),[2] a gold standard to validate NIRS cerebral monitors. ECMO centers using the cephalad catheter often have an in-line monitor (Statsat, Gish Biomedical, Rancho Santa Margarita CA, USA) to measure $SjvO_2$ continuously, which allows for high temporal resolution comparison of $SjvO_2$ to NIRS.

1.3. Experimental Setup

The prototype CAS NIRS monitor used a laptop computer to collect and process NIRS data as well as to record data from a pulse oximeter (Radical, Masimo Corp., Irvine CA, USA) attached to the neonate's foot and from the Statsat $SjvO_2$ monitor, if available. Data were sampled every three seconds and displayed on the laptop computer. Periodic drawing of $SjvO_2$ blood samples were analyzed by a co-oximeter to verify the Statsat values, as well as to provide further data for NIRS validation. NIRS monitoring sessions of ECMO patients were 2 to 10 days long, so a large quantity of high temporal resolution data was collected. This allowed for detection of brain oxygenation changes that may occur naturally without disturbing the subject or altering the ECMO procedure. During the weaning phase of ECMO, inspired oxygen (FiO_2) was sometimes increased to assess the viability of the neonate's lungs, which increased $SjvO_2$ by about 20%.

1.4. Data Analysis

Calibration of the CAS NIRS system and determination of the optimum tissue small-vessel venous to arterial blood volume (Kv:Ka) ratio is an iterative process using linear and non-linear regression techniques, as well as by trial and error, to best fit Eq. (5) to Eq. (6). Since the Kv:Ka ratio has never been thoroughly studied in humans, especially neonates, we empirically estimated the Kv:Ka ratio during different brain oxygenation transitions. To do this, segments of data from each subject were analyzed to identify the best candidates for determination of the Ka:Kv ratio. Data segments that show changes of $SjvO_2 \geq 10\%$ with trend changes of SnO_2 that track well with SaO_2 and $SjvO_2$ over several hour periods were selected for analysis. Multivariate linear regression techniques following that of Brun et al.[3] and Watzman et al.[4] were utilized to resolve Kv:Ka. Once the optimum NIRS calibration was found, Bland-Altman analysis was applied to compare NIRS SnO_2 with the $SmvO_2$ reference of Eq. (6) and NIRS $SnvO_2$ to $SjvO_2$.

2. RESULTS

Figure 1 and Figure 2 show the linear regression results for NIRS SnO_2 and $SnvO_2$, respectively, from the combination of different data segments from three subjects that show brain oxygenation changes over several hour periods. Figures 1 and 2 included a total of 50.7 hours (~60,000 samples) of data, consisting of 20.4 hours from Subject N1, 5.9 hours from Subject N2, and 24.4 hours from Subject N3. The range of SaO_2, as determined by a pulse oximeter, was 86%-100%. The range of $SjvO_2$, as determined by the Statsat cephalad venous monitor was 41.2%-88.3%. NIRS SnO_2 ranged from 55.3%-99.7% and $SnvO_2$ ranged from 38.7%-99.6%. The NIRS-interrogated brain venous to arterial compartment ratio Kv:Ka was estimated to be 70:30 from analysis of candidate transitional brain oxygenation data segments for each subject, as shown in Table 1. Bland-Altman analysis (bias ± 2*standard deviation) of SnO_2 was 1.01 ± 5.19, and $SnvO_2$ was 1.46 ± 7.36. A representative, 10-hour recording of NIRS SnO_2 and $SnvO_2$ along with SaO_2 and $SjvO_2$ are shown in Figures 3 and 4. NIRS and $SjvO_2$ monitoring appeared more sensitive to brain oxygenation changes compared with pulse oximetry.

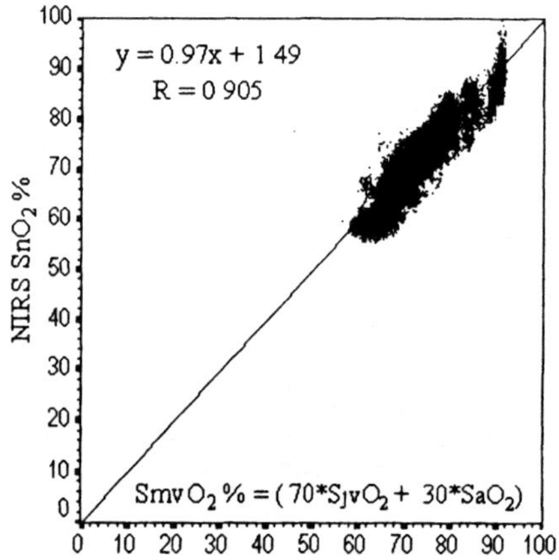

Figure 1. Correlation of NIRS SnO_2 vs. $SmvO_2$.

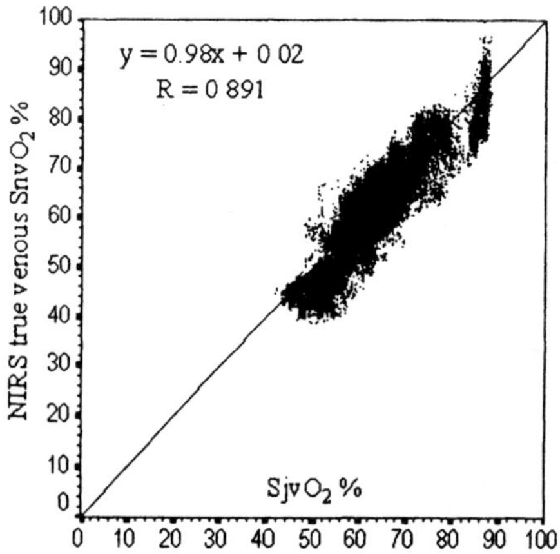

Figure 2. Correlation of NIRS $SnvO_2$ vs. $SjvO_2$.

It was found that prior methods[3, 4] to determine Kv:Ka ratio with confidence were insufficient because different results were found using the two methods. To elaborate, Brun et al.[3] related relative oxygenation changes (i.e. $\Delta SnO_2) = B_1 * SvO_2 + B_2 * SaO_2 + B_3$, whereas $Kv = B_1 / (B_1+B_2)$, $Ka = B_2 / (B_1+B_2)$, and B_3 is the Y-intercept of the multivariate liner regression equation. This technique was used to calculate the "Relative Method" results in Table 1. The Relative Method was also useful to obtain an initial Kv:Ka ratio estimate for an uncalibrated NIRS monitor using ΔSnO_2, provided that the Kv:Ka ratio was consistent. Watzman et al.[4] related absolute oxygen saturation (i.e. $SnO_2) = Kv * SvO_2 + Ka * SaO_2$, where $Ka + Kv = 1$, without a Y-intercept (Table 1 "Absolute Method"). We found that only in some data segments the determined Kv:Ka ratios from the two methods were similar, provided that the Y-intercept was close to zero. We concluded that a data segment with a near-zero Y-intercept indicated that the Kv:Ka ratio was either relatively constant or a good estimation of the ratio average, accounting for some variations.

Table 1. Analysis of NIRS-interrogated venous to arterial compartment (Kv:Ka) ratio.

Subject ID	Data file ID	Data length hours	Brain oxygen change	Kv from Absolute Method	Kv from Relative Method	Relative Method: intercept	Regression correlation R^2	Estimated Ave. Kv (range)
N1	N1-cal1	10.0	^FiO_2	.730	.676	-5.24	.852	.67 -.73
N1	N1-cal2	10.4	^$SjvO_2$	.703	.722	2.09	.831	.70 -.73
N2	N2-cal	5.9	random	.743	.713	-5.76	.708	.71 -.75
N3	N3-cal1	5.4	^FiO_2	.601	.546	-6.28	.876	.54 -.60
N3	N3-cal2	6.1	various	.502	.804	23.5	.826/.837	unknown
N3	N3-cal2a	3.2	^FiO_2	.545	.548	0.32	.897	.54 -.55
N3	N3-cal2b	2.9	random	.644	.682	3.78	.642	.64 -.69

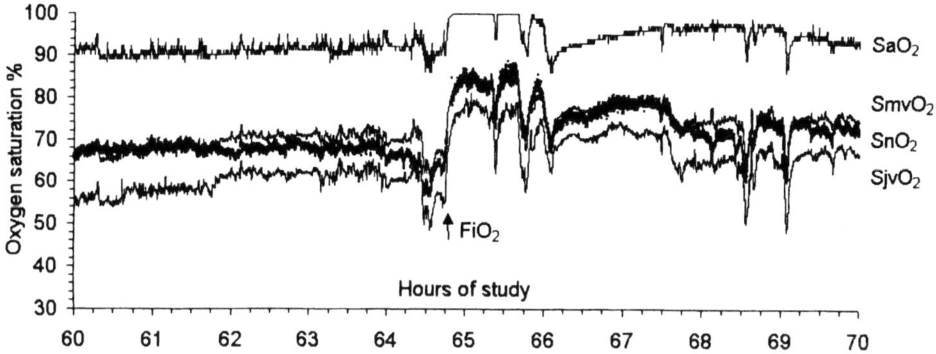

Figure 3. Representative result from Subject N1 during brief increase of FiO_2 showing relationship of NIRS SnO_2 to $SmvO_2 = [.70{*}SjvO_2 + .30{*}SaO_2]$. (From data ID "N1-cal1"). SaO_2: upper solid thin line, $SjvO_2$: lowest solid thin line, $SmvO_2$: middle solid thin line, and NIRS SnO_2: thick dotted line.

To interpret the values in Table 1, the estimated average venous compartment contribution, Kv, was somewhere between the values obtained from the Relative Method and Absolute Method, whereas the closer the Kv values were to each other, the higher confidence we had in the results. The average Kv for Subject N1 was found to be 0.67 to 0.73 (mean Kv:Ka ratio 70:30), and the average Kv for Subject N2 was 0.71 to 0.75 (mean Kv:Ka ratio 73:27). However, for Subject N3, the Y-intercepts for some data segments were high, and the Kv values determined by the two methods were far apart. For example, one data segment, "N3-cal2" (see Table 1), had a high Y-intercept and well-separated Kv values from the two methods. Interestingly, when this data segment was partitioned at a time where the brain oxygenation appeared to be affected by different causes to form two smaller data segments called "N3-cal2a" (^FiO_2 hyperoxia event) and "N3-cal2b" (random oxygenation changes), two different Kv values were found with near-zero Y-intercepts. Along with the results from data "N3-cal1" the Kv:Ka ratio for subject N3 was interpreted to range from 54:46 to 69:31 with a high degree of certainty.

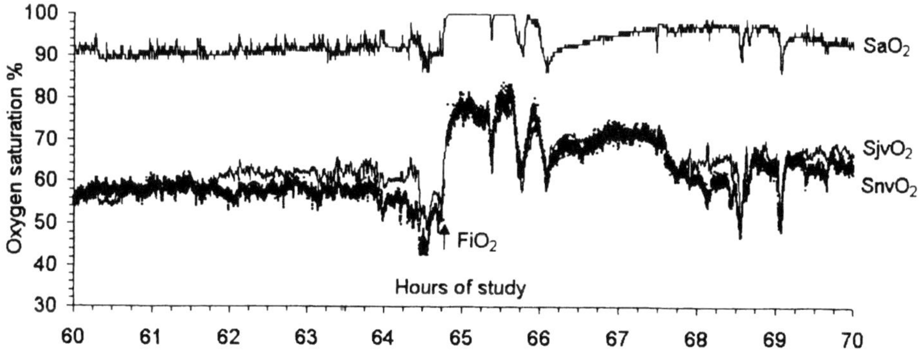

Figure 4. NIRS true venous $SnvO_2$ result from Subject N1 during brief increase of FiO_2. (From data ID "N1-cal1"). SaO_2: upper solid thin line, $SjvO_2$: lower solid thin line, and NIRS $SnvO_2$: thick dotted line.

If we took the mean Kv:Ka ratio of all three subjects (N1: 70:30; N2: 73:27; and N3: 61.5:38.5), the overall mean Kv:Ka result was 68:32. Since we had a small sample size (n = 3), we used a value of 70:30 to correlate NIRS to $SjvO_2$ and SaO_2 to generate Figures 1 and 2 for our preliminary results. The analyzed data segments tabulated in Table 1 were included in the NIRS correlation results, including 12.9 hours of other data from Subject N3, where the Kv:Ka ratio could not be determined. If the Kv:Ka ratio varies, the possible error in determination of SnO_2, assuming that the regional NIRS value tracks well with the global O_2 saturation values, can be evaluated by the expression:

$$\text{Error}\ (SnO_2) = \Delta SmvO_2 = \Delta Kv * (SaO_2 - SjvO_2) \tag{8}$$

Notice that the higher the SaO_2-$SjvO_2$ difference, the higher the possible error in SnO_2. Under normal conditions ($SaO_2 = 100$, $SjvO_2 = 70$), if ΔKv changed 0.1, the error in SnO_2 would be 3%. The SaO_2-$SjvO_2$ difference included in the NIRS correlation results shown in Figures 1 and 2 ranged from 11.6% to 52.2%.

3. DISCUSSION

Internal jugular venous $SjvO_2$ is a global brain oxygenation parameter, while NIRS parameters, depending on the placement of the sensor, represent regional brain oxygenation. Our preliminary results show that NIRS-measured oxygen saturation generally tracks well with global brain measurements, which suggest that a regional measurement can approximate global brain oxygenation. We demonstrated that NIRS-determined true venous oxygen saturation, $SnvO_2$, is possible by combining NIRS with a pulse oximeter, provided that the venous to arterial compartment Kv:Ka ratio does not vary to a large degree within a subject or between subjects. However, some brain pathologies, such as stroke, regional brain death, and carotid artery stenosis, may de-couple the relationship between regional and global oxygenation parameters, as well as alter the Kv:Ka ratio to a large degree. Cerebral hemodynamic changes such as hyper-capnia and hypocapnia may also influence the Kv:Ka ratio.[4] Inhomogeneous systemic arterial oxygen saturation could occur in neonates with ductus arteriosus or septal defects since arterial and venous blood may not be fully mixed in the brain compared with the lower extremities. Therefore, knowledge of the patient's condition is important in interpreting NIRS measurements. Also surgical procedures that briefly occlude the carotid artery for insertion of a shunt or cannulae, as done for veno-arterial ECMO, could disrupt the Kv:Ka ratio, especially if brain hemisphere collateral perfusion is poor.

An accurate estimate of the Kv:Ka ratio depends on how well NIRS regional measurements track global measurements. High temporal resolution monitoring of brain oxygenation changes aid in the assessment, compared with discrete blood sampling alone. Good Kv:Ka ratio estimates emerge from data analyzed by multivariate linear regression that show a near zero Y-intercept, backed by comparable results from linear regression analysis with a forced zero Y-intercept. High Y-intercepts and lack of agreement between the regression methods indicate that the Kv:Ka ratio may have changed during the study. Extracerebral tissue interference may hinder accurate Kv:Ka ratio estimation due to increased scattering of NIRS data to reference brain oxygenation measurements. In conclusion, the preliminary results of the CAS neonate NIRS monitor are encouraging. Analysis of data from more neonate patients may reveal the general mean and range of the Kv:Ka ratio and its impact on NIRS interpretation.

4. ACKNOWLEDGEMENTS

Study supported by NIH-NINDS R44NS39723.

REFERENCES

1. P. B. Benni, Method for non-invasive spectrophotometric blood oxygenation monitoring, U.S. Patent No. 6,456,862, USPTO, (2002).
2. R. Pettignano, M. Labuz, T. Gauthier, J. Huckaby, and R. Clark, The use of cephalad cannulae to monitor jugular venous oxygen content during extracorporeal membrane oxygenation, *Crit. Care* **1**(3), 95-99 (1997).
3. N. C. Brun, A. Moen, K. Borch, O. D. Saugstad, and G. Greisen, Near-infrared monitoring of cerebral tissue oxygen saturation and blood volume in newborn piglets, *Am. J. Physiol.* **273**(2 Pt. 2), H682-H686 (1997).
4. H. M. Watzman, C. D. Kurth, L. M. Montenegro, J. Rome, J. M. Steven, and S. C. Nicolson, Arterial and venous contributions to near-infrared cerebral oximetry, *Anesthesiol.* **93**(4), 947-953 (2000).

INTERACTIONS BETWEEN HIF-1 AND Jab1: BALANCING APOPTOSIS AND ADAPTATION

Outline of a working hypothesis

Mona Larsen, Anja Høg, Eva L. Lund, and Paul E. G. Kristjansen

Abstract: When cells experience hypoxia, they either die by apoptosis or adapt to the hypoxic conditions by a series of compensatory mechanisms. Hypoxia inducible factor-1 (HIF-1) is a transcription factor involved in both processes, but the exact mechanisms regulating whether the cells survive (adapt) or perish by apoptosis are largely unknown.

We hypothesize that the balancing between apoptosis and adaptation is governed by a triangular feedback system involving the α-subunit of HIF-1, p53, and jun activating binding protein 1 (Jab1). Jab1 and p53 bind competitively to the same domain on HIF-1α resulting in either stabilization or degradation of HIF-1α, respectively. Moreover, p53 is stabilized by binding to HIF-1α, whereas its interaction with Jab1 targets p53 for degradation. Thus as a consequence we propose that the ratio between p53 and Jab1 determine whether a hypoxic induction of HIF-1 results in apoptosis or adaptation, with Jab1 as the factor promoting adaptation. On this background we consider Jab1 an interesting molecular target for anticancer therapy.

1. INTRODUCTION

When cells are exposed to hypoxia, they either die or adapt to the hypoxic conditions. Both processes involve upregulation of the transcription factor hypoxia inducible factor-1 (HIF-1), which consists of an α and a β subunit. This has been documented in several studies, and concordantly the pattern of tumor localization in humans suffering from von Hippel-Lindau (VHL) disease confirms this dual role of HIF-1α. These patients develop highly vascularized tumors due to a biallelic mutation in the VHL gene. VHL is essential for degradation of HIF-1α during normoxia, hence mutational inactivation of VHL causes expression of HIF-1α during normoxia.[1] This occurs in carriers of a monoallelic mutation in the VHL gene, which is a germ-line mutation. Therefore biallelic mutations must occur in any cell type, yet the neoplasms predominantly occur in the retina, the kidneys, and in the central nervous system. This suggests that over-expression of HIF-1α in most human tissues induces cell death.[2] We are currently investigating factors that govern whether HIF-1 induces adaptation or apoptosis, based on a hypothesis that the effect is dependent on whether HIF-1α binds or interacts with p53 protein or jun activating binding protein 1 (Jab1).

2. BACKGROUND

2.1. Oxygen Supply in Tumors

During growth of solid tumors, oxygen supply by diffusion becomes inadequate and scattered regions of hypoxia develop. The hypoxia triggers angiogenesis and the tumor develops its own vascular beds. The tumor vessels are unevenly distributed with enhanced flow resistance and other defects resulting in an irregular blood-flow. Thus the oxygen and nutrition supply is insufficient and removal of CO_2, lactate, and other waste products becomes inefficient. This causes the internal milieu in a tumor to become hypoxic and acidic. Half of all solid tumors have a median oxygen partial pressure (Po_2) of 1.5% (10 mm Hg), which is less than 10% of the oxygen supply to normal tissue.[3] Oxygen supply is crucial for cellular homeostasis, and the mammalian cell has a variety of regulation mechanisms to compensate for a low oxygen tension and to restore oxygen supply. HIF-1 is established as a key-player in these mechanisms.[4]

2.2. Hypoxia Inducible Factor-1

HIF-1 is a transcription factor highly expressed in cells during oxygen-deprivation. The HIF-1β subunit is constitutively expressed, and HIF-1 gene activation is regulated through the expression/degradation cycle of the HIF-1α protein (Fig. 1). HIF-1 activates genes with a hypoxia response element (HRE) in their promoter, enhancer, or untranslated regions. Expression of the HIF-1 activated genes promote angiogenesis, erythropoiesis, iron homeostasis, glycolysis, NO-expression, and cell proliferation.[5, 6] Recently, it has been acknowledged that HIF-1 also participates in initiation of hypoxic apoptosis.[7] So far it has been established that HIF-1 activates expression of NIX and BNIP3,[8] which are pro-apoptotic proteins, and that HIF-1α also plays a role in stabilization and activation of p53.[9] This stabilization may depend on the phosphorylation of HIF-1α. The phosphorylation status of HIF-1α changes during long-term hypoxia (>48 hours), where dephosphorylation occurs. The dephosphorylated HIF-1α stabilizes p53 and hypoxia-induced apoptosis may be promoted this way.[10] This stabilization of p53 happens through a protein-protein interaction between HIF-1α and p53, which inhibits the Hdm2 degradation of p53. In addition, dephosphorylated HIF-1α also prevents nuclear accumulation of Hdm2.[9-11] The mechanism controlling the phosphorylation status of HIF-1α during hypoxia is not yet established.

2.2.1. Regulation of HIF-1 Expression

HIF-1α expression is regulated post-translationally, meaning that HIF-1α is constitutively expressed but degraded during normoxia and stabilized during hypoxia/anoxia.[12] The normoxic degradation of HIF-1α occurs through proteolysis in the ubiquitin-proteasome system, initiated by an oxygen-dependent proline hydroxylation of HIF-1α, which is generated by HIF-1α prolyl-4-hydroxylase (HIF-PH).[13] The hydroxylation promotes binding of the VHL protein, which is a subunit of the E3 ligase that ubiquitinates HIF-1α, and thereby targets HIF-1α for degradation. During hypoxia, degradation of HIF-1α in the proteasome is abolished, as the VHL no longer binds to HIF-1α. This leads to an instant increase in the expression of HIF-1α (Figure 1).

An asparagine in the trans-activation domain of HIF-1α is hydroxylated during normoxia. This is prevented during hypoxia, and, combined with an increased phosphorylation in this domain, the capacity of HIF-1α as a transcription factor is enhanced.[14-16] The phosphorylation is to some extent executed by p44 MAP kinases[14, 15] (Figure 1).

Even though the VHL-mediated degradation of HIF-1α is blocked during hypoxia, HIF-1α expression is still regulated during hypoxia. It has been shown that HIF-1α has protein-protein interaction with either Jab1 or p53 during hypoxia. Jab1 and p53 compete in this binding, since Jab1 and p53 interact with the same domain on HIF-1α.[17] But, as binding of p53 to HIF-1α causes HIF-1α degradation facilitated by a HIF-1α ubiquitination by the Hdm2-complex,[18-20] binding of Jab1 stabilizes HIF-1α and increases the capacity of HIF-1 as a transactivator[17] (Figure 1). Therefore, the ratio of Jab1:p53 is likely to play a crucial role in determining the direction of the cellular response to acute hypoxia.

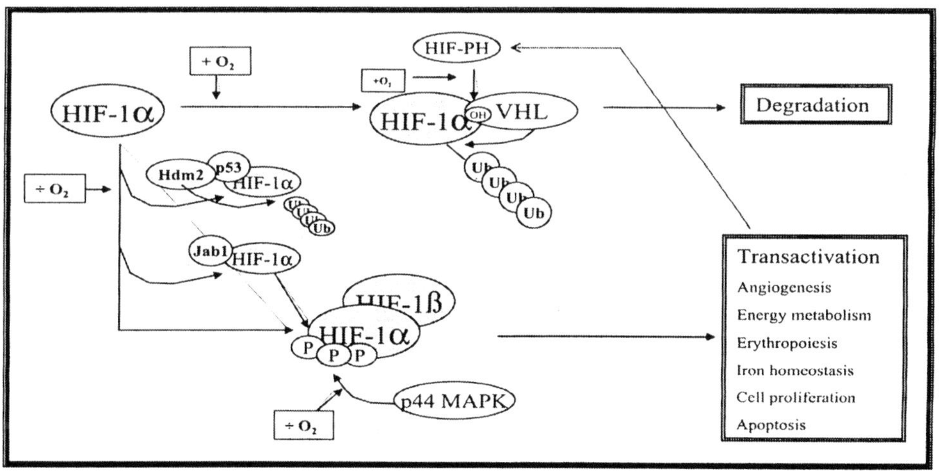

Figure 1. Regulation of HIF-1α. The dotted line illustrates an auto-regulated mechanism, increasing degradation of HIF-1α after re-oxygenation.

2.3. Jun Activating Binding Protein 1

Jab1 was first identified by F. Claret et al. as a protein that interacts with c-Jun and JunD; this interaction promotes cell cycling through stabilization of complexes of c-Jun or JunD with AP-1 sites, potentiating them as trans-activators.[21] Jab1 exists and functions both as a monomer and as a subunit in the COP9 signalosome and for this reason it is also called COP9 signalosome subunit 5 (CSN5).[22] The COP9 regulates ubiquitin activated proteolyses,[23] such as the degradation of p53.[24] Jab1 is the subunit in COP9 primarily responsible for this - probably as a kinase.[24, 25]

Furthermore, Jab1 binds cyclin-dependent kinase inhibitor p27 ($p27^{kip1}$) in the nucleus and, due to a leucine-rich nuclear export signal sequence on Jab1, the complex is exported to the cytoplasm where $p27^{kip1}$ is degraded.[26, 27] This $p27^{kip1}$ degradation contributes to the role of Jab1 as a cell cycle enhancer because $p27^{kip1}$ is a mitotic inhibitor that abrogates the G_1/S-phase transition. Moreover, expression of $p27^{kip1}$ induced during hypoxia is essential for hypoxic cell cycle arrest.[28, 29] There is contradicting evidence on whether this induction is HIF-1α dependent.[28, 29] The cellular localization of Jab1 is affected by the VHL status. In cells with wild type VHL, Jab1 is localized in the cytosol, in contrast to VHL mutated cells that have a nuclear localization of Jab1. This has a down-stream effect on $p27^{kip1}$; the usual accumulation of $p27^{kip1}$ in the nucleus during serum starvation causing cell cycle arrest does not occur in VHL-mutated cells. Jab1 and VHL interact indirectly, probably through HIF-1α and/or through a protein called TRC8.[30] Finally, as mentioned previously, Jab1 interacts with HIF-1α and induces the transactivity of HIF-1.

2.4. p53

During hypoxia, p53 has been found to be upregulated. This upregulation is likely to play a role in hypoxia induced apoptosis,[31-34] but not for the hypoxia-induced cell cycle arrest.[28, 29, 31] However, this is still somewhat controversial.[35] The upregulation of p53 may occur through a HIF-1α-induced stabilization that inhibits degradation of p53.[9, 11] In other instances, looking at Hepa1 cells, p53 is not stabilized by HIF-1α,[36] and some investigators cannot even recognize a significant up-regulation of p53 during hypoxia.[28] Finally, several reports have shown that p53 expression during hypoxia causes degradation of HIF-1α via a facilitation by p53 of the ubiquitination of HIF-1α[18-20] (Figures 1 and 2).

In summary, HIF-1α, Jab1, and p53 seem to regulate the cell during hypoxia in a triangular feedback pattern, which is depicted in Figure 2.

3. THE ROLE OF Jab1 DURING HYPOXIA – A HYPOTHESIS

In light of the information above, we have established a working hypothesis on a mechanism that may control whether HIF-1/hypoxia induces hypoxic adaptation or hypoxic apoptosis. The hypothesis is that the interaction of Jab1 with HIF-1α promotes adaptation to hypoxia, in contrast to the interaction of p53 and HIF-1α at the same domain, which promotes apoptosis. This is based on the role of Jab1 as promoter of

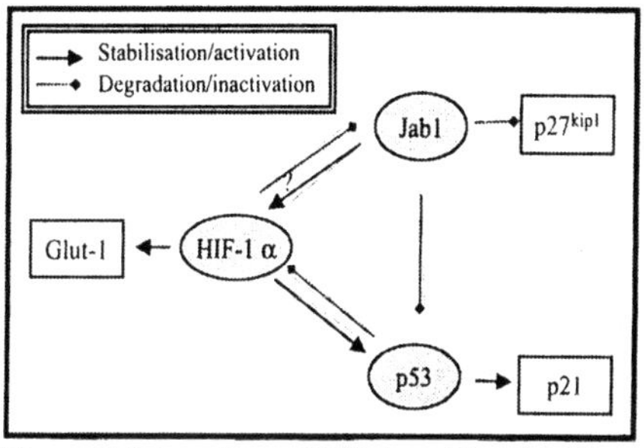

Figure 2. Interactions of HIF-1α, p53 and Jab1 under hypoxia.

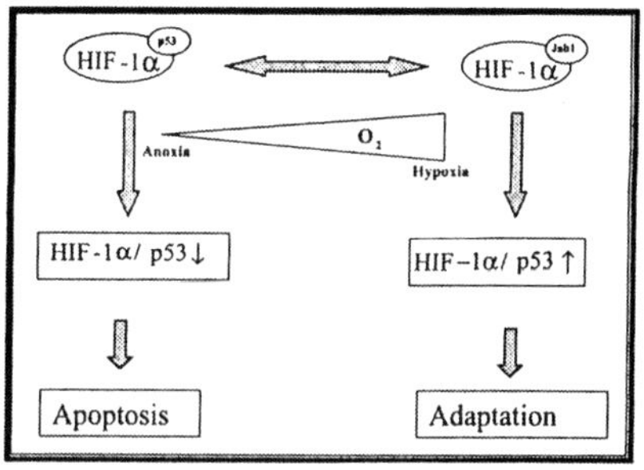

Figure 3. Effect of hypoxia depending on whether HIF-1α binds to p53 or Jab1.

HIF-1α stability and transactivation, whereas p53 leads to HIF-1α degradation and decreases transactivation. Correspondingly, HIF-1α stabilizes p53, promoting apoptosis in this way. The binding of either Jab1 or p53 may also control the dephosphorylation of HIF-1α during an extended period of hypoxia. As mentioned, Jab1 is a kinase and could be responsible for HIF-1α phosphorylation, but if Jab1 is displaced from HIF-1α after severe hypoxia due to an increased expression of p53, the phosphorylation cannot be maintained (Figure 3).

3.1. The Relevance of the Triad for Cancer

3.1.1. The Role of Each Factor in Tumors

3.1.1a. HIF-1α. HIF-1α expression cannot be detected in normal tissues, whereas the protein is expressed in many cancer forms[37, 38] and in their metastases.[39] The over-expression of HIF-1α in tumors is a result of the abnormal and often hypoxic microenvironment in the tumors, but it can also be related to genetic changes such as loss-of-function mutations of tumor-suppressor genes (e.g. p53, VHL, or PTEN) and activation of oncogenes (e.g. H-ras).[37, 40] HIF-1α over-expression in tumors correlates with tumor progression,[41] formation of metastasis, and resistance to radiotherapy, and therefore also to poor prognosis.[42] HIF-1α over-expression also correlates with an increased angiogenesis, resistance to carboplatin and etoposide chemotherapy, and increased invasiveness.[43-45]

3.1.1b. Jab1. Jab1 expression is associated with poor prognosis in various cancer types,[46] $p27^{kip1}$ expression is inversely correlated with Jab1,[26] and finally, expression of $p27^{kip1}$ is associated with a good prognosis.[47]

3.1.1c. p53. Approximately one half of solid tumors have mutations in the p53 gene.[48] p53 deficient tumors are characterized by an increased neovascularization and growth, probably caused by the lack of the negative effect of p53 on HIF-1α.[11, 18] This may also be the reason why p53 deficient tumors are more resistant to antiangiogenic therapy.[49]

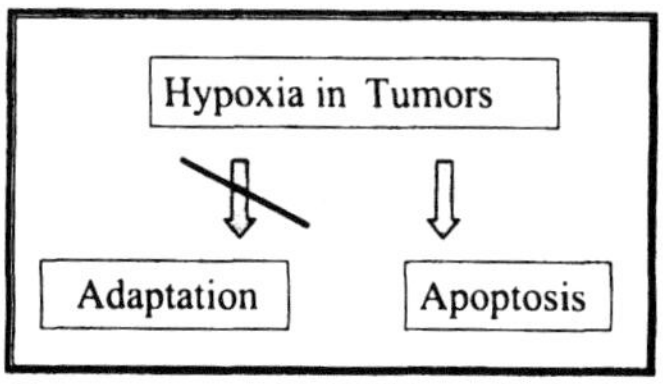

Figure 4. Consequences of inhibition of Jab1 in hypoxic tumors.

3.2. Perspectives of the Hypothesis

The fact that HIF-1α is correlated to poor prognosis makes HIF-1α a potential diagnostic or therapeutic target in cancer. Furthermore, HIF-1 enhances the malignant phenotype by promoting adaptation. But it also appears that HIF-1α mutated cells could have a lower tendency to go into apoptosis under hypoxic stress and have an accelerated growth compared with wild type controls,[7] possibly due to lack of stabilization of p53. In this light, inhibition of HIF-1α may not be the optimal therapeutic target.

Instead, an inhibition of Jab1 may have a desirable antitumor effect because Jab1, through HIF-1, promotes the ability of the cells to adapt to hypoxia and enhances the ability of cancer cells to survive and proliferate. Inhibition of Jab1 will therefore abrogate HIF-1's ability to promote adaptation to hypoxia, without inhibiting HIF-1's pro-apoptotic effects, as this most likely happens through a stabilization of p53. This stabilization may even happen to a greater extent when Jab1 expression is decreased both because Jab1 no longer is there to prevent binding of p53 to HIF-1α and because the direct Jab1 degradation of p53 will be abolished. In this way, anti-Jab1 treated cancer cells should have a greater propensity to enter apoptosis in the adverse environment of a solid tumor. In theory, Jab1 could be a better therapeutic target in cancer than HIF-1α (Figure 4).

On the other hand, success of the proposed anti-Jab1 treatment could at least partly be dependent on the p53 proficiency of the tumors, a proficiency which is not present in 50% of all solid tumors.[48] However, the idea that dependency is on an intact p53 is still purely speculative. Another problem is that—in contrast to HIF-1α—Jab1 is expressed in many tissue types, which may increase the risk of side effects (unpublished data).

4. APPENDIX: WORK IN PROGRESS

In order to test our hypothesis, the following series of investigations is underway to explore hypoxia induced apoptosis, and to investigate the role of HIF-1, Jab1, and p53 in this process. The first round of experiments has focused on:

- Defining the degree of hypoxia where hypoxic apoptosis is the predominant response rather than hypoxic adaptation.
- Measuring the expression of Jab1, HIF-1α, and p53, along with their downstream proteins, $p27^{kip1}$, Glut-1, and p21, under various degrees of hypoxia.
- Investigating whether the degree of hypoxia, where apoptosis takes over, is influenced by blocking the expression of Jab1, HIF-1α, or p53, or by corresponding over-expression.

REFERENCES

1. M. Ivàn, and W. G. Kaelin, Jr., The von Hippel-Lindau tumor suppressor protein, *Curr. Opin. Genet: Dev.* **11**, 27-34 (2001).
2. K. Kondo and W. G. Kaelin, Jr., The von Hippel-Lindau tumor suppressor gene, *Exp. Cell Res.* **264**, 117-125 (2001).
3. J. M. Brown, Exploiting the hypoxic cancer cell: mechanisms and therapeutic strategies, *Mol. Med. Today* **6**, 157-162 (2000).
4. G. L. Semenza, HIF-1 and mechanisms of hypoxia sensing, *Curr. Opin. Cell Biol.* **13**, 167-171 (2001).
5. G. L. Semenza, Hypoxia-inducible factor 1: control of oxygen homeostasis in health and disease, *Pediatr. Res.* **49**, 614-617 (2001).
6. G. L. Semenza, HIF-1: mediator of physiological and pathophysiological responses to hypoxia, *J. Appl. Physiol* **88**, 1474-1480 (2000).
7. P. Carmeliet, Y. Dor, J. M. Herbert, D. Fukumura, K. Brusselmans, M. Dewerchin, M. Neeman, F. Bono, R. Abramovitch, P. Maxwell, C. J. Koch, P. Ratcliffe, L. Moons, R. K. Jain, D. Collen, E. Keshert, and E. Keshet, Role of HIF-1alpha in hypoxia-mediated apoptosis, cell proliferation and tumour angiogenesis, *Nature* **394**, 485-490 (1998).
8. R. K. Bruick, Expression of the gene encoding the proapoptotic Nip3 protein is induced by hypoxia, *Proc. Natl. Acad. Sci. USA* **97**, 9082-9087 (2000).
9. W. G. An, M. Kanekal, M. C. Simon, E. Maltepe, M. V. Blagosklonny, and L. M. Neckers, Stabilization of wild-type p53 by hypoxia-inducible factor 1alpha, *Nature* **392**, 405-408 (1998).
10. H. Suzuki, A. Tomida, and T. Tsuruo, Dephosphorylated hypoxia-inducible factor 1alpha as a mediator of p53-dependent apoptosis during hypoxia, *Oncogene* **20**, 5779-5788 (2001).
11. D. Chen, M. Li, J. Luo, and W. Gu, Direct interactions between HIF-1 alpha and Mdm2 modulate p53 function, *J. Biol. Chem.* **278**, 13595-13598 (2003).
12. S. Salceda, I. Beck, V. Srinivas, and J. Caro, Complex role of protein phosphorylation in gene activation by hypoxia, *Kidney Int.* **51**, 556-559 (1997).
13. P. Jaakkola, D. R. Mole, Y. M. Tian, M. I. Wilson, J. Gielbert, S. J. Gaskell, A. Kriegsheim, H. F. Hebestreit, M. Mukherji, C. J. Schofield, P. H. Maxwell, C. W. Pugh, and P. J. Ratcliffe, Targeting of HIF-alpha to the von Hippel-Lindau ubiquitylation complex by O2-regulated prolyl hydroxylation, *Science* **292**, 468-472 (2001).
14. E. Minet, T. Arnould, G. Michel, I. Roland, D. Mottet, M. Raes, J. Remacle, and C. Michiels, ERK activation upon hypoxia: involvement in HIF-1 activation, *FEBS Lett.* **468**, 53-58 (2000).
15. D. E. Richard, E. Berra, E. Gothie, D. Roux, and J. Pouyssegur, p42/p44 mitogen-activated protein kinases phosphorylate hypoxia-inducible factor 1alpha (HIF-1alpha) and enhance the transcriptional activity of HIF-1, *J. Biol. Chem.* **274**, 32631-32637 (1999).
16. D. Lando, D. J. Peet, D. A. Whelan, J. J. Gorman, and M. L. Whitelaw, Asparagine hydroxylation of the HIF transactivation domain a hypoxic switch, *Science* **295**, 858-861 (2002).
17. M. K. Bae, M. Y. Ahn, J. W. Jeong, M. H. Bae, Y. M. Lee, S. K. Bae, J. W. Park, K. R. Kim, and K. W. Kim, Jab1 interacts directly with HIF-1alpha and regulates its stability, *J. Biol. Chem.* **277**, 9-12 (2002).
18. R. Ravi, B. Mookerjee, Z. M. Bhujwalla, C. H. Sutter, D. Artemov, Q. Zeng, L. E. Dillehay, A. Madan, G. L. Semenza, and A. Bedi, Regulation of tumor angiogenesis by p53-induced degradation of hypoxia-inducible factor 1alpha, *Genes Dev.* **14**, 34-44 (2000).
19. M. V. Blagosklonny, W. G. An, L. Y. Romanova, J. Trepel, T. Fojo, and L. Neckers, p53 inhibits hypoxia-inducible factor-stimulated transcription, *J. Biol. Chem.* **273**, 11995-11998 (1998).
20. L. O. Hansson, A. Friedler, S. Freund, S. Rudiger, and A. R. Fersht, Two sequence motifs from HIF-1alpha bind to the DNA-binding site of p53, *Proc. Natl. Acad. Sci. USA* **99**, 10305-10309 (2002).
21. F. X. Claret, M. Hibi, S. Dhut, T. Toda, and M. Karin, A new group of conserved coactivators that increase the specificity of AP-1 transcription factors, *Nature* **383**, 453-457 (1996).
22. S. F. Kwok, R. Solano, T. Tsuge, D. A. Chamovitz, J. R. Ecker, M. Matsui, and X. W. Deng, Arabidopsis homologs of a c-Jun coactivator are present both in monomeric form and in the COP9 complex, and their abundance is differentially affected by the pleiotropic cop/det/fus mutations, *Plant Cell* **10**, 1779-1790 (1998).
23. M. H. Glickman, D. M. Rubin, O. Coux, I. Wefes, G. Pfeifer, Z. Cjeka, W. Baumeister, V. A. Fried, and D. Finley, A subcomplex of the proteasome regulatory particle required for ubiquitin-conjugate degradation and related to the COP9-signalosome and eIF3, *Cell* **94**, 615-623 (1998).
24. D. Bech-Otschir, R. Kraft, X. Huang, P. Henklein, B. Kapelari, C. Pollmann, and W. Dubiel, COP9 signalosome-specific phosphorylation targets p53 to degradation by the ubiquitin system, *EMBO J.* **20**, 1630-1639 (2001).

25. M. Seeger, R. Kraft, K. Ferrell, D. Bech-Otschir, R. Dumdey, R. Schade, C. Gordon, M. Naumann, and W. Dubiel, A novel protein complex involved in signal transduction possessing similarities to 26S proteasome subunits, *FASEB J.* **12**, 469-478 (1998).
26. K. Tomoda, Y. Kubota, and J. Kato, Degradation of the cyclin-dependent-kinase inhibitor p27Kip1 is instigated by Jab1, *Nature* **398**, 160-165 (1999).
27. K. Tomoda, Y. Kubota, Y. Arata, S. Mori, M. Maeda, T. Tanaka, M. Yoshida, N. Yoneda-Kato, and J. Y. Kato, The cytoplasmic shuttling and subsequent degradation of p27Kip1 mediated by Jab1/CSN5 and the COP9 signalosome complex, *J. Biol. Chem.* **277**, 2302-2310 (2002).
28. L. B. Gardner, Q. Li, M. S. Park, W. M. Flanagan, G. L. Semenza, and C. V. Dang, Hypoxia inhibits G1/S transition through regulation of p27 expression, *J. Biol. Chem.* **276**, 7919-7926 (2001).
29. N. Goda, H. E. Ryan, B. Khadivi, W. McNulty, R. C. Rickert, and R. S. Johnson, Hypoxia-inducible factor 1 alpha is essential for cell cycle arrest during hypoxia, *Mol. Cell Biol.* **23**, 359-369 (2003).
30. R. M. Gemmill, L. T. Bemis, J. P. Lee, M. A. Sozen, A. Baron, C. Zeng, P. F. Erickson, J. E. Hooper, and H. A. Drabkin, The TRC8 hereditary kidney cancer gene suppresses growth and functions with VHL in a common pathway, *Oncogene* **21**, 3507-3516 (2002).
31. T. G. Graeber, J. F. Peterson, M. Tsai, K. Monica, A. J. Fornace, Jr., and A. J. Giaccia, Hypoxia induces accumulation of p53 protein, but activation of a G1-phase checkpoint by low-oxygen conditions is independent of p53 status, *Mol. Cell Biol.* **14**, 6264-6277 (1994).
32. T. G. Graeber, C. Osmanian, T. Jacks, D. E. Housman, C. J. Koch, S. W. Lowe, and A. J. Giaccia, Hypoxia-mediated selection of cells with diminished apoptotic potential in solid tumours, *Nature* **379**, 88-91 (1996).
33. E. M. Hammond, N. C. Denko, M. J. Dorie, R. T. Abraham, and A. J. Giaccia, Hypoxia links ATR and p53 through replication arrest, *Mol. Cell Biol.* **22**, 1834-1843 (2002).
34. E. M. Hammond, M. J. Dorie, and A. J. Giaccia, ATR/ATM targets are phosphorylated by ATR in response to hypoxia and ATM in response to re-oxygenation, *J. Biol. Chem.* **278**(14), 12207-12213 (2003).
35. M. Achison and T. R. Hupp, Hypoxia attenuates the p53 response to cellular damage, *Oncogene* **22**, 3431-3440 (2003).
36. R. H. Wenger, G. Camenisch, I. Desbaillets, D. Chilov, and M. Gassmann, Up-regulation of hypoxia-inducible factor-1alpha is not sufficient for hypoxic/anoxic p53 induction, *Cancer Res.* **58**, 5678-5680 (1998).
37. K. L. Talks, H. Turley, K. C. Gatter, P. H. Maxwell, C. W. Pugh, P. J. Ratcliffe, and A. L. Harris, The expression and distribution of the hypoxia-inducible factors HIF-1alpha and HIF-2alpha in normal human tissues, cancers, and tumor-associated macrophages, *Am. J. Pathol.* **157**, 411-421 (2000).
38. K. L. Sondergaard, D. A. Hilton, M. Penney, M. Ollerenshaw, and A. G. Demaine, Expression of hypoxia-inducible factor 1alpha in tumours of patients with glioblastoma, *Neuropathol. Appl. Neurobiol.* **28**, 210-217 (2002).
39. H. Zhong, A. M. De Marzo, E. Laughner, M. Lim, D. A. Hilton, D. Zagzag, P. Buechler, W. B. Isaacs, G. L. Semenza, and J. W. Simons, Overexpression of hypoxia-inducible factor 1alpha in common human cancers and their metastases, *Cancer Res.* **59**, 5830-5835 (1999).
40. G. L. Semenza, Hypoxia, clonal selection, and the role of HIF-1 in tumor progression, *Crit Rev. Biochem. Mol. Biol.* **35**, 71-103 (2000).
41. P. Birner, M. Schindl, A. Obermair, C. Plank, G. Breitenecker, and G. Oberhuber, Overexpression of hypoxia-inducible factor 1alpha is a marker for an unfavorable prognosis in early-stage invasive cervical cancer, *Cancer Res.* **60**, 4693-4696 (2000).
42. D. M. Aebersold, P. Burri, K. T. Beer, J. Laissue, V. Djonov, R. H. Greiner, and G. L. Semenza, Expression of hypoxia-inducible factor-1alpha: a novel predictive and prognostic parameter in the radiotherapy of oropharyngeal cancer, *Cancer Res.* **61**, 2911-2916 (2001).
43. M. I. Koukourakis, A. Giatromanolaki, E. Sivridis, C. Simopoulos, H. Turley, K. Talks, K. C. Gatter, and A. L. Harris, Hypoxia-inducible factor (HIF1A and HIF2A), angiogenesis, and chemoradiotherapy outcome of squamous cell head-and-neck cancer, *Int. J. Radiat. Oncol. Biol. Phys.* **53**, 1192-1202 (2002).
44. A. Unruh, A. Ressel, H. G. Mohamed, R. S. Johnson, R. Nadrowitz, E. Richter, D. M. Katschinski, and R. H. Wenger, The hypoxia-inducible factor-1 alpha is a negative factor for tumor therapy, *Oncogene* **22**, 3213-3220 (2003).
45. P. Birner, B. Gatterbauer, G. Oberhuber, M. Schindl, K. Rossler, A. Prodinger, H. Budka, and J. A. Hainfellner, Expression of hypoxia-inducible factor-1 alpha in oligodendrogliomas: its impact on prognosis and on neoangiogenesis, *Cancer* **92**, 165-171 (2001).
46. L. Sui, Y. Dong, M. Ohno, Y. Watanabe, K. Sugimoto, Y. Tai, and M. Tokuda, Jab1 expression is associated with inverse expression of p27(kip1) and poor prognosis in epithelial ovarian tumors, *Clin. Cancer Res.* **7**, 4130-4135 (2001).

47. M. A. Kouvaraki, G. Z. Rassidakis, L. Tian, R. Kumar, C. Kittas, and F. X. Claret, Jun activation domain-binding protein 1 expression in breast cancer inversely correlates with the cell cycle inhibitor p27(Kip1), *Cancer Res.* **63**, 2977-2981 (2003).
48. A. J. Levine, p53, the cellular gatekeeper for growth and division, *Cell* **88**, 323-331 (1997).
49. J. L. Yu, J. W. Rak, B. L. Coomber, D. J. Hicklin, and R. S. Kerbel, Effect of p53 status on tumor response to antiangiogenic therapy, *Science* **295**, 1526-1528 (2002).

PAST, PRESENT, AND FUTURE OF OXYGEN IN CANCER RESEARCH

Paul Okunieff, Bruce Fenton, and Yuhchyau Chen

Abstract: The first pathologists, oncologists, and medical physicists were aware that tumors were populated by an aberrant vasculature. The classic observations of Thomlinson and Gray in the 1950's established that O_2 diffusion distances caused tumor to grow in cords. Tumor necrosis was observed surrounding a Krogh cylinder of viable tumor. That work helped explain earlier work by Warburg, who demonstrated a predisposition for tumors to favor anaerobic respiration, and it became the basis for 5 decades of subsequent research aimed at improving tumor oxygenation at the time of radiation. The role of O_2 in modifying radiation response was attributed exclusively to the reactive free radicals that can be formed when O_2 is present. These radicals produce approximately three-fold more irreparable double strand breaks in DNA.

Subsequently it became clear that tumor had nutritional insufficiencies in addition to hypoxia. Ischemic regions are hypoglycemic, acidotic, have poor penetration of drugs, increased interstitial pressure, and altered immunological states. Ischemic regions can have intermittent reflow and associated redox stress. The relative impact of O_2 compared to these associated phenomenon, and the degree to which hypoxia causes or follows these associated physiologic stresses, have been studied in detail. ISOTT scientists are responsible for much of the elucidation of the specific effects of O_2, ADP/ATP ratios, hypoglycemia, and acidosis on tumor responses to radiation and hyperthermia. Many questions still remain.

1. INTRODUCTION

The first pathologists, oncologists, and medical physicists were aware that tumors were populated by an aberrant vasculature. The classic observations of Thomlinson and Gray in the 1950's established that O_2 diffusion distances caused tumor to grow in cords about their vasculature.[1] Tumor necrosis was observed surrounding a Krogh cylinder of viable tumor.[2] That work helped explain earlier work by Warburg,[3] who demonstrated a

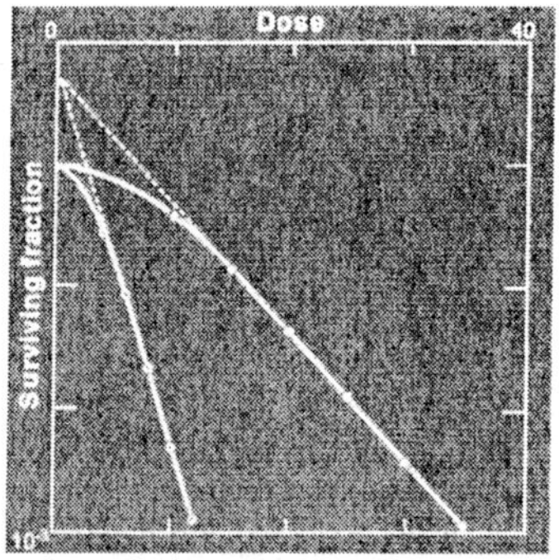

Figure 1. The impact of oxygen on the radiation response of tumors outperforms that of any other known sensitizer. For example, at 1000 cGy, only about 1/2 of the tumor cells in an experiment might lose reproductive integrity. If the cells are oxygenated at levels over 10 mm Hg, the cell kill increases so that only 1 in 200 cells survive.

predisposition for tumors to favor anaerobic respiration, and it became the basis for 5 decades of research aimed at improving tumor oxygenation at the time of irradiation. For most of a half century, the reduction of radiation-induced cell kill that occurs in the presence of low O_2 was attributed exclusively to the decreased number of reactive free radicals that can be formed. These radicals produce approximately three-fold more irreparable double strand breaks in DNA in the presence of oxygen, and that number of increased radicals corresponds to the three-fold difference in dose response (Figure 1).[4] This dogma remains the major explanation of the oxygen effect, but we now know it does not fully explain it.

The role that oxygen plays in tumor metabolism is clear. Tumor has nutritional insufficiencies that go hand in hand with hypoxia.[5, 6] Hypoxia, in the presence of adequate circulation, has very different effects than does ischemia, wherein oxygen and other nutrients and waste removal add to the toxic environment. Ischemic regions are hypoglycemic, acidotic, have poor penetration of drugs, increased interstitial pressure, and altered immunological states.[7, 8] Ischemic regions can have intermittent reflow and associated redox stress.[9] The relative impact of O_2 compared to these associated phenomenon, and the degree to which hypoxia causes or follows these associated physiologic stresses, has been studied extensively.[7, 8, 10-12] ISOTT scientists are responsible for much of the elucidation of the specific effects of O_2, ATP/ADP, hypoglycemia, and acidosis on tumor responses.[13-19] Many questions remain regarding the interaction between oxygen and tumor metabolism (Figure 2).

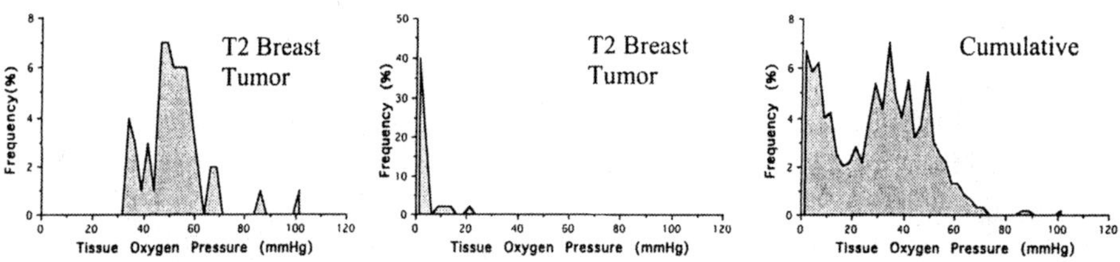

Figure 2. Human tumors of the same size and type can have very different oxygen distributions. For example, two women with T2 (approximately 2-3 cm) tumors were found to have extremely different oxygen profiles, measured just before resection. In one patient, no readings were below 10 mm Hg, and in the second over 85% of the readings were under 10 mm Hg. This difference in oxygen distribution is sufficient to explain the heterogeneity of cure rates for radiated breast cancers.[20, 21]

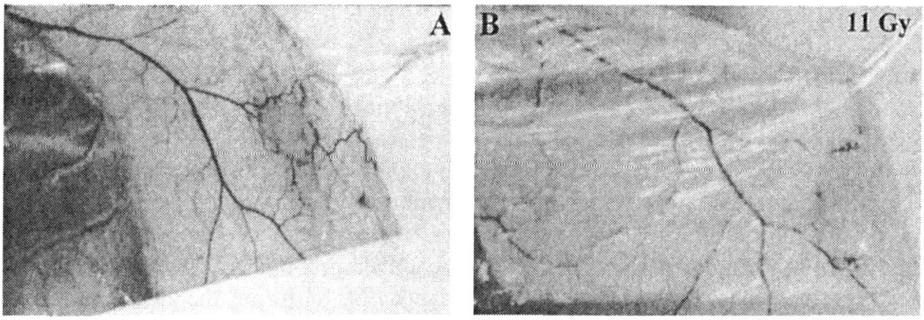

Figure 3. Radiation powerfully reduces local angiogenesis. The effect is long-lasting and can be irreversible. Shown is the effect of radiating mouse skin, and then implanting tumor. **A:** In the non-irradiated skin, there is dramatic angiogenesis. **B:** In the skin given 11 Gy 1 week before the tumor was injected, there is almost no angiogenesis.

More recently, the molecular impact of O_2 on gene expression has been of greater interest than its redox chemistry. Hypoxia is associated with apoptosis, selection of oncogenes, and production of angiogenic growth factors (Figure 3). Hypoxic tumors have an increased propensity to metastasize in addition to a decreased response to radiation, immunotherapy, and most chemotherapy. Molecules that induce similar molecular signal pathways are considered oxygen-mimetic, much in the way that drugs that had similar redox chemistry were commonly considered oxygen-mimetic. This molecular approach to understanding oxygen biology has led to clearly overly simplistic beliefs: for example, that hypoxia and Cobalt exposure have the same biological properties because both induce hypoxia inducible factor 1 alpha (HIF1α).[18] Nevertheless, the observation that O_2 impacts molecular pathways has breathed new life into hypoxia research. Ultimately, the redox chemistry and molecular biological aspects of oxygen metabolism and biology must merge.

As was true when ISOTT scientists studied the relative impact of oxygen, glucose, and acidosis on tumor metabolism, modern ISOTT scientists must study the relative impact of oxygen on cell signal pathways critical to tumor progression and response. The process will be very complicated. For example, of the known genes of the human genome,[22, 23] it is difficult to identify a single one that would not respond directly or indirectly to acute or chronic changes in oxygen concentration. Sorting out the large number of irrelevant responses from the clinically important reaction in tumor biology will be key to understanding the role of oxygen in tumor progression and resistance to therapy. There will be a large number of complex and interactive molecular signaling pathways, and it will be critical to define those pathways that are actually O_2-specific. It is possible that in the end we will find that the most important clinical measurement is simply the O_2 concentration of the tissue, ending at the same point we started in the 1950's. Hopefully, if this is the case, we will have developed simple non-invasive methods of measuring tissue oxygenation.

2. IMPACT OF OXYGEN ON RADIATION DAMAGE TO DNA

At radiation doses used to treat humans with cancer, the energy deposited (2 to 3 Gy per day, and about 30 to 60 Gy total) is minimal. For example, 2 Gy, in the absence of temperature regulation by blood flow, will raise tissue temperature by 0.002°C. The same dose causes thousands of easily repaired single strand DNA breaks, and a dozen or two double strand breaks.[24] The latter are less well repaired and, while rarely causing metabolic dysfunction of a cell, they regularly inactivate the cellular reproductive integrity since chromosomes will not segregate properly at mitosis. The redox chemistry at the DNA level is largely responsible for the oxygen effect. In the absence of oxygen, DNA that is ionized can easily find a replacement electron from a neighboring hydrogen on a water molecule. In the presence of oxygen, the organic DNA radical instead reacts with oxygen to form a peroxide. The peroxide can then go on to react with other organic molecules, creating a covalent molecular change that deforms the DNA and is not repairable. This phenomenon can then render the cell reproductively inactivated. The role of oxygen in this process likely explains most of the oxygen effect. DNA repair is extremely efficient after irradiation. Well over 99% of single strand DNA damage is repaired in the presence of oxygen, and likewise at least 95% of double strand breaks are repaired. Oxygen is needed for full repair to be accomplished. Thus oxygen plays a role in both increasing and repairing DNA damage after irradiation. The DNA repair processes after irradiation can be error-prone, though error-free repair is possible for tetraploid and aneuploid cells. As mentioned previously, oxygen plays a role in the regulation of the approximately 18 different known DNA repair mechanisms; even a very small change in the frequency or fidelity of repair would have profound effects on radiation-related tumor cell killing: for example, if the repair rate for double strand breaks were to decrease to 90%, radiation would be more than twice as effective.

3. OXYGEN MIMETIC DRUGS

Oxygen mimetic drugs are those that can undergo single electron transfer. In the presence of oxygen, oxygen mimetic drugs undergo futile cycling[25]; in the absence of oxygen, instead the molecules will irreversibly covalently bind to the radiation-induced DNA, protein, and lipid radical. To reproductively inactivate cells, the drug must gain access to the nucleus and must have sufficiently low toxicity to allow the required concentration in the blood and tissue. It must also not be metabolized. Since most normal tissues are not hypoxic, several drugs meet this requirement. The most useful drugs have been nitroimidazoles. Some normal tissues, prominently including the liver and brain, are relatively low in oxygen,[8] and, accordingly, these drugs can produce hepatotoxicity and neurotoxicity.

A second class of oxygen mimetics are useful for studying molecular biological effects like those caused by oxygen. They have almost none of the oxygen effects expected *in vivo*, and thus are only an *in vitro* tool. These are drugs that induce oxygen-related gene expression. For example, while Nickel itself has almost no effect on tumor response to radiation, Nickel induces HIF1α and is useful for studying the role that HIF1α plays on tumor progression, growth, and apoptosis.[26]

HIF1α may be misnamed, since it responds to many stimulae other than hypoxia. Furthermore, most molecules judged to be oxygen markers (HIF1α, Glut1, VEGF) immunostain for hypoxia in tumors (Table 1), but do not appear to specifically stain hypoxic cells *in vivo*. More likely, relevant, molecularly oxygen-related biomarkers will take one of two forms: those that impact specific DNA repair pathways under different

Table 1. Candidate molecular markers for detection of hypoxia in tumors.

VEGF, FGF1, FGF2	Angiogenesis factors that are upregulated by hypoxia. They are commonly upregulated by other stress, including trauma. Tumors also constituatively increase these molecules in order to improve angiogenesis and augment metastatic potential
Glut 1	Glycolytic intermediate upregulated when aerobic glycolysis is reduced due to low oxygen availability. Can be constituatively upregulated in tumors (Warberg hypothesis) and thus is not universally useful.
Involulcrin, CA-9, HIF-1, NDRG1, Hyaluronidase	Proteins that seem to accumulate in tissues or cells that are hypoxic. Evidence that they are expressed in tissues by hypoxic cells and evidence that expression corresponds to distance from blood vessels has rarely been carefully tested.

oxygen conditions, and those that alter expression of nuclear antioxidant enzymes and of redox metabolites.

4. MINIMALLY OR NON-INVASIVE *IN VIVO* DIRECT OR INDIRECT OXYGEN MEASUREMENTS

The optimal *in vivo* measurement method of oxygen does not exist: specifically, a method that can directly, non-invasively, and volumetrically measure oxygen in any tissue environment, including microregions.[27] There are, however, some that come close and others that might have some clinical utility.

4.1. Oxygen Electrode Measurements

The advent of automated oxygen electrodes that can be placed into a tumor, automatically repositioned in multiple tumor regions, and provide histograms of near-accurate oxygen partial pressures has allowed human tumor oxygen measurements. Using these systems, it has been shown that oxygen distribution can explain the response heterogeneity of human head and neck cancers and of human breast cancers.[20, 21, 28, 29] Low oxygen partial pressure can also explain the resistance of human gliomas to radiation treatments.[30] Electrode measurements have also demonstrated that hypoxia is associated with, if not a cause of, tumor metastatic progression.[31-35] The precision of electrodes has been questioned,[36] but the difference from true oxygen tension, particularly at the very low or very high levels, does not seem to be clinically significant. The electrode techniques have obvious disadvantages. The electrode is only of use for tumors that are superficial. Electrodes also require invasive instrumentation of the tumor, a potential clinical misadventure.

4.2. Imaging Techniques

There are imaging techniques that specifically measure oxygen-related phenomenon, or those that measure physiological parameters expected to correlate indirectly with tissue oxygen partial pressures. Among the direct techniques, there are electron spin resonance (ESR) techniques, nuclear magnetic resonance spectroscopy (MRS) techniques, and positron emission tomography techniques (PET).[37-43] ESR can be performed using implanted or infused spin traps,[25] and while it only allows for a point measurement, importantly, ESR can be utilized for simple serial measurements during a therapeutic intervention. Infused ESR contrast agents are in development. It is not yet clear if they will be useful, but limited animal data are interesting. Imaging of tissues more than a few centimeters deep will be technically difficult due to penetration of the radiofrequencies and other limitations that occur due to the low magnetic field strengths.[25] Very interesting is the use of fluorinated hydrocarbons and MRS. Here a concurrent measurement of oxygen, temperature, and pH might be possible.[44] The requirement for infusion of a large volume of contrast agent, and the accumulation of that agent in macrophage (and thus long biological half-life) are of concern. Since the resonance frequency of Fluorine is similar to that of hydrogen, it is likely that resonance coils used for proton spectroscopy can be easily adapted for Fluorine spectroscopy. Positron emission tomography techniques currently rely on drugs that bind proteins in hypoxic tumor regions; they are thus indirect, imperfect measures.[45, 46] Theoretically, designer compounds meant to visualize RNA or proteins related to hypoxia could be developed. MRS can also be used to indirectly measure tissue physiology as a surrogate for oxygen. For example, ATP, ADP, inorganic phosphate, and other high energy phosphates can be detected. The relaxation rates of the molecules are also oxygen sensitive.[47] Thus the method allows for a variety of opportunities for estimation of oxygen-related metabolism and concentration. High field clinical magnets are becoming commonplace and thus this modality might become useful in the near future. Dynamic magnetic resonance imaging and computerized tomographic imaging methods both can measure tumor blood flow. Since oxygen is usually the limiting nutrient for tumor metabolism, blood flow (and thus oxygen delivery rate) can be a useful measure of adequate oxygenation.[48] It is necessarily an indirect estimator of oxygenation, but is easily performed (but not often employed) using modern clinical imaging technologies.

4.3. Near Infrared Spectroscopy Techniques

The physical and spectral chemistry of oxygen is probably not just fortuitous. Indeed the large number of methods available to measure oxygen exist because molecular oxygen has almost unique chemical processes. For example, it is why nuclear and electron spin resonance relaxation changes occurs as a function of tissue oxygenation, why chemical electrode measurements are nearly oxygen specific, why specific enzymes and metabolites are altered in function based on oxygen concentrations in the tissues, and why near infrared spectroscopy is likewise quite specific for the partial pressure of molecular oxygen. Near infra red spectroscopy in its simplest form, as a peripheral blood gas monitor is already the most logistically important technology for measuring oxygen in the clinic. Multiple color wavelengths can be used with laser or filtered light technologies. Much of the pioneering work to advance this technology for measurement of oxygen profiles was originally presented at ISOTT. These include both invasive and non-invasive in-vivo methods.[38, 49-52] Invasive electrode techniques are also available, and some can concurrently measure temperature and pH.[53, 54] Most of the invasive work however is in animal models. Non-invasive methods are available for some deep tissue measurements and are used to measure brain oxygenation in premature or sick infants, wherein the skull is rather transparent.[49, 50, 52] Optimal design and implementation of these non-invasive techniques remains an open research subject. Adopting these technologies for the evaluation of deeper structures, such as tumors in adults, remains elusive. The problems of scatter, absorption, and reflection of light may ultimately not be solvable for deep applications.

5. MOLECULAR MARKERS FOR HYPOXIA

It is difficult to image a protein expression level that would not be altered indirectly by ischemia or hypoxia. In identifying a molecule as a hypoxia marker, therefore, there is huge potential for identifying markers that are not at all quantitatively or specifically related to oxygen concentration. There are several characteristics that, at minimum, any endogenous, oxygen-dependent protein must have in order to be a useful marker clinically; it is necessary that the cells that are hypoxic are making the marker; the marker should remain intracellular or at least not diffuse away from the hypoxic region; the marker must also be stable enough to be measured in histological sections of tumor (Figure 4).

A powerful method to validate the oxygen dependence is increased expression of a molecule at increased distance from the nearest blood vessel.[55] There is debate as to whether HIF1α and related proteins meet this requirement. What is clear, however, is that evidence of the utility of HIF1α as a hypoxia marker is difficult to reproduce consistently. One explanation is the short half-life of this compound. NDRG1, in several tumor models, does meet minimum requirements of validation.[56] NDRG1 sometimes is constituatively expressed, and in those tumors is of limited utility as an oxygen marker.

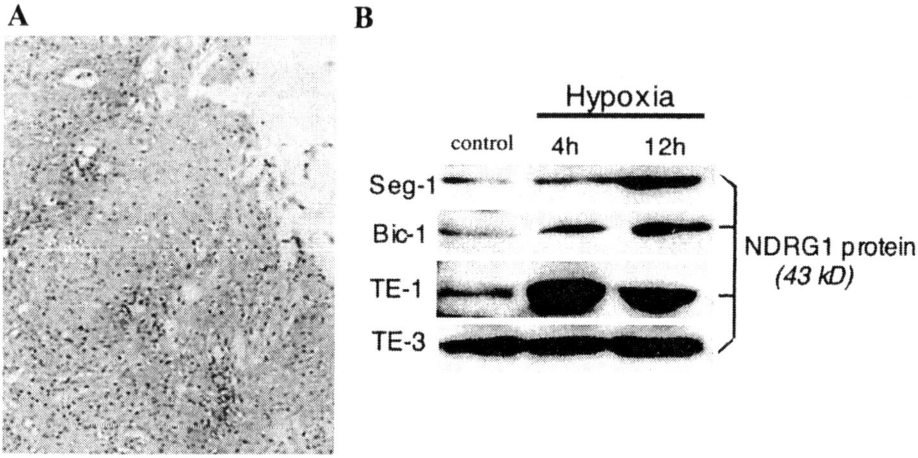

Figure 4. There are many molecular signals and reactions to hypoxia and to oxygen. Few are fully specific. **A**: While VEGF is over-expressed in tissues that contain hypoxia, the proximity of staining and blood vessels suggests that it is not actually staining the hypoxic cells themselves. The mechanism of this discordance includes diffusion of the material away from the cells that produce it. **B**: Other markers are clearly expressed by hypoxic cells, but in some tumors are also expressed by cells that are not hypoxic. In the example, among 4 human esophageal tumor lines, three respond to hypoxia with an increase in NDRG1, one has constituative production of NDRG1 and cannot upregulate when made hypoxic.

There will probably also be tumors that do not produce NDRG1, even under hypoxic conditions. For those as well, NDRG1 will not be clinically useful. Optimally, a biological molecule would have an oxygen response curve that included hypoxia and euoxia, and could be quantitatively calibrated. Unfortunately, it is highly unlikely that any such molecule will be discovered.

6. CONCLUSION

Remarkably, after 60 years of research, the measurement of oxygen concentration in human tumors remains a technical problem. There is an enormous body of literature identifying oxygen as a most important modifier of tumor growth, progression, and resistance to therapy. Add to that new anti-cancer drugs aimed at inhibition of specific growth and angiogenesis factors. The action of these must be measured in terms of limiting the nutritional component, oxygen. Developing logistically and technically reliable methods of measuring tissue oxygen is an imperative. Substantial opportunities exist in the future to exploit new technologies, computer imaging methodologies, and molecular biology. Ultimately, for any of the oxygen measurement techniques to be fully clinically useful, they will have to provide valid estimates of the tissue oxygenation state, if not accurate measurements of oxygen concentration or partial pressures.

REFERENCES

1. R. H. Thomlinson, Changes of oxygenation in tumors in relation to irradiation, *Front. Radiat. Ther. Oncol.* **3**, 109-112 (1968).
2. K. Groebe, and P. Vaupel, Evaluation of oxygen diffusion distances in human breast cancer xenografts using tumor-specific in vivo data: Role of various mechanisms in the development of tumor hypoxia, *Int. J. Radiat. Oncol. Biol. Phys.* **15**(3), 691-697 (1988).
3. O. Warburg, Uber den Stoffwechsel der Carcinomzellen, *Klin. Wschr.* **4**, 534-536 (1925).
4. J. P. Freyer, K. Jarrett, S. Carpenter, and M. R. Raju, Oxygen enhancement ratio as a function of dose and cell cycle phase for radiation-resistant and sensitive CHO cells, *Radiat. Res.* **127**, 297-307 (1991).
5. E. K. Rofstad, P. DeMuth, B. M. Fenton, and R. M. Sutherland, 31P nuclear magnetic resonance spectroscopy studies of tumor energy metabolism and its relationship to intracapillary oxyhemoglobin saturation status and tumor hypoxia, *Cancer Res.* **48**, 5440-5446 (1988).
6. P. Vaupel, C. Schaefer, and P. Okunieff, Intracellular acidosis in murine fibrosarcomas coincides with ATP depletion, hypoxia, and high levels of lactate and total Pi, *NMR Biomed.* **7**, 128-136 (1994).
7. P. Vaupel, H. P. Fortmeyer, S. Runkel, and F. Kallinowski, Blood flow, oxygen consumption, and tissue oxygenation of human breast cancer xenografts in nude rats, *Cancer Res.* **47**, 3496-3503 (1987).
8. P. Vaupel, F. Kallinowski, and P. Okunieff, Blood flow, oxygen and nutrient supply, and metabolic microenvironment of human tumors: a review, *Cancer Res.* **49**, 6449-6465 (1989).
9. B. A. Teicher, Hypoxia and drug resistance, Cancer Metastasis Rev. **13**(2), 139-168 (1994).
10. I. Lee, Y. Boucher, T. J. Demhartner, and R. K. Jain, Changes in tumour blood flow, oxygenation and interstitial fluid pressure induced by pentoxifylline, *Br. J. Cancer* **69**, 492-496 (1994).
11. J. Biaglow, M. Dewirst, D. Leeper, R. Burd, and S. Tuttle, Factors controlling oxygen utilization, *Adv. Exp. Med. Biol.* 317-324 (2005).
12. P. Vaupel, A. Mayer, S. Briest, and M. Höckel, Hypoxia in breast cancer: role of blood flow, oxygen diffusion distances, and anemia in the development of oxygen depletion, *Adv. Exp. Med. Biol.* 333-342 (2005).
13. P. Vaupel, P. Okunieff, and L. J. Neuringer, Blood flow, tissue oxygenation, pH distribution, and energy metabolism of murine mammary adenocarcinomas during growth, *Adv. Exp. Med. Biol.* **248**, 835-846 (1989).
14. P. Vaupel, K. Schlenger, and M. Höckel, Blood flow and tissue oxygenation of human tumors: an update, *Adv. Exp. Med. Biol.* **317**, 139-152 (1992).

15. M. Tamura, O. Hazeki, S. Nioka, B. Chance, and D. S. Smith, The simultaneous measurements of tissue oxygen concentration and energy state by near-infrared and nuclear magnetic resonance spectroscopy, *Adv. Exp. Med. Biol.* **222**, 359-363 (1988).
16. K. Erickson, R. D. Braun, D. Yu, J. Lanzen, D. Wilson, D. M. Brizel, T. W. Secomb, J. E. Biaglow, and M. W. Dewhirst, Effect of longitudinal oxygen gradients on effectiveness of manipulation of tumor oxygenation, *Cancer Res.* **63**(15), 4705-4712 (2003).
17. S. Nioka, D. S. Smith, B. Chance, H. V. Subramanian, S. Butler, and M. Katzenberg, Oxidative phosphorylation system during steady-state hypoxia in the dog brain, *J. Appl. Physiol.* **68**(6), 2527-2535 (1990).
18. J. T. Erler, C. J. Cawthorne, K. J. Williams, M. Koritzinsky, B. G. Wouters, C. Wilson, C. Miller, C. Demonacos, I. J. Stratford, and C. Dive, Hypoxia-mediated down-regulation of Bid and Bax in tumors occurs via hypoxia-inducible factor 1-dependent and -independent mechanisms and contributes to drug resistance, *Mol. Cell. Biol.* **24**(7), 2875-2889 (2004).
19. B. Chance, S. Nioka, W. Warren, and G. Yurtsever, Mitochondrial NADH as the bellwether of tissue O_2 delivery, *Adv. Exp. Med. Biol.* 17-22 (2005).
20. P. Okunieff, E. P. Dunphy, M. Höckel, D. J. Terris, and P. Vaupel, The role of oxygen tension distribution on the radiation response of human breast carcinoma, *Adv. Exp. Med. Biol.* **345**, 485-492 (1994).
21. P. Okunieff, M. Höckel, E. P. Dunphy, K. Schlenger, C. Knoop, and P. Vaupel, Oxygen tension distributors are sufficient to explain the local response of human breast tumors treated with radiation alone, *Int. J. Radiat. Oncol. Biol. Phys.* **26**, 631-636 (1993).
22. S. Istrail, G. G. Sutton, L. Florea, A. L. Halpern, C. M. Mobarry, R. Lippert, et al., Whole-genome shotgun assembly and comparison of human genome assemblies, *Proc. Natl. Acad. Sci. USA*, **101**(7), 1916-1921 (2004).
23. J. C. Venter, M. D. Adams, E. W. Myers, P. W. Li, R. J. Mural, G. G. Sutton, et al., The sequence of the human genome, *Science* **291**(5507), 1304-1351 (2001).
24. G. Taucher-Scholz, J. A. Stanton, M. Schneider, and G. Kraft, Induction of DNA breaks in SV40 by heavy ions, *Adv. Space Res.* **12**(2-3), 73-80 (1992).
25. J. T. Hwang, M. M. Greenberg, T. Fuchs, and K. S. Gates, Reaction of the hypoxia-selective antitumor agent tirapazamine with a C1'-radical in single-stranded and double-stranded DNA: the drug and its metabolites can serve as surrogates for molecular oxygen in radical-mediated DNA damage reactions, *Biochemistry* **38**(43), 14248-14255 (1999).
26 H. Cangul, Hypoxia upregulates the expression of the NDRG1 gene leading to its overexpression in various human cancers, *BMC Genet.* **5**(1), 27 (2004).
27. H. Swartz, and J. Dunn, The difficulties in comparing in vivo oxygen measurements: turning the problems into virtues! *Adv. Exp. Med. Biol.* 295-302 (2005).
28. D. M. Brizel, G. S. Sibley, L. R. Prosnitz, R. L. Scher, and M. W. Dewhirst, Tumor hypoxia adversely affects the prognosis of carcinoma of the head and neck, *Int. J. Radiat. Oncol. Biol. Phys.* **38**, 285-289 (1997).
29. P. Okunieff, J. de Bie, E. P. Dunphy, D. J. Terris, and M. Höckel, Oxygen distributions partly explain the radiation response of human squamous cell carcinomas, *Br. J. Cancer* **27**, S185-S190 (1996).
30. R. Rampling, G. Cruickshank, A. Lewis, S. A. Fitzsimmons, and P. Workman, Direct measurement of pO2 distribution and bioreductive enzymes in human malignant brain tumors, *Int. J. Radiat. Oncol. Biol. Phys.* **29**, 427-431 (1994).
31. D. M. Brizel, S. P. Scully, J. M. Harrelson, L. J. Layfield, J. M. Bean, L. R. Prosnitz, and M. W. Dewhirst, Tumor oxygenation predicts for the likelihood of distant metastases in human soft tissue sarcoma, *Cancer Res.* **56**, 941-943 (1996).
32. D. M. Brizel, G. L. Rosner, L. R. Prosnitz, and M. W. Dewhirst, Patterns of variability of tumor oxygenation in human soft tissue sarcomas, cervical carcinomas, and lymph node metastases, *Int. J. Radiat. Oncol. Biol. Phys.* **32**, 1121-1125i (1995).
33. M. Höckel, K. Schlenger, B. Aral, M. Mitze, U. Shaffer, and P. Vaupel, Association between tumor hypoxia and malignant progression in advanced cancer of the uterine cervix, *Cancer Res.* **56** 4509-4515 (1996).
34. M. Höckel, C. Knoop, B. Vorndran, E. Baussmann, M. Mitze, P. G. Knapstein, and P. Vaupel, Intratumoral pO2 predicts survival in advanced cancer of the uterine cervix, *Radiother. Oncol.* **26**, 45-50 (1993).
35. P. Okunieff, I. Ding, P. Vaupel, and M. Höckel, Evidence for and against hypoxia as the primary cause of tumor aggressiveness, *Adv. Exp. Med. Biol.* **510**, 69-75 (2003).
36. B. M. Seddon, D. J. Honess, B. Vojnovic, G. M. Tozer, and P. Workman, Measurement of tumor oxygenation: in vivo comparison of a luminescence fiber-optic sensor and a polarographic electrode in the p22 tumor, *Radiat. Res.* **155**(6), 837-846 (2001).
37. G. Ilangovan, A. Bratasz, and P. Kuppusamy, Non-invasive measurement of tumor oxygenation using embedded microparticulate EPR spin probe, *Adv. Exp. Med. Biol.* 67-74 (2005).

38. Y. S. Sakata, O. Y. Grinberg, S. Grinberg, R. Springett, and H. M. Swartz, Simultaneous NIR-EPR spectroscopy of rat brain oxygenation, *Adv. Exp. Med. Biol.* 357-362 (2005).
39. N. Khan, H. Hou, P. Hein, R. J. Comi, J. C. Buckey, O. Grinberg, I. Salikhov, S. Y. Lu, H. Wallach, and H. M. Swartz, Black magic and EPR oximetry: from lab to clinical trials, *Adv. Exp. Med. Biol.* 119-126 (2005).
40. H. Hou, O. Y. Grinberg, S. A. Grinberg, N. Khan, J. F. Dunn, and H. M. Swartz, Cerebral PtO_2 acute hypoxia, and volatile anesthetics in the rat brain, *Adv. Exp. Med. Biol.* 179-186 (2005).
41. R. D. Shonat, and A. S. Norige, Developing strategies for three-dimensional imaging of oxygen tension in the rodent retina, *Adv. Exp. Med. Biol.* 173-178 (2005).
42. G. Schears, J. Creed, T. Zaitseva, S. Schultz, D. F. Wilson, and A. Pastuszko, Cerebral oxygenation during repetitive apnea in newborn piglets, *Adv. Exp. Med. Biol.* 1-8 (2005).
43. E. Takahashi, T. Takano, A. Numata, N. Hayashi, S. Okano, O. Nakajima, Y. Nomura, and M. Sato, Genetic oxygen sensor: GFP as an indicator of intracellular oxygenation, *Adv. Exp. Med. Biol.* 39- 44 (2005).
44. Y. Song, K. L. Worden, X. Jiang, D. Zhao, A. Constantinescu, H. Liu, and R. P. Mason, Tumor oxygen dynamics: comparison of 19F MR EPI and frequency domain NIR spectroscopy, *Adv. Exp. Med. Biol.* **530**, 225-236 (2003).
45. L. Bentzen, S. Keiding, M. Nordsmark, L. Falborg, S. B. Hansen, J. Keller, O. S. Nielsen, and J. Overgaard, Tumour oxygenation assessed by 18F-fluoromisonidazole PET and polarographic needle electrodes in human soft tissue tumours, *Radiother. Oncol.* **67**(3), 339-344 (2003).
46. J. S. Rasey, W. J. Koh, M. L. Evans, L. M. Peterson, T. K. Lewellen, M. M. Graham, and K. A. Krohn, Quantifying regional hypoxia in human tumors with positron emission tomography of [18F]fluoromisonidazole: a pretherapy study of 37 patients, *Int. J. Radiat. Oncol. Biol. Phys.* **36**(2), 417-428 (1996).
47. P. Okunieff, T. Tokuhiro, P. Vaupel, and L. J. Neuringer, Interaction of oxygen partial pressure and energy metabolism with the relaxation rate of inorganic phosphate: a 31P NMR study, *Adv. Exp. Med. Biol.* **277**, 95-105 (1990).
48. F. Kallinowski, K. H. Schlenger, S. Runkel, M. Kloes, M. Stohrer, P. Okunieff, and P. Vaupel, Blood flow, metabolism, cellular microenvironment, and growth rate of human tumor xenografts, *Cancer Res.* **49**(14), 3759-3764 (1989).
49. P. B. Benni, B. Chen, F. D. Dykes, S. F. Wagoner, M. Heard, A. J. Tanner, T. L. Young, K. Rais-Bahrami, O. Rivera, and B. Short, Validation of the CAS neonatal NIRS system by monitoring VV-EMCO patients, *Adv. Exp. Med. Biol.* 195-202 (2005).
50. C. E. Elwell, J. R. Henty, T. S. Leung, T. Austin, J. H. Meek, D. T. Delphy, and J. S. Wyatt, Measurement of CMRO2 in neonates undergoing intensive care using near infrared spectroscopy, *Adv. Exp. Med. Biol.* 263-268 (2005).
51. F. A. Howe, J. P. Connelly, S. P. Robinson, R. Springett, and J. R. Griffiths, The effects of tumor blood flow and oxygenation modifiers on subcutaneous tumours as determined by NIRS, *Adv. Exp. Med. Biol.* 75-82 (2005).
52. K. von Siebenthal, M. Keel, J.-C. Fauchère, V. Dietz, D. Haensse, U. Wolf, U. Helfenstein, O. Bänziger, H. U. Bucher, and M. Wolf, Variability of cerebral hemoglobin concentration in very preterm infants during the first 6 hours of life, *Adv. Exp. Med. Biol.* 91-98 (2005).
53. M. Urano, Y. Chen, J. Humm, J. A. Koutcher, P. Zanzonico, and C. Ling, Measurements of tumor tissue oxygen tension using a time-resolved luminescence-based optical oxylite probe: comparison with a paired survival assay, *Radiat. Res.* **158**(2), 167-173 (2002).
54. T. Jarm, G. Sersa, and D. Miklavcic, Oxygenation and blood flow in tumors treated with hydralazine: evaluation with a novel luminescence-based fiber-optic sensor, *Technol. Health Care* **10**(5), 363-380 (2002).
55. B. M. Fenton, S. F. Paoni, B. Grimwood, and I. Ding, Varied response of spontaneous tumors to antiangiogenic agents, *Adv. Exp. Med. Biol.* 59-66 (2005).
56. I. Ding, P. Okunieff, K. Salnikow, W. Liu, and B. Fenton, A new intrinsic hypoxia marker in esophageal cancer, *Adv. Exp. Med. Biol.* **540**, 227-233 (2003).

BENZOATE HYDROXYLATION:
A measure of oxidative stress in divers

Gernot Gronow, Wataru Kähler, Andreas Koch, and Norbert Klause

Abstract: Hyperoxia may facilitate the formation of reactive oxygen species. Recent experiments indicated signs of oxidative stress after 3.5 h hyperoxic diving. We analyzed in the urine of healthy, 100% O_2-breathing male volunteers before and after 45 min sea-water diving (170 kPa) or 30 min resting at 280 kPa in a pressure chamber (HBO) for sub-fractions of hydroxybenzoate (HB), monohydroxybenzoate (MHB), and of dihydroxybenzoate (DHB). Measurements were performed by HPLC and electrochemical or UV-detection. Additionally, urinary concentrations of thiobarbituric acid-reactive substances (TBARS) and of creatinine (CREA) were analyzed by standard colorimetric assays. During HBO treatment, TBARS, DHB, 2,4-DHB, and 3,4-DHB increased significantly. MHB and CREA did not change. 2,4- and 3,4-DHB-alterations correlated with changes in TBARS. Diving induced urine dilution and stimulated oxygen consumption. Urinary TBARS and HB rose significantly higher during diving at 170 kPa than during HBO at 280 kPa. A different pattern in urinary sub-fractions of DHB could be observed in divers: 2,6 > 2,3 > 2,5 > 3,4. Changes in 2,6- and 2,5-DHB correlated significantly with alterations in TBARS. 2,6-DHB probably indicated renal oxidant stress similar to previously described animal experiments. It is concluded that analyzing urinary HB may provide a sensitive measure to quantify and qualify oxidant stress in divers.

1. INTRODUCTION

A growing number of amateur and professional divers prefer to breathe oxygen at high concentration. At increased O_2 pressure (hyperoxia), however, tissue integrity is threatened by the formation of reactive oxygen species (ROS). Cellular survival may then be limited by the duration of exposure and applied partial pressure.[1] It is well documented that prolonged hyperoxia may induce in man tissue alterations such as lung edema or an increased excitability of the central and neuromuscular system.[2, 3] Little is known about the effect of brief periods of hyperoxia.

Recently, Bearden et al.[4] could demonstrate that 4.6 m below surface divers breathing 82% O_2 developed hematological signs of oxidative stress within 3.5 hours. The erythrocyte osmotic fragility and the plasma content of thiobarbituric acid-reactive substances (TBARS) rose significantly. In another, quite different experiment by Brichta and co-workers,[5] repetitive hyperbaric oxygenation (HBO, 5 days, 1 h/d) in a pressure chamber also induced hematological alterations. The authors observed a decrease in the number of red blood cells and in hemoglobin concentration, as well as a nearly two-fold increase in plasma lipid peroxidation products. By contrast, in the time course of HBO-treatment, urinary chemiluminescence, another parameter of lipid peroxidation, fell by about 32%. Brichta and co-workers[5] concluded that hyperoxic vasoconstriction in the kidney may have prevented the release of lipid peroxidation products into the urine. They did not measure urinary TBARS or hydroxylated benzoate (HB). HB may be formed by the reaction of salicylate or benzoate with the hydroxyl radical OH˙, leading to 2,3-dihydroxybenzoate (2,3-DHB) as a main reaction product.[6] Recent animal experiments provided evidence that oxidative stress may also stimulate the release of 2,6-dihydroxybenzoate (2,6-DHB).[7]

The aim of the present study was, therefore, to test in man whether or not a short period (30–45 min) of hyperoxia would raise urinary TBARS and HB, specifically 2,6-DHB. Two different conditions of oxidative stress were applied. Young male volunteers inhaled 100% oxygen either in a pressure chamber (HBO) or during seawater diving with a close-circuit oxygen rebreathing apparatus.

2. METHODS

We chose two groups of healthy male subjects, one consisted of 28 trained divers (age 29 ± 4.5 years, body weight 83.4 ± 3.4 kg), and one of 10 untrained volunteers (age 27 ± 3.6 years, body weight 79.2 ± 4.6 kg). Both groups had to breathe 100% oxygen at commonly used conditions: the divers for 45 min at 170 kPa, and the HBO group for 30 min at 280 kPa. Divers had to swim 7 m below seawater level, and the volunteers rested in a pressure chamber. Urine samples were analyzed immediately before and after the experimental period. We measured the formation of thiobarbituric acid-reactive substances (TBARS) as well as of hydroxybenzoate (HB), monohydroxybenzoate (MHB) and dihydroxybenzoate (DHB). HB, MHB, DHB and fractions of 2,3-, 2,4-, 2,5-, 3,4- 3,5- and of 2,6-DHB were analyzed by HPLC and electrochemical or UV-detection.[7] TBARS were measured by a colorimetric assay[8] modified by a two-wavelength method using the 550 nm-absorption to correct 535 nm-readings for urochrome interference. 1,1,3,3-tetra-ethoxypropane (TEP) served as the reference for TBARS. Creatinine was measured by a standard colorimetric assay. Values were compensated for urinary volume changes on the basis of excreted creatinine (scheme in Fig. 1). Statistical significance was calculated by Student *t* test. Differences were considered significant if the *P* value was less than 0.05.

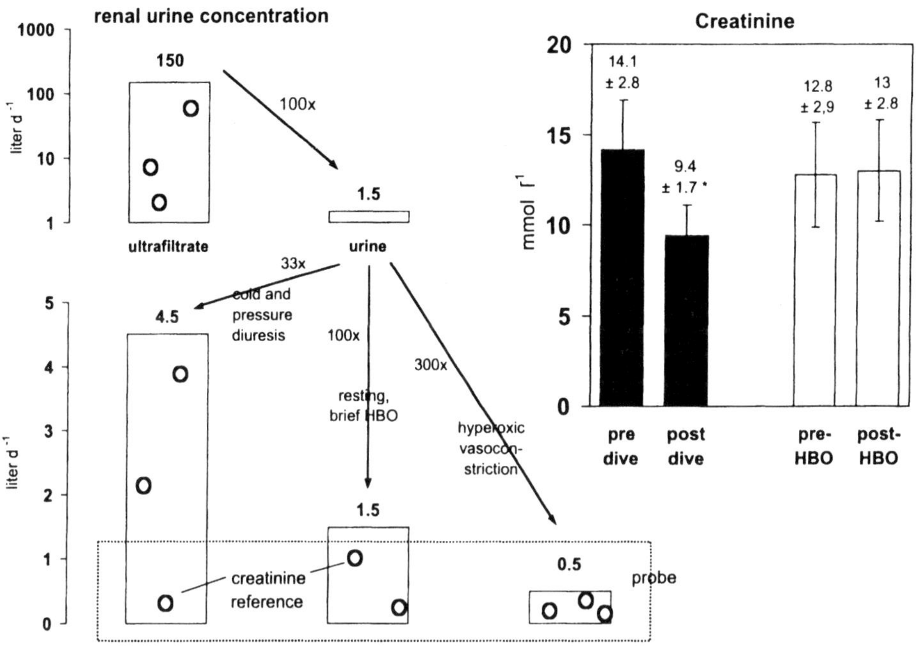

Figure 1. *Left:* Compensation of differences in urine volume by using creatinine as reference. *Right:* Urinary creatinine concentration in man breathing 100% oxygen during 45 min seawater diving (170 kPa, n = 28, black columns) or 30 min hyperbaric oxygenation (HBO, 280 kPa, n = 10, white columns). Mean ± SD

3. RESULTS

Filtered plasma constituents are usually concentrated in the urine (Fig. 1). Seawater diving-induced urine dilution was indicated by a significant (33%) decrease in creatinine concentration (Fig. 1, right panel, black columns); by contrast, brief hyperbaric oxygenation (HBO) had no effect (open columns). The concentration of thiobarbituric-reactive substances (TBARS, Fig. 2, upper left panel) and of hydroxybenzoate (HB, Fig. 2, lower left panel) increased significantly in both groups, by 172.9% ± 31.4% in divers, and by 148.3% ± 24.1% in the HBO group. TBARS in control groups were significantly different. Professional divers had, prior to the experiments, about 75% higher urinary TBARS-values than the HBO group. Accordingly, we found a higher pre-dive urinary content of hydroxybenzoate (HB, + 87%). Increases in HB (Fig. 2, lower left panel) were moderate in post-HBO (+17.4 ± 10.3%, open columns) and substantial in the post-dive group (+43.7% ± 11.3%, black columns). The latter consumed 3.3 ± 0.2 l min^{-1} oxygen, while members of the HBO group consumed about 90% less (0.4 ± 0.18 l min^{-1}).

No 2-monohydroxybenzoate = salicylate could be detected in the urine of both groups. Among remaining monohydroxybenzoates (MHB), the subspecies 3-MHB and 4-MHB could not be identified separately and were, without presence of salicylate (2-MHB), summarized as MHB in Figure 2 (upper right panel). Moreover, no measurable concentration of MHB was found in the urine of the pre- and post-HBO group (open columns). In contrast, urinary MHB and dihydroxybenzoate (DHB) levels of divers were

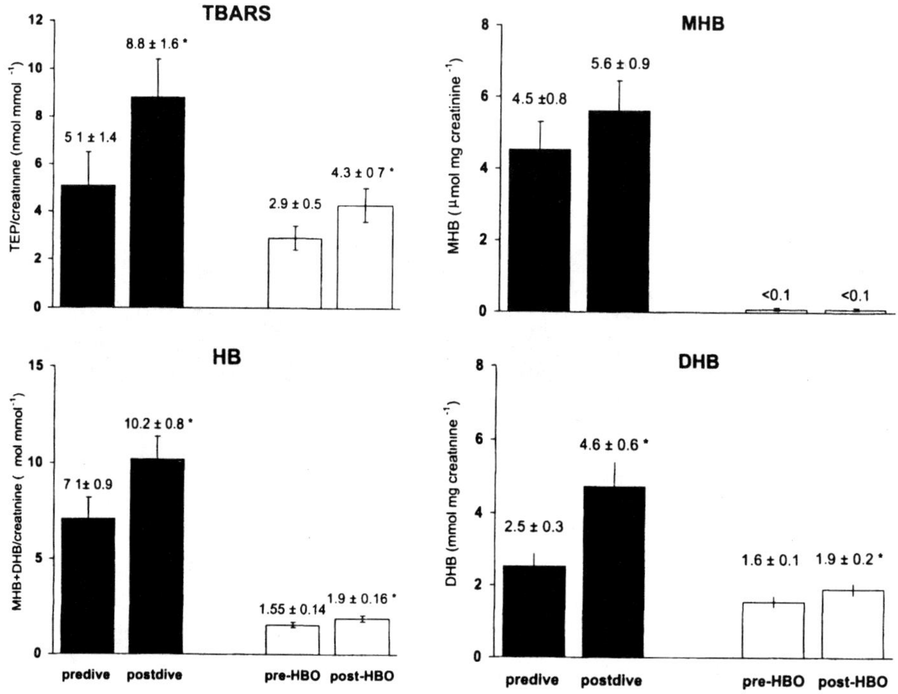

Figure 2. Urinary concentration of thiobarbituric acid-reactive substances (TBARS), hydroxylated benzoate (HB), monohydroxybenzoate (MHB), and of dihydroxybenzoate (DHB) in man breathing 100% oxygen during 45 min seawater diving (170 kPa, n = 28, black columns) or 30 min hyperbaric oxygenation (HBO, 280 kPa, n = 10, white columns). Mean ±SD

significantly elevated (Fig. 2, lower right panel, black columns). Post-dive MHB levels rose moderately, and post-dive DHB levels rose significantly by +87.8% ± 28.6%. Urinary DHB in the post-HBO group (open columns) rose by a smaller, but significant amount (+22.6% ± 10.2%). Accordingly, in the post-dive group (with the exception of salicylate), all hydroxylation sites of benzoate were present (Fig. 3, black columns). Preferred positions were (in decreasing order) 2,6- (+170%), 2,3- (+115%), 2,5- (+86%), 3,4- (+78%), 3,5- (+48%), and 2,4-DHB (+34%). Increases in 2,5-, 2,6- and 3,4-DHB were significant. In the post-HBO group, DHB fractions rose, with the exception of 2,3-DHB (-10%) (Fig. 3, open columns). Preferred hydroxylation sites were 2,4- (+147%) and 3,4-DHB (+75%). Despite large inter-individual variations, significant linear correlations could be calculated for changes in urinary concentration of TBARS and certain fractions of hydroxybenzoate (Fig. 3, right panel). In divers, 2,5-DHB and 2,6-DHB rose in accordance with increasing TBARS (upper right panel), and in the HBO group, changes in 2,4-DHB and 3.4-DHB correlated significantly with TBARS (Fig. 3, lower right panel).

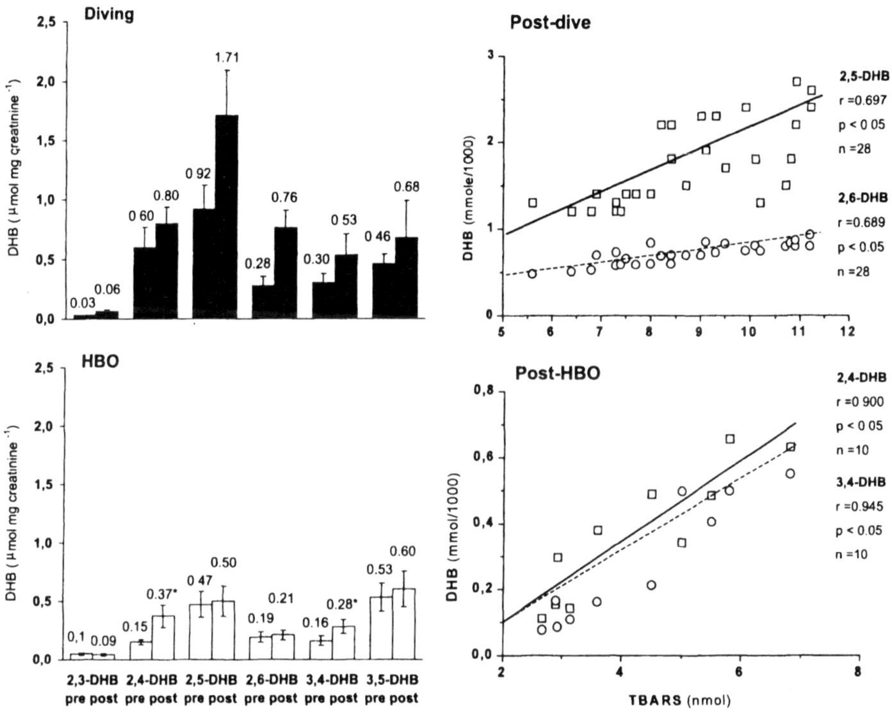

Figure 3. *Left:* Urinary concentration of fractions of 2,3-, 2,4-, 2,5-, 2,6-, 3,4-, and 3,5 dihydroxybenzoate (DHB) in man breathing 100% oxygen during 45 min seawater diving (170 kPa, n = 28, black columns) or 30 min HBO-treatment (280 kPa, n = 10, white columns). Mean ± SD

4. DISCUSSION

Tolerance of pulmonary tissue to reactive oxygen species (ROS) is limited by exposure time and applied pressure.[1, 9] In medical care, for example, 100% O_2 at 50 kPa may be safely inhaled for 24 h. The increased pressure below water surface reduces this time interval. At 300 kPa, for example, pulmonary tolerance against 100% O_2 is limited to 3.5 hours.[10] Recent experiments provide evidence that, even within this time interval, 82% O_2 at 145 kPa induced pulmonary and hematological signs of oxidative stress in man.[4] One hour, repetitive, hyperbaric oxygen treatment (HBO) was also reported to increase plasma peroxides, but urine lipid peroxidation seemed to decline.[5]

We investigated the effect of moderate hyperoxia in a shorter time period, either induced by 45 min diving (100% O_2 at 170 kPa) or by 30 min resting in a pressure chamber (HBO-treatment, 100% O_2 at 280 kPa). Parameters of oxidative stress were analyzed in the urine. Variations in urinary volume by hyperoxic vasoconstriction[11] or by cold or pressure diuresis[12, 13] were compensated by using filtered creatinine as a reference (Fig. 1, left panel). Our data provided no evidence for an altered renal excretion of creatinine during HBO treatment (Fig. 1, right panels, open columns). Thus, a renal retention of plasma peroxides, as assumed by Brichta et al.,[11] appears to be unlikely according to our HBO experiments. Instead, we found a significant increase in tested parameters of oxidative stress, i. e. in urinary thiobarbituric acid-reactive substances (TBARS) and in benzoate hydroxylation products (Fig. 2, left panel). These data are in accordance with the observation that 30 min hyperoxia also stimulated *n*-pentene excretion in man.[14] More likely, five days of repetitive HBO-exposure in the experiments of Brichta et al.[5] may have stimulated renal adaptation to oxidative stress by induction of protective enzymes such as superoxide dismutase (SOD) and catalase, as well as by an increased provision of cytoprotective ROS-scavengers.[9]

A candidate for trapping ROS in the kidney is benzoate, a xenobiotic. Benzoate is a natural constituent in human food, and it is also used as food preserver. Under normal conditions, benzoate is conjugated in the liver and kidney with glycine. The end product, hippuric acid (benzoyl-glycine), is excreted in the urine. Under conditions of oxidative stress, however, hippurate may be metabolized in the kidney to benzoate and glycine, thus providing two cytoprotective compounds: glycine to stabilize cell membranes,[15, 16] and benzoate to trap free hydroxyl radicals. It has previously been demonstrated in animal experiments that during different types of oxidative stress (i.e., post-anoxic reoxygenation, post-ischemic reperfusion, or the presence of H_2O_2) rat kidney cells released hydroxylated benzoate, at the expense of hippurate.[7, 17] Accordingly, in our experiments oxidative stress stimulated the urinary excretion of hydroxybenzoate (HB). Among tested sub-fractions, monohydroxybenzoate (MHB) rose insignificantly in divers, while the HBO group had no detectable urinary MHB-content (Fig. 2, right panel, open columns). By contrast, both groups excreted substantial amounts of dihydroxybenzoate (DHB), the divers about twice as much as the HBO group. Consistently, increased benzoate hydroxylation during diving was accompanied by an elevated oxygen consumption.

The high urinary amounts of DHB enabled us to discriminate between different sites of benzoate hydroxylation. It has been postulated for thermodynamic reasons that a preferred site of hydroxyl radical attack may be position 3- and, mediated in part by the cytochrome P_{450}-system, position 5-.[6, 18] According to this rule, mainly 2,3-DHB and 2,5-DHB should have been generated in our experiments. By contrast, diving and HBO treatment stimulated the excretion of nearly all fractions of DHB: in divers, sub-fractions (in decreasing order) 2,6 > 2,3 > 2,5 > 3,4, and, in the HBO group, sub-fractions 2,4 > 3,4 (Fig. 3, left panel). According to thermodynamic rules, absolute 2,5-DHB-excretion was high in both groups, but 2,3-DHB represented only less than 1% of total HB. Obviously, *in vivo*, not only thermodynamic rules but also hyperoxia-driven pathways generated hydroxyl radicals and directed hydroxylation to most if not all free positions of the benzoate ring.[19]

At normal PO_2, biological signals may already induce some hydroxyl radical generation via enzyme systems.[20] However, in view of the high O_2 consumption observed for divers and the demonstrated tight correlation between the appearance of urinary lipid peroxidation products (TBARS) and certain DHB-fractions (Fig. 3, right panel), it appears reasonable to assume that in our experiments the predominant driving force for benzoate hydroxylation was hyperoxia. Obviously, diving 45 min at a lower pressure (170 kPa) induced more pronounced signs of oxidative stress than resting 30 min at a higher pressure (280 kPa, Fig. 2). The higher urinary level of HB in divers probably originated from an increased oxygen consumption, amplified by additional variables such as muscular exercise and cold. A preferred hydroxylation pathway among tested HB appeared to be DHB to the expense of MHB and 2,3-DHB (Fig. 3). Further experiments are necessary to clarify whether parallel increases in certain sub-fractions of DHB and lipid peroxides indicated individual tissue responses, especially 2,6-DHB in divers and 2,4-DHB in the HBO group (Fig. 3, right panel). An increase in 2,6-DHB, for example, may have been typical for oxidative stress in the kidney.[7] It is concluded that moderate hyperoxia may stimulate in man urinary accumulation of TBARS and of hydroxylated benzoate (HB). Analyzing urinary sub-fractions of HB may provide an additional, sensitive measure to quantify and qualify oxidative stress in divers.

REFERENCES

1. J. A. Kazzaz, J. Xu, T. A. Palaia, L. Mantell, A. M. Fein, and S. Horowitz, Cellular oxygen toxicity: oxidant injury without apoptosis, *J. Biol. Chem.* **271**, 15182-15186 (1996).
2. A. Y. Shaikh, J. Yu, Y. Wu, L. He, and C. Y. Hsu, Melatonin protects bovine cerebral endothelial cells from hyperoxia induced DNA damage and death, *Neurosci. Lett.* **229**, 193-197 (1997).
3. R. L. Morton, D. Ikle, and C. W. White, Loss of lung mitochondrial aconitase activity due to hyperoxia in bronchopulmonary dysplasia in primate, *Am. J. Physiol.* **274**, L127-L134 (1998).
4. S. E. Bearden, S. N. Cheuvront, A. Ring, and E. M. Haymes, Oxidative stress during a 3,5-hour exposure to 120 kPa(a) PO_2 in human divers, *Undersea Hyper. Med.* **26**, 159-164 (1999).
5. M. Brichta, L. Hock, J. Plöse, H. Kappus, R. Beneke, and K. Behn, Lipid-peroxide related hemodilution during repetitive hyperbaric oxygenation, *Adv. Exp. Med. Biol.* **345**, 189-194 (1994)
6. P. T. Diaz, Z. W. She, W.- B. Davis, and T. L. Clanton, Hydroxylation of salicylate by the in vitro diaphragm: evidence for hydroxyl radical production during fatigue, *J. Appl. Physiol.* **75**, 540-545 (1993).
7. M. Mályusz, W. Kähler, and G. Gronow, Hippurate metabolism as a hydroxyl radical trapping mechanism in the rat kidney, *Kidney Blood Pres. Res.* **24**, 149-158 (2001).
8. H. Ohkawa, N. Ohishi, and K. Yagi, Assay for lipid peroxides in animal tissues by thiobarbituric acid reaction, *Anal. Biochem.* **95**, 351-358 (1979).
9. J. M. C. Gutteridge, and B. Halliwell, Free radicals and antioxidants in the year 2000: a historical look to the future, *Ann. NY Acad. Sci.* **899**, 136-147 (2000).
10. J. M. Clark, R. M. Jackson, C. J. Lambertsen, R. Gelfand, W. D. Hiller, and M. Unger, Pulmonary function in men after oxygen breathing at 3 ATA for 3.5 h, *J. Appl. Physiol.* **71**, 878-885 (1991).
11. K. H. Muhvich, M. R. Piano, R. A. M. Myers, J. L. Ferguson, and L. Marzella, Hyperbaric oxygenation decreases blood flow in normal and septic rats, *Undersea Biomed. Res.* **19**, 31-40 (1992).
12. J. Anderson and E. Schagatay, Arterial oxygen desaturation during apnea in humans, *Undersea Hyperb. Med.* **25**, 21-35 (1998).
13. L. E. Wittmers, Pathophysiology of cold exposure, *Minn. Med.* **84**, 30-36 (2001).
14. S. Morita, S. M. Snider, and Y. Inada, Increased n-pentene excretion in humans: a consequence of pulmonary oxygen exposure, *Anesthesiol.* **64**, 730-733 (1986).
15. J. M. Weinberg, J. A. Davis, M. Abarzua, and T. Kiani, Relationship between cell ATP and glutathione content and protection by glycine against hypoxic tubular cell injury, *J. Lab. Clin. Med.* **113**, 612-623 (1989).
16. G. Gronow, N. Klause, and M. Mályusz, Support of hypoxic renal cell volume regulation by glycine, *Adv. Exp. Med. Biol.* **277**, 705-712 (1990).
17. G. Gronow, M. Moussavian, and M. Mályusz, Effect of hydroxyl radical scavengers in renal cortical cells, *Adv. Exp. Biol. Med.* **487**, 345-351 (1999).
18. R. A. Floyd, R. Henderson, J. J. Watson, and P. K. Wong, Use of salicylate with high pressure liquid chromatography and electrochemical detection (LCED) as a sensitive measure of hydroxyl free radicals in adriamycin-treated rats, *J. Free Radic. Biol. Med.* **225**, 221-235 (1986).
19. M. Benedetti-Strolin, and K. F. Tipton, Involvement of monooxygenases and amino oxidase in hydroxyl radical generation in vivo, *Neurobiol.* **7**, 123-134 (1999).
20. M. Paolini, A. Antelli, L. Pozetti, D. Spetlova, P. Perocco, L. Valgimigli, G. F. Pedulli, and G. Cantelli-Forti, Induction of cytochrome P450-enzymes and overgeneration of oxygen radicals in betacarotene supplemented rats, *Carcinogen.* **22**, 1483-1495 (2001).

MITOCHONDRIAL NADH AS THE BELLWETHER OF TISSUE O_2 DELIVERY

Britton Chance, Shoko Nioka, Warren Warren, and Gunay Yurtsever

Abstract: It is proposed that the redox state of mitochondrial NADH[1] will complement blood gas analysis for measuring the health and welfare of human tissues. Use of arterial oxygen saturation levels (SaO_2), especially as assayed by the Nellcor instrument, has spread almost everywhere in medicine despite the fact that hypoxia of internal organs, liver, kidney, brain, pancreas, etc. is not well indicated by peripheral digital oxygenation. Indeed, there is an implied liability in the failure to infer central oxygenation from peripheral values. Near infrared (NIR) sensing of deep tissue saturation of hemoglobin (StO_2) requires multi-wavelength, multi-site measurement of both absorption and scattering properties by time or frequency domain NIR methods. Corrections for underlying water and lipid absorptions can be made so that the correct value for, and saturation oh hemoglobin are obtained.

Nevertheless, the significance of blood oxygen saturation, even localized to particular organs, can be questioned from the standpoint of what is the critical value of the desaturation from which the tissue can recover[2]; for example, in the case of cortical neurons where stroke, compression ischemia, etc. cause O_2 lack, this value becomes of significant clinical importance in both the brain and the spinal chord. These approaches are actively pursued and the possibility of subsurface redox state measurement in human tissues may eventually emerge as the quantitative metric of tissue metabolic state and of hypoxic stress.

The great flexibility and versatility of the fast, economical and "tetherless" nature of opto-electronic technology is appropriate to the manifold challenges of neuronal function as currently measured by intrinsic signals and soon to be studiable by extrinsic signals of metabolism and electrophysiological functions.

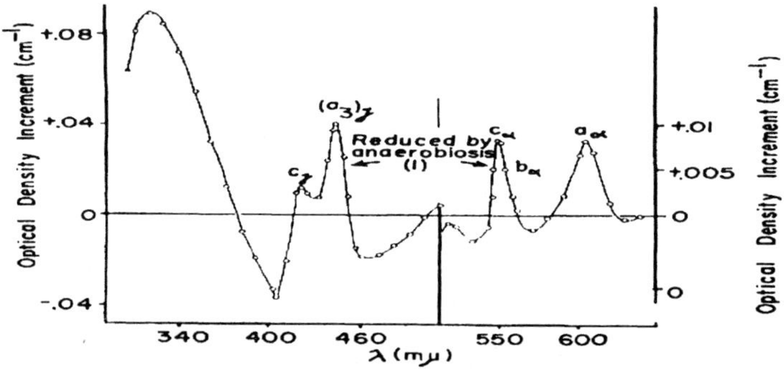

Figure 1. The first spectroscopic recordings of the oxidized-reduced components of the respiratory chain which revealed not only cytochromes in the alpha band region but also the Soret bands and, unknown to Keilin, the NADH peak and flavin trough.

1. INTRODUCTION

The invention of the dual-wavelength spectrophotometer[3] permitted delineation of the cytochromes of the highly scattering suspension of intact mitochondria as never before. The salient features of the oxidized minus reduced difference spectra[4] corroborated what the spectroscope and Keilin's keen eyes saw, except for the flavin trough and the NADH peak that were quite novel. The functionality of neither of these could have been studied, as indeed Keilin had already verified for the cytochromes.[5] He vividly portrayed the disappearance and reappearance of cytochrome bands in yeast cell suspensions as observed with the microspectroscope as evidence for their functionality. But, surprisingly, Keilin, the cell biologist, overlooked the fact that the biologist's "cell autofluorescence" also changed in the yeast, as did the cytochrome absorption bands. The dual wavelength spectrometer linked NADH and flavin absorption and fluorescence to the respiratory chain and were the subject of intense experimentation in the Chance laboratory from 1955 onwards[5, 6] (Figure 1).

Transference of optical techniques from isolated mitochondrial preparations to tissues[7] was to be the focus of the Chance laboratory activity for the next several decades. The first and foremost studies were "function activational," where the absorption and fluorescence changes of NADH were observed on stimulation. The activation of muscle function caused a rise of ADP and PI concentrations and oxidation of NADH together with changes in other carriers of the respiratory chain, "crossovers."[8] The spectroscopic study of Lardy and Wellman's tightly coupled guinea pig mitochondrial liver preparations showed the four stages of metabolic activity[9] (endogenous, starved, fully active, and hypoxic). These "states" formed the basis for many studies of functional activity of isolated mitochondria and mitochondria in many intact tissues: liver, brain, kidney, adrenal, skeletal muscle, etc.[10, 11, 12] and served as markers of metabolic intensity and patency of energy coupling.

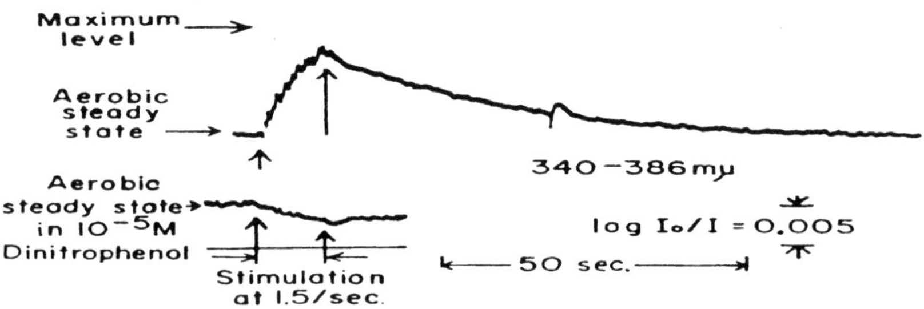

Figure 2. Functional activation of mitochondrial NADH oxidation in stimulation of sartorius muscle and of glycolytic NADH reduction in uncoupled mitochondria.

2. FUNCTIONAL ACTIVATION

The first experiment to demonstrate the functionality of the novel "pyridine nucleotide" absorption band was carried out with Clancy Connelly on strips of the frog sartorius muscle bathed in Ringers in a special holder and electrically stimulatable in the dual wavelength light beams.[13] The observation of a several percent decrease of absorption due to NADH oxidation in muscle mitochondria on stimulating the frog sartorius muscle (Figure 2) was, to Chance and Connelly, just as exciting as Keilin's microspectroscopic observations on the disappearance of the absorption bands of cytochromes on oxygenation of yeast.

The ADP activation of muscle mitochondria was the focus of muscle biochemistry due to the activities of R. E. Davis, Mommaerts, Fleckenstein, and Kushmeric.[14, 15] Furthermore, a test that was not available to David Keilin, namely uncoupling with dinitrophenol, afforded a typical control of muscle studies. The oxidative response of NADH to stimulation was not only ablated by such an uncoupling agent, and, in fact, just the opposite occurred: a small reduction of NAD occurred. This was later attributed to glycolytic activity based on studies of the mainly glycolytic tissue electroplax of electrophorus electricus.[16]

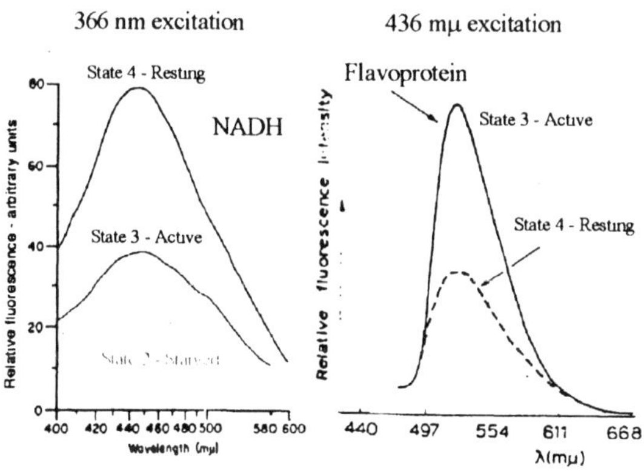

Figure 3. The fluorescence emission spectra of NADH and flavoprotein using 366 and 436 nm excitation wavelengths, respectively, with responses of these components of the respiratory chain to metabolic activation where NADH is oxidized and flavoprotein is reduced permitting calculation of the redox ratio.

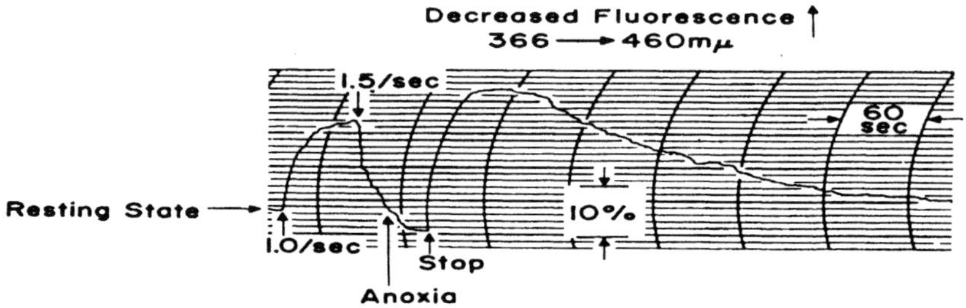

Figure 4. The use of NADH fluorescence to quantify functional activity in the frog's *sartorius* muscle showing oxidation on functional activation as the fluorescent signal of NADH decreases.

Chance and Baltscheffsky[17] followed up Warburg's observation that the pyridine nucleotides were fluorescent and found that, indeed, mitochondria not only showed a strong fluorescence emission band at 450 nm, but also that this fluorescence emission band was linked to changes of metabolic activity observed in the cytochrome chain with the dual wavelength spectrophotometer, tying the NADH pool (the major absorption band of the mitochondria) closely to cytochrome function (Figure 3).

Armed with the highly sensitive fluorescent label of mitochondrial function in tissues, nearly all available tissues were studied, even the salt gland of the herring gull as an example of a high NADH, high mitochondrial tissue as contrasted with the electric organ of electrophorus, which was characterized by a paucity of mitochondria and glycolytic function.[17]

The fluorometric method turned out to be extremely convenient and sensitive, not requiring tissue transmission, being able to be measured by reflectance or "remission," and thus affording a veritable "window" into metabolic activity of cells and tissues, particularly in the brain. The traces of Figure 2 are duplicated with a high fidelity using NADH fluorescence[18] (Figure 4).

3. CRITICAL OXYGEN TENSION AND O_2 GRADIENTS

The NADH fluorescence signal served as a primary key to mitochondrial function and to tissue oxygen gradients in the brain, which has persistently resisted valid studies of the copper component of cytochrome oxidase due to oxyhemoglobin interference. The mitochondrial NADH response to oxygen can be compared with that of hemoglobin in the capillary bed as measured optically in the visible and the NIR regions, as well as the functional activity in brain by EEG. Thus, for the first time and most importantly, it was possible to quantify the oft-postulated oxygen gradient between hemoglobin deoxygenation and NAD reduction as measured fluorometrically. We found that hemoglobin can be deoxygenated to over 80% before the NADH reduction begins and energy coupling is compromised[19] (Figure 5). Furthermore, the electrical activity measured with monopolar EEG[20] (Figure 6) terminated not at the start of NAD reduction, but beyond the midpoint, corresponding to 85% reduction of NAD showing how energy coupling is maintained in hypoxic states through the rapidity of the cytochrome oxidase-oxygen reaction, together with the rapid electron transport through the chain to maintain oxygenation of NADH and the energy coupling in the mitochondrial membranes at low O_2 concentration.

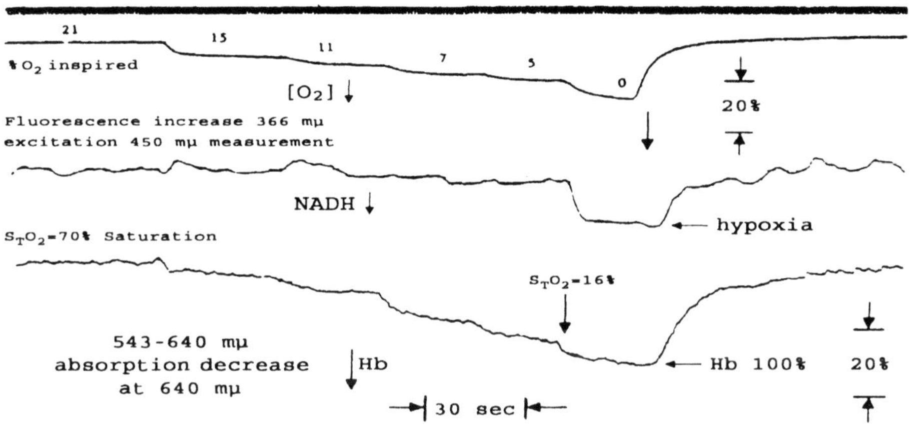

Figure 5. The resistance of the mitochondrial redox state to severe hypoxia. The middle trace does not show a significant change above the noise level until the FIO_2 is below 5% and the hemoglobin has been desaturated to only 16% based upon the normoxic saturation of 70%.

Not satisfied with the development of fluorometric technology for measuring NADH in cells and tissues, attention was paid to the flavoprotein signal, well-known for the fluorescence of riboflavin, but in this case highly selective for the α-ketoglutarate and pyruvate dehydrogenases, key enzymes of Krebs' citric acid cycle. Here, to our great delight, it was the oxidized form that was fluorescent and the reduced form non-fluorescent. Thus, for the first time the ratio of flavoprotein to NADH signals gave a direct readout of the redox state of the mitochondria, a quantity which was to embellish our studies of metabolic control in normal cells and tissues[7] and to shed a new and important light on cancer cell metabolism.[21] In the latter case, contrary to Warburg's idea of respiratory deficiency, our studies of many lines of Ascites cells (in collaboration with Hauska) in the early 1950s belied the Warburg-Burk hypothesis of respiratory deficiency and replaced it with the more flexible and likely to be valid hypothesis of inadequate tremor of oxygen to the tumor mitochondria,[22] which, perhaps due to the chaotic nature

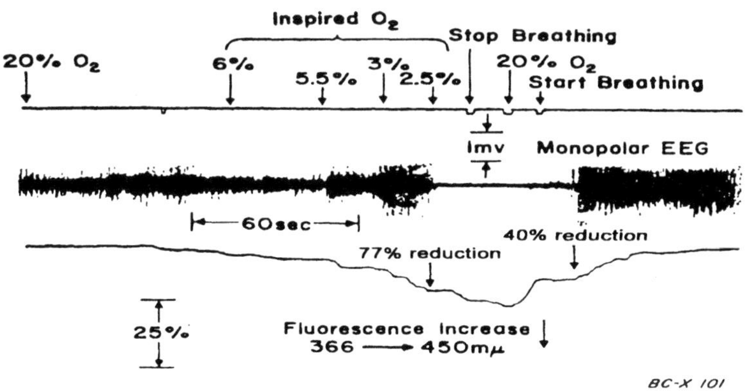

Figure 6. Correlation of NAD reduction with functional activity in the rat brain showing that the electrical activity of the brain ceases when NADH is reduced by 77% and is reactivated when the reduction is 44%.

of the vascular bed (extreme angiogenesis), failed to deliver adequate oxygen to the tumor mitochondria. The ratio of flavoprotein and NADH fluorescence also afforded for the first time a measure of mitochondrial redox state independent of the number of mitochondria involved and was found to be relatively insensitive to changes in hemoglobin,[1] which is a great boon to many studies, particularly cancer, as mentioned previously.

The previous figures suggest a high O_2 affinity of the NADH signal. The use of luminous bacteria to detect the oxygen affinity of the cytochrome chain was exploited by my student, Fred Schindler, who perfected the method by using liter volumes of suspended yeast cells or mitochondria to which microliter additions of aerated buffer were added to precisely calibrate the method. The luminosity of the luminous bacteria followed the decrease of oxygen in respiring mitochondria, as shown in Figure 7, so that the cytochrome oxidase reaction continued until the oxygen concentration was less than 10^{-10} molar oxygen.[23] In later studies with the late and lamented Carlos Saronio, the explanation was provided that cytochrome oxidase transfers a pair of electrons to oxygen immediately upon reaction, forming a ferryl ion, making it a fast, essentially irreversible process.[24]

It turns out that the reaction of most hemoproteins with ligands with oxygen and indeed peroxide is an irreversible reaction involving very rapid electron transfer to the substrate, usually forming a peroxi-intermediate, a ferryl ion. Strange as it may seem, most recent data from Takashi Yonetani (personal communication) suggest that ligation of hemoglobin with oxygen involves structural changes that, while not involving a higher oxidation state of the ion, involve a different structure of the protein. This suggests that biology has created enzymatic structures that achieve very rapid velocity constants for the reaction with ligands that are essentially irreversible reactions, so that the reverse reaction does not deteriorate the ligand binding efficiency. This is illustrated in Figure 8, where an essentially irreversible reaction of cytochrome a_3 and oxygen is shown (i.e. no back reaction) so that the steady state depends upon the oxygen concentration and the Michaelis-Menten affinity depends upon the turnover number (k3) and not upon the

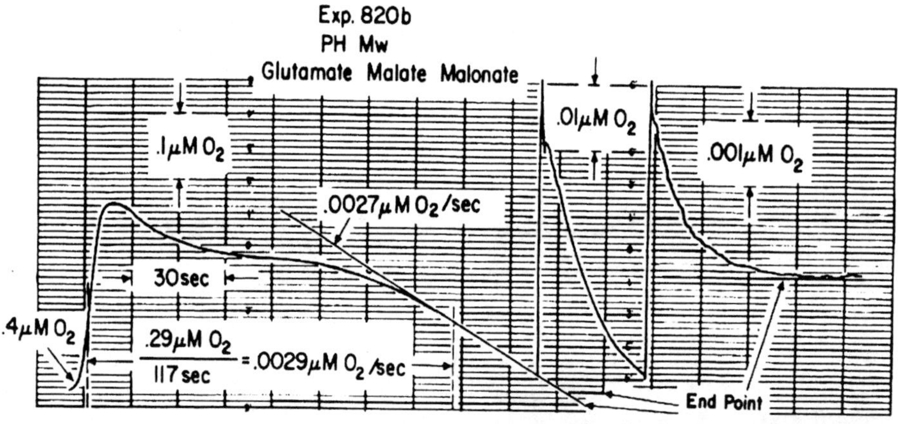

Figure 7. Schindler's remarkable diagram of the use of bacterial luminescence to record the kinetics of oxygenization with a linear range 0.1 μmolar oxygen as measured by independent calibration of a precisely known volume of oxygen added to one liter of a mitochondrial suspension.

Basic Chemistry

$$O_2 + a_3^{2+} \xrightarrow{k_1} a_3^{3+} + O_2^- \quad (1)$$

$$c^{2+} + a_3^{3+} \xrightarrow{k_3} a_3^{2+} + c^{3+} \quad (2)$$

$$\text{sum} \quad O_2 + c^{2+} \longrightarrow c^{3+} + O_2^- \quad (3)$$

$$\frac{da_3^{3+}}{dt} = k_1 [O_2] a_3^{2+} - k_3 c^{2+} a_3^{3+} \quad (4)$$

For Steady State:

$$k_1 [O_2] a_3^{2+} = k_3 c^{2+} a_3^{3+} \quad (5)$$

$$\frac{a_3^{3+}}{a_3^{2+}} = \frac{k_1}{k_3} \frac{[O_2]}{c^{2+}} \quad (6)$$

The redox state of a_3 depends upon O_2

The velocity relationship follows for Michaelis Menten kinetics:

$$\frac{V}{V_M} = \frac{1}{1 + \frac{k_3}{k_1 [O_2]}} \quad (7)$$

$$\text{where } K_M = \frac{k_3}{k_1} \text{ and } k_3' = k_3 c^{2+} \quad (8)$$

for $k_3 = 10\ \text{sec}^{-1}$ and $k_1 = 10^8\ M^{-1} \times \text{sec}^{-1}$

$$K_M = 10^{-7}\ M\ O_2 \quad (9)$$

Figure 8. The chemical equation for a very simplified reaction of cytochrome oxidase and O_2 where no reversibility is shown, but instead a steady state flux at 10 per second gives an effective Km 10^{-7} molar oxygen where 10 per second is a middle respiration rate.[25]

reversible reaction of cytochrome oxidase and oxygen. This further extension of the general observation was made in studies of peroxidase with H_2O_2 where indeed computer simulation of the enzyme mechanism, which did not require the Michaelis-Menten dissociation of the enzyme substrate compound, fitted the kinetics to a high degree of accuracy over the whole course from the initial reaction of enzyme with substrate through its turnover number to its release after a cycle of enzyme activity.[26] Thus, irreversibility seems to be the rule in this class of hemoproteins.

4. THE OVERLAP OF MYOGLOBIN AND HEMOGLOBIN SPECTRA

In experiments on the cat soleus carried out by Glenn Millikan, a simple colorimeter was used to transmit light of two colors through the isolated soleus muscle in the cat.[27] Glenn made the first kinetic recordings of the function of myoglobin in stimulated muscle, showing its rapid deoxygenation and reoxygenation. In that experiment, and in all subsequent experiments, the distinction between myoglobin and hemoglobin was not possible optically (except for a small difference in the visible region exploited by Clyde Barlow[28]). However, in order to be certain of the myoglobin signal as compared to hemoglobin, the proton signal of myoglobin has been exploited in Jack Leigh's laboratory[29] and, more recently, in Tom Jue's laboratory.[30] However, the combined measurement of absorbency of hemoglobin plus myoglobin and the NMR measurement of myoglobin alone has been accomplished and gives the striking result of Figure 9 (unpublished data), where the NIR data show continued deoxygenation with stimulation of the muscle at 5 per second. The progressive changes of the NIR signal proceeds with no change of the NMR proton spectrum of myoglobin until the animal has been caused to breathe pure nitrogen, causing immediate response of NMR signal of myoglobin and of

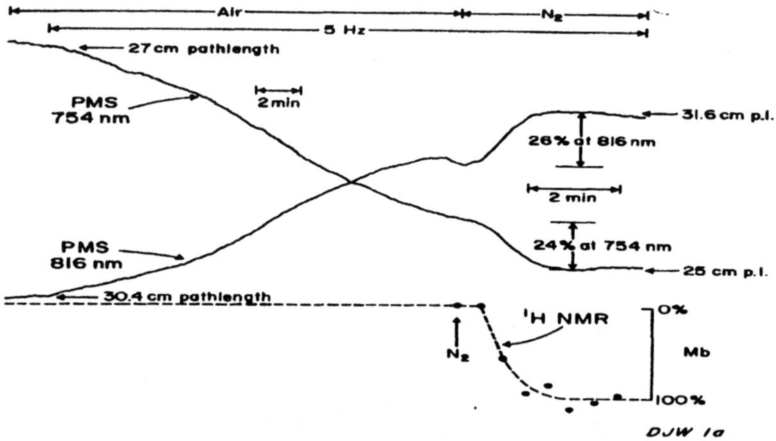

Figure 9. Correlation of frequency of 100 MHz NIR (PMS) and ¹H NMR of Hb (PMS) and Mb ¹HNNR kinetics.

the NIR traces as well. Thus, this particular preparation was able to maintain myoglobin completely oxygenated throughout the period of stimulation until N_2 hypoxia occurred, leading to the general observation that myoglobin is oxygenated during most of the stimulation interval, and can therefore function effectively for any shuttle phenomenon that is postulated (if it were deoxygenated, of course, shuttling would serve no purpose). This demonstrates two clear conclusions: (1) that myoglobin can indeed act as a store against hypoxic stress; (2) that if myoglobin has a shuttle capacity, then it is in the oxygenation state, and therefore effective to carry out this function.

On this basis, the profiles for oxygen utilization in muscle and brain take the form shown in Figure 10, where hemoglobin with a K_l of approximately 30 µmolar, myoglobin with a K_l of approximately 4 µmolar, and cytochrome oxidase with a K_m of 10^{-7} molar or less, spaces the affinities as shown for the resting or low activity state as in Figure 10. However, with the higher activation of electron transport, the tissue O_2 gradients, as in the perfused heart, coalesce these profiles so that the apparent affinity of hemoglobin, myoglobin, and cytochrome oxidase becomes very similar, yet hemoglobin

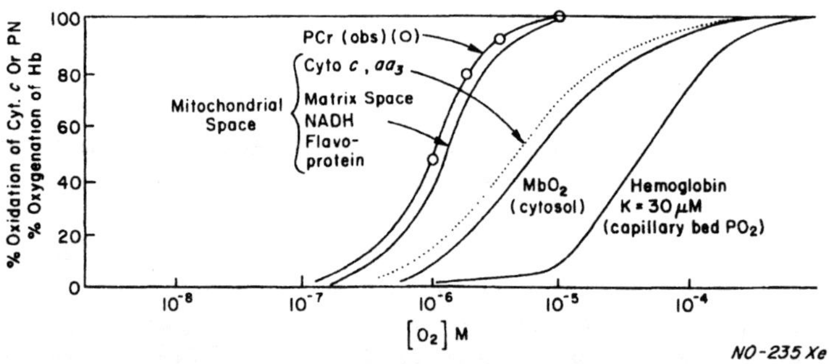

Figure 10. The resting state profiles for hemoglobin, myoglobin, and cytochrome at a value of 10-20. This indicates well the profiles for PCr decrease as a result of mitochondrial insufficiencies.[31]

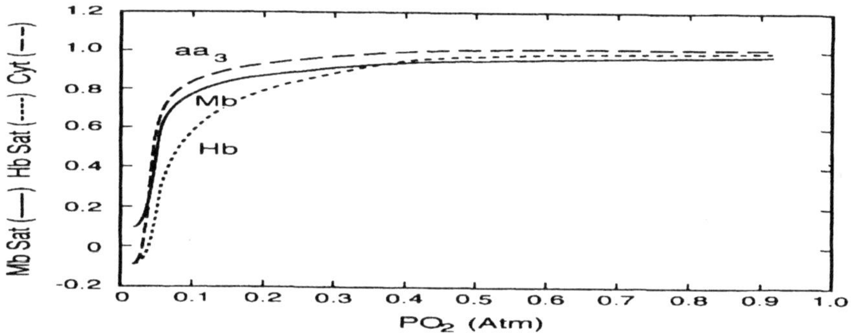

Figure 11. The effect upon tissue-oxygen gradients of high reparation rates upon the profiles for hemoglobin, myoglobin, and cytochrome are coalesced so that the difference of oxygen affinities is now largely subsumed by the oxygen gradient between the hemoglobin, myoglobin, and cytochrome oxidase.

can be somewhat deoxygenated before myoglobin[32] (Figure 11). These profiles suggest indeed that myoglobin will be kept oxygenated over most of the course of hemoglobin deoxygenation and cytochrome will be maintained oxidized until most of the myoglobin is deoxygenated. Activation of high levels of electron transport, as in the heart cardiac tissue, will coalesce these gradients at high respiration rates. This could be evidenced by stimulation of the dog leg in the above experiment at a high rather than low rate, that indeed one could expect some myoglobin deoxygenation when hemoglobin is largely deoxygenated (Figure 9).

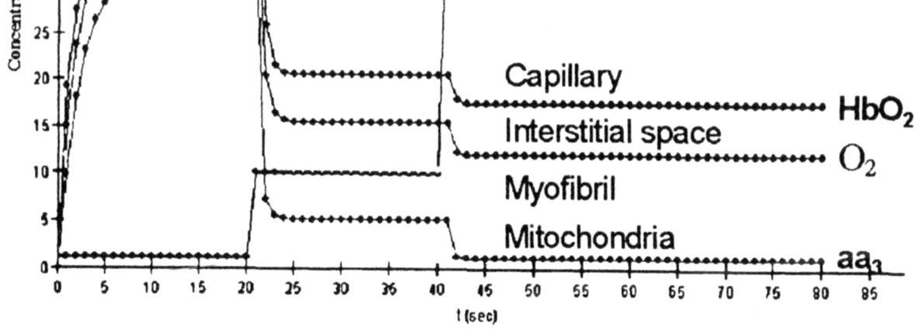

Figure 12. Simulation of the distribution of oxygen concentrations in a four compartment model for three different cellular respiration rates VO_2 2→10→50 μmolar/sec.

5. SIMULATION

The data available suggest a simulation of a four-compartment model representing (1) the capillary bed, (2) the extracellular space, (3) the myofibrils containing myoglobin, and (4) the mitochondrial space, and finally, the cytochrome oxidase activity, each depending upon the oxygen gradients and the oxygen extraction effectiveness (Figure 12). This model gives a qualitative picture of the relative response in the various compartments and can clearly separate the myoglobin and cytochrome oxidase responses.[33]

6. THE FUTURE

The fact that NADH communicates between the citric acid cycle and the mitochondrial compartments gives it a special role in indicating whether or not there is enough available NAD to interact with the citric acid cycle enzymes and allow the cycle to function, providing the reducing equivalents necessary for energy conservation in the respiratory chain. Thus, NAD/NADH is in a unique position to determine the effectiveness of oxidative metabolism. However, the absorption band at 340-350 nm for excitation and 450 nm for emission renders non-invasive optical methods ineffective in studying this component in the internal organs. In order to circumvent this difficulty, two approaches are under consideration.

The first approach is measurement of fluorescence of NADH and flavoprotein in internal tissues or body cavities: for example, the cheek pouch. Since modern optoelectronics affords miniaturization of bulky components, we have designed and are ready to construct a "cheek pouch fluorometer" (Figure 13) with telemetered output that would indicate the metabolic redox state of the internal tissues and therefore could represent the general state of oxygenation of body tissues in terms of the flavoprotein-to-pyridine nucleotide ratio. And with the use of LEDs that are emitting in the near-ultraviolet (360 nm and 450 nm), excitation of NADH and flavoprotein fluorescence can readily be obtained in extremely small-sized light sources. Furthermore, with silicon diode detectors of the time-shared FP/PN signals, it is readily possible to give an amplified output suitable for telemetry to a nearby receiver, remotely indicating the redox ratio of the untethered body tissue.

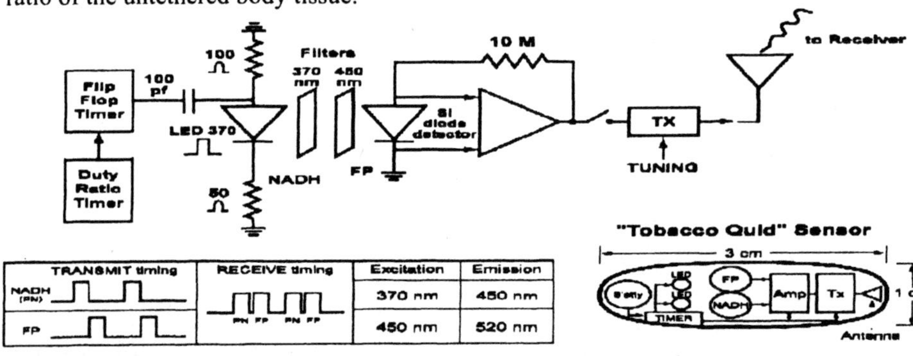

Figure 13. Illustrating one possibility to measure flavoprotein and NADH florescence in the human subject by creation of a buccal pouch fluorometer to measure redox ratio. The device is expected to be 3 cm by 1 cm with a radio transmission out to a 1 m distance to a recorder using time shared 370 and 450 nm excitation for the measurement of fluorescent signals through suitable secondary filters.

Alternatively, penetration of the skin may be possible with a unique concept for measuring fluorescence with our ultrafast pulse-shaping technique developed in W. Warren's laboratory,[34] in which advantage is taken of the fact that very high peak power chirped pulses evoke non-linear optical effects, causing frequency doubling of the excitation light from 720 to 360 nm and exciting fluorescence deep within the tissue due to the propagation of the 720 nm light, providing that the pulse is appropriately modulated; at the present time, the depth is estimated to be almost 1 mm with possibilities for deeper penetration, as is well known from the NIR studies.

The problem is, however, only half way solved, because the emission of NADH at 450 nm will be readily absorbed by the tissues, which incapacitates the method. Therefore, instead of measuring the fluorescence emission, the absorption of the excitation at 720 nm is proposed. While this is small, due to the large concentration of NADH in mitochondria, a small increment of absorption is to be expected when NADH in subsurface mitochondria of tissues is studied.[3] Furthermore, the method is directly applicable to flavoprotein, where the excitation would be 900 nm with corresponding changes in the absorption of the excitation wavelength. We expect, however, that, due to the smaller concentration of flavin and its somewhat smaller absorption band (Figure 1), this will be more difficult. Nevertheless, the goal is to measure the FP/PN ratio deep within the tissue directly by multiphoton technology. It is furthermore expected that the method will exploit dual wavelength technology as was used in the frog muscle studies of Figure 2 (i.e. 374 nm was used as a suitable reference wavelength for NADH and 540 nm as a suitable reference for flavoprotein).

In sum, the study of mitochondrial fluorescent compounds has been most fruitful. The signals are large, their correlation with hemoglobin and with functional activity is readily demonstrated, and the key role of NADH and flavoprotein in coupling between the citric acid cycle and the mitochondrial respiratory chain is a clear and strong one.

REFERENCES

1. B. Chance, B. Schoener, R. Oshino, F. Itshak, and Y. Nakase, Oxidation-reduction ratio studies of mitochondria in freeze-trapped samples, *J. Biol. Chem.* **254**, 4764-4771 (1979).
2. S. E. Nicklin, I. A. A. Hassan, Y. A. Wickramasinghe, and S. A. Spoencer, The light still shines, but not that brightly? The current status of perinatal near infrared spectroscopy, *Arch. Dis. Child Fetal. Neonatal* **88**, F263-F268 (2003).
3. B. Chance, Rapid and sensitive spectrophotometry. III. A double beam apparatus, *Rev. Sci. Ins.* **22**(8), 634-638 (1951).
4. B. Chance, and G. R. Williams, Respiratory enzymes in oxidative phosphorylation. II. Difference spectra, *J. Biol. Chem.* **217**(1), 395-407 (1955).
5. M. Dixon, R. Hill, and D. Keilin, The Absorption Spectrum of the Component c of Cytochrome. Proceedings of the Royal Society of London. Series B, Containing Papers of a Biological Character, Vol. 109, No. 760. (Sep. 1, 1931), pp. 29-34.
6. B. Chance, Spectrophotometric and kinetic studies of flavoproteins in tissues, cell suspensions, mitochondria and their fragments, in: *Flavins and Flavoproteins*, edited by E.C. Slater (Elsevier, New York, 1996) pp. 496-510.
7. B. Chance, Optical method, *Ann. Rev. Biophys. Biophysical Chem.* **20**, 1-30 (1991).
8. B. Chance, and G. R. Williams, Respiratory enzymes in oxidative phosphorylation. II. The steady state, *J. Biol. Chem.* **217**(1), 409-427 (1955).
9. H. A. Lardy, and H. Wellman, Oxidative phosphorylations – role of inorganic phosphate and acceptor systems in control of metabolic rates, *J. Biol. Chem.* **195**(1), 215-224 (1952).
10. B. Chance, F. Jobsis, B. Schoener, and P. Cohen, Intracellular oxidation –reduction states in vivo, *Science* **137**(3529), 499-508 (1962).

11. B. Chance, B. Schoener, and V. Legallais, Metabolically linked changes in fluorescence emission spectra of cortex of rat brain, kidney and adrenal gland, *Nature* **195**(4846), 1073-1076 (1962).
12. B. Chance, and B. Schoener, A correlation of absorption and fluorescence changes in ischemia of rat liver in vivo, *Biochemische Zeitschrift* **341**(4), 340 (1965).
13. B. Chance, and C. M. Connelly, Method for the estimation of the increase in concentration of adenosine diphosphate in muscle sarcosomes following a contraction, *Nature* **179**(4572), 1235-1237 (1957).
14. W. F. H. M. Momaerts, The biochemistry of muscle, *Ann. Rev. Biochem.* **23**, 381-404 (1954).
15. A. Fleckenstein, J. Janke, R. E. Davies, and H. A. Krebs, Chemistry of muscle contraction-contraction of muscle without fission of adenosine triphosphate or creatine phosphate, *Nature* **174**(4441), 1081-1083 (1954).
16. X. Aubert, B. Chance, and R. D. Keynes, Optical studies of biochemical events in electric organ of electrophorus, Proceedings of the Royal Society of London Series B- Biological Sciences, **160**(979), 211 (1964).
17, B. Chance, and B. Baltscheffsky, Respiratory enzymes in oxidative phosphorylation. Binding of intramitochondrial reduced pyridine nucleotide, *J. Biol. Chem.* **233**(3), 736-739 (1958).
18. B. Chance, Continuous recording of intracellular reduced pyridine nucleotide changes in skeletal muscle in vivo, *Texas Reports on Biology and Medicine* **22**(1), 836-841 (1964).
19. B. Chance, B. Schoener, and F. Schindler, The Intracellular Oxidation-Reduction State, Oxygen in the Animal Organism: *Proceedings of a Symposium in London*, edited by Dickens and Niel (Pergamon Press, London 1964).
20. B. Chance, and B. Schroener, Correlation of oxidation-reduction chances of intracellular reduced pyridine nucleotide and chances in electroencephalogram of rat in anoxia, *Nature* **195**(4845), 956-958 (1962).
21. N. Ramanujam, R. Richards-Kortum, S. Thomsen, A. Mahadevan-Jansen, M. Follen, and B. Chance, Low temperature fluorescence imaging of freeze-trapped human cervical tissues, *Optics Express* **8**(6), 335-343 (2000).
22. B. Chance, and L. N. Castor, Some patterns of the respiratory pigments of ascites tumors of mice, *Science* **116**(3008), 200-202 (1955).
23. F. J. Schindler, Oxygen kinetics in the cytochrome oxidase-oxygen reaction, Dissertations: University of Pennsylvania, 115-118 (1964).
24. C. Saronio, Initial phases in reaction of cytochrome-oxidase with oxygen at very low temperatures in intact mitochondria, Federation Proceedings **5**, 331289-331289 (1974).
25. B. Chance, G. R. Williams, W. F. Holmes, and J. Higgins, Respiratory enzymes in oxidative phosphorylation. V. A mechanism for oxidative phosphorylation, *J. Biol. Chem.* **217**, 439-452 (1955).
26. B. Chance, The kinetics of the enzyme-substrate compound of peroxidase, *J. Biol. Chem.* **151**, 553–577 (1943).
27. G. A. Millikan, Experiments on muscle haemoglobin in vivo; The instantaneous measurement of muscle metabolism, *Proceedings of the Royal Society of London.* Series B, Biological Sciences, **123**(831), 218-241 (1937).
28. C. H. Barlow, K. A. Kelly, and K. J. Kelly, Simultaneous determination of hemoglobin and myoglobin oxygen binding curves by spectral curve fitting, *Appl. Spectrosc.* **46**(5), 758-763 (1992).
29. E. A. Noyszewski, E. L. Chen, R. Reddy, Z. Y. Wang, and J. S. Leigh, A simplified sequence for observing deoxymyoglobin signals in vivo: Myoglobin excitation with dynamic unexcitation and saturation of water and fat, *Magn. Res. Med.* **38**(5), 788-792 (1997).
30. T. K. Tran, N. Sailasuta, R. Hurd, and T. Jue, Spatial distribution of deoxymyoglobin in human muscle: an index of local tissue oxygenation NMR, *Biomedicine*, **12**(1), 26-30 (1999).
31. B. Quistorff, B. Chance, and A. Hunding, An experimental model of the Krogh tissue cylinder: Two-dimensional quantitation of the oxygen gradient, in: *Oxygen Transport to Tissue-III*, edited by I. A. Silver (Plenum Press, New York, 1978), pp. 127-133.
32. J. J. Kelly, K. A. Kelly, S. A. Hartley, and C. H. Barlow, Myoglobin oxygen binding curves determined by phosphorescence quenching of palladium porphyrin, *Appl. Spectrosc.* **45**(7), 1177-1182 (1991).
33. P. Huang, B. Chance, X. Wang, R. Kime, S. Nioka, and E. M. Chance, Modeling of oxygen diffusion and metabolism from capillary to muscle, *Adv. Exp. Med. Biol.* **540**, 325-330 (2002).
34. H. S. Tan, E. Schreiber, and W. S. Warren, High-resolution indirect pulse shaping by parametric transfer, *Optics Letters*, **27**(6), 439-441 (2002).

PROCESS SCALE-UP STUDIES FOR PROTEIN C SEPARATION USING IMAC

Huiping Wu and Duane F. Bruley

Abstract: Protein C (PC) is the pivotal anticoagulant and antithrombotic in the human coagulation cascade. PC deficiency can disturb the blood hemostasis and cause thrombosis, inhibiting oxygen transport to tissue, and resulting in major medical problems such as deep vein thrombosis (DVT). The current treatment can cause bleeding and other major medical problems. PC circulates in the blood as a zymogen and is only activated when and where it is needed. PC is a safe anticoagulant without harmful side effects.

A combination of ion-exchange chromatography and IMAC IDA-Cu was studied for the relatively large scaled PC separation from Cohn fraction IV-1. Almost half of the active PC was recovered by using this process. In future work, we will verify the linearity of the IMAC column scale-up. This process can be used to produce PC from Cohn fraction IV-1 at large quantities and low cost to treat PC-deficient patients.

1. INTRODUCTION

Protein C (PC) is the pivotal anticoagulant and antithrombotic in the human blood coagulation cascade. It is a member of the homologous vitamin K-dependent (VKD) family, also consisting of coagulation protein factors VII, IX, X, proteins M, Z, and prothrombin. PC is a trace protein at a normal concentration of 4 μg/ml in human blood; however, serious problems can occur when the PC level in the blood is lowered. Patients deficient in PC are at risk of deep vein thrombosis (DVT) and other clotting complications. When oxygen transfer to the tissue is reduced or stopped, irreversible damage can occur leading to extremity amputation, organ failure, and death.

At the present time, heparin and coumadin are used to treat PC deficiency and other hypercoaguable conditions. The disadvantage of these drugs is that both have dangerous side effects, including excessive internal bleeding, stroke, or major organ failure.

Previous animal testing and clinical trials indicate that PC is an effective anticoagulant/antithrombotic for many medical indications without harmful side effects. When considering that the PC-deficient population is 1 in 300 people, the benefit of having inexpensive PC available to medical patients is enormous. PC can be obtained from human blood plasma. Because of the complexity of the source material, immunoaffinity chromatography[1] is used for the separation. However, because of the high cost of the monoclonal antibodies, this technology is very expensive. A new separation technology should be investigated for low cost PC production.

Immobilized metal affinity chromatography (IMAC) utilizes protein metal binding properties for separation. IMAC is dependent upon the presence of adsorption centers created in a hydrophilic stationary support.[2] The adsorption centers consist of two components, a chelator and a metal ion. The chelator, such as iminodiacetic acid (IDA), is linked to a matrix via a spacer arm by covalent bonds. The metal ion, usually the borderline metal ion, such as Cu^{2+}, Ni^{2+}, Zn^{2+}, or Co^{2+}, is immobilized to the chelator by coordinate bonds.[3] The histidine, cysteine, and tryptophan of a protein have metal binding affinity. Protein can be released from the support gel by reducing the binding affinity between the immobilized metal ion and the adsorbed protein, such as changing pH, and it can be displaced by a competitive agent, which is similar to the amino acid residue involved in metal binding.

IMAC is inexpensive and operates under mild conditions so that proteins do not denature.[4] Also, there is no immunogenic contamination, as compared with immuno-affinity chromatography. IMAC has been used to study PC separation from human blood plasma derivatives,[5-7] and IMAC has been used successfully to separate two homologous vitamin K-dependent blood proteins.[8] Therefore, a process with an IMAC element should be able to achieve low cost PC cocktail production.

2. MATERIALS AND METHODS

2.1. Materials

- Cohn fraction IV-1 paste was supplied by the American Red Cross (ARC) Plasma Derivatives Laboratory (Rockville, MD). It contained ethanol (between 2%-5%) and trace amounts of acetate and citrate ions (from an original plasma pool). The main protein components were albumin, alpha-1 antitrypsin, alpha-2 macroglobulin, antithrombin III, ceruloplasmin, transferrin, prothrombin, factor IX, factor X, haptoglobulin, apo A-1, alpha-1 lipoprotein, GC-globulins, alpha-1 acid glycoprotein, and PC. PC was at approximately 100 mg per kg of paste.
- Protac activator for PC activation was obtained from American Diagnostica Inc. (Hauppauge, NY).
- S-2366 chromogenic substrate for enzymatic activity assay of activated PC was obtained from Kabi Diagnostica (Franklin, OH).
- Fresh, frozen plasma that was used as a PC standard for PC activity assay and a protein C-albumin mixture was supplied by the ARC.
- All reagents were obtained from Sigma (St. Louis, MO) or Aldrich (Milwaukee, WI), unless otherwise specified.
- DEAE Sepharose fast flow and HiTrap chelating Sepharose columns were obtained from Pharmacia Biotech (Piscataway, NJ).

- Immulon II 96 well flat-bottomed microtiter plates were obtained from Dynatech Laboratories Inc. (Chantilly, VA), which were used for PC amidolytic activity assays.
- A Pharmacia assembly of equipment from Pharmacia (Piscataway, NJ) was used to perform chromatography separation of proteins.
- An EL 340 Biokinetics Reader obtained from Biotech Instruments (Winooski, VT) was used for PC amidolytic activity assay optical readings.
- A HP 8452A diode array spectrophotometer from Hewlett Packard (Palo Alto, CA) was used for protein UV absorption ($A_{280\ nm}$-$A_{320\ nm}$) measurements.

2.2. Methods

2.2.1. *DEAE Chromatography of Cohn Fraction IV-1*

The Cohn fraction IV-1 paste was dissolved in sodium citrate buffer by constant stirring with a magnetic stirrer. After centrifugation at 8000 rpm for 15 minutes, the supernatant was loaded onto the DEAE column, which had been equilibrated with equilibration buffer. The column was then washed with the same buffer to remove as many contaminants as possible. Then the column was washed with a wash buffer to remove lightly bound contaminants. Finally, the bound proteins were eluted with an elution buffer.

2.2.2. *DEAE Eluate Chromatography Using IMAC Column*

The HiTrap chelating Sepharose column was washed with water, then charged with Cu^{2+}. The unbound metal ion was washed out of the column with water. The column was charged with imidazole buffer and then equilibrated with equilibration buffer. The DEAE eluate was loaded onto an IDA-Cu column. Then the column was washed with the same equilibration buffer to remove the lightly bound contaminants. The bound PC was eluted with an elution buffer.

2.2.3. *PC Amidolytic Activity Assay*

PC activity was measured using protac activator and S-2366 chromogenic substrate by adapting the method of Odegaard et al.[9] This activator rapidly and specifically catalyzed the conversion of PC into an active form. Chromogenic substrate was used to measure the activated PC.

3. RESULTS AND DISCUSSION

An ion exchange DEAE column was used for Cohn fraction IV-1 volume reduction and mass capture of the VKD proteins.

3.1. DEAE Equilibration Buffer and First Wash Buffer

The majority of the contaminant proteins in Cohn fraction IV-1, such as α-1 antitrypsin and human albumin, cannot remain in DEAE gel in sodium citrate buffer.

Therefore, sodium citrate was used as equilibration buffer as well as first wash buffer to perform DEAE column operation. Since the higher the sodium citrate concentration, the more dissolved the proteins, a suitable concentration of sodium citrate was found to be efficient for DEAE equilibration and wash.

3.2. DEAE Column Second Wash Buffer Study

After an overnight wash with the first wash buffer to remove loosely bound contaminants, the DEAE column was then washed with the second wash buffer. For ion exchange chromatography, when sodium chloride concentration increases, the bound proteins can be released from the gel. In order to remove lightly bound contaminants as much as possible and still retain VKD proteins bound to the column, a series of sodium chloride concentrations were studied and a suitable concentration was chosen as the second DEAE wash buffer to wash the DEAE column.

3.3. DEAE Elution Buffer Study

Since VKD proteins bound to the DEAE column tightly, a series of relatively high concentrations of sodium chloride were used to study DEAE column elution. A suitable concentration of sodium chloride was determined to be able to elute VKD proteins efficiently. By using equilibration buffers, first wash buffer, second wash buffer, and elution buffer, a DEAE column could separate Cohn fraction IV-1 with 86% active PC recovery and 32-fold increase in PC purity.

After systematic investigations of different chelator and metal ion combinations, an IDA-Cu column was determined to be suitable for DEAE eluate separation.

3.4. Determination of IDA-Cu Wash Buffer

Since PC binds to the IDA-Cu column tightly, it could not be released from the column by lowering pH. A buffer containing imidazole was necessary for desorption. After study, a buffer containing a low concentration of imidazole was found to be able to release the major contaminant protein, prothrombin, which remained PC-bound in the column; therefore, this concentration of imidazole was used for the IDA-Cu column equilibration and wash buffer.

3.5. Determination of IDA-Cu Elution Buffer

An IDA-Cu column was equilibrated with equilibration buffer containing a low concentration of imidazole. DEAE eluate was loaded onto the column. The column was then washed with the same buffer. Then, a buffer containing a different concentration of imidazole was used to elute the bound PC. The stronger the elution buffer, the more the tightly bound contaminants would be eluted. After many studies, a suitable concentration of imidazole was considered to be an efficient buffer for PC elution from an IDA-Cu column. When an IDA-Cu column was used to separate PC from DEAE eluate by using equilibration buffer, wash buffer, and elution buffer, PC purity was increased 4-fold, with 83% active PC recovery.

Table 1. Protein C recovery.

Material	Volume, ml	PC recovery DEAE	PC recovery IDA-Cu
Cohn IV-1	1600	71.1%	
DEAE eluate	360		66.1%
IDA-Cu eluate	230		

3.6. Scale-Up of the Process

This process was scaled-up by using a 120 ml (d = 5.0 cm, h = 6.1 cm) DEAE column and a 20 ml (d = 2.6 cm, h = 3.8 cm) IDA-Cu column. Approximately 52 grams of Cohn fraction IV-1 was used as source material for the DEAE run. Protein C activity assay showed that the PC recovery for this step was 71.1% (Table 1).

The DEAE eluate was loaded onto the IDA-Cu column. The column was then washed with the wash buffer to remove loosely bound contaminants. The bound protein C was eluted by an elution buffer. Protein C activity assay indicated the protein C recovery of this step was 66.1% (Table 1).

4. CONCLUSIONS

A process combining ion exchange and IMAC has been developed to separate Cohn fraction IV-1 in order to produce a PC cocktail. An ion exchange DEAE column is used to remove most of the contaminants from Cohn fraction IV-1, thus increasing the PC ratio as high as 32-fold, with approximately 86% activity recovery. The IMAC IDA-Cu column then increases the PC ratio approximately 4-fold with approximately 83% PC recovery. Therefore, the combination of DEAE and IDA-Cu can produce a PC cocktail with more than 100-fold PC ratio increase. It can be scaled-up easily, and the scale-up is linear. This process can be scaled-up further to produce large quantities of protein C at low cost to treat protein C-deficient patients.

5. ACKNOWLEDGMENTS

The authors would like to express appreciation for the National Science Foundation grant No. CTS-9904465. Thanks also go to Dr. William N. Drohan and Dr. Annemarie H. Ralston for their assistance, and to the American Red Cross for their support in the form of materials for the study.

REFERENCES

1. W. H. Velander, T. Morcol, D. B. Clark, D. Gee, and W. N. Drohan, Technological challenges for large-scale purification of protein C, in: *Advances in Applied Biotechnology Series 11: Protein C and Related Anticoagulants*, edited by D. F. Bruley and W. N. Drohan (Portfolio Publications Co., Texas, 1990).
2. J. J. Winzerling, B. Patrick, and J. Porath, How to use immobilized metal ion affinity chromatography, *Methods: A Companion to Methods in Enzymology* **4**, 4-13 (1992).

3. L. Andersson, Fraction of human serum proteins by immobilized metal affinity chromatography, *J. Chromat.* **15**, 167-174 (1984).
4. J. W. Wong, R. L. Albright, and N. H. Wang, Immobilized metal ion affinity chromatography (IMAC) chemistry and bioseparations applications, *Separ. Purific. Methods* **20**(1), 49-106 (1991).
5. H. Wu, D. F. Bruley, K. A. Kang, and W. N. Drohan, Protein C separation from human plasma Cohn fraction IV-1 using immobilized metal affinity chromatography, *Topical Conference on Separation Science and Technologies-Proceeding of 1997 Annual AIChE Meeting*, 1443-1448 (1997).
6. H. Wu, D. F. Bruley, and K. A. Kang, Protein C separation from human blood plasma Cohn Fraction IV-1 using immobilized metal affinity chromatography, *Adv. Exp. Med. Biol.* **454**, 697-704 (1998).
7. S. S. Tadepalli, D. F. Bruley, K. A. Kang, and W. Drohan, Immobilized metal affinity chromatography process identification and scale-up for protein C production, *Adv. Exp. Med. Biol.* **428**, 31-43 (1997).
8. H. Wu, and D. F. Bruley, Homologous blood protein separation using immobilized metal affinity chromatography-protein C separation from prothrombin with application to the separation of factor IX and prothrombin, *Biotechnol. Prog.* **15**(5), 928-931 (1999).
9. O. R. Odegaard, K. Try, and T. R. Anderson, Protein C: an automated activity assay, *Haemostasis* **17**, 109-113 (1987).

HYPOXIA-INDUCED ALTERATIONS IN HYALURONAN AND HYALURONIDASE

Feng Gao, Paul Okunieff, Zeqiu Han, Ivan Ding, Luping Wang, Weimin Liu, Jiaying Zhang, Shanmin Yang, Jinguo Chen, Charles B. Underhill, Sunghee Kim, and Lurong Zhang

Abstract: Hyaluronan (HA), a large negatively-charged polysaccharide, is a major component of vessel basal membrane. HA is expressed by a variety of cells, including tumor and endothelial cells. We hypothesized that HA could be up-regulated by hypoxia to enhance vessel formation. To determine the effect of hypoxia on the production of HA, tumor cells were treated with either media alone (control) or a hypoxia inducer (CoCl or NaN_3) for 24 h. The level of HA in the media was then measured by ELISA. The results showed that both CoCl and NaN_3 induced the production of HA. Since the low molecular weight form of HA (SMW) possesses pro-angiogenic properties, we investigated whether hypoxia-induced HA can be processed into SMW. Under hypoxic conditions, the activity of hyaluronidase, the enzyme responsible for degrading HA, was measured by an ELISA-like assay. The activity of hyaluronidase was shown to be up-regulated by hypoxia and, further, could carry out the function of processing HA into SMW. In addition, the hypoxic areas of tumor tissues were stained strongly with biotinylated HA-binding proteins, indicating that the level of HA was high compared to the oxic areas. This study demonstrates that hypoxia can stimulate the production of HA and the activity of hyaluronidase, which may promote angiogenesis as a compensation mechanism for hypoxia.

1. INTRODUCTION

Hyaluronan (HA) is a large, negatively-charged polysaccharide consisting of repeated units of N-acetylglucosamine and glucuronic acid.[1] It is widely distributed and is a major component of most extracellular matrices.[1] Due to its high negative charge and hydrophilic character, HA can absorb a large amount of water, which allows it to exert a swelling pressure on surrounding tissue. HA is synthesized by both normal and tumor cells and is highly regulated. For example, cells in mitotic phase synthesize significant amounts of HA, presumably to provide space for the two daughter cells.[2-5] An early study showed that if HA synthesis is blocked by an inhibitor, then the cells became arrested in mitosis.[6] Furthermore, the synthesis of HA is dramatically elevated in the initial phase of wound healing.[7, 8] The rate of HA synthesis is influenced by a number of factors in a dose-dependent manner[9]: inflammatory mediators, growth factors, activators of adenylate cyclase,[10] insulin-like growth factor-1 (IGF-1), and platelet-derived growth factor (PDGF). Interleukin-1, tumor necrosis factors-alpha and -beta, and transforming growth factor-beta 1 can also stimulate the synthesis of HA.[11-13] However, beyond this, little is known about the effects of stress factors, such as hypoxia, on the synthesis of HA.

Recent studies have demonstrated that HA is involved in angiogenesis. In particular, low molecular weight HA is a potent angiogenic factor.[14, 15] It has been shown that hyaluronidase (HAase) in lysosomes is responsible for the degradation of HA,[14-17] digesting the high molecular weight HA (several thousand repeats of disaccharide) into low molecular weight HA (less than 25 repeats of disaccharide). Again, it is unclear whether HAase is regulated by hypoxia.

In this study, the effect of hypoxia on both the production of HA and the activity of HAase were examined. The results of this study provide new evidence for the regulation of HA and HAase and suggest that these molecules act synergistically to produce a pro-angiogenic factor for counteracting hypoxia to avoid further tissue damage.

2. METHODS

2.1. Reagents and Cells

T47D human breast cancer cells were obtained from the Tumor Cell Bank of Lombardi Cancer Center at Georgetown University (Washington DC) and maintained in 10% calf serum and 90% DMEM (Dulbecco's Modification of Eagle's Medium). Cobalt chloride ($CoCl_2$) and sodium azide (NaN_3) were purchased from Sigma (St. Louis, MO). The biotinylated HA-binding protein was prepared in our laboratory, as described previously.[18]

2.2. Treatment

T47D cells (2×10^4) were cultured in 1 ml of media in 24-well plates and grown to 80% confluence. After changing to fresh media, the cells were treated without (control) or with 100 μM of $CoCl_2$ or 1 mM of NaN_3 (in triplicate) for 24 h. Then the media from each well was collected, and the cell layer was extracted in 0.3 ml of lysis buffer (PBS containing 0.1% sodium deoxycholate, 0.5 mg/ml leupetin, 1 mg/ml pepstatin, and 0.2 mM phenylmethyl-sulfonyl fluoride). The samples were frozen at -20°C until analyzed for HA and HAase.

2.3. ELISA for HA

A modified enzyme-linked assay[19] was used. Briefly, a high-bound ELISA plate (Falcon, Lincoln Park, NJ) was coated with 100 μg/ml of hyaluronan (Sigma, St. Louis, MO) in PBS at room temperature overnight, then blocked with 10% calf serum and 90% PBS. Then 100 μl of media sample was mixed with 50 μl of 50 μg/ml biotinylated-HA binding protein (b-HABP) at 37°C for 1 h, then transferred to the HA-coated ELISA plate. The unbound b-HABP remaining in the sample mixture could then bind to the HA-coated plate and was detected by incubation with 0.5 μg/ml of peroxidase-labeled streptavidin followed by 0.5 mg/ml of peroxidase substrate, consisting of H_2O_2 and azinobis (3-ethyl-benzthiazoline sulfonic acid) in 0.1 M sodium citrate, pH 5.0. The plate was read at A_{405} and the concentration of HA in the samples was calculated from a standard curve.

2.4. ELISA-like Assay for HAase

Using a high-bound ELISA plate at room temperature, 100 μl of 100 μg/ml of HA in PBS was applied, incubated overnight, and then the unbound HA was washed away with PBS. After blocking with 3% bovine albumin, 100 μl of the cell lysate containing 100 μg of protein was added to the each well, incubated at 37°C for 2 h, then the digested, detached HA was washed away. The HA remaining bound to the plate was detected by biotinylated-HA binding protein, as described above. The relative activity of HAase in equal amounts of lysate was determined from a standard curve constructed with a known concentration (0, 0.15, 0.31, 0.62, 1.25, 2.5, 5, 10 U/ml) of bovine testicular HAase.

2.5. HA Degradation Assay

Two μg of ^{3}H-labeled high molecular weight HA (prepared in our laboratory using cultured rat fibrosarcoma cells) was added to T47D cells treated without (control) or with 100 μM of $CoCl_2$, 1 mM of NaN_3, or NaN_3 plus a 50-fold excess of HA, and incubated for 48 h. Then 200 μl of 20 mg/ml pronase E was added to each well and further incubated for 4 h. The media was collected, and the lysates were frozen and thawed 5 times in 0.2 ml water. After centrifuging down the cell debris, each supernatant was mixed with its media, and applied to a Centricon 30 (Amicon, Danvers, MA). When centrifuged, the high molecular weight HA was retained on the top of the filter membrane, while the low molecular weight HA passed through. The media was collected from the bottom chamber, mixed with 6 volumes of scintillation fluid, and processed in a β-counter to determine the amount of low molecular weight HA in each well.

2.6. Histochemical Staining for Tissue HA

Five μm sections of tumors derived from KHT fibrosarcoma cells grown in C3H/He mice[18] were de-paraffinized and incubated in a 10% hydrogen peroxide solution for 10 min to eliminate endogenous peroxidase activity. After thorough washing, the sections were incubated with PBS (control) or affinity-purified, biotinylated HA-binding proteins (1:20) followed by peroxidase labeled-streptavidin (1:250) and the peroxidase substrate,

3-amino-9-ethyl-carbazole (AEC) with H_2O_2, which gives a red reaction product. The samples were then counter-stained with hematoxylin (blue) and preserved with CrystalMount (Biomedia, Foster City, CA). The negative controls were processed with the same procedure, but the phosphate buffer was used to replace the HA-binding protein in the initial staining.

2.7. Statistical Analysis

The mean and standard error were calculated from the raw data, and then subjected to the Student *t* test. The data for the media alone group and the $CoCl_2$- or NaN_3-treated groups were compared. A P value of $< .05$ was regarded as statistically significant.

3. RESULTS

3.1. Hypoxia Induces the Production of HA

When the T47D cells were treated with media alone (control), 100 μM of $CoCl_2$, or 1 mM of NaN_3, the optimal doses for these chemical inducers of hypoxia for 24 h,[20, 21] the amount of HA in the conditioned media increased as determined by quantitative ELISA (Fig. 1). Similar results were obtained from three separate experiments under the same conditions.

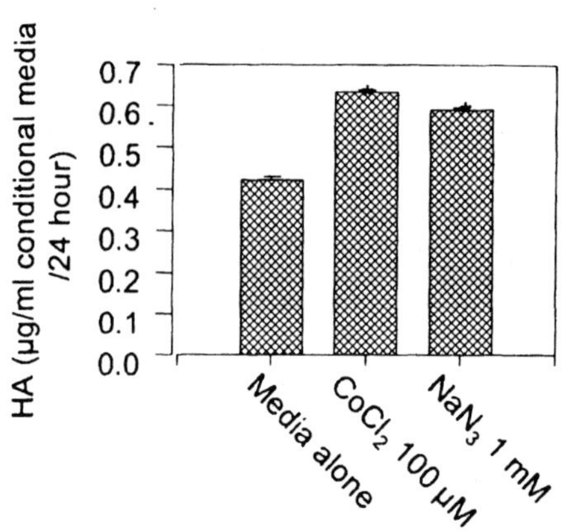

Figure 1. The hypoxia-induced production of HA in conditioned media of T47D cells. The cells were treated with media alone (control), 100 μM of $CoCl_2$, or 1 mM of NaN_3 for 24 h, then the HA was measured in the media by an HA-ELISA (* = $P < .05$)

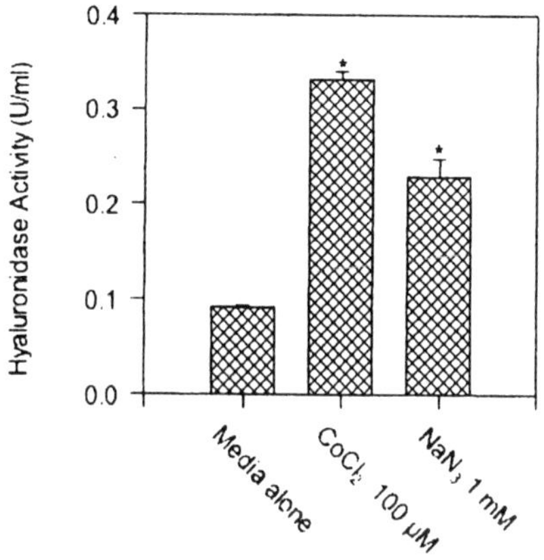

Figure 2. The activity of HAase was enhanced by hypoxia. Lysates from cells treated with media alone (control), $CoCl_2$ or NaN_3 for 24 h were analyzed for the activity of HAase with an ELISA-like assay. The result showed that HAase activity was dramatically increased (* = $P < .01$).

3.2. Hypoxia Enhances the Activity of HAase

To determine if hypoxia altered the levels of HAase, equal amounts of lysate from cells treated with PBS vehicle, 100 μM of $CoCl_2$ or 1 mM of NaN_3 for 24 h, were added to plates coated with HA. The digested HA then was washed away and the remaining undigested HA was measured with an ELISA-like assay. The results (Fig. 2) showed that the activity of HAase in lysates of cells exposed to chemically induced hypoxia was higher than in the control cells ($P < .01$).

3.3. Hypoxia Induces the Degradation of HA

Since low molecular weight HA possesses angiogenic activity and an increase in HAase activity was detected in the lysates of hypoxic cells, we tested whether there was a similar increase in low molecular weight HA. For this we treated the cultured cells with ^{3}H-HA (high molecular weight) and used size-specific membranes (Centricon 30) to separate the high from low molecular weight HA. Figure 3 shows that the HA was processed into the low molecular weight form at a much higher rate by the hypoxic cells compared with the control cells. This effect could be blocked by the addition of an excess of unlabelled HA, indicating the assay was specific.

3.4. HA Increases in the Hypoxic Areas of KHT Tumors

KHT is a C3H/He mouse fibrosarcoma cell line that is very sensitive to hypoxia.[22] In addition, it produces much more HA than human T47D cells, as evidenced by the staining of HA in tumors (data not shown). These two characteristics offered us the opportunity to examine the *in vivo* effects of hypoxia on HA. The tumors were harvested from mice, processed in paraffin, and stained with biotinylated HA-binding proteins. Normally, the hypoxic areas are located near necrotic regions some distance from normal

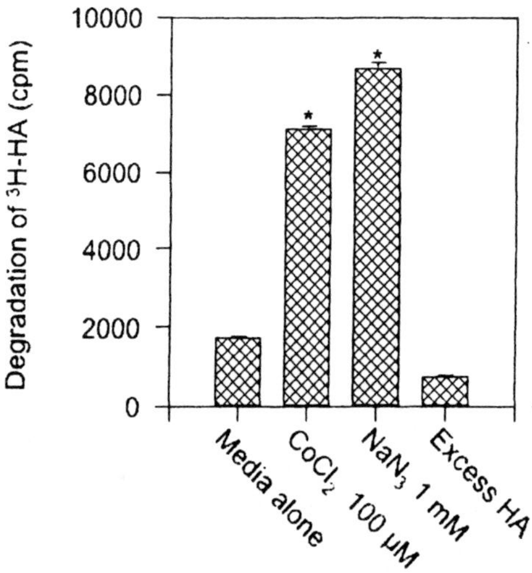

Figure 3. The degradation of HA was enhanced by hypoxia. Two µg of high molecular weight ^{3}H-HA was added to the cells treated with either media alone (control), 100 µM of $CoCl_2$, 1 mM of NaN_3, or NaN_3 plus 50-fold excess of HA for 48 h; the low molecular weight HA was obtained by passage through a Centricon 30. The radioactivity of low molecular weight HA was counted with a ß-counter.

oxygenated areas. Figure 4 shows that a significant increase in HA staining (in red) was associated with areas of hypoxia as compared with other regions.

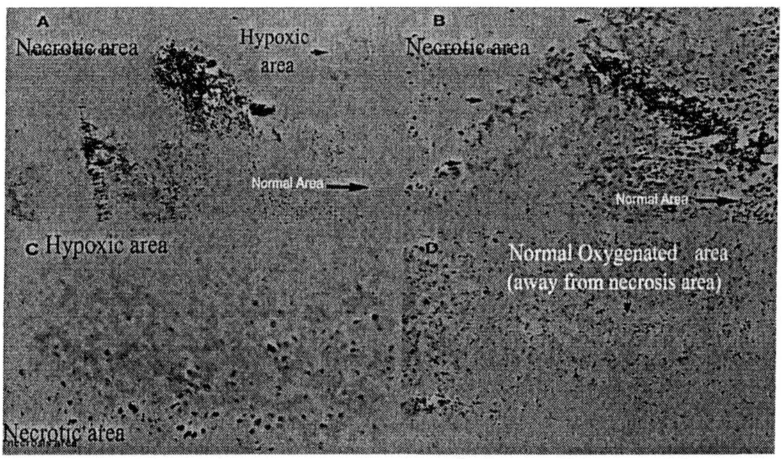

Figure 4. HA staining in the hypoxia-sensitive KHT mouse fibrosarcoma. Sections of the tumor were stained with biotinylated HA-binding proteins (1:20) followed by peroxidase labeled-streptavidin (1:250) and AEC substrate. **A** and **B**: HA in hypoxic areas stained strongly (red), while the distant, normal oxygenated areas had a much lower intensity of staining (X 100); **C**: hypoxic areas with strong HA staining (X 200); **D**: the staining in normal areas distant from necrosis and hypoxic area was greatly reduced (X 100).

4. DISCUSSION

In this study, we have demonstrated that chemical inducers of hypoxia could up-regulate the production of HA and the activity of HAase (Figures 1 and 2), which resulted in a significant increase in the expression of low molecular weight HA in the hypoxic cells (Fig. 3). This *in vitro* phenomenon was confirmed *in vivo* by histochemical staining of KHT tumors, in which strong staining of HA was apparent in areas of hypoxia (Fig. 4).

We believe that this is a common phenomenon that occurs with hypoxia induced by different conditions, such as tumor hypoxia due to the unbalanced growth of tumor cells relative to their blood supply, or hypoxia due to a wound that cuts off the blood supply. Under these conditions, cytokines, chemokines, and interleukins are quickly produced and many are involved in HA synthesis regulation. In addition, a Hox transcription factor has recently been discovered to be functionally associated with the production of HA.[23] Therefore, the regulation of HA is likely to be a complex mechanism involving multiple inputs.

It has been demonstrated that HA is increased immediately after wounding, and peaks a few days prior to the appearance of collagens.[7, 8] The biological function of HA may include the following: 1) it provides a new matrix through which fibroblast or endothelial cells migrate into a wound area to repair it; and 2) it generates a new angiogenic factor to stimulate new vessel formation, which is critical for tissue repair and survival.

The enhanced activity of HAase may act synergistically with the increased levels of HA to produce angiogenic, low molecular weight HA. The optimal pH for lysosome HAase is 5 to 6,[24] which can be easily achieved under hypoxic conditions due to the accumulation of acidic metabolic products at the level of both cells and tissues. Thus, hypoxia-induced low pH could account for the increased level of HAase activity. We believe that the increased HA and HAase activity play a dual role in angiogenesis, providing a matrix for the endothelial cells to migrate in and as a stimulator of proliferation. In our future studies, we will conduct experiments to compare the number of CD31-positive cells (representing endothelial cells) between the hypoxic and normoxic areas, which have high levels or relatively low levels of HA, respectively.

While several molecules, such as HIF (hypoxia induced factor[25]) and VEGF (vascular endothelium growth factor[26]), have been identified as hypoxia-related factors, we have demonstrated here for the first time that HA and HAase are also regulated by hypoxia, which may promote angiogenesis and counteracts the effects of hypoxia.

5. ACKNOWLEDGMENTS

This work was supported by grants from the U. S. Army Med. Res. & Mat. Command (DAMD17-00-1-0081 and DAMD17-01-1-0708) and NCI/NIH (CA71545).

REFERENCES

1. T. C. Laurent, Biochemistry of hyaluronan, *Acta Otolaryngol.* **442**, 7-24 (1987).
2. N. Main, Analysis of cell-growth phase-related variation in hyaluronate synthase activity of isolated plasma-membrane fractions of cultured human skin fibroblasts, *Biochem. J.* **237**, 333-342 (1986).

3. M. Tomida, H. Koyama, and T. Ono, Induction of hyaluronic acid synthetase activity in rat fibroblasts by medium change of confluent cultures, *J. Cell Physiol.* **86**, 121-130 (1975).
4. L. Hronowski, and T. P. Anastassiades, The effect of cell density on net rates of glycosaminoglycan synthesis and secretion by cultured rat fibroblasts, *J. Biol. Chem.* **255**, 10091-10099 (1980).
5. K. Matuoka, M. Namba, and Y. Mitsui, Hyaluronate synthetase inhibition by normal and transformed human fibroblasts during growth reduction, *J. Cell Biol.* **104**(4), 1105-1115 (1987).
6. M. Brecht, U. Mayer, E. Schlosser, and P. Prehm, Increased hyaluronate synthesis is required for fibroblast detachment and mitosis, *Biochem. J.* **239**, 445-450 (1986).
7. M. Stern, B. Schmidt, T. B. Dodson, R. Stern, and L. B. Kaban, Fetal cleft lip repair in rabbits: histology and role of hyaluronic acid, *J. Oral Maxillofac. Surg.* **50**(3), 263-268 (1992).
8. B. A. Mast, R. F. Diegelmann, T. M. Krummel, and I. K. Cohen, Scarless wound healing in the mammalian fetus, *Surg. Gynecol. Obstet.* **174**(5), 441-451 (1992).
9. Y. Imai, R. Odajima, Y. Inoue, and Y. Shishiba, Effect of growth factors on hyaluronan and proteoglycan synthesis by retroocular tissue fibroblasts of Graves' ophthalmopathy in culture, *Acta Endocrinol.* **126**(6), 541-552 (1992).
10. T. C. Laurent, and J. R. E. Fraser, The properties and turnover of hyaluronan, in: *Functions of the Proteoglycans*, edited by D. Evered, and J. Whelan (Wiley, Chichester, 1986), pp. 9-29.
11. D. M. Butler, G. F. Vitti, T. Leizer, and J. A. Hamilton, Stimulation of the hyaluronic acid levels of human synovial fibroblasts by recombinant human tumor necrosis factor alpha, tumor necrosis factor beta (lymphotoxin), interleukin-1 alpha, and interleukin-1 beta, *Arthritis Rheum.* **31**(10), 1281-1289 (1988).
12. A. Tengblad, U. B. G. Laurent, K. Lilja, R. N. Cahill, A. Engstrom-Laurent, J. R. Fraser, H. E. Hansson, and T. C. Laurent, Concentration and relative molecular mass of hyaluronate in lymph and blood, *Biochem. J.* **236**, 521-525 (1986).
13. H. D. Haubeck, R. Kock, D. C. Fischer, E. Van de leur, K. Hoffmeister, and H. Greiling, Transforming growth factor beta1, a major stimulator of hyaluronan synthesis in human synovial lining cells, *Arthritis Rheum.* **38**(5), 669-677 (1995).
14. D. C. West, I. N. Hampson, F. Arnold, and S. Kumar, Angiogenesis induced by degradation products of hyaluronic acid, *Science* **228**(4705), 1324-1326 (1985).
15. D. C. West, and S. Kumar, Hyaluronan and angiogenesis, *Ciba Foundat. Symp.* **143**, 187-201 (1989).
16. G. Kreil, Hyaluronidases – a group of neglected enzymes, *Protein Sci.* **4**(9), 1666-1669 (1995).
17. T. C. Laurent, and J. R. Fraser, Hyaluronan, *FASEB J.* **6**(7), 2397-2404 (1992).
18. C. B. Underhill, and L. Zhang, Analysis of hyaluronan using biotinylated hyaluronan-binding proteins, *Methods Mol. Biol.* **137**, 441-447 (2000).
19. N. Liu, F. Gao, Z. Han, X. Xu, C. B. Underhill, and L. Zhang, Hyaluronan synthase 3 overexpression promotes the growth of TSU prostate cancer cells, *Cancer Res.* **61**(13), 5207-5214 (2001).
20. C. R. Rose, S. G. Waxman, and B. R. Ransom, Effects of glucose deprivation, chemical hypoxia, and simulated ischemia on Na+ homeostasis in rat spinal cord astrocytes, *J. Neurosci.* **18**(10), 3554-3562 (1998).
21. J. P. Piret, D. Mottet, M. Raes, and C. Michiels, $CoCl_2$, a chemical inducer of hypoxia-inducible factor-1, and hypoxia reduce apoptotic cell death in hepatoma cell line HepG2, *Ann. NY Acad. Sci.* **973**, 443-447 (2002).
22. D. W. Siemann, I. M. Johansen, and M. R. Horsman, Radiobiological hypoxia in the KHT sarcoma: prediction using the Eppendorf histograph, *Int. J. Radiat. Oncol. Biol. Phys.* **40**(5), 1171-1176 (1998).
23. J. A. Mack, S. R. Abramson, Y. Ben, J. C. Coffin, J. K. Rothrock, E. V. Maytin, V. C. Hascall, C. Largman, and E. J. Stelnicki, Hoxb13 knockout adult skin exhibits high levels of hyaluronan and enhanced wound healing, *FASEB J.* **17**(10), 1352-1354 (2003).
24. J. Chayen, and L. Bitensky, Lysosomal enzymes and inflammation with particular reference to rheumatoid diseases, *Ann. Rheum. Dis.* **30**(5), 522-536 (1971).
25. E. Minet, G. Michel, J. Remacle, and C. Michiels, Role of HIF-1 as a transcription factor involved in embryonic development, cancer progression and apoptosis, *Int. J. Mol. Med.* **5**(3), 253-259 (2000).
26. A. L. Harris, Hypoxia – a key regulatory factor in tumour growth, *Nat. Rev. Cancer* **2**(1), 38-47 (2002).

RATE OF DECREASE OF PO_2 FROM AN ARTERIOLE WITH ARRESTED FLOW

Roland N. Pittman, Aleksander S. Golub, and William F. Schleicher

Abstract: When flow to a region is arrested, the amount of oxygen contained within the stationary blood decreases at a rate dependent on the oxygen utilization of the surrounding tissue. We used phosphorescence quenching microscopy to measure arteriolar PO_2 in the mesentery of male Sprague-Dawley rats. Flow was quickly stopped (< 1 s) by occluding the microvessels using an inflatable Saran bag attached to the microscope objective. The rate of decline in PO_2 following occlusion yielded a calculated initial flux of oxygen out of the vessel lumen of 8.0×10^{-7} ml O_2 cm^{-2} sec^{-1}. An upper limit on the oxygen consumption of the arteriolar wall was calculated by assuming that all of the oxygen in the lumen was consumed by the wall at the initial rate. This value was 2.5×10^{-3} ml O_2 cm^{-3} sec^{-1} and is an overestimate since the oxygen consumption of the nearby parenchymal cells was neglected. The calculated maximum oxygen consumption of the wall is more than an order of magnitude smaller than that reported previously for arterioles in the rat mesentery (6.5×10^{-2} ml O_2 cm^{-3} sec^{-1}). We conclude that oxygen consumption of the arteriolar wall is similar to previous values for other vascular tissues.

1. INTRODUCTION

Duling and Berne[1] reported a longitudinal decrease in blood oxygenation, using PO_2 microelectrodes, in the arteriolar network of the hamster cheek pouch and the hamster and rat cremaster muscle. Subsequent studies confirmed the longitudinal gradient in hemoglobin oxygen saturation (SO_2) also in the hamster retractor muscle,[2-4] hamster dorsal skin fold,[5] rat liver,[6] rat pancreas,[7] rat mesentery,[8] and rat cremaster muscle.[9]

There are several potential destinations for the oxygen that diffuses from an arteriole: the oxygen could be consumed by parenchymal cells surrounding the arteriole; it could diffuse to nearby microvessels[10-14]; or it could be consumed by the cells (i.e., endothelial and vascular smooth muscle cells) of the arteriolar wall itself. Initially, this last possibility was not thought to be a major sink for oxygen since the arteriolar wall is thin compared with the surrounding tissue, and there was no indication that the vessel wall consumes oxygen at a rate much greater than that of the parenchymal tissue.

The diffusion of oxygen from arterioles can be quantified for the steady state by demanding mass balance for oxygen in an unbranched arteriolar segment. The diffusive flow of oxygen from the lumen of the segment should be equal to the difference between the convective inflow and outflow of oxygen in the arteriole. This mass balance for oxygen can be written as:

$$QO_2^{DIFF} = QO_2^{CONV}\,(\text{inflow}) - QO_2^{CONV}\,(\text{outflow}) \qquad (1)$$

where QO_2^{DIFF} is the diffusive flow of oxygen across the arteriolar wall and QO_2^{CONV} is the convective flow of oxygen through the arteriole.

The flux of oxygen ($JO_2 = QO_2^{DIFF}$/luminal surface area) calculated from the experimental measurements (i.e., diameter, velocity, hemoglobin concentration, SO_2) was an order of magnitude higher than the value predicted by a standard computational model of oxygen diffusion and consumption. Popel et al.[12, 13] proposed that Krogh's diffusion coefficient, K ($= \alpha\ DO_2$), in perfused tissue might be an order of magnitude higher than that in excised, unperfused tissue, the environment used for most determinations of diffusion coefficient. Measurements by Meng et al.[15] and Bentley et al.[16] indicated that K could be larger by a factor of two than was generally accepted, but the order of magnitude discrepancy was not resolved by these results.

This apparent discrepancy still eludes explanation. Could it be due to an experimental artifact or incorrect interpretation of the data? Could the theory be incorrect, either in its mathematical formulation or by the incorporation of invalid assumptions in the computational model? These issues have been considered previously by Pittman[17] and Vadapalli et al.[18]

Based on their reports of a large transmural PO_2 gradient, Tsai et al.[8, 19] have proposed that the explanation for this apparent discrepancy is a large oxygen consumption of the arteriolar wall that is about 100 times that of resting skeletal muscle. Their data were obtained in the rat mesentery from measurements of PO_2 using the phosphorescence quenching technique; however, the view of Tsai et al.[8, 19] is contrary to the preponderance of evidence on oxygen consumption of vascular tissue as compiled by Vadapalli et al.[18] These authors estimated the maximum oxygen consumption of the vessel wall based on its mitochondrial content: $M_{mt} = 5 \times 10^{-3}$ ml O_2 cm^{-3} s^{-1}. All published values for vascular cells and *in vitro* tissue are smaller than M_{mt}, whereas the value calculated by Tsai et al.[8, 19] is an order of magnitude larger than M_{mt}.

Since there is no obvious reason why the oxygen consumption of the arteriolar wall should be so high, it is natural to seek an alternative approach to estimate the oxygen consumption of the arteriolar wall. Under normal circumstances, blood flow through the microcirculation provides adequate oxygenation of tissues. However, when flow to a region is arrested (e.g., occlusion of arterial inflow), the amount of oxygen contained within the now stationary blood decreases at a rate dependent on the oxygen utilization of the surrounding tissue (i.e., vessel wall and parenchymal cells). Thus, by monitoring the disappearance of oxygen in the lumen of an arteriole in which blood flow has been quickly stopped, an upper estimate of vessel wall oxygen consumption can be made.

2. METHODS

2.1. Experimental

Male Sprague-Dawley rats (N = 6) were anesthetized with an i.p. injection of sodium pentobarbital (6.5 mg/100 g) and the mesentery was prepared for observations using intravital microscopy. The trachea was cannulated with PE-240 tubing to achieve a patent airway and a femoral vein was cannulated with PE-90 tubing with a PE-50 tip for infusion of supplemental anesthetic and phosphor. The animal and the mesentery were placed on a heated platform[20] that maintained both the animal's core temperature and that of the mesentery at 37°C. The mesentery was covered with a single layer of Saran film to minimize gas exchange with the atmosphere.

In order to rapidly arrest blood flow in the mesentery, the microscope objective (Neofluar 40X/0.75, Zeiss) was outfitted with a small (25 mm diam.), inflatable Saran bag. When the bag was inflated with air to 50 mmHg, using an electronic pressure control unit, it gently pressed uniformly on the mesenteric tissue beneath the objective, and microvascular flow throughout the preparation (i.e., not just local arteriolar occlusion) was stopped within one second. Saran film was used because it was thin, transparent, and practically impermeable to oxygen from the atmosphere. Presumably, the modest tissue deformation caused by inflation of the bag had a negligible effect on cellular metabolism of the arteriole and adjacent tissue.

Oxygen tension (PO_2) within arterioles was measured using the time domain version of the phosphorescence quenching method implemented for intravital microscopy.[21] A Xenon flash lamp excited the phosphorescent probe, Pd meso-tetra (4-carboxyphenyl) porphyrin, that was bound to bovine serum albumin (Oxygen Enterprises, Philadelphia, PA). The probe was injected into the animal's circulation (0.5 ml of 10 mg/ml probe concentration) about 15 minutes before the measurements began. PO_2 was measured continuously before occlusion, during arrest of the circulation, and after restoration of flow. The size of the excitation region was adjusted for each measurement to be a square with side length equal to the vessel diameter. The excitation (i.e., flash) frequency was 1 Hz, and one PO_2 value was calculated for each phosphorescence decay curve using the fitting procedure described previously.[21, 22]

The experimental protocol started with the selection of an arteriole for study that had a relatively straight section at least 100 μm long. The sequence of events, including arrest of flow, was videotaped for later analysis of internal and external arteriolar diameter. Blood flow was quickly arrested by inflating the Saran bag between the objective and the mesentery. PO_2 measurements were made continuously at a rate of one measurement per second until the end of the recovery period.

2.2. Calculations

The approach to estimate an upper limit on the oxygen consumption of the arteriolar wall is based on observing the time course of the disappearance of oxygen from the lumen of an arteriole in which blood flow has been stopped. The amount of oxygen in the lumen, AO_2, is given by:

$$AO_2 = \pi R_i^2 \, L \, C_{Hb} \, [Hb] \, SO_2 \tag{2}$$

where R_i is the internal radius of the arteriole, L is the length of the segment, C_{Hb} is the oxygen binding capacity of the hemoglobin in the red blood cells, [Hb] is the hemoglobin concentration in the segment, and SO_2 is the oxygen saturation of the hemoglobin. The small amount of oxygen dissolved in the plasma is neglected. The rate of disappearance of oxygen from lumen following stoppage of flow is given by:

$$-dAO_2/dt = \pi R_i^2 \, L \, C_{Hb} \, [Hb] \, dSO_2/dt \tag{3}$$

The slope of the oxygen dissociation curve for rat blood can be used to express the time rate of change of SO_2 in terms of the measured PO_2 time course:

$$dSO_2/dt = (\partial SO_2/\partial PO_2) \, dPO_2/dt \tag{4}$$

The rate of oxygen consumption by the wall is $M \pi (R_e^2 - R_i^2) L$, where M is the volume-specific consumption and R_e is the external radius of the arteriole. If one assumes that all of the oxygen disappearing from the lumen is consumed by the wall, then one can obtain an estimate for M that is an upper limit. Using the initial (maximum) rate of PO_2 decrease to obtain an estimate of wall oxygen consumption during normal perfusion (i.e., corresponding to the extrapolated value just before flow arrest) yields the following expression:

$$M_{upper} = - \{C_{Hb} \, [Hb]/[(R_e/R_i)^2 - 1]\} \, (\partial SO_2/\partial PO_2) \, (dPO_2/dt)_{initial} \tag{5}$$

3. RESULTS

Arterioles in two different branching orders were utilized for these measurements. The smaller arterioles had an internal diameter of 25 ± 1 μm (N = 6), and the larger arterioles had an internal diameter of 46 ± 4 μm (N = 12). The initial rate of decline in PO_2 was 0.36 and 0.29 mmHg sec^{-1}, respectively, for the smaller and larger arterioles in the mesentery. The time course of the decrease in PO_2 following arrest of the circulation in the mesentery is presented in Figure 1.

An upper limit on the oxygen consumption of the arteriolar wall was calculated using the equation above for M_{upper}. The values obtained were 2.1×10^{-3} and 3.0×10^{-3} ml O_2 cm^{-3} sec^{-1}, respectively, for the smaller and larger arterioles of the mesentery.

4. DISCUSSION

The oxygen consumption of the cells making up the arteriolar wall determined with our approach yielded values similar to previous ones for other vascular tissues.[18] As we had anticipated, the assumptions used to calculate wall oxygen consumption resulted in

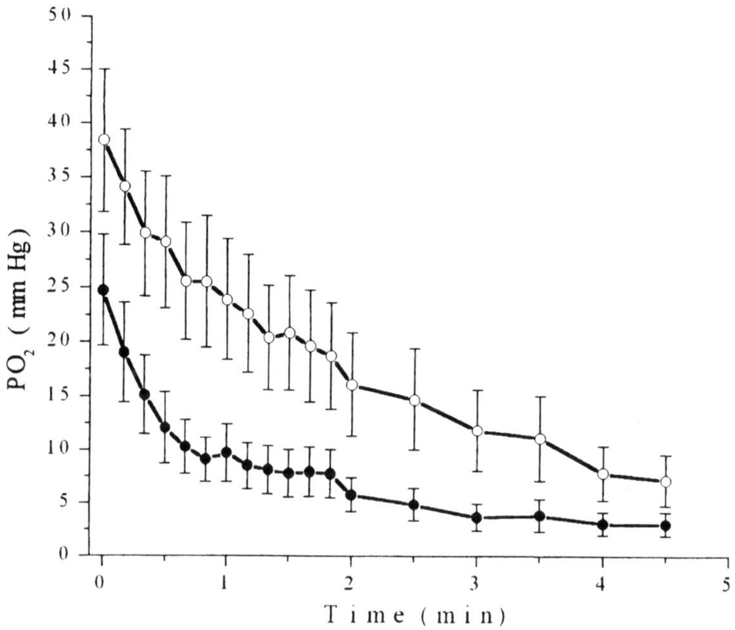

Figure 1. Plot of mean (± standard error) arteriolar PO_2 values versus time beginning with onset of arrested blood flow. The filled circles represent data from small arterioles, and the open circles represent data from large arterioles.

an overestimate, since the consumption of the parenchymal cells beyond the wall, where some of the oxygen must have diffused, was neglected. Furthermore, it is interesting to note that the upper limits of arteriolar wall oxygen consumption that we found are below the maximum value reported by Vadapalli et al.[18] for this tissue, based on its observedmitochondrial content. In our present study, the calculated maximum values of arteriolar wall oxygen consumption are thus more than an order of magnitude (about 25 times) smaller than that reported for the rat mesentery (6.5×10^{-2} ml O_2 cm^{-3} sec^{-1}) by Tsai et al.,[8] where their value was based on considerations of oxygen mass balance in perfused microvessels *in vivo.*

The reason for the 25-fold difference between our result and that of Tsai et al.[8] is unknown. Both studies were carried out on the rat mesentery and both measured PO_2 using the phosphorescence quenching method. It is possible that differences in the implementation of the phosphorescence method to measure PO_2 could be involved, although the exact nature of the difference is not clear. We have previously expressed concern regarding the potential consumption of oxygen by the method itself,[23] especially in regard to PO_2 determinations at sites with stationary fluid (e.g., the interstitium) not supported by robust carriers of oxygen, such as red blood cells. We are left with the unsatisfying circumstance that the order of magnitude discrepancy between apparent and predicted oxygen diffusion from arterioles remains to be reconciled. It does not appear that a large oxygen consumption by the arteriolar wall is a viable explanation; perhaps the answer lies elsewhere as discussed previously by Vadapalli et al.[18]

5. ACKNOWLEDGEMENTS

This work was supported by grant HL18292 from the National Heart, Lung and Blood Institute.

REFERENCES

1. B. R. Duling, and R. M. Berne, Longitudinal gradients in periarteriolar oxygen tension. A possible mechanism for the participation of oxygen in the local regulation of blood flow, *Circ. Res.* **27**, 669-678 (1970).
2. L. Kuo, and R. N. Pittman, Effect of hemodilution on oxygen transport in arteriolar networks of hamster striated muscle, *Am. J. Physiol.* **254**, H331-H339 (1988).
3. D. P. Swain, and R. N. Pittman, Oxygen exchange in the microcirculation of hamster retractor muscle, *Am. J. Physiol.* **256**, H247-H255 (1989).
4. L. Kuo, and R. N. Pittman, Influence of hemoconcentration on arteriolar oxygen transport in hamster striated muscle, *Am. J. Physiol.* **259**, H1694-H1702 (1990).
5. H. Kerger, I. P. Torres Filho, M. Rivas, R. M. Winslow, and M. Intaglietta, Systemic and subcutaneous microvascular oxygen tension in conscious Syrian golden hamsters, *Am. J. Physiol.* **267**, H802-H810 (1995).
6. N. Tateishi, N. Maeda, and T. Shiga, A method for measuring the rate of oxygen release from single microvessels, *Circ. Res.* **70**, 812-819 (1992).
7. A. Seiyama, S. Tanaka, H. Kosaka, and T. Shiga, O_2 transfer from single microvessels to acinar cells in secretin-stimulated pancreas of rat, *Am. J. Physiol. Heart Circ. Physiol.* **270**, H1704-H1711 (1996).
8. A. G. Tsai, B. Friesenecker, M. C. Mazzoni, H. Kerger, D. G. Buerk, P. C. Johnson, and M. Intaglietta, Microvascular and tissue oxygen gradients in the rat mesentery, *Proc. Natl. Acad. Sci.* **95**, 6590-6595 (1998).
9. H. Kobayashi, and N. Takizawa, Oxygen saturation and pH changes in cremaster microvessels of the rat, *Am. J. Physiol. Heart Circ. Physiol.* **270**, H1453-H1461 (1996).
10. M. L. Ellsworth, C. G. Ellis, A. S. Popel, and R. N. Pittman, Role of microvessels in oxygen supply to tissue, *News Physiol. Sci.* **9**, 119-123 (1994).
11. M. L. Ellsworth, and R. N. Pittman, Arterioles supply oxygen to capillaries by diffusion as well as by convection, *Am. J. Physiol.* **258**, H1240-H1243 (1990).
12. A. S. Popel, R. N. Pittman, and M. L. Ellsworth, The rate of oxygen loss from arterioles is an order of magnitude higher than expected, *Am. J. Physiol.* **256**, H921- H924 (1989).
13. D. P. V. Weerappuli, R. N. Pittman, and A. S. Popel, Effect of convection in capillaries on oxygen removal from arterioles in striated muscle, *J. Theor. Biol.* **147**, 275-288 (1990).
14. T. W. Secomb, and R. Hsu, Simulation of oxygen transport in skeletal muscle: diffusive exchange between arterioles and capillaries, *Am. J. Physiol. Heart Circ. Physiol.* **267**, H1214-H1221 (1994).
15. H. Meng, T. B. Bentley, and R. N. Pittman, Oxygen diffusion in hamster striated muscle: comparison of in vitro and near in vivo conditions, *Am. J. Physiol.* **263**, H35-H39 (1992).
16. T. B. Bentley, H. Meng, and R. N. Pittman, Temperature dependence of oxygen diffusion and consumption in mammalian striated muscle, *Am. J. Physiol.* **264**, H1825-H1830 (1993).
17. R. N. Pittman, Influence of microvascular architecture on oxygen exchange in skeletal muscle, *Microcirc.* **2**, 1-18 (1995).
18. A. R. Vadapalli, R. N. Pittman, and A. S. Popel, Estimating oxygen transport resistance of the microvascular wall, *Am. J. Physiol. Heart Circ. Physiol.* **279**, 657-671 (2000).
19. A. G. Tsai, P. C. Johnson, and M. Intaglietta, Oxygen gradients in the microcirculation, *Physiol. Rev.* **83**, 933-963 (2003).
20. A. S. Golub, and R. N. Pittman, Thermostatic animal platform for intravital microscopy of thin tissues, *Microvasc. Res.* **66**, 213-217 (2003).
21. L. Zheng, A. S. Golub, and R. N. Pittman, Determination of PO_2 and its heterogeneity in single capillaries, *Am. J. Physiol.* **271**, H365-H372 (1996).
22. A. S. Golub, A. S. Popel, L. Zheng, and R. N. Pittman, Analysis of phosphorescence decay in heterogeneous systems using distributions of quencher concentration, *Biophys. J.* **73**, 452-465 (1997).
23. R. N. Pittman, A. S. Golub, A. S. Popel, and L. Zheng, Interpretation of phosphorescence quenching measurements made in the presence of oxygen gradients, *Adv. Exp. Med. Biol.* **454**, 375-383 (1998).

MEASUREMENT OF $CMRO_2$ IN NEONATES UNDERGOING INTENSIVE CARE USING NEAR INFRARED SPECTROSCOPY

Clare E. Elwell, Julian R. Henty, Terence S. Leung, Topun Austin, Judith H. Meek, David T. Delpy, and John S. Wyatt

Abstract: Greater understanding of the rate of oxygen delivery and uptake in sick preterm and term infants undergoing intensive care is an important aim of brain-orientated neonatal medicine. Near infrared spectroscopy (NIRS) is a continuous, non-invasive and portable technique which can be used to measure cerebral blood flow (CBF) in infants. It is also possible to use spatially resolved spectroscopy to measure absolute mean cerebral oxygen saturation ($SmcO_2$). The aim of this study was to investigate the derivation of cerebral metabolic rate for oxygen ($CMRO_2$) from these two measurements. Nine preterm infants were studied, of median (range) gestational age 25 (23-37) weeks. A NIRO300 was used to measure CBF and $SmcO_2$ simultaneously over the right and left hemisphere. Median (range) left and right cerebral hemisphere values for $CMRO_2$ were 0.95 (0.79–1.53) ml $100g^{-1}.min^{-1}$ and 0.88 (0.69–1.46) ml $100g^{-1}.min^{-1}$, respectively. No significant difference was seen between the left- and right-sided values. These values are similar to median (range) values previously reported in infants using positron emission tomography or more invasive NIRS methods. Further work is necessary to define limits on the use of this technique, particularly in the assumption of the venous:arterial compartment volume ratio across different infants.

1. INTRODUCTION

Advances in neonatal medicine, leading to increased survival rates among extremely preterm and sick term infants, have led to a focus on the causes of long-term neurological deficit in babies who have undergone neonatal intensive care. Central to the premise of brain-orientated neonatal medicine is the development of reliable, non-invasive, and cotside techniques for the assessment of cerebral circulation and metabolism. Cerebral hypoxia and ischemia are important causes of neurodisability, but most current techniques only provide information about the level of oxygen delivery to the cerebral tissue. Measurements of cerebral tissue metabolism, which would provide information about the balance of cerebral oxygen delivery and demand, may prove to be of much greater clinical relevance. In adults, measurements of cerebral metabolic rate for oxygen ($CMRO_2$) have been used to relate the severity of cerebral ischaemia to long-term cerebral deficit.[1] No such correlation has yet been definitively described in the neonate.

The key to any measurement which will influence clinical decision making is its reliability, ease of use, and, specifically in neonates, its non-invasive nature. $CMRO_2$ has previously been measured in neonates using positron emission tomography (PET),[2] ^{133}Xe clearance,[3] and near infrared spectroscopy (NIRS) methods.[4] To varying degrees, each of the manoeuvres or interventions in these studies has ethical implications which may limit the number of infants in which measurements can be performed. In this paper, we describe a new method for the minimally invasive measurement of regional $CMRO_2$ in neonates undergoing intensive care using a combination of conventional NIRS and spatially resolved spectroscopy methods.

2. BACKGROUND

2.1. Measurement of $CMRO_2$

A generic expression for $CMRO_2$ can be given by:

$$CMRO_2(ml\ O_2.100g^{-1}.\mathrm{min}^{-1}) = CBF \cdot OEF \cdot O_2{}^{art} \qquad (1)$$

where CBF is cerebral blood flow (ml of blood. $100g^{-1}$ tissue.min^{-1}), OEF is the oxygen extraction fraction, and $O_2{}^{art}$ is the arterial oxygen content (ml of O_2.ml $blood^{-1}$). This expression is central to the measurement of $CMRO_2$ by PET, where CBF can be measured using the injection of ^{15}O labelled H_2O, and OEF is measured using inhalation of ^{15}O labelled O_2, as performed on neonates in a study by Altman et al.[2] Arterial blood sampling is also required for the measurement of $O_2{}^{art}$.

Where direct measurement of OEF is not possible, the expression for $CMRO_2$ can be expanded to:

$$CMRO_2(ml.O_2.100g^{-1}.\mathrm{min}^{-1}) = CBF \cdot (SaO_2 - SvO_2) \cdot K \cdot tHb \qquad (2)$$

where SaO_2 is the arterial oxygen saturation (fractional value); SvO_2 is the venous oxygen saturation (fractional value); K is the oxygen combining power of haemoglobin (Hb) (1.306 ml.g^{-1})[5]; tHb is the haemoglobin concentration in blood (g.ml^{-1}). This expression does not incorporate dissolved oxygen, which is a reasonable omission given that within the range of haemoglobin concentrations seen in neonates this would only account for less than 1% of the blood oxygen content. In 1993, Skov et al.[3] measured absolute $CMRO_2$ by measuring CBF using ^{133}Xe clearance and SvO_2 using NIRS and a head down tilt manoeuvre. Yoxall et al.[4] used entirely NIRS-based methods and measured CBF using the oxygen bolus technique[6] and SvO_2 by employing partial jugular venous occlusion. Recent technical developments (described below) have led to the advent of systems which provide an online continuous measurement of absolute mean cerebral tissue oxygen saturation, $SmcO_2$, which is a mean tissue saturation across all vascular compartments in the tissue of interest. The relative contributions to this measurement of the venous and arterial oxygen saturations can be estimated with the venous volume fraction ($V_{ven} \approx 0.75$) using the equation:

$$SmcO_2 = V_{ven} \cdot SvO_2 + (1 - V_{ven}) \cdot SaO_2 \tag{3}$$

Using this relationship, a new expression for $CMRO_2$ can be derived that is independent of SvO_2:

$$CMRO_2\ (ml.100g^{-1}.\min^{-1}) = CBF \cdot \frac{(SaO_2 - SmcO_2)}{V_{ven}} \cdot K \cdot tHb \tag{4}$$

2.2. Near Infrared Spectroscopy Techniques

Conventional NIRS or differential spectroscopy provides a continuous measurement of the quantified changes in concentration of oxy (HbO_2) and deoxyhaemoglobin (HHb). Skov et al.[3] and Yoxall et al.[4] measured the changes in HbO_2 and total haemoglobin concentration (Hbtot) (HbO_2 + HHb) during manoeuvres designed to alter venous blood volume only (i.e. head down tilt and partial jugular venous occlusion). SvO_2 is then easily computed from the ratio of the change in HbO_2 to the change in tHb. Spatially resolved spectroscopy (SRS) depends upon measuring light attenuation as a function of spacing across a series of detectors. This attenuation slope is then fitted to a modified diffusion equation, and assuming a wavelength dependence of scattering, scaled absorption coefficients for HbO_2 and HHb can be derived. From this, the tissue oxygenation index (TOI) can be computed.[7] For measurements in the neonatal brain, this parameter provides a continuous, absolute, non-invasive measurement of the mean cerebral tissue saturation ($SmcO_2$), without the need for any physiological manipulation.

CBF can be measured using NIRS by employing a small change in oxyhaemoglobin concentration as an intravascular tracer.[8] The arterial input of this bolus into the brain is measured using pulse oximetry and the cerebral response function is measured using NIRS via the change in cerebral HbO_2 concentration. A modified Fick principle is then used to calculate CBF in absolute units of $ml.100g^{-1}.min^{-1}$. The aim of this study is to use the measurement of $SmcO_2$ and CBF to investigate the derivation of $CMRO_2$ using Eq. (4).

3. METHODS

Nine infants were studied, all of whom were receiving ventilatory support. The median (range) gestational age was 25 weeks (23-37) and their median (range) birth weight was 800 g (512-2676). The median age (range) at study was 3 days (1-43) post-delivery. Ethical permission for the study was obtained from the local ethics committee, and informed consent was obtained from one or both parents of each infant prior to the study. A NIRO300 spectrometer (Hamamatsu Photonics K.K.) was used to measure CBF and $SmcO_2$ simultaneously over the right and left hemisphere. The optodes were placed over each parietal region with an interoptode spacing of 4 cm. Optical density data were collected at 6 Hz. Arterial oxygen saturation (SaO_2), mean arterial blood pressure, partial pressure of carbon dioxide, and heart rate were simultaneously recorded. For each infant, between 2 and 5 successful measurements of CBF were made. $SmcO_2$ was measured continuously throughout each study and CBF measurements were made after collecting stable $SmcO_2$ data. tHb was measured for each baby from a venous blood sample and a differential pathlength factor of 5.13 was assumed.[9] $CMRO_2$ was calculated using Eq. (4), assuming a venous volume fraction of 0.75 (see Discussion).

4. RESULTS

Table 1 lists the results of the mean $CMRO_2$ measurements for each infant and includes a clinical summary. Median (range) right and left cerebral hemisphere values for $CMRO_2$ were found to be 0.95 (0.74 – 1.52) ml 100 $g^{-1}.min^{-1}$ and 0.90 (0.66 – 1.52) ml 100 $g^{-1}.min^{-1}$, respectively. Using a paired *t* test, no significant difference was seen between the left and right sided values. Figure 1 shows the relationship between the average of the left and right sided $CMRO_2$ values for each infant and (a) gestational age (in completed weeks) at birth and (b) post-menstrual age (in days) at the time of study. Using linear regression analysis, a positive correlation was found between $CMRO_2$ and gestational age ($P < .01$) and postmenstrual age ($P < .02$).

Table 1. Summary of the $CMRO_2$ results and clinical details of the infants studied.

Gestational Age at Birth (completed weeks)	Age at Study (Days)	$CMRO_2$ ($ml.100g^{-1}.min^{-1}$)		Clinical Summary
		Right	Left	
23	1	0.86	0.88	HMD
23	43	0.76	0.88	PDA
24	1	0.87	0.89	HMD
24	2	0.80	0.74	HMD
25	8	0.98	0.69	HMD
28	16	1.23	1.17	HMD
31	2	1.46	1.46	Asphyxia
36	5	0.95	1.31	Asphyxia
37	1	1.53	-	Asphyxia

HMD: hyaline membrane disease, PDA: patent ductus arteriosus.

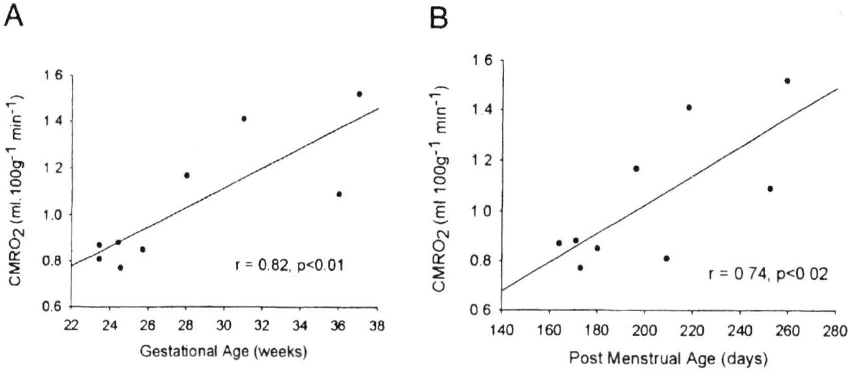

Figure 1. Relationship of $CMRO_2$ with (a) gestational age at birth in completed weeks and (b) post-menstrual age at the time of study in days. The mean of the left and right sided $CMRO_2$ values for each infant was used in the regression analysis.

5. DISCUSSION

We describe a minimally invasive technique for the measurement of regional $CMRO_2$ in neonates undergoing intensive care. The median (range) values we report of 1.03 (0.72 – 1.52) $ml.100g^{-1}.min^{-1}$ are within the range of those quoted by other studies using more invasive methods. Altman et al.[2] used PET to measure $CMRO_2$ in a group of 11 infants and reported values of median 0.4 $ml.100g^{-1}.min^{-1}$ (range 0 – 1.3). Skov et al.[3] reported values of 1.4 $ml.100g^{-1}.min^{-1}$ in 10 asphyxiated term babies and a lower value of 1.0 $ml.100g^{-1}.min^{-1}$ in nine preterm infants with hyaline membrane disease. An NIRS based study by Yoxall et al.[4] reported median values of 0.47 $ml.100g^{-1}.min^{-1}$ (range 0.17 – 1.60) from a group of 20 neonates of median gestational age 27 weeks (range 24 – 41) (values corrected to a DPF of 5.13 and K of 1.304 $ml.g^{-1}$ to allow direct comparison with those reported in the current study) and an increase of $CMRO_2$ with gestational age. The data from all of these studies show that the vast majority of reported values for $CMRO_2$ in neonates are below the value of 1.3 $ml.100g^{-1}.min^{-1}$ assumed to be the lower limit of viability of the adult brain. This lowered level of cerebral metabolism has been attributed to reduced energetic demand in the neonatal brain and decreased levels of functional activation.

The use of continuous, online $SmcO_2$ measurements for the estimate of $CMRO_2$ reported in this paper negates the need for any physiological manoeuvre such as a head down tilt or partial jugular venous occlusion. Central to our method, however, is the assumption of venous volume fraction, V_{ven}, which, in the absence of any published neonatal values, we have assumed to be 0.75 (the value generally employed for studies in adults[10]). However, a number of partial venous jugular occlusion measurements was performed on one of the infants in our study, which allowed the estimation of SvO_2. Combining these data with simultaneous SaO_2 and $SmcO_2$ measurements allowed the calculation of V_{ven} in this infant from the following equation:

$$V_{ven} = \frac{SaO_2 - SmcO_2}{SaO_2 - SvO_2} \tag{5}$$

The mean value of V_{ven} calculated in this single infant was 0.70. Further work is obviously required to determine the range of V_{ven} values to be expected in neonates of different ages and with varying pathology.

To date, the number of infants in which $CMRO_2$ has been measured has been limited by the use of physiological manoeuvres which may not always be appropriate to perform repeatedly on infants undergoing intensive care. When CBF is measured using a NIRS system capable of delivering an absolute measure of mean tissue saturation, the method described in this paper allows the automatic calculation of $CMRO_2$. It is therefore hoped that a large database of $CMRO_2$ values can be acquired in a wide range of infants to allow normal ranges to be defined and the clinical and prognostic significance of such measurements to be assessed.

6. ACKNOWLEDGEMENTS

We thank the parents and staff of the Neonatal Unit, University College Hospitals, London. This work was supported by the Wellcome Trust, Hamamatsu Photonics KK, and the United Kingdom Engineering and Physical Sciences Research Council.

REFERENCES

1. W. J. Powers, R. L. Grubb, D. Darriet, and M. E. Raichle, Cerebral blood flow and cerebral metabolic rate of oxygen for cerebral function in humans, *J. Cereb. Blood Flow Metab.* **5**, 600-608 (1985).
2. D. I. Altman, J. M. Perlman, J. J. Volpe, and W. J. Powers, Cerebral oxygen metabolism in newborns, *Pediatrics* **92**, 99-104 (1993).
3. L. Skov, O. Pryds, G. Griesen, and H. Lou, Estimation of cerebral venous saturation in newborn infants by near infrared spectroscopy, *Pediatr. Res.* **33**, 52-55 (1993).
4. C. W. Yoxall, and M. Weindling, Measurement of cerebral oxygen consumption in the human neonate using near infrared spectroscopy: cerebral oxygen consumption increases with advancing gestational age, *Pediatr. Res.* **44**(3), 283-290 (1998).
5. I. C. Gregory, The oxygen and carbon monoxide capacities of foetal and adult blood, *J. Physiol.* **236**, 625 (1974).
6. A. D. Edwards, J. S. Wyatt, C. E. Richardson, D. T. Delpy, M. Cope, and E. O. Reynolds, Cotside measurement of cerebral blood flow in ill newborn infants by near infrared spectroscopy, *Lancet* **2**(8614), 770-771 (1988).
7. S. J. Matcher, P. Kirkpatrick, K. Nahid, M. Cope, and D. T. Delpy, Absolute quantification methods in tissue near-infrared spectroscopy, *Proc. SPIE* **2389**, 486-495 (1995).
8. C. E. Elwell, M. Cope, A. D. Edwards, J. S. Wyatt, D. T. Delpy, and E. O. R Reynolds, Quantification of adult cerebral haemodynamics by near infrared spectroscopy, *J. Applied Physiol.* **77**, 2753-2760 (1994).
9. A. Duncan, J. Meek, M. Clemence, C. E. Elwell, L. Tyszczuk, M. Cope, and D. T. Delpy, Optical pathlength measurements on adult head, calf and forearm and the head of the newborn infant using phase resolved optical spectroscopy, *Phys. Med. Biol.* **40**, 295-304 (1995).
10. G. McHedlishvili, *Arterial behaviour and blood circulation in brain* (Plenum Press, New York, 1986).

THE PATHWAYS OF OXYGEN IN BRAIN I

Delivery and metabolism of oxygen

Albert Gjedde

Abstract: Flow-metabolism coupling in brain is different from flow-metabolism coupling in other vascular beds. In the classic description of Krogh,[1] the capillary bed is a system of parallel tubes serving cylinders of tissue known as Krogh's cylinders. This simple arrangement yielded a quantitative expression of oxygen delivery to the tissue. However, in brain tissue, the arrangement is so disorderly that no prediction of oxygen tensions in the tissue is possible.[2]

Only two claims of the capillary bed in the brain appear to be indisputable, i.e., the capillaries have a common arterial source and a common venous terminus, and their density is proportional to the average regional rates of metabolism at steady-state. The following revision of the mechanism of flow-metabolism coupling in brain arose from the simple assumption, first introduced by Erwin R. Weibel in *The Pathway for Oxygen,*[3] that every segment of the capillary bed "feeds" the same amount of brain tissue, i.e., that every fraction of the tissue is served by commensurate fractions of capillary density and oxygen diffusibility and accounts for the same fraction of the total oxygen consumption.

1. THEORY

1.1. Blood-Brain Transfer of Oxygen

Let the oxygen extraction fraction along the axial dimension of the capillary bed (z, ranging from 0 to 1) be proportional to the fraction of capillary segments cumulatively served by the blood flowing from the arterial inlet,

$$\varepsilon_{O_2}(z) = z \frac{J_{O_2}}{F C_{O_2}^{\mathrm{art}}} \tag{1}$$

where the oxygen extraction fraction is $\varepsilon_{O_2}(z)$, J_{O_2} is the measured CMR_{O_2}, F is the measured blood flow, and $C_{O_2}^{art}$ is the measured arterial oxygen concentration. The extraction of oxygen from brain capillaries causes the capillary oxygen tension to decline from the arterial oxygen tension at the arterial end of capillaries to the venous oxygen tension at the venous end. The segmental oxygen tension as a function of z is then given by the remaining oxygen, which in turn is given by the segmental extraction fraction, according to the Hill equation of the relationship between oxygen tension and oxygen-hemoglobin dissociation,

$$p_{O_2}^{cap}(z) = \frac{P_{50}^{cap}}{w(z)} \sqrt[h]{\frac{1}{\varepsilon_{O_2}(z)} - 1} \tag{2}$$

where the oxygen tension as a function of position in the capillary is $p_{O_2}^{cap}(z)$, P_{50}^{cap} is the half-saturation tension of oxygen, h is the Hill coefficient, and $w(z)$ is the correction for prior arterial desaturation.†

The oxygen diffusibility, which accounts for the extraction, is given by the ratio between the oxygen flux and the difference between the pressures of oxygen in that

† The term $w(z)$ corrects for the degree of desaturation of hemoglobin that exists prior to the entry into the capillary. In normoxia, the effect is small.[4] The fractional remainder of the arterial oxygen is given by the ratio,

$$1 - \varepsilon_{O_2}(z) = \frac{1 + \left[\dfrac{P_{50}^{art}}{P_{O_2}^{art}}\right]^{h_1}}{1 + \left[\dfrac{P_{50}^{cap}}{p_{O_2}^{cap}}\right]^{h_2}} \tag{3}$$

where h_1 is the Hill coefficient in arterial blood (2.84) and P_{50}^{art} is the half-saturation tension in arterial blood (26 mm Hg), while h_2 is the Hill coefficient in capillaries (3.5 when corrected for Bohr shift) and P_{50}^{cap} is the half-saturation tension in capillaries (35 mm Hg when corrected for Bohr shift). This equation rearranges to,

$$p_{O_2}^{cap}(z) = \frac{P_{50}^{cap} \sqrt[h_2]{\dfrac{1}{\varepsilon_{O_2}(z)} - 1}}{\sqrt[h_2]{\dfrac{1}{\varepsilon_{O_2}(z)}\left(\dfrac{P_{50}^{art}}{P_{O_2}^{art}}\right)^{h_1} + 1}} = P_{50}^{cap} \sqrt[h_2]{\frac{1 - \varepsilon_{O_2}(z)}{\left(\dfrac{P_{50}^{art}}{P_{O_2}^{art}}\right)^{h_1} + \varepsilon_{O_2}(z)}} \tag{4}$$

in which the denominator has the value 1.02 for an extraction of 20% at an arterial oxygen tension of 114 mm Hg (our standard) and the value of 1.16 at a tension of 52 mm Hg for an extraction of 20%. Since the fraction of physically dissolved oxygen is about 2% in normoxia (but less in hypoxia), the effective denominator is lower (as the physically dissolved oxygen "spares" some of the bound oxygen during extraction).

segment of the capillary bed and the pressure in the mitochondria of the tissue segment served by the segment of capillary bed,

$$l(z) = \frac{J_{O_2}}{p_{O_2}^{cap}(z) - P_{O_2}^{mit}} \tag{5}$$

where the apparent oxygen diffusibility is $l(z)$ as a function of the segmental position. The true diffusibility of oxygen depends on the distribution of mitochondria in the tissue. If the distribution is even, the true diffusibility is twice $l(z)$ on the average. If the mitochondria are placed at the end of the diffusion path, $l(z)$ is an estimate of the real diffusibility of oxygen. The term $p_{O_2}^{cap}(z)$ is the tension in the capillary segment, which was assumed to equal the arterial oxygen tension $P_{O_2}^{art}$ and the venous oxygen tension P_{VO_2}. $P_{O_2}^{mit}$ is the minimum mitochondrial oxygen tension, which we claim, as an initial assumption, to be approximately uniform in brain.

The equations show that the apparent diffusibility increases regularly from the arterial to the venous end of the capillary bed. The diffusibility depends on the surface area of the capillary segment and the average distance from the segment to the mitochondria served by the segment.

The linearity of the extraction fraction as a function of the capillary segmentation implies that the entire axial dimension can be projected onto a single transaxial plane at the position $z = 0.5$, where the extraction fraction is half of the totally extracted fraction ($\varepsilon_{O_2}(0.5) = E_{O_2}/2$). The weighted average capillary oxygen tension of all segments of the capillary bed is then,[5]

$$P_{O_2}^{cap} = \frac{P_{50}^{cap}}{\Omega} \sqrt[h]{\frac{2}{E_{O_2}} - 1} \tag{6}$$

where Ω is the correction for prior arterial desaturation, when the negligible component of physically dissolved oxygen is ignored.[§]

The corresponding weighted average oxygen diffusibility is,

$$L = \frac{J_{O_2}}{P_{O_2}^{cap} - P_{O_2}^{mit}} \tag{8}$$

[§] The term Ω is the average capillary correction for the degree of desaturation of hemoglobin that exists prior to entry into the capillary. In agreement with Eq. (4), it obeys the expression,

$$\Omega = \sqrt[h_2]{\frac{2}{E_{O_2}} \left(\frac{P_{50}^{art}}{P_{O_2}^{art}} \right)^{h_1} + 1} \tag{7}$$

which has the value 1.02 for an extraction of 40% at an arterial oxygen tension of 114 mm Hg (our standard), the value of 1.16 for an extraction of 40% at a tension of 52 mm Hg, and 1.12 for an extraction of 60% at a tension of 52 mm Hg.

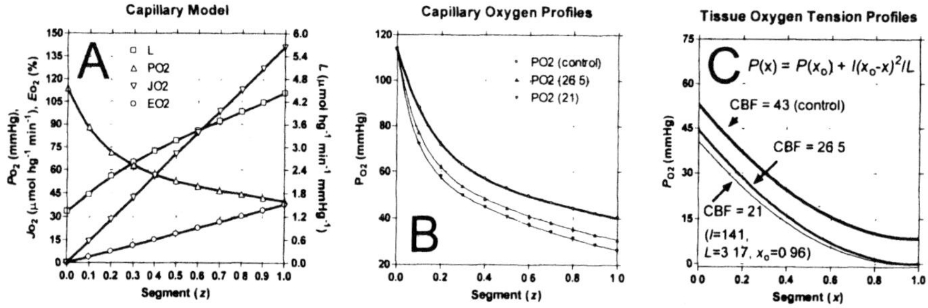

Figure 1. Oxygen tension, extraction fraction, and oxygen diffusibility as a function of axial segmentation of cortical capillary bed (z direction), calculated for baseline (A) and two oligemia states (B) of current study. Also shown in C are the tissue oxygen tension profiles (C) for volume fraction x at $z = 0.5$ for the baseline and oligemia states shown in B. *Panel* A: Abscissa: Axial segments of capillary bed (z). Ordinate: Capillary oxygen extraction fraction $(\varepsilon_{O_2}(z))$, tension $(p_{\mathrm{capo}_2}(z))$, and diffusibility $(l(z))$ as functions of axial segment. J_{O_2} is the measured $CMRO_2$ (141 µmol hg^{-1} min^{-1} at the baseline), F the measured blood flow (43 ml hg^{-1} min^{-1}), $C_{O_2}^{\mathrm{art}}$ the measured arterial oxygen concentration (8.7 mM), P_{50}^{cap} the calculated half-saturation tension of oxygen (35 mm Hg, when corrected for the Bohr shift), and h the Hill coefficient (3.5 when corrected for Bohr shift). $P_{\mathrm{capo}_2}(0)$ was assumed to equal the arterial oxygen tension $P_{O_2}^{\mathrm{art}}$ (114 mm Hg). *Panel P:* Abscissa: Axial segment of capillary bed (z). Ordinate: Capillary oxygen tensions $(p_{O_2}^{\mathrm{cap}}(z))$ as function of axial segment, calculated as explained in A for baseline (CBF 43 ml hg^{-1} min^{-1}), and two oligemia states (CBF 26.5 and 21 ml hg^{-1} min^{-1}), chosen to be just at and just below the limit of adequate oxygen delivery. The numerical solution of the three equations with the three unknowns $\varepsilon_{O_2}(z)$, $p_{O_2}^{\mathrm{cap}}(z)$, and $x_O(z)$ yielded the capillary oxygen tension profile for insufficient oxygen supply. *Panel C:* Abscissa: Tissue volume fraction (x). Ordinate: Oxygen tension profiles $(p_{O_2}(x))$ as a function of tissue volume fraction for three cases of panel B, using equation indicated in panel. The end-path oxygen tension was 8.5 mm Hg.

where J_{O_2} is the tissue's metabolic rate of oxygen (CMR_{O_2}), and L is the oxygen diffusibility at $z = 0.5$, when $\varepsilon_{O_2}(z)$ rises linearly. It follows that the oxygen consumption is given by,

$$J_{O_2} = L\left[P_{50}^{\mathrm{cap}}\sqrt[h]{\frac{2}{E_{O_2}} - 1} - P_{O_2}^{\mathrm{mit}}\right] \tag{9}$$

2.2. Metabolism of Oxygen

The minimum mitochondrial oxygen tension represents the balance between the diffusion of oxygen to the site of metabolism and the reaction of oxygen with cytochrome *c* oxidase, according to the simple Michaelis-Menten expression,[6-8]

$$J_{O_2} = \frac{J_{\max} P_{O_2}^{\mathrm{mit}}}{P_{50}^{\mathrm{cytox}} + P_{O_2}^{\mathrm{mit}}} \tag{10}$$

where P_{50}^{cytox} is the apparent half-saturation tension of the oxygen reaction with cytochrome oxidase, and J_{max} is maximum velocity of the cytochrome oxidase reaction. Equation (10) eliminates $P_{O_2}^{\text{mit}}$ from Eq. (9) and yields the formula, which prescribes the rate of blood flow needed to supply the tissue with a given amount of oxygen,‡

$$F = \frac{J_{O_2}}{2\, C_{\text{hb}}}\left(1+\left[\frac{P_{50}^{\text{cap}}}{P_{a_{O_2}}}\right]^h\right)\left(1+\left(\left[1+\frac{L\, P_{50}^{\text{cytox}}}{J_{\text{max}}-J_{O_2}}\right]\left[\frac{J_{O_2}}{L\, P_{50}^{\text{cap}}}\right]\right)^h\right) \qquad (11)$$

where F is the blood flow to the tissue (per unit weight or volume), J_{O_2} is the average oxygen metabolism of the tissue, C_{hb} is the arterial hemoglobin concentration, L is the oxygen diffusibility, P_{50}^{cap} is the half-saturation oxygen tension of hemoglobin (equal to 35 mm Hg in human beings, when corrected for the Bohr shift), h is Hill's coefficient (equal to 3.5, also when corrected for the Bohr shift) and $P_{O_2}^{\text{art}}$ is the arterial oxygen tension. The power of h predicts that blood flow must change more than the rate of oxygen consumption when oxygen metabolism rises.

3. RESULTS AND DISCUSSION

Equation (11) is the quantitative expression of the flow-metabolism couple. The basis for the equation is shown in Figure 1. Panel A shows that the oxygen extraction as a function of capillary segment rises linearly. Panels B and C show that blood flow limits the oxygen delivery below a threshold of 20-25 ml hg^{-1} min^{-1}, corresponding to an oxygen extraction of 60%. Below this rate of blood flow, the oxygen supply is not sufficient to satisfy the demand.

We validated the flow-metabolism formula for cases of visual cortex activation,[9, 13] cortical motor activation,[10] and hypothermia,[12] shown in panels A-C of Figure 2. Panel B of Figure 2 also includes data published by Kastrup et al.[11] In panels A and B of Figure 2, the curves indicate the prediction provided by Eq. (11) for human cortical gray matter having an oxygen diffusibility of approximately 4 μmol hg^{-1} min. mm Hg^{-1}, while panel C predicts the relationship in progressive hypothermia of porcine cortical tissue with a slightly higher oxygen diffusibility of approximately 5 μmol hg^{-1} min. mm Hg^{-1} because of the smaller size of pig brains. The couple downregulates blood flow rates when the energy demand of neurons falls at low temperatures.[12]

Fundamentally similar models of non-linear flow-metabolism coupling were formulated in recent years.[3, 5, 14-16] The inverse relation between oxygen extraction fraction and oxygen tension of the capillary bed is central to these revisions of the

‡ Calculation of L from Eq. (8) corrects for the degree of hypoxemic desaturation of arterial blood, which exists prior to entry into the capillary, assuming that the magnitude of L is adjusted to correct for desaturation by a process of recruitment. Equation (11) shows that the combined desaturation prior to capillary entry and commensurate adjustment of the blood/brain oxygen diffusion capacity serve to render the baseline magnitude of L valid for all degrees of desaturation.

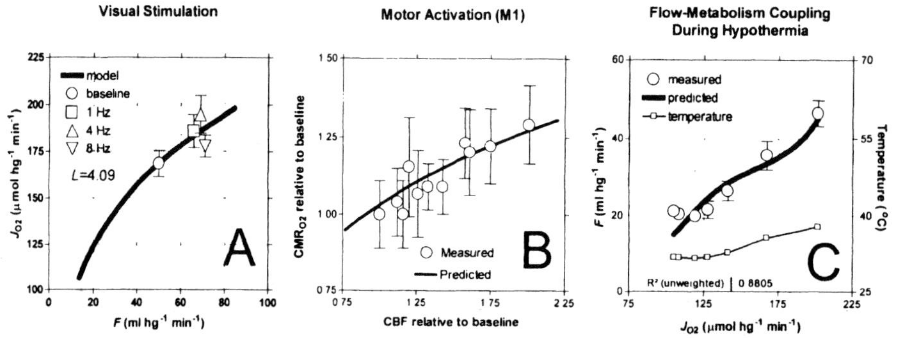

Figure 2. Validation of non-linear flow-metabolism couple for cases of brain tissue activation and deactivation. In panels A and B, heavy curves indicate prediction for human cortical gray matter with oxygen diffusibility of approximately 4 µmol hg^{-1} min. mm Hg^{-1}. Panel C predicts behavior for porcine cortical tissue with slightly higher oxygen diffusibility of approximately 5 µmol hg^{-1} min. mm Hg^{-1}. *Panel A:* Abscissa: Cortical blood flow. Ordinate: Oxygen consumption in human visual cortex activation by visual stimulation (modified from Vafaee et al.[9]). *Panel B:* Abscissa: Cortical blood flow. Ordinate: Oxygen consumption in human motor cortex activation by finger motion (modified from Gjedde et al.[10] and Kastrup et al.[11]). *Panel C:* Abscissa: Oxygen consumption. Ordinate: Cortical blood flow and cerebral core temperatures during hypothermia of porcine cortex (modified from Sakoh & Gjedde[12]).

classical flow-metabolism couple. The inverse relation is necessary because every increase of the extraction fraction represents a reduction of the partial pressure of oxygen. The non-linear flow-metabolism couple predicted and received subsequent confirmation of the well-known tolerance of brain oxidative metabolism to low arterial oxygen tension.[4]

4. ACKNOWLEDGEMENTS

The Institute of Experimental Clinical Research (University of Aarhus), the Scandinavia-Japan Sasakawa Foundation, the Medical Research Council of Canada and Denmark, and the National Science Foundation of Denmark (center of excellence grant for "Center of Functionally Integrative Neuroscience" at Aarhus University Hospitals) supported these studies.

REFERENCES

1. A. Krogh, The number and distribution of capillaries in muscles with calculations of the oxygen pressure head necessary for supplying the tissue, *J. Physiol.* **52**, 405-415 (1919).
2. C. Y. Wang, and J. Bassingthwaighte, Capillary supply regions, *Math. Biosci.* **173**, 103-114 (2001).
3. E. R. Weibel, *The Pathway for Oxygen* (Harvard University Press, Cambridge, 1984).
4. A. Gjedde, Cerebral blood flow change in arterial hypoxemia is consistent with negligible oxygen tension in brain mitochondria, *NeuroImage* **17**, 1876-1881 (2002).
5. M. S. Vafaee, and A. Gjedde, Model of bloodbrain transfer of oxygen explains nonlinear flow metabolism coupling during stimulation of visual cortex, *J. Cereb. Blood Flow Metab.* **20**, 747-754 (2000).
6. E. Gnaiger, R. Steinlechner-Maran, G. Mendez, T. Eberl, and R. Margreiter, Control of mitochondrial and cellular respiration by oxygen, *J. Bioenerg. Biomembr.* **27**, 583-596 (1995).

7. E. Gnaiger, B. Lassnig, A. Kuznetsov, G. Rieger, and R. Margreiter, Mitochondrial oxygen affinity, respiratory flux control and excess capacity of cytochrome c oxidase, *J. Exp. Biol.* **201**, 1129-1139 (1998).
8. C. Guilivi, Functional implications of nitric oxide produced by mitochondria in mitochondrial metabolism, *Biochem. J.* **332**, 673-679 (1998).
9. M. S. Vafaee, E. Meyer, S. Marrett, T. Paus, A. C. Evans, and A. Gjedde, Frequencydependent changes in cerebral metabolic rate of oxygen during activation of human visual cortex, *J. Cereb. Blood Flow Metab.* **19**, 272-277 (1999).
10. A. Gjedde, S. Marrett, M. Sakoh, and M. Vafaee, Model of oxygen delivery to brain tissue in vivo explains beneficial effect of hypothermia in ischemia, in: *Brain Activation and CBF Control*, edited by M. Tomita, I. Kanno, and E. Hamel (Elsevier, Tokyo, 2002), pp. 223-229.
11. A. Kastrup, G. Kruger, T. Neumann-Haefelin, G. H. Glover, and M. E. Moseley, Changes of cerebral blood flow, oxygenation, and oxidative metabolism during graded motor activation, *NeuroImage* **15**, 74-82 (2002).
12. M. Sakoh, and A. Gjedde, Neuroprotection in hypothermia linked to redistribution of oxygen in brain, *Am. J. Physiol.* **285**, H17-H25 (2003).
13. A. Gjedde, and S. Marrett, Glycolysis in neurons, not astrocytes, delays oxidative metabolism of human visual cortex during sustained checkerboard stimulation in vivo, *J. Cereb. Blood Flow Metab.* **21**, 1384-1392 (2001).

THE PATHWAYS OF OXYGEN IN BRAIN II

Competitions for cytochrome c oxidase and NOS are keys to flow-metabolism coupling

Albert Gjedde

Abstract: It has been well-known for many years that cerebral oxygen consumption remains constant during moderate changes of blood flow, as measured during hypo- or hypercapnia or indomethacin administration. Current models of flow-metabolism coupling link blood-brain transfer of oxygen to oxygen metabolism in mitochondria.[2-6] The resulting quantitative relations between flow and metabolism reveal that a close link between diffusion and metabolism prevents the enzyme from maintaining a constant oxygen consumption when blood flow changes, unless the enzyme's affinity towards oxygen is adjusted commensurately.

1. INTRODUCTION

It has been well-known for many years (e.g., James et al.[1]) that cerebral oxygen consumption remains constant during moderate changes of blood flow, as measured during hypo- or hypercapnia or indomethacin administration. Current models of flow-metabolism coupling link blood-brain transfer of oxygen to oxygen metabolism in mitochondria.[2-6] The resulting quantitative relations between flow and metabolism reveal that a close link between diffusion and metabolism prevents the enzyme from maintaining a constant oxygen consumption when blood flow changes, unless the enzyme's affinity towards oxygen is adjusted commensurately.

The quantitative relation given by the author in this volume is,

$$F = \frac{J_{O_2}}{2\,C_{hb}}\left(1+\left[\frac{P_{50}^{cap}}{P_{a_{O_2}}}\right]^h\right)\left(1+\left(\left[1+\frac{L\,P_{50}^{cytox}}{J_{max}-J_{O_2}}\right]\left[\frac{J_{O_2}}{L\,P_{50}^{cap}}\right]\right)^h\right) \tag{1}$$

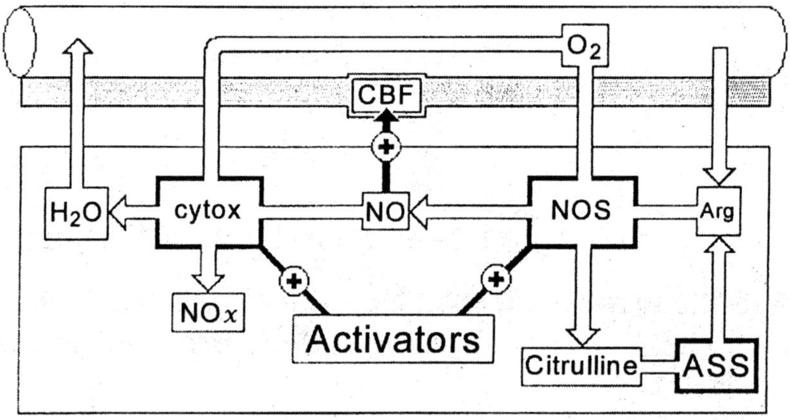

Figure 1. Model of NO synthesis from oxygen and competition of NO and oxygen for metabolism by cytochrome oxidase.[7, 8] At the core of the model is competition for cytochrome oxidation by flow-related competitor, which is likely to be NO. The model predicts that affinity of cytochrome oxidase for oxygen declines when blood flow rate is elevated because of a direct or indirect link to flow-related competitor. NO and NOS refer to flow-related competitor and its synthesizing enzyme, CBF is cerebral blood flow, Arg is arginine, cytox is cytochrome oxidase, and ASS is argininosuccinate synthase.

where F is the blood flow to the tissue (per unit weight or volume), J_{O_2} is the average oxygen metabolism of the tissue, C_{hb} is the arterial hemoglobin concentration, L is the oxygen diffusibility, P_{50}^{cap} is the half-saturation oxygen tension of hemoglobin (equal to 35 mm Hg in human beings, when corrected for the Bohr shift), h is Hill's coefficient (equal to 3.5, also when corrected for the Bohr shift), and $P_{O_2}^{art}$ is the arterial oxygen tension.

It is known that the apparent affinity, or P_{50}^{cytox}, declines when the turnover rate of the enzyme rises, but this mechanism does not explain the maintenance of constant oxygen consumption in the absence of a turnover change.

Many blood flow stimulators act by means of nitric oxide (NO) including carbon dioxide and hydrogen ions, which are products of brain metabolism.[9] This chapter tests the hypothesis that the interaction of oxygen and a hypothetical flow-related competitor for cytochrome oxidase, which is likely to be nitric oxide (NO), links the actions of cytochrome oxidase and nitric oxide synthase (NOS). The effects of this linkage are consistent with current knowledge of the properties of flow-metabolism coupling.

2. MODEL

The model of hypothetical oxygen and NO interactions at cytochrome oxidase and NO in mitochrondria[7, 8] is shown in Figure 1. In addition to Eq. (1), which relates the blood flow and cytochrome oxidation rates, the model incorporates the following linked interactions: first, increase of the normalized concentration of the flow-related competitor (Δx) raises the P_{50} of cytochrome oxidase,

$$P_{50}^{\text{cytox}} = P_{\text{M}}^{\text{cytox}}\left(1+\Delta x_i\right) \tag{2}$$

where $P_{\text{M}}^{\text{cytox}}$ is the inherent (highest) affinity. Second, the rate of blood flow is a function of the magnitude of the flow-related competitor,

$$F = F_o e^{k\ \Delta x_i} \tag{3}$$

where F_o is the baseline flow and k is a scaling factor. Third, the change of the flow-related competitor is the difference between its total level and a baseline,

$$\Delta x_i = x_i - x_o \tag{4}$$

where Δx_i is the change of the flow-related inhibitor, and x_i and x_o are the baseline and total levels of the flow-related competitor. Fourth, the level of the hypothetical flow-related competitor is determined by the relative strengths of synthesis by NOS and metabolism by cytochrome oxidase,

$$Sx_i = x_{\max} S_{\arg} S_{O_2} \tag{5}$$

where Sx_i is the competitor's occupancy of cytochrome oxidase. NOx represents the metabolites of the flow-related competitor, $x_{\max}$ is the ratio between the maximum reaction rates of NOS and cytochrome oxidase, and $S_{\arg}$ and S_{O_2} are the arginine and oxygen occupancies of NOS.

3. METHODS

The linked Eqs. (1)-(5) were solved numerically for changes of cerebral blood flow resulting from changes of NOS $(x_{\max})$ or cytochrome oxidase activities ($J_{\max}$). To permit the numerical solution, the enzyme occupancies of Eq. (5) were further expanded to incorporate the effects of increased turnover on substrate affinity and competition among multiple paths of metabolism. Thus,

$$Sx_i = \frac{x_i}{x_{\text{M}}+x_i} \tag{6}$$

where x_i is the normalized concentration of the flow-related competitor, and x_{M} the corresponding half-saturation concentration. The arginine occupancy was given by,

$$S_{\arg} = \frac{x_{\arg}\left(R - Sx_i\right)}{1 + x_{\max} + x_{\arg}\left(R - Sx_i\right)} \tag{7}$$

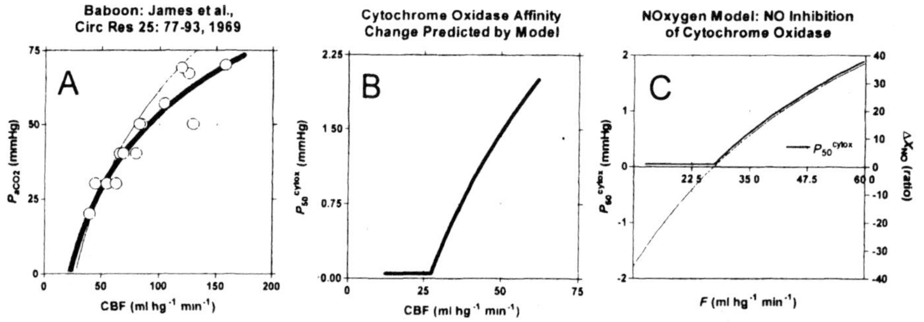

Figure 2. Evidence of flow-related inhibitor action in flow-metabolism coupling. Panel A shows summary of relationships between cerebral blood flow rates (CBF) in baboon and arterial carbon dioxide tensions, reported by James et al.[1] The heavy curve shows regression to published values; thin curve shows line presented in panel B. Panel B shows change of cytochrome oxidase affinity for oxygen P_{50}^{cytox} required to maintain constant oxygen consumption during extrinsically imposed flow change in generic human brain with baseline values listed in Table 1. Panel C shows accompanying change of hypothetical flow-related competitor.

where x_{arg} is the normalized arginine concentration, and $R - Sx_i$ represents the combined influences of recirculation and recruitment of blood-brain arginine transfer,

$$R = 0.6F\left(1 - e^{-1.5\,PS/F}\right)/\,25.8 \tag{8}$$

Table 1. Baseline literature values and range of variables and constants used in analysis of affinity-dependent model of flow-metabolism couple.[10] The terms "NO" and "NOS" represent the hypothetical flow related competitor.

Substrate	*Enzyme/Protein/Process*	*Variable*	*Baseline (Range)*	*Unit*
Oxygen	Hemoglobin	C_{hb}	9	µmol/ml
		$P_{\mathrm{O_2}}^{\mathrm{art}}$	114	mm Hg
		P_{50}^{cap}	35	mm Hg
		h	3.5	ratio
	Diffusibility	L	3.2	µmol/hg/min/mm Hg
	Cytochrome *c* oxidase	J_{max}	159 (142-250)	µmol/hg/min
		$P_{\mathrm{M}}^{\mathrm{cytox}}$	0.05	mm Hg
		$J_{\mathrm{O_2}}$	141	µmol/hg/min
	NOS	$P_{\mathrm{M}}^{\mathrm{NOS}}$	1	mm Hg
		x_{max}	20.5 (0-120)	ratio
		x_{arg}	1	ratio
		PS	200	ml/hg/min
NO	Arteriolar dilatation	F_o	26.91	ml/hg/min
		k	0.0213	ratio
	Cytochrome *c* oxidase	x_{M}	200	ratio
		x_o	30	ratio

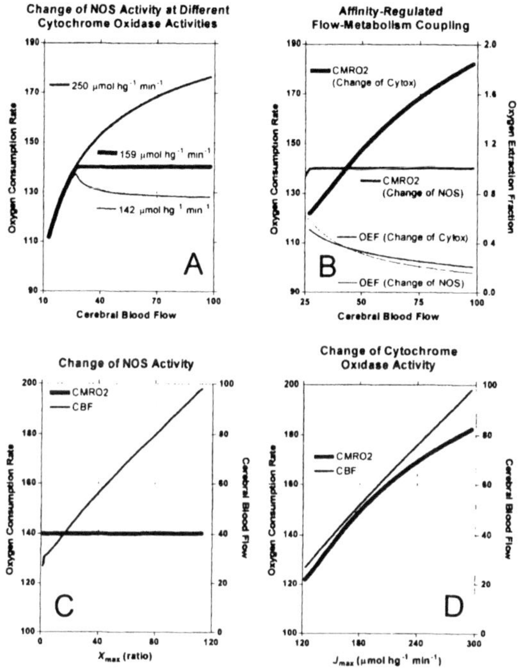

Figure 3. Analysis of affinity-dependent model of flow metabolism couple. Panel A shows dependence of oxygen consumption rate on cerebral blood flow. Curves show oxygen consumption rates for three maximum cytochrome reaction rates listed next to line symbols. Panel B shows metabolism-induced change of blood flow (heavy curve) and constant oxygen consumption rate in presence of extrinsically imposed flow change (light curve). Panel C shows influence of extrinsically imposed change of maximum NOS reaction rate on oxygen consumption rate (heavy curve) and cerebral blood flow (light curve). Panel D shows influence of maximum rate of cytochrome oxidase reaction on oxygen consumption rate (heavy curve) and cerebral blood flow (light curve).

and increased competition with the breakdown of the flow-related competitor (S_{X_i}). The recruitment effect was based on the known permeability of the blood-brain barrier to basic amino acids. The oxygen occupancy was given by,

$$S_{O_2} = \frac{x_{O_2}}{1 + x_{\max} + x_{O_2}} \tag{9}$$

where x_{O_2} is the relative oxygen tension, equal to the $P_{O_2}^{\text{mit}} / P_{\text{M}}^{\text{NOS}}$ ratio, and $x_{\max}$ accounts for the effect of enzyme turnover on the affinity.

4. RESULTS

The apparent insensitivity of oxygen consumption to flow change above a certain threshold is explained by the effect of the hypothetical adjustment of cytochrome oxidase affinity. The hypothesis claims that the affinity of the enzyme undergoes the inverse change of blood flow in the presence the enzyme competitor, as shown in Figure 2.

The analysis summarized in Figure 3 reveals that changes of the maximum activity of cytochrome oxidase (J_{max}) affect oxygen consumption only at blood flow rates above the threshold, which depends to an extent on the rate of blood flow (Panel A). Note the paradoxical increase of oxygen consumption induced by lowering of blood flow when cytochrome oxidase activity is low. This observation matches experimental findings by Nemoto et al.[11] in patients with oligemia of the brain, and it raises the question of how changes of cytochrome oxidase activity, induced by changes of cellular metabolism, in turn lead to the appropriate changes of blood flow required to match the necessary blood flow to the change of metabolic demand. The solution to the problem of the apparent insensitivity of blood flow changes to changes of cellular metabolism, shown in panel B of Figure 3, emerges from the suggestion of reciprocal inhibition of the synthesis or release of the hypothetical flow-related inhibitor exerted by the level of oxygen in the mitochondria.

Oxygen inhibits the synthesis of the flow-related inhibitor, which in turn inhibits cytochrome oxidase. Increased consumption of oxygen must then release the synthesis of the flow marker from the inhibition exerted by oxygen. The hypothetical function of such reciprocity is predicted in panels C and D of Figure 3. Purely for simplicity, nitric oxide synthase (NOS) is named as one possible synthesizer of the hypothetical flow-related inhibitor. Panel D of Figure 3 shows that the reciprocal inhibition causes changes of cytochrome oxide activity that lead to simultaneous changes of blood flow and oxygen consumption, while changes of the affinity of the flow marker synthesis lead to changes only of blood flow. The result is a mechanism explaining the known properties of flow-metabolism coupling.

5. DISCUSSION

The current explanation of the non-linear coupling of cerebral blood flow to oxygen consumption during neuronal excitation holds that blood flow must rise more than oxygen consumption to compensate for the lack of an oxygen reserve in brain mitochondria. This hypothesis fails to explain how oxygen consumption is maintained during moderate reductions of cerebral blood flow.

In the model of flow-metabolism coupling presented in Figure 1, two enzymes (cytochrome oxidase and NOS) compete for the same substrate (oxygen), while two substrates (oxygen and NO) compete for the same enzyme (cytochrome oxidase). The model adds the influence of NO on blood flow and the influence of blood flow on oxygen delivery, as expressed in Eq. (1). Independent variations of cytochrome oxidase and NOS activity solved the model equations for dependent variations of blood flow and metabolism. The solutions predicted parallel changes of oxygen consumption and blood flow for a change of cytochrome axidase activity, as shown in panels B and D of Figure 3. However, the model predicted a change only of blood flow when NOS activity varied, as shown in panel C of Figure 3. This pattern is consistent with earlier and more recent observations on the effect of NO on the mechanism of flow-metabolism coupling.[12, 13]

6. ACKNOWLEDGMENTS

The author acknowledges the support of the Institute of Experimental Clinical Research, University of Aarhus, Denmark. Additional support was generously provided by the Scandinavia-Japan Sasakawa Foundation, the Medical Research Councils of Canada and Denmark, and the National Science Foundation of Denmark (center of excellence grant to the Center of Functionally Integrative Neuroscience at Aarhus University Hospitals).

REFERENCES

1. I. M. James, R. A. Millar, and M. J. Purves, Observations on the extrinsic neural control of cerebral blood flow in the baboon, *Circ. Res.* **25**, 77-93 (1969).
2. A. Gjedde, S. Ohta, H. Kuwabara, and E. Meyer, Is oxygen diffusion limiting for bloodbrain transfer of oxygen? in: *Brain Work and Mental Activity, Alfred Benzon Svmposium 31*, edited by N. A. Lassen, D. H. Ingvar, M. E. Raichle, and L. Friberg (Munksgaard, Copenhagen, 1991), pp. 177-184.
3. A. Gjedde, The relation between brain function and cerebral blood flow and metabolism, in: *Cerebrovascular Disease*, edited by H. H. Batjer (Lippincott-Raven, Philadelphia, 1997), pp. 23-40.
4. R. B. Buxton, and L. R. Frank, A model for the coupling between cerebral blood flow and oxygen metabolism during neural stimulation, *J. Cereb. Blood Flow Metab.* **17**, 64-72 (1997).
5. F. Hyder, R. G. Shulman, and D. L. Rothman, A model for the regulation of cerebral oxygen delivery, *J. Appl. Physiol.* **85**, 554-564 (1998).
6. M. S. Vafaee, and A. Gjedde, Model of blood-brain transfer of oxygen explains nonlinear flow-metabolism coupling during stimulation of visual cortex, *J. Cereb. Blood Flow Metab.* **20**, 747-754 (2000).
7. C. Guilivi, J. J. Poderoso, and A. Boveris, Production of nitric oxide by mitochondria, *J. Biol. Chem.* **273**, 11038-11043 (1998).
8. S. L. Effering, T. M. Sarkela, and C. Guilivi, Biochemistry of mitochondrial nitric-oxide synthase, *J. Biol. Chem.* **277**, 38079-38086 (2002).
9. A. Villringer, and U. Dirnagl, Coupling of brain activity and cerebral blood flow: basis of functional neuroimagiag, *Cerebrovasc. Brain Metab. Rev.* **7**, 240-276 (1995).
10. A. Gjedde, S. Marrett, and M. Vafaee, Oxidative and nonoxidative metabolism of excited neurons and astrocytes, *J. Cereb. Blood Flow Metab.* **22**, 1-14 (2002).
11. E. M. Nemoto, H. Yonas, H. Kuwabara, R. R. Pindsola, D. Sashin, C. C. Meltzer, J. C. Price, and Y. Chang, Detection of stage II compromised cerebrovascular reserve by xenon CT cerebral blood flow with acetazolamide and oxygen extraction fraction by positron emission tomography, in: *Brain Imaging Using PET*, edited by M. Senda, and P. Herscovitch (Elsevier, San Diego, 2002).
12. P. J. Goadeby, H. Kaube, and K. L. Hoskin, Nitric oxide synthesis couples cerebral blood flow and metabolism, *Brain Res.* **595**, 167-170 (1992).
13. O. Z. Chi, X. Liu, and H. R. Weiss, Effects of inhibition of neuronal nitric oxide synthase on NMDA-induced changes in cerebral blood flow and oxygen consumption, *Exp. Brain Res.* **148**, 256-260 (2003).

EXOGENOUS AND ENDOGENOUS MARKERS OF TUMOUR OXYGENATION STATUS

Definitive markers of tumour hypoxia?

Kaye J. Williams, Catriona A. Parker, and Ian J. Stratford

Abstract: Hypoxia is a physiological abnormality that has been detected in all solid tumours analysed to date. Studies using polarographic needle electrodes have shown an unequivocal link between the extent of tumour hypoxia and poor treatment outcome. The practical limitations of polarographic needle electrodes have warranted investigation into alternative strategies enabling routine assessment of tumour hypoxia in the clinical setting. This review focuses on the clinical evaluation of exogenous and endogenous markers of tumour hypoxia that may fulfil this role.

1. INTRODUCTION

Hypoxia arises in tumours as a consequence of a poor and disorganised blood supply. The rapid growth of tumour cells exceeds that of the underlying capillary network. Regions of chronic hypoxia ensue where tumour cells lie at a distance from a perfused capillary that exceeds the limits of oxygen diffusion. In addition, high interstitial pressure within tumours can result in vascular occlusion leading to areas of acutely hypoxic cells surrounding the temporarily closed vessels. Importantly, hypoxic cells within tumours remain viable and are either quiescent or proliferate at a markedly lower rate than well-oxygenated tumour cells.[1, 2] Furthermore, hypoxic cells are resistant to radiotherapy and to some forms of chemotherapy.[3] A prevailing belief exists that these radio- and chemo-resistant cells may act as foci for tumour re-growth following treatment. In addition, hypoxia has been linked with a more aggressive disease phenotype in both the experimental and clinical setting.[3-5] These observations have incited numerous studies to establish the prognostic significance of hypoxia in cancer therapy with the aim of providing a rational basis for treatment stratification dependent upon tumour oxygenation status.

2. CLINICAL SIGNIFICANCE OF TUMOUR HYPOXIA

It was first recognised in 1955 that human tumours contain micro-regions of potentially hypoxic cells.[6] The implications from a radiobiological point of view were immediately recognised given the prior elucidation of oxygen concentration as a key determinate of radiation response.[7] In spite of these early observations, robust and direct measurements of tumour oxygenation were not possible until the development of polarographic needle electrodes in the late 1980's. Following the establishment of routine protocols for tumour oxygen electrode measurements, clear demonstrations of the importance of pre-treatment oxygenation status on the outcome of radiotherapy have ensued. Hypoxia has been found to be a poor prognostic indicator for radiotherapy outcome in carcinoma of the breast[8] and cervix,[9, 10] head and neck cancer,[11] soft tissue sarcoma,[12] and squamous cell carcinoma lymph node metastases.[13] The observations of Höckel and co-workers in 1996[5] confounded the initial thoughts that hypoxia linked with poor prognosis following radiotherapy only as a consequence of the inherent resistance of cells under low oxygen tension to radiation treatment. They found that hypoxia was an equally poor prognostic indicator for cervical carcinoma treated with surgery alone. These observations correlated with studies in murine tumours demonstrating that hypoxia could promote a more aggressive tumour phenotype through the promotion of genomic instability and the direct selection of cells with reduced apoptotic potential.[4] Indeed, hypoxia coupled with low apoptotic index defines a subset of patients with highly aggressive carcinomas of the cervix.[14]

3. EXOGENOUS AND ENDOGENOUS MARKERS OF TUMOUR HYPOXIA

Clearly, the use of polarographic needle electrodes has enabled the true importance of tumour hypoxia to be established in the clinical setting. However, the use of oxygen electrodes is restricted to accessible cancers thereby preventing their wide scale application to all tumour types. In addition, the electrode consumes oxygen by electrochemical reduction, and hence the probe has to be moved step-wise through a tissue to avoid an under estimation of the oxygen tension. This prevents the analysis of temporal changes in oxygen tension at a single location that may be important in determining the efficacy of treatments aimed at increased tumour oxygenation. The OxyLite™ (Oxford Optronix, UK) is a fibre optic oxygen sensor that enables temporal changes in oxygen tension to be measured[15] and gives comparable assessment of tumour oxygenation to that determined by polarographic needle electrodes in experimental models.[16] Although overcoming one limitation of the oxygen electrode, the OxyLite™ still requires an accessible tumour.

Exogenous and endogenous markers are currently under evaluation as alternative approaches to enable an assessment of tumour hypoxia. Their use is potentially applicable to all solid neoplasms and precludes the requirement for specialised equipment and procedures. Exogenous markers are essentially bioreductive drugs that can be administered systemically. Conditions of low oxygen tension (below 10 mm Hg) facilitate the bioactivation of these agents resulting in adduct formation that can be detected in tumour sections or disaggregated tumour cells using the appropriate antibodies. Endogenous markers of tumour hypoxia are normal cellular proteins whose expression is linked with exposure to hypoxic stress. Again, expression can be evaluated in standard biopsy material using immunohistochemical techniques. Of importance is to establish whether these markers are indeed definitive indicators of tumour hypoxia.

3.1. Exogenous Markers

Two bioreductive markers combined with immunohistochemistry have been developed for clinical use, pimonidazole[17] and EF-5.[18] Both are 2-nitroimidazoles. Pre-clinical studies demonstrated excellent correlations between marker binding and "classical" methodologies used to identify hypoxic cells.[19, 20] Clinical evaluation of these markers has principally been undertaken in squamous cell carcinoma of the head and neck and in carcinoma of the cervix. The results of a multi-centre comparative analysis of pimonidazole binding and polarographic needle electrode measurements have been recently reported.[21] Eighty-six patients with primary carcinoma of the uterine cervix were recruited from centres in Manchester, Aarhus, and Vancouver. Patients were infused with pimonidazole and then oxygen electrode measurements were made 10 to 24 hours later. Biopsies were taken and the extent of pimonidazole binding was analysed using a semi-quantitative scoring system. No significant correlation was found between the extent of pimonidazole binding and the percentage of readings below 10, 5, or 2.5 mm Hg obtained with the oxygen electrode. A number of factors could contribute to this lack of concordance. Oxygen electrodes measure areas of acute and chronic hypoxia, whereas pimonidazole is likely to identify a greater proportion of chronically hypoxic cells. In addition, oxygen electrode measurements have the potential to overestimate the level of hypoxia through either compression artefacts or through the measurement of necrotic areas as anoxic. Poor perfusion can potentially interfere with pimonidazole dissemination that may lead to an underestimation of hypoxia. However, in differentiated tissues there is potential for pimonidazole binding to overestimate hypoxia, since binding can occur in the absence of hypoxia.[22] In this case, scoring could be refined on the basis of differentiation in the sample.

The lack of correlation between pimonidazole binding and the current gold-standard method for measuring hypoxia in tumours may be somewhat disappointing. However, given the subtleties in the precise hypoxic tumour populations that may be detected with pimonidazole compared with the oxygen electrode, it may have been expected. The key issue is whether pimonidazole binding is of predictive value in a clinical scenario where hypoxia is a known adverse indicator for outcome. This was found to be the case in a recent study where pimonidazole binding was shown to predict for treatment outcome in head and neck cancer.[23] Stratification of patients by the median value of pimonidazole binding demonstrated that those patients with more hypoxic tumours had significantly poorer locoregional control and disease free-survival 2 years after treatment. The patients in this study were potential candidates for inclusion in a Phase II trial with accelerated radiotherapy combined with carbogen and nicotinamide (ARCON). Carbogen and nicotinamide treatments are aimed at improving tumour oxygenation and hence increasing the efficacy of radiotherapy. Patients were then stratified according to both hypoxic fraction and treatment (ARCON *versus* radiotherapy, surgery, or both [non-ARCON]). The difference in outcome dependent upon hypoxic fraction was significantly reduced in the ARCON patients. For those patients with hypoxic fractions above the median, locoregional control at 2 years was approximately 70% in the ARCON group versus 20% in the non-ARCON group. High pimonidazole binding appeared to identify a cohort of patients that would benefit from a treatment regimen specifically designed to improve oxygenation, giving the clearest indication to date of the potential utility of exogenous markers for hypoxia in the clinical setting.

3.2. Endogenous Markers

The cost of exogenous markers may be prohibitive for their wide spread application in all clinical situations. To circumvent this issue there has been a great deal of interest in the potential use of endogenous markers of tumour hypoxia. These are proteins whose expression is increased as a consequence of hypoxic exposure. In most cases, this is orchestrated by the transcription factor hypoxia inducible factor-1 (HIF-1). HIF-1 is composed of two subunits, HIF-1$\alpha$and HIF-1β. In the presence of oxygen, the αsubunit is constantly degraded. This is achieved via a post-translational modification of the αsubunit that enables the von Hippel-Lindau protein to bind and target HIF-1α to the proteosome. The prolyl-hydroxylase enzymes required for this process are dependent on oxygen and hence are inhibited under hypoxia. This allows HIF-1α to accumulate and bind to the constitutive βsubunit forming active HIF-1.[24]

Of the numerous genes that are induced by HIF-1, much attention recently has been focussed upon the use of carbonic anhydrase 9 (CA9) and glucose transporter 1 (Glut-1) as endogenous markers for tumour hypoxia. Both play critical roles in the cellular adaptation to hypoxic stress. In the absence of oxygen, anaerobic metabolism ensues. Carbonic anhydrases reversibly catalyse the hydration of carbon dioxide to form carbonic acid. Tumour-derived CA9 may be particularly important in maintaining carbon dioxide diffusion gradients across cell membranes that facilitate the efflux of hydrogen ions generated through anaerobic metabolism.[25] Increased Glut-1 expression enables a greater uptake of glucose into cells to compensate for the poorer ATP generation per glucose molecule in anaerobic vs. oxidative metabolism. CA9 and Glut-1 appear to be the predominant tumour-expressed members of their individual protein families, and the over-expression of both proteins has been associated with poor prognosis in some cancers.[25-27] With the recognition that CA9 and Glut-1 are HIF-1 targets,[28-30] the aim is to establish whether the tumour-specific expression of these two proteins reflects the extent of hypoxia in clinical tumours.

Three related studies have correlated CA9 and Glut-1 expression with oxygen electrode measurements, pimonidazole binding, and radiotherapy outcome in cervical cancer.[31-33] CA9 expression correlated strongly with oxygen electrode measurements, but had no significant relationship with pimonidazole binding.[31, 33] In contrast, Glut-1 staining shows a highly significant correlative relationship with pimonidazole binding and a weaker, but significant, relationship with oxygen electrode measurements.[32, 33] This may suggest that the dynamics of CA9 and Glut-1 up-regulation differ. CA9 expression may be induced more rapidly in response to hypoxia, thereby allowing a better correlation with the oxygen electrode that measures both acute and chronic hypoxia. This is perhaps supported by the fact that a positive correlation between CA9 and pimonidazole binding has been reported in head and neck cancer when the pimonidazole was administered 2 hours prior to biopsy.[23] In the case of the cervical carcinoma studies, stratifying patients according to presence or absence of staining was of prognostic significance for metastases-free survival for both CA9[31] and Glut-1,[32] with patients showing no expression having a better prognosis. In addition, absence of CA9 expression predicted for patients with a better prognosis in terms of disease specific survival.[31] In contrast, in the head and neck cancer study, although CA9 expression correlated with pimonidazole binding, CA9 alone was of no predictive value.[23]

It may be the case that individual endogenous markers for hypoxia have utility in specific disease types. Indeed down-regulation of CA9 has been observed in stomach and oesophageal cancer compared with the level of expression observed in normal tissue or pre-neoplastic lesions.[25] Similarly, evaluation of CA9 expression following aerobic or hypoxic exposure reveals a non-uniform response in a panel of human tumour cell lines (Fig. 1).

The underlying mechanistic basis for the differences in CA9 expression in the cell line panel is unclear but may reflect differences in SP1/SP3 activity[35] or promoter methylation status[36] that have been shown to positively and negatively affect CA9 expression, respectively.

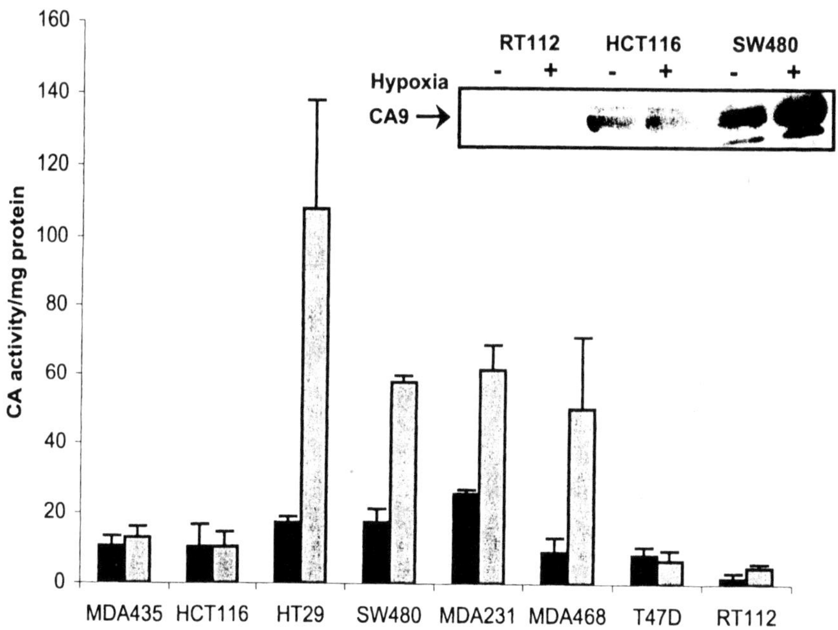

Figure 1. Carbonic anhydrase (CA) activity in membrane extracts of cells derived from a melanoma metastasis to the breast (MDA435), colon carcinomas (HCT116, HT29, SW480), breast carcinomas (MDA231, MDA468, T47D), and a bladder carcinoma (RT112), following overnight exposure to air (black bars) or hypoxia (grey bars). CA activity was determined using the Wilbur-Anderson method.[34] The inlay shows Western analysis of CA9 protein expression in representative cell lines from the panel.

Rather than looking at downstream targets of the HIF-1 protein, the expression of the oxygen regulated HIF-1α subunit has been investigated as a putative endogenous marker of tumour hypoxia. As with CA9 and Glut-1, HIF-1α is commonly over-expressed in cancer.[37] The extent of expression has been associated with increasing tumour grade and is generally higher in metastases than primary tumours.[37] Support that HIF-1α may relate to hypoxic fraction perhaps comes from the observations that HIF-1α over-expression is a poor prognostic indicator for radiotherapy outcome in cervix[38, 39] and oropharyngeal[40] cancer. Pre-clinical studies have been undertaken to evaluate the extent of co-localisation between of HIF-1α expression and the binding of exogenous hypoxic markers. *In vitro* analyses in glioblastoma cells supported a strong correlation between HIF-1α expression and pimonidazole binding with decreasing oxygen concentration.[41] However, when the glioma cells were grown as xenografts in nude mice, the concordance between HIF-1α expression and pimonidazole binding was much weaker.[41] Similar data were obtained in cervical cancer xenografts. Co-localisation was apparent, but the area positive for HIF-1α exceeded that for the hypoxic marker EF-5.[42] There are several potential explanations for this difference, not least are differences in the absolute oxygen requirements for HIF-1α stabilisation and exogenous marker binding and the time-course of these two phenomena (< 5 minutes vs. 30-60 minutes).

On the basis of the time-course of HIF-1α stabilisation, one might assume that HIF-1α expression would be a reasonable marker for acute hypoxia. A follow-up study to those conducted in cervical carcinoma xenografts would support this point. HIF-1α expression was analysed in biopsies taken from cervical carcinoma patients and the extent of expression was found to correlate with pre-treatment oxygen electrode measurements.[43] Although, in some cases, the patterns of HIF-1α expression relative to blood vessels may have been consistent with both acute (HIF-1α adjacent to presumably non-perfused vessels) and chronic hypoxia (HIF-1α expression distal to vessels), in other samples, homogeneous HIF-1α staining was apparent. Interestingly, HIF-1α was of no prognostic significance in this study using disease-free survival as an end-point.[43] HIF-1α expression, pimonidazole binding, and iododeoxyuridine uptake as a marker of vessel perfusion have been simultaneously analysed in a series of head and neck cancer patients.[44] Here, a clear correlation was seen between the diffusion-limited fraction of the tumour and the areas binding pimonidazole. However, the co-localisation of pimonidazole binding and HIF-1αexpression was low. Again, a homogeneous pattern of HIF-1α expression was reported for a number of samples that would not be conducive to HIF-1α acting as an appropriate marker for either acute or chronic hypoxia in these cases.[44]

This inevitably raises the question of whether there are means other than exposure to hypoxia by which HIF-1α expression can be achieved within a tumour. Indeed, multiple pathways characteristic to neoplastic transformation can impact upon HIF-1α expression, either by directly inhibiting the degradation of the protein (for example, mutation in the von Hippel-Lindau protein[45]) or by increasing HIF-1α synthesis, as has been associated with the activation of tyrosine kinase receptor-mediated cascades.[46]

The implication from these observations is that HIF-1α expression may be a better marker for aggressive disease in some cancer types rather than as a marker for tumour hypoxia *per se*. Importantly, although HIF-1α may be expressed in a manner independent of hypoxia in some tumours, this does not necessarily mean that the same is true of all HIF-1 regulated proteins. The expression of both CA9 and Glut-1 is not solely regulated by HIF-1. Perturbations in cell redox that occur as a consequence of hypoxia are also important regulators of the expression of both of these proteins.[25, 30] Furthermore, over-expression of HIF-1α alone may be insufficient to achieve inappropriate activation of the HIF-1 transcription factor. In addition to the oxygen dependent modifications of HIF-1α that normally target the protein for degradation, hydroxylation of an asparagine residue in the C-terminal domain of the protein facilitates binding of an inhibitory complex.[47] This prevents the recruitment of the p300 cofactor that is an obligate requirement for the optimal HIF-1 transactivation. Indeed, the interaction between p300 and the HIF-1 complex is facilitated by redox-factor 1 in a cell redox sensitive manner.[48] Thus, although aerobic HIF-1α expression may be achieved, there are multiple mechanisms in place to limit the potential for transactivation of HIF-1 responsive genes in the absence of exposure to hypoxic stress.

4. SUMMARY

The prognostic significance of hypoxia as an adverse indicator for treatment outcome is now well established. With the development of therapeutic strategies to target tumour hypoxia with the aim of enhancing the efficacy of conventional treatments, there is a clear need to identify patients that would benefit from such an approach. Although the polarographic oxygen electrode has been an invaluable tool in establishing the role hypoxia plays in clinical response, the development of hypoxic markers with wide-scale application is needed. Initial clinical evaluation of the exogenous markers pimonidazole, and EF-5 are looking very promising. In addition, studies utilising endogenous markers such as CA9 and Glut-1 have provided a substantial body of evidence to support their use as markers of tumour hypoxia. In contrast, HIF-1α expression may be of limited utility as an indicator of hypoxic fraction HIF-1α over-expression in cancer can be achieved by numerous hypoxia- and redox-independent mechanisms. A greater understanding of the biology of hypoxia-regulated gene expression will facilitate the appropriate use of endogenous markers in the clinical setting. With the rapid advancements being made in this field, the potential for routine screening of tumour hypoxia and individualisation of therapy based on this assessment may not be too distant.

5. ACKNOWLEDGEMENTS

The authors acknowledge the financial support of the Medical Research Council and Cancer Research UK. The CA9 antibody used was kindly donated by Silvia Pastorekova, Slovak Academy of Sciences, 84246 Bratislava, Slovak Republic. Thanks are also due to Rachel Airley and Rachel Cowen for helpful discussions in the preparation of this manuscript.

6. REFERENCES

1. L. Webster, R. J. Hodgkiss, and G. D. Wilson, Cell cycle distribution of hypoxia and progression of hypoxic tumour cells in vivo, *Br. J. Cancer* **77**, 227-234 (1998).
2. R. E. Durand, and J. A. Raleigh, Identification of nonproliferating but viable hypoxic tumour cells in vivo, *Cancer Res.* **58**, 3547-3550 (1998).
3. B. A. Teicher, Hypoxia and drug resistance, *Cancer Metastasis Rev.* **13**, 39-68 (1994).
4. T. G. Graeber, C. Osmanian, T. Jacks, D. E. Housman, S. J. Koch, S. W. Lowe, and A. J. Giaccia, Hypoxia-mediated selection of cells with diminished apoptotic potential in solid tumours, *Nature* **379**, 88-91 (1996).
5. M. Höckel, K. Schlenger, B. Aral, M. Mitze, U. Schaffer, and P. Vaupel, Association between tumour hypoxia and malignant progression in advanced cancer of the uterine cervix, *Cancer Res.* **56**, 4509-4515 (1996).
6. R. H. Thomlinson, and L. H. Gray, The histological structure of some human lung cancers and the possible implications for radiotherapy, *Br. J. Cancer* **9**, 539-549 (1955).
7. L. H. Gray, A. D. Conger, M. Ebert, S. Hornsey, and O. C. Scott, Concentration of oxygen dissolved in tissue at the time of irradiation as a factor in radiotherapy, *Br. J. Radiol.* **26**, 638-648 (1953).
8. P. Okunieff, M. Höckel, E. P. Dunphy, K. Schlenger, C. Knoop, and P. Vaupel, Oxygen tension distributions are sufficient to explain the local response of human breast tumours treated with radiation alone, *Int. J. Radiat. Oncol. Biol. Phys.* **26**, 631-636 (1993).
9. M. Höckel, C. Knoop, K. Schlenger, B. Vorndran, E. Baussmann, M. Mitze, P. G. Knapstein, and P. Vaupel, Intratumoural pO_2 predicts survival in advanced cancer of the uterine cervix, *Radiother. Oncol.* **26**, 45-50 (1993).
10. A. W. Fyles, M. Milosevic, R. Wong, M. C. Kavanagh, M. Pintilie, A. Sun, W. Chapman, W. Levin, L. Manchul, T. J. Keane, and R. P. Hill, Oxygenation predicts radiation response in patients with cervix cancer, *Radiother. Oncol.* **48**,149-156 (1998).
11. M. Nordsmark, and J. Overgaard, A confirmatory prognostic study on oxygenation status and regional control in advanced head and neck squamous cell carcinoma treated by radiation therapy, *Radiother. Oncol.* **57**, 39-43 (2000).
12. D. M. Brizel, S. P. Scully, J. M. Harrelson, L. J. Layfield, J. M. Bean, L. R. Prosnitz, and M. W. Dewhirst, Tumour oxygenation predicts for the likelihood of distant metastasis in human soft tissue sarcoma, *Cancer Res.* **56**, 941-943 (1996).
13. R. A. Gatenby, H. B. Kessler, J. S. Rosenblum, L. R. Coia, P. J. Moldofsky, W. H. Hartz, and G. J. Broder, Oxygen distribution in squamous cell carcinoma metastases and its relationship to outcome of radiation therapy, *Int. J. Radiat. Oncol. Biol. Phys.* **14**, 831-838 (1988).
14. M. Höckel, K. Schlenger, S. Höckel, and P. Vaupel, Hypoxic cervical cancers with low apoptotic index are highly aggressive, *Cancer Res.* **59**, 4525-4528 (1999).
15. J. Bussink, J. H. A. M. Kaanders, A. M. Strik, and A. J. van der Kogel, Effects of nicotinamide and carbogen on oxygenation in human tumour xenografts measured with luminescence based fibre optic probes, *Radiother. Oncol.* **57**, 21-30 (2000).
16. B. M. Seddon, D. J. Honess, B. Vojnovic, G. M. Tozer, and P. Workman, Measurement of tumour oxygenation: *In vivo* comparison of a luminescence fibre optic sensor and a polarographic electrode in the p22 tumour, *Radiat. Res.* **155**, 837-846 (2001).
17. J. A. Raleigh, D. P. Calkin-Adams, L. H. Rinker, C. A. Ballenger, M. C. Weissler, W. C. Fowler, D. B. Novotny, and M. A. Varia, Hypoxia and vascular endothelial growth factor expression in human squamous cell carcinomas using pimonidazole as a hypoxic marker, *Cancer Res.* **58**, 3765-3768 (1998).
18. S. M. Evans, S. Hahn, D. R. Pook, W. T. Jenkins, A. A. Chalian, P. Zhang, C. Stevens, R. Weber, G. Weinstein, I. Benjamin, N. Mirza, M. Morgan, S. Rubin, W. G. McKenna, E. M. Lord, and C. J. Koch, Detection of hypoxia in human squamous cell carcinoma by EF5 binding, *Cancer Res.* **60**, 2018-2024 (2000).
19. J. A. Raleigh, S. C. Chou, G. E. Arteel, and M. R. Horsman, Comparisons among pimonidazole binding, oxygen electrode measurements, and radiation response in C3H mouse tumours, *Radiat. Res.* **151**, 580-589 (1999).
20. P. L. Olive, R. E. Durand, J. A. Raleigh, C. Luo, and C. Aquino-Parsons, Comparison between the comet assay and pimonidazole binding for measuring tumour hypoxia, *Br. J. Cancer* **83**, 1525-1531 (2000).
21. M. Nordsmark, J. Loncaster, C. Aquino-Parsons, S. C. Chou, M. Ladekarl, H. Havsteen, J. C. Lindegaard, S. E. Davidson, M. Varia, C. West, R. Hunter, J. Overgaard, and J. A. Raleigh, Measurements of hypoxia using pimonidazole and polarographic oxygen-sensitive electrodes in human cervix carcinomas, *Radiother. Oncol.* **67**, 35-44 (2003).

22. H. L. Janssen, F. J. Hoebers, D. Sprong, L. Goethals, K. J. Williams, I. J. Stratford, K. M. Haustermans, A. J. Balm, and A. C. Begg, Differentiation-associated staining with anti-pimonidazole antibodies in head and neck tumours, *Radiother. Oncol.* **70**, 91-97 (2004).
23. J. H. A. M. Kaanders, K. I. E. M. Wijffels, H. A. M. Marres, A. S. E. Ljungkvist, L. A. M. Pop, F. J. A. van den Hoogen, P. C. M. de Wilde, J. Bussink, J. A. Raleigh, and A. J. van der Kogel, Pimonidazole binding and tumour vascularity predict for treatment outcome in head and neck cancer, *Cancer Res.* **62**, 7066-7074 (2002).
24. C. W. Pugh, and P. J. Ratcliffe, The von Hippel-Lindau tumour suppressor, hypoxia-inducible factor-1 (HIF-1) degradation and cancer pathogenesis, *Semin. Cancer Biol.* **13**, 83-89 (2003).
25. C. P. S. Potter, and A. L. Harris, Diagnostic, prognostic and therapeutic implications of carbonic anhydrases in cancer, *Br. J. Cancer* **89**, 2-7 (2003).
26. R. S. Haber, A. Rathan, K. R. Weiser, A. Pritsker, S. H. Itzkowitz, C. Bodian, G. Slater, A. Weiss, and D. E. Burnstein, Glut-1 glucose transporter expression in colorectal carcinoma, *Cancer* **83**, 34-40 (1998).
27. M. Younes, R. W. Brown, M. Stephenson, M. Gondo, and P. T. Cagle, Overexpression of Glut-1 and Glut-3 in stage 1 non small cell lung carcinoma is associated with poor survival, *Cancer* **80**, 1046-1051 (1997).
28. C. C. Wycoff, N. J. P. Beasley, P. H. Watson, K. J. Turner, J. Pastorek, A. Sibtain, G. D. Wilson, H. Turley, K. L. Talks, P. H. Maxwell, C. W. Pugh, P. J. Ratcliffe, and A. L. Harris, Hypoxia-inducible expression of tumour associated carbonic anhydrases, *Cancer Res.* **60**, 7075-7083 (2000).
29. B. L. Ebert, J. D. Firth, and P. J. Ratcliffe, Hypoxia and mitochondrial inhibitors regulate expression of glucose transporters via distinct cis-acting sequences, *J. Biol. Chem.* **270**, 29083-29089 (1995).
30. K. J. Williams, B. A. Telfer, R. E. Airley, H. P. Peters, M. R. Sheridan, A. J. van der Kogel, A. L. Harris, and I. J. Stratford, A protective role for HIF-1 in response to redox manipulation and glucose deprivation: implications for tumourigenesis, *Oncogene* **21**, 282-290 (2002).
31. J. A. Loncaster, A. L. Harris, S. E. Davidson, J. P. Logue, R. D.Hunter, C. C. Wycoff, J. Pastorek, P. J. Ratcliffe, I. J. Stratford, and C. M. L. West, Carbonic anhydrase (CAIX) expression, a potential new intrinsic marker of hypoxia: correlations with tumour oxygen measurements and prognosis in locally advanced carcinoma of the cervix, *Cancer Res.* **61**, 6394-6399 (2001).
32. R. Airley, J. Loncaster, S. Davidson, M. Bromley, S. Roberts, A. Patterson, R. Hunter, I. Stratford, and C. West, Glucose transporter Glut-1 expression correlates with tumour hypoxia and predicts metastasis free survival in advanced carcinoma of the cervix, *Clin. Cancer Res.* **7**, 928-934 (2001).
33. R. Airley, J. Loncaster, J. Raleigh, A. L. Harris, S. E. Davidson, R. D. Hunter, C. M. L. West, and I. J. Stratford, Glut-1 and CAIX as intrinsic markers of hypoxia in carcinoma of the cervix: relationship to pimonidazole binding, *Int. J. Cancer* **104**, 85-91 (2003).
34. K. Wilbur, and N. Anderson, Electrometric and colorimetric determination of carbonic anhydrase, *J. Biol. Chem.* **176**, 147-154 (1948).
35. S. Kaluz, M. Kaluzova, and E. Stanbridge, Expression of the hypoxia marker carbonic anhydrase IX is critically dependent on SP1 activity. Identification of a novel type of hypoxia-responsive enhancer, *Cancer Res.* **63**, 917-922 (2003).
36. M. Cho, H. Uemura, S-C. Kim, Y. Kawada, K. Yoshida, Y. Hirao, N. Konishi, S, Saga, and K. Yoshikawa, Hypomethylation of the MN/CA9 promoter and upregulated MN/CA9 expression in human renal cell carcinoma, *Br. J. Cancer* **85**, 563-567 (2001).
37. H. Zhong, A. M. De Marzo, E. Laughner, M. Lim, D. A. Hilton, D. Zagzag, P. Buechler, W. B. Isaacs, G. L. Semenza, and J. W. Simons, Overexpression of hypoxia-inducible factor 1alpha in common human cancers and their metastases, *Cancer Res.* **59**, 5830-5835 (1999).
38. B. Bachtiary, M. Schindl, R. Potter, B. Dreier, T. H. Knocke, J. A. Hainfellner, R. Horvat, and P. Birner, Overexpression of hypoxia inducible factor 1α indicates diminished response to radiotherapy and unfavourable prognosis in patients receiving radical radiotherapy for cervical cancer, *Clin. Cancer Res.* **9**, 2234-2240 (2003).
39. P. Burri, V. Djonov, D, Aebersold, K. Lindel, U. Studer, H. J. Altermatt, L. Mazzucchelli, R. H. Greiner, and G. Gruber, Significant correlation of hypoxia inducible factor 1α with treatment outcome in cervical cancer treated with radical radiotherapy, *Int. J. Radiat. Oncol. Biol. Phys.* **56**, 494-501 (2003).
40. D. M. Aebersold, P. Burri, K. T. Beer, J. Laissue, V. Djonov, R. H. Greiner, and G. L. Semenza, Expression of hypoxia inducible factor 1α: a novel predictive and prognostic parameter in the radiotherapy of oropharyngeal cancer, *Cancer Res.* **61**, 2911-2916 (2001).
41. D. Vordermark, and J. M. Brown, Evaluation of hypoxia inducible factor 1α (HIF-1α) as an intrinsic marker of tumour hypoxia in U87 MG human glioblastoma: in vitro and xenograft studies, *Int. J. Radiat. Oncol. Biol. Phys.* **56**, 1184-1193 (2003).

42. V. Vukovic, H. K. Haugland, T. Nicklee, A. J. Morrison, and D. W. Hedley, Hypoxia inducible factor 1α is an intrinsic marker for hypoxia in cervical cancer xenografts, *Cancer Res.* **61**, 7394-7398 (2001).
43. H. K. Haugland, V. Vukovic, M. Pintilie, A.W. Fyles, M. Milosevic, R. P. Hill, and D. W. Hedley, Expression of hypoxia inducible factor 1αin cervical carcinomas: correlation with tumour oxygenation, *Int. J. Radiat. Oncol. Biol. Phys.* **53**, 854-861 (2002).
44. H. L. K. Janssen, K. M. G. Haustermans, D. Sprong, G. Blommestijn, I. Hofland, F. J. Hoebers, E. Blijweert, J. A. Raleigh, G. L. Semenza, M. A. Varia, A. J. Balm, M-L. F. van Velthuysen, P. Delaere, R. Sciot, and A. C. Begg, HIF-1A, pimonidazole and iododeoxyuridine to estimate hypoxia and perfusion in human head and neck tumors, *Int. J. Radiat. Oncol. Biol. Phys.* **54**, 1537-1549 (2002).
45. P. H. Maxwell, M. S. Wiesener, G-W. Chang, S. C. Clifford, E. C. Vaux, M. E. Cockman, C. C. Wycoff, C. W. Pugh, E. R. Maher, and P. J. Ratcliffe, The tumour suppressor protein VHL targets hypoxia inducible factors for oxygen-dependent proteolysis, *Nature* **399**, 271-275 (1999).
46. E. Laughner, P. Taghavi, K. Chiles, P. C. Mahon, and G. L. Semenza, HER2 (neu) signaling increases the rate of hypoxia-inducible factor 1alpha (HIF-1alpha) synthesis: novel mechanism for HIF-1-mediated vascular endothelial growth factor expression, *Mol. Cell Biol.* **21**, 3995-4004 (2001).
47. D. Lando, J. J. Gorman, M. L. Whitelaw, and D. J. Peet, Oxygen-dependent regulation of hypoxia-inducible factors by prolyl and asparaginyl hydroxylation, *Eur. J. Biochem.* **270**, 781-790 (2003).
48. M. Ema, K. Hirota, J. Mimura, H. Abe, J. Yodoi, K. Sogawa, L. Poellinger, and Y. Fujii-Kuriyama, Molecular mechanisms of transcription activation by HLF and HIF1alpha in response to hypoxia: their stabilization and redox signal-induced interaction with CBP/p300, *EMBO J.* **18**, 1905-1914 (1999).

THE DIFFICULTIES IN COMPARING *IN VIVO* OXYGEN MEASUREMENTS:

Turning the problems into virtues

Harold M. Swartz and Jeff Dunn

Abstract: There has been rapid development of effective new tools that provide information on oxygenation *in vivo* and an increased recognition of how valuable such information can be. Consequently, there also has been considerable interest in comparing and evaluating the accuracy and usefulness of the different types of measurements.

The various types of measurements usually do not measure the same thing. They may measure pO_2 or $[O_2]$ or something less directly related, such as hemoglobin saturation. They may make measurements in different compartments (e.g. intracellular, extracellular, vascular) in the volume that they sample, the time span over which they average, the local perturbation that they may cause, etc. They also differ in their sensitivity, accuracy, ability to measure repetitively.

However, these potentially confounding and confusing differences can be made into an outstanding virtue, if their nature is considered carefully. Then a proper model can relate them to each other. The ability to relate the various measurements to each other can be a powerful tool to test the validity of models that attempt to explain fully the distribution of oxygen in real systems and the factors that affect this. We then could have a major advancement in our understanding of oxygen transport in tissues, with an ability to determine accurately the effects of physiological and pathophysiological perturbations on oxygenation at all levels of cells and tissues *in vivo*.

1. INTRODUCTION

There has been both rapid development of effective new tools that provide information on oxygenation *in vivo* and an increased recognition of the value of such information.[1] Consequently, there also has been considerable interest in comparing and evaluating the accuracy and usefulness of the different types of measurements.[2-5] Such information seems essential for proper interpretation of the measurements that are obtained with these elegant techniques.

When such comparisons are made, there appear to be considerable discrepancies among the various techniques that reflect differences that are more than simply differences in accuracy, potentially raising questions as to their validity. On the other hand, if the differences in these measurements are real and are taken into account, then it should be possible to construct and test a model for oxygen distribution in tissues that uses the different types of information to test and improve the model. It should be especially useful to use the model to predict the changes that will occur with various types of perturbations, such as altering the amount of oxygen to which the system is exposed, changing perfusion patterns. This paper considers the ways in which the different types of measurements can vary and how these differences could be exploited to obtain a much more comprehensive understanding of oxygen in tissues.

2. THE DIFFICULTIES IN COMPARING *IN VIVO* OXYGEN MEASUREMENTS

The classical approach is to use one technique as a gold standard and then to compare another technique to it. This straightforward approach, however, has potential pitfalls that can result in misleading conclusions. The first, and perhaps most obvious, is what is the "gold standard"? What properties does it have, in addition to historical precedence, which makes it appropriate? In an *in vivo* study, it is unlikely that a single type of measurement will be clearly the most accurate under most circumstances, and therefore a single gold standard is unlikely. The various methods also may differ in their sensitivity, accuracy, and ability to measure repetitively. Clearly one needs to consider carefully the strengths and weaknesses of each type of measurement, including the "gold standard," and try to gain insights into the uncertainties.

But differences in sensitivity and accuracy are not the biggest problem. For real measurements of oxygen *in vivo*, the most problematic aspect is that the various types of measurements usually do not measure the same thing. This is because various techniques usually differ significantly in one or more of the following aspects:

1. the parameter that they measure
2. the spatial dimension that they probe directly
3. the time resolution of the measurements
4. the compartment in which the measurement is made
5. the perturbations that occur as a consequence of the measurement

2.1. The Parameter that is Measured

This is one of the most important variables among different measurement techniques. They may measure oxygen directly (as either pO_2 or $[O_2]$),[6-9] or they may measure something that is less directly related to these fundamental parameters, such as hemoglobin saturation.[10-12] Some important and useful techniques, such as measurements of perfusion, are even less directly related to direct measurements of oxygen, but often are compared directly with direct measurements of oxygen.[13, 14]

The measurement of pO_2 vs. $[O_2]$ can provide very different results, unless the location of the probe is known and taken into account. Because of the high lipid solubility of oxygen, a measurement of $[O_2]$, even in a system that is in full equilibrium, should give very different values among regions with different lipophilicity.[15] The concentration of oxygen in membranes will be 6–10 times higher in lipid rich areas such as membranes because of the differential partitioning of oxygen. On the other hand, an accurate measurement of pO_2 will provide the same value regardless of the lipophilicity of the site where the measurement is made. This does not mean that measurement of pO_2 is inherently superior. For example, if the parameter of interest requires knowledge of the $[O_2]$, as would be the case for some types of oxidative damage where oxygen is a direct reactant, then a direct measurement of the concentration of oxygen in the compartment of interest would be desirable. This type of information also could be provided by a measurement of pO_2, combined with a knowledge of the solubility of oxygen in the region of interest.

The methods that are usually considered to measure the concentration of oxygen actually often are based on the collision rate of oxygen with the probe, which usually is a function of both the concentration of oxygen and the diffusion rates of oxygen and the probe.[15] Thus, the interpretation of such data as reflecting the concentration of oxygen assumes knowledge of the diffusion rates in the compartment of interest.

The methods that measure the pO_2 directly include oxygen electrodes,[6] $^{17}O_2$ (by NMR),[16] and EPR oximetry based on particulates.[8, 17] The methods that are considered to measure $[O_2]$ include optical oxygen quenching methods,[18] NMR methods based on relaxation times (^{19}F and protons and some $^{17}O_2$ methods),[19] and EPR oximetry based on soluble free radicals.[20-23]

Many frequently employed methods do not measure oxygen directly but, instead, a parameter that may be related to oxygen. These include methods based on concentrations of metabolites,[24] measurements of hemoglobin[10-12] or myoglobin saturation,[25] and measurements of hemodynamic parameters, such as perfusion[14] and diffusion.[26] These reflect very important physiological and biochemical parameters of proven experimental and clinical value. While such measurements can be very valuable, additional assumptions are needed to relate them to the actual pO_2 or $[O_2]$ in tissues. In some cases, as with the saturation of myoglobin, these involve relatively small assumptions about intracellular pH and the myoglobin/oxygen binding curve.[25] In others, such as with BOLD MRI, the assumptions are extensive, involving arterial/venous volume ratios, blood volume, and changes in p50 and pH through the capillary bed.[27] On the other hand, an appropriate model of oxygen distribution in tissue should enable these parameters to be related to the actual oxygen levels in tissues under many circumstances.

2.2. The Spatial Dimension that they Probe Directly

The spatial dimension that is probed directly is another critical variable, and one about which there frequently is confusion.[28] Most measurements differ in the volume that they sample to produce the measurement. A common question is "how does measurement of X relate to the measurements with the 'gold standard'?" (which often is the oxygen electrode). The problem is that the distribution of oxygen in real tissues is very heterogeneous, sometimes varying significantly across dimensions smaller than a whole cell. Therefore, unless appropriate modeling and calculations are employed, the direct comparisons of different techniques can be very misleading because they are sampling different amounts of a heterogeneous system. The problem is difficult but not insurmountable: for example, Braun et al.[4] have illustrated how one can compare two techniques (OxyLite and oxygen electrode) that sample quite different spatial volumes. The challenges in making such comparisons include both recognizing that this is an important issue and obtaining an adequate database with which to make the comparison. Comparisons of oxygen electrodes with other techniques may be especially difficult because the very small volume sampled by the oxygen electrode will reflect heterogeneity that is not sensed by techniques that sample much larger volumes. Valid comparisons then require that one obtain a large number of data points with the electrodes, but that may not be practical because of the perturbations that can be caused by such sampling. It may be more effective to use a model to compare the methods, taking into account the volumes that are sampled by each method, the heterogeneity that is expected to be present, and the probability that the oxygen electrode will have sparse sampling of the total heterogeneity.

2.3. The Time Resolution of the Measurements

The local amount of oxygen varies temporally as well as spatially. This is especially evident in studies that follow the circulation in capillaries.[29, 30] Therefore, if different methods have different sampling times, they may give quite different results even though both are quite accurate. The variations in sampling time are both intrinsic (i.e. particular methods have different minimum times required to obtain a data point) and operational (i.e. to improve sensitivity some type of time averaging is employed), and therefore need to be considered for the particular conditions under which the measurements have been made.

2.4. The Compartment in which the Measurement is Made

The various methods may make measurements in specific compartments (e.g. intracellular for selectively localizing probes, interstitial for macroscopic probes physically introduced into tissues, vascular for methods based on hemoglobin) or over the entire sample, or in different compartments randomly (e.g. the Eppendorf). Within the vascular system, the methods may differ in the weighting that they give among arteries, veins, and capillaries. They therefore may report quite different values from the same region even though the measurements per se are accurate. Again, with proper modeling, these differences should be able to be taken into account.

2.5. The Perturbations that Occur as a Consequence of the Measurement

Every real measurement has some potential for perturbing the system that it measures. The perturbations arising from various methods vary with the method, the sites where the measurements are made, and the amount and timing of repeated measurements. For example, the introduction of a physical probe, such as an oxygen electrode or a fiber optic, must cause some local perturbation. The perturbation may vary with the interval between the physical perturbation and the time of the measurements. For example, the methodology used with a stepped introduction of a small oxygen electrode is based on the expectation that the acute perturbations caused by its passage can be minimized by rapid measurements after a short step back to relieve local pressure. On the other hand, leaving the electrode in place for longer times sometimes is not considered to be a good approach because of reactions to the presence of this foreign object. Methods such as EPR oximetry with particulates or NMR fluorine-based oximetry often use a different approach in which the measurements are deferred until some time later, to allow for dissipation of perturbations from the insertion of the material.

3. TURNING THE PROBLEMS INTO VIRTUES

If the nature of these potentially confounding and confusing differences is considered carefully, it becomes clear that these problems can be made into outstanding virtues because then a proper model can relate them to each other. Most importantly, the ability to relate the various measurements to each other can be a powerful tool to test the validity of models that attempt to explain fully the distribution of oxygen in real systems and the factors that affect this. The modeling would need to be very complex and comprehensive, including consideration of the physiological and anatomical factors that affect the distribution and utilization of oxygen and also the basis and operation of the various methods. Careful consideration of compartmentalization may play a key role in this modeling.

We can readily measure the arterial and venous pO_2. Some combination of intravascular reporting agents may provide more extensive information on pO_2 in the microvasculature. These include NIR or MRI assessment of hemoglobin saturation[10-12] or EPR measurements with intravascular agents.[23] It is possible that some fluorinated MR agents may be taken up in the endothelium, providing data on the vessel boundaries. Electrodes and fiber optic-based measurements, such as the OxyLite, provide data in the extravascular and pericellular spaces.[6, 7] Measurements of intracellular pO_2 can be made with methods such as soluble EPR agents[31, 32] and myoglobins.[25] Finally, the intramitochondrial pO_2 may be inferred through a knowledge of the cytochrome oxidation states.[33, 34]

The development of such a model will require a cooperative effort between modelers, measurers, and physiologists. This will be difficult, but, if successful, we then could have a major advancement in our understanding of oxygen transport in tissues, with an ability to determine accurately the effects of physiological and pathophysiological perturbations in oxygenation at all levels of cells and tissues *in vivo*. The fact that the various methods also differ in their sensitivity, accuracy, and ability to measure repetitively also can be considered in an appropriate model, including provision for relating the uncertainties in the measurements of each technique. This also would add to the potential information that can be obtained. The result could be a much more complete picture of the oxygen distribution in tissue and how it changes under various conditions.

This approach inherently would require iteration, in which the data are obtained and analyzed with a specific model, and then the model would be modified to take into account discrepancies observed when the measuring conditions are changed. Optimistically, the final product would be a model that accurately predicts the measurements that would be obtained with any particular technique and that can be used to describe fully the distribution of oxygen under any plausible condition.

4. THE PROPOSAL

These considerations lead to the conclusion that a carefully designed set of studies utilizing an appropriate set of measurement techniques, combined with the development of a model of the factors that can affect the distribution of oxygen, should be planned and carried out. It could provide the data and concepts needed for a very full understanding of oxygen in tissues. This would be a formidable undertaking, requiring careful planning and excellent methodology, employing many different techniques in pairs or larger groups. In order to obtain and analyze this large and broad set of information, it will require the interaction of a number of investigators with different and complementary expertise. This could be done through the interactions of the community of experts that comprise the membership of ISOTT.

5. ACKNOWLEDGEMENTS

This work was supported by PO1 EB002180, "Measurement of pO_2 in Tissues In Vivo and In Vitro," and used the facilities of P41 EB002032, "EPR Center for the Study of Viable Systems."

6. REFERENCES

1. H. M. Swartz, and J. F. Dunn, Measurements of oxygen in tissues: overview and perspectives on methods, *Adv. Exp. Med. Biol.* **540**, 1-12 (2003).
2. D. R. Collingridge, W. K. Young, B. Vojnovic, P. Wardman, E. M. Lynch, S. A. Hill, and D. J. Chaplin, Measurement of tumor oxygenation: a comparison between polarographic needle electrodes and a time-resolved luminescence-based optical sensor, *Radiat. Res.* **147**, 329-334 (1997).
3. B. M. Seddon, D. J. Honess, B. Vojnovic, G. M. Tozer, and P. Workman, Measurement of tumor oxygenation: in vivo comparison of a luminescence fiber-optic sensor and a polarographic electrode in the p22 tumor, *Radiat. Res.* **155**, 837-846 (2001).
4. R. D Braun, J. L. Lanzen, S. A. Snyder, and M. W. Dewhirst, Comparison of tumor and normal tissue oxygen tension measurements using OxyLite or microelectrodes in rodents, *Am. J. Physiol. Heart Circ. Physiol.* **280**(6), H2533-H2544 (2001).
5. M. C. Kavanagh, A. Sun, Q. Hu, and R. P. Hill, Comparing techniques of measuring tumor hypoxia in different murine tumors: Eppendorf pO_2 histograph, [3H] misonidazole binding and paired survival assay, *Radiat. Res.* **145**, 491-500 (1996).
6. D. W. Lubbers, The meaning of the tissue oxygen distribution curve and its measurement by means of Pt electrodes, *Prog. Respir. Res.* **3**, 112-123 (1969).
7. C. I. Nwaigwe, M. A. Roche, O. Grinberg, and J. F. Dunn, Effect of hyperventilation on brain tissue oxygenation and cerebrovenous pO_2 in rats, *Brain Res.* **868**, 150-156 (2000).
8. E. L. Rolett, A. Azzawi, K. J. Liu, M. N. Yongbi, H. M. Swartz, and J. F. Dunn, Critical oxygen tension in rat brain: a combined ^{31}P-NMR and EPR oximetry study, *Am. J. Physiol.* **279**(1), R9-R16 (2000).
9. J. F. Glockner, S.-W. Norby, and H. M. Swartz, Simultaneous measurement of intracellular and extracellular oxygen concentrations using a nitroxide-liposome system, *Magn. Reson. Med.* **29**(1), 12-18 (1993).
10. D. T. Delpy, M. C. Cope, E. B. Cady, J. S. Wyatt, P. A. Hamilton, P. L. Hope, S. Wray, and E. O. Reynolds, Cerebral monitoring in newborn infants by magnetic resonance and near infrared spectroscopy, *Scand. J. Clin. Lab. Invest. Suppl.* **188**, 9-17 (1987).
11. S. Ogawa, T. M. Lee, A. R. Kay, and D. W. Tank, Brain magnetic resonance imaging with contrast dependent on blood oxygenation, *Proc. Natl. Acad. Sci. USA*, **87**(24), 9868-9872 (1990).
12. J. F. Dunn, J. A. O'Hara, Y. Zaim-Wadghiri, H. Lei, M. E. Meyerand, O. Y. Grinberg, H. Hou, P. J. Hoopes, E. Demidenko, and H. M. Swartz, Changes in oxygenation of intracranial tumors with carbogen: a BOLD MRI and EPR oximetry study, *J. Magn. Reson. Imaging* **16**(5), 511-521 (2002).

13. T. Q. Duong, C. Iadecola, and S. G. Kim, Effect of hyperoxia, hypercapnia, and hypoxia on cerebral interstitial oxygen tension and cerebral blood flow, *Magn. Reson. Med.* **45**(1), 61-70 (2001).
14. H. Lei, O. Grinberg, C. I. Nwaigwe, H. G. Hou, H. Williams, H. M. Swartz, and J. F. Dunn, The effects of ketamine-xylazine anesthesia on cerebral blood flow and oxygenation observed using nuclear magnetic resonance perfusion imaging and electron paramagnetic resonance oximetry, *Brain Res.* **913**(2), 174-179 (2001).
15. G. Bacic, T. Walczak, F. C. Demsar, and H. M. Swartz, Electron spin resonance imaging of tissues with lipid-rich areas, *Magn. Res. Med.* **8**, 209-219 (1988).
16. D. Fiat, J. Dolinsek, J. Hankiewicz, M. Dujovny, and J. Ausman, Determination of regional cerebral oxygen consumption in the human: ^{17}O natural abundance cerebral magnetic resonance imaging and spectroscopy in a whole body system, *Neurol. Res.* **15**(4), 237-248 (1993).
17. K. J. Liu, P. Gast, M. Moussavi, S. W. Norby, N. Vahidi, T. Walczak, M. Wu, and H. M. Swartz, Lithium phthalocyanine: a probe for electron paramagnetic resonance oximetry in viable biological systems, *Proc. Natl. Acad. Sci. USA* **90**, 5438-5442 (1993).
18. S. A. Vinogradov, L. W. Lo, W. T. Jenkins, S. M. Evans, C. Koch, and D. F. Wilson, Noninvasive imaging of the distribution in oxygen in tissue in vivo using near-infrared phosphors, *Biophys. J.* **70**(4), 1609-1617 (1996).
19. R. P. Mason, A. Constantinescu, S. Hunjan, D. Le, E. W. Hahn, P. P. Antich, C. Blum, and P. Peschke, Regional tumor oxygenation and measurement of dynamic changes, *Radiat. Res.* **152**(3), 239-249 (1999).
20. M. C. Krishna, S. Subramanian, P. Kuppusamy, and J. B. Mitchell, Magnetic resonance imaging for in vivo assessment of tissue oxygen concentration, *Semin. Radiat. Oncol.* **11**(1), 58-69 (2001).
21. T. I. Smirnova, A. I. Smirnov, R. B. Clarkson, and R. L. Belford, Accuracy of oxygen measurements in T2 (line width) EPR oximetry, *Magn. Reson. Med.* **33**(6), 801-810 (1995).
22. J. L. Zweier, S. Thompson-Gorman, and P. Kuppusamy, Measurement of oxygen concentration in the intact beating heart using electron paramagnetic resonance spectroscopy:a technique for measuring oxygen concentration in situ, *J. Bioenerg. Biomembr.* **23**, 855-871 (1991).
23. H. M. Swartz, and H. Halpern, EPR studies of living animals and related model systems (in vivo EPR), in: *Spin Labeling: The Next Millennium*, edited by L. J. Berliner (Plenum Publishing, New York, 1998), pp 367-404.
24. J. F. Dunn, S. Frostick, G. E. Adams, I. J. Stratford, N. Howells, G. Hogan, and G. K. Radda, Induction of tumour hypoxia by a vasoactive agent. A combined NMR and radiobiological study, *FEBS Letters* **249**(2), 343-347 (1989).
25. T. K. Tran, N. Sailasuta, U. Kreutzer, R. Hurd, Y. Chung, P. Mole, S. Kuno, and T. Jue, Comparative analysis of NMR and NIRS measurements of intracellular pO_2 in human skeletal muscle, *Am. J. Physiol.* **276**(6 Pt 2), R1682-1690 (1999).
26. F. Calamante, M. F. Lythgoe, G. S. Pell, D. L. Thomas, M. D. King, A. L. Busza, C. H. Sotak, S. R. Williams, R. J. Ordidge, and D. G. Gadian, Early changes in water diffusion, perfusion, T1, and T2 during focal cerebral ischemia in the rat studied at 8.5 T, *Magn. Reson. Med.* **41**(3), 479-485 (1999).
27. C. Baudelet, and B. Gallez, How does blood oxygen level-dependent (BOLD) contrast correlate with oxygen partial pressure (pO_2) inside tumors? *Magn. Reson. Med.* **48**(6), 980-986 (2002).
28. H. M. Swartz, J. Dunn, O. Grinberg, J. O'Hara, and T. Walczak, What does EPR oximetry with solid particles measure -- and how does this relate to other measures of pO_2? *Adv. Exp. Med. Biol.* **428**, 663-670 (1997).
29. M. W. Dewhirst, E. T. Ong, G. L. Rosner, S. W. Rehmus, S. Shan, R. D. Braun, D. M. Brizel, and T. W. Secomb, Arteriolar oxygenation in tumour and subcutaneous arterioles: effects of inspired air oxygen content, *Br. J. Cancer Suppl.* **27**, S241-S246 (1996).
30. A. G. Hudetz, J. D. Wood, B. B. Biswal, I. Krolo, and J. P. Kampine, Effect of hemodilution on RBC velocity, supply rate, and hematocrit in the cerebral capillary network, *J. Appl. Physiol.* **87**(2), 505-509 (1999).
31. O. Y. Grinberg, P. E. James, and H. M. Swartz, Are there significant gradients of pO_2 in cells? *Adv. Exp. Med. Biol.* **454**, 415-423 (1998).
32. P. Morse, and H. Swartz, Measurement of intracellular oxygen concentration using the spin label TEMPOL, *Magn. Reson. Med.* **2**, 114-127 (1985).
33. A. Mayevsky, Brain NADH redox state monitored in vivo by fiber optic surface fluorometry, *Brain Res.* **319**(1), 49-68 (1984).
34. D. F. Wilson, and W. L. Rumsey, Factors modulating the oxygen dependence of mitochondrial oxidative phosphorylation, *Adv. Exp. Med. Biol.* **222**, 121-131 (1988).

PRELIMINARY STUDY OF SIMULTANEOUS MULTI-ANTICOAGULANT DEFICIENCY DIAGNOSIS BY FIBER OPTIC MULTI-ANALYTE BIOSENSOR

Liang Tang and Kyung A. Kang

Abstract: Protein C (PC), protein S (PS), antithrombin III, and plasminogen are four important anticoagulants in blood plasma. Deficiency of any of these biomolecules may lead to thrombo-embolic complications including lung embolism, heart attack, and stroke. A multi-factor sensing system is beneficial for identifying the cause of abnormal blood clotting more effectively, rapidly, and cost-effectively. As an initial effort toward simultaneous multi-anticoagulant detection, a PC and PS dual-sensing system has been under development in our research group.

A fiberoptic PC biosensor utilizing fluorophore-mediated sandwich immunoassay was already developed for rapid (~5 minutes) PC deficiency diagnosis. After a single PS sensor was developed for the PS deficiency diagnosis, the two sensors were connected in series to form a dual-sensing system. The cross-reactivity between the analytes and the sensors was found to be minimal. For easier sensing operation, a mixture of fluorophore-linked anti-PC and anti-PS was applied. The results showed that the mixture can be used with a slight signal reduction. When PC and PS was mixed in a sample, the signal intensity was decreased by approximately 5% for both sensors. A study is currently being performed to overcome the signal reduction by increasing the flow velocity and incubation time.

1. INTRODUCTION

Hemostasis maintains blood fluidity within blood vessels (anticoagulation) and, at the same time, prevents blood loss by immediately sealing defects (coagulation) on the injured vessel wall.[1] When the anticoagulation capacity is impaired, venous thrombo-embolism (VTE) complications including thrombosis, lung embolism or heart attack may occur due to the hindrance in oxygen and nutrient transport to tissues.[2]

Table 1. Frequency of hemostatic defects leading to VTE.

Hemostatic Defect	Freq. in General Population	Freq. in VTE Patients
APC resistance*	3%-7%	20%-40%
AT deficiency	0.02%	5%-6%
PC deficiency	0.2%-0.4%	5%-6%
PS deficiency	0.1%	5%-6%
PLG deficiency	-	1%

* A single point mutation of the coagulant Factor V.[6]

The annual incidence of VTE is approximately 1 in 1000 persons, and 50,000-100,000 deaths occur each year in the US.[3] Table 1 lists the inherited hemostatic defects that cause VTE.[4] Among them, the most frequent (20%-40%) is the resistance to activated protein C (APC resistance). It is mainly caused by a single point gene mutation of human Factor V, a coagulation factor: the mutated factor does not become inactivated by the activated protein C (APC).[5, 6] The other defects, deficiencies of PC, protein S (PS), antithrombin III (AT), and plasminogen (PLG), share a similar nature: insufficient amount of anticoagulant proteins in plasma. Altogether, these defects account for about 15%-20% of patients with VTE. Compared with the subjects with APC resistance, their risk of thrombosis is higher and their frequency of deep vein thrombosis is more (88% in PC, 90% in AT, 100% in PS deficiency, and 57% in APC resistance).[7] Therefore, accurate diagnostic tools for these anticoagulant deficiencies are invaluable. In addition, a multi-factor sensing system is very beneficial not only to identify the actual cause more effectively and quickly, but also to reduce costs for diagnoses.

Due to the low level of anticoagulants in plasma (~nM), sensing methods for these factors need to be highly attuned. In addition, the presence of many homologues to these proteins in plasma requires a highly specific assay. Currently, the diagnosis for these factor deficiencies is performed by enzyme linked immunosorbent assay (ELISA). Although very accurate, ELISA is time-consuming (hours to a day), expensive, and technically complicated. A fiber optic biosensor, performing a sandwich, fluorophore-mediated immunoassay on the fiber surface, can provide an accurate, rapid, cost-effective, and user-friendly deficiency diagnostic method. The detector for this sensing system is a compact fluorometer, capable of monitoring four optical fiber sensors simultaneously.

As an initial effort toward the development of a simultaneous multi-anticoagulant detection system, the feasibility of a PC and PS dual-sensing system was explored using a two-channel fiber optic biosensing system. A fiber optic PC biosensor has already been developed for rapid (~5 minutes) PC deficiency diagnosis.[8-15] PS is a co-factor for PC activation and, therefore, an important anticoagulant. The heterozygous, PS-deficient patients have 15%-50% of the normal level of functional PS (10 μg/ml), which is 1.5-5 μg/ml (20-70 nM).[1]

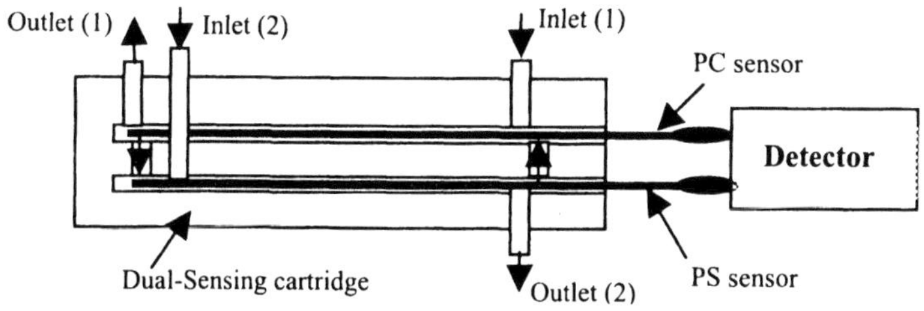

Figure 1. Schematic diagram of the dual-sensing unit.

2. MATERIALS, INSTRUMENTS, AND METHODS

2.1. Materials and Instruments

Human PC (MW = 62,000) and two different kinds of mouse monoclonal anti-human PC were provided by the American Red Cross (Rockville, MD). Human PS (MW = 70,000) and different types of mouse monoclonal anti-human PS were purchased from Haematologic Tech (Essex Junction, VT). Cy5 reactive dye was from Amersham Pharmacia Biotech (Uppsala, Sweden). Fluorometer, Analyte 2000™, and quartz optical fibers were obtained from Research International (Monroe, WA).

2.2. Methods

Optical fibers 6 cm in length were chemically treated, and, primary monoclonal antibodies (1° Mabs) against PC and PS were immobilized on the surface of PC and PS fibers, respectively, via an avidin-biotin bridge.[11] Then, the fiber was inserted into a glass chamber (100 ml) and hot glued to form a functional sensor unit.[11] For the dual-sensing unit, a PC and a PS sensor were placed in series (Fig. 1). Conjugation of Cy5 with secondary monoclonal antibodies (2° Mabs) was performed according to manufacturer's instructions. PC and PS samples were prepared in the emulated human plasma (103 mg-HSA/ml- phosphate buffered saline).[13] During the assay, samples (1 ml) and Cy5-2° Mab were circulated through the sensing chamber at a pre-determined flow velocity for 0.5 and 2 min, respectively.[14]

3. RESULTS AND DISCUSSION

Each data point in figures is presented with the mean value and standard deviation of multiple measurements (> 3 times) on at least two sensors, unless otherwise stated.

3.1. PS Biosensor Development

The molar concentration of the PS sensing range (20-70 nM) is greater than that of the PC (8-40 nM). Therefore, the incubation times for the PC sample and Cy5-2° Mab (0.5/2 min) were expected to be sufficient for the PS sensing. The effect of the flow

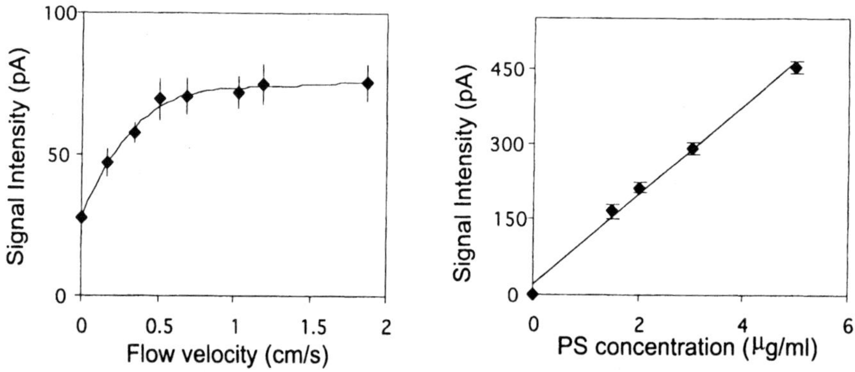

Figure 2. **A**: Effect of flow velocity on PS sensor performance (1.5 µg/ml PS). **B**: Sensitivity of the PS sensor at a flow velocity of 0.5 cm/s.

velocity on the PS sensing performance was investigated to determine the minimal flow velocity.

PS samples at 1.5 mg/ml (the lower sensing limit) and Cy5-2° Mab-PS were applied (i.e., circulated) at various flow velocities between 0.1 and 2 cm/s in a sensor unit. As Figure 2A shows, the signal intensities increased linearly up to the flow velocity of 0.5 cm/s (increase by 153%, compared with the static conditions). As the velocity became higher than 0.5 cm/s, the signal increase slowly tapered, showing the changes in the reaction kinetics from the mass-transfer-limited to the reaction-limited. The minimum effective velocity for the PS sensing was determined to be 0.5 cm/s. This was less than the optimal velocity of the PC sensing (0.7 cm/s for 1 mg/ml PC),[15] possibly because of the higher PS sensing range.

Using this minimum effective velocity, the sensitivity of the PS sensor in the sensing target range was tested (Fig. 2B). The signal intensity increased linearly, with the PS concentration over the range tested with a correlation coefficient of 0.99.

3.2. Effect of the Mixture of Two Second Antibodies on Sensing Performance

For dual factor sensing after the sample incubation, Cy5-linked antibody against PC (Cy5-2° Mab-PC) and Cy5-linked antibody against PS (Cy5-2° Mab-PS) need to be introduced into the respective sensor chambers. This can be done by either applying each antibody to the respective chamber, as shown in Figure 1 (inlet 1 and outlet 1 for PC sensor; inlet 2 and outlet 2 for PS sensor), or by running a mixture of two different Cy5-2° Mabs through both sensors [Fig. 1 (inlet 1 and outlet 2)], if Cy5-2° Mabs do not cross-react. The second case was tested first because it requires an easier design for the fluid direction manipulation. The mixture was applied to each sensor to investigate the extent of cross-reactivity.

Figure 3A shows the signal intensities generated by a PC sensor using only Cy5-2° Mab-PC (■) and the Cy5-2° Mab mixture (◊). Both measurements were performed by a single PC sensor to eliminate possible fiber-to-fiber sensing variation. The signal intensities generated by the mixture slightly decreased, but the reduction was within 6%, with a similar standard deviation for both (5%-10%). A possible cause for the signal

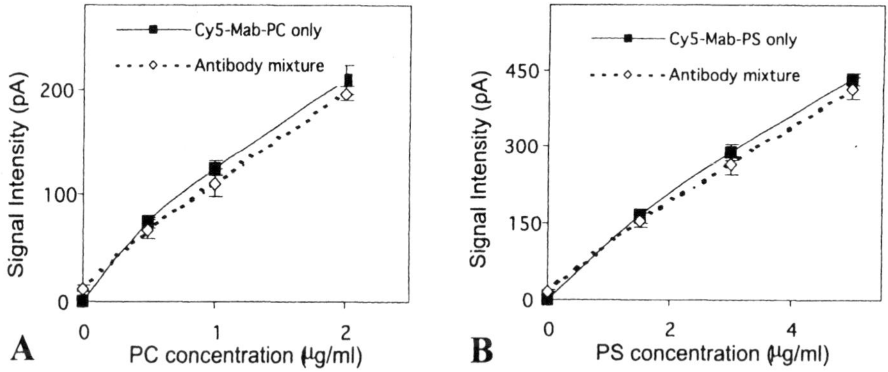

Figure 3. **A**: Effect of the Cy5-2° Mab mixture on the PC sensor performance. **B**: Effect of the Cy5-2° Mab mixture on the PS sensor performance.

reduction could be that the presence of another type of molecule causes a slower mass transfer rate. The background signal of the Cy5-2° Mab mixture slightly increased (10-15 pA) for all tested concentrations, compared with that of Cy5-2° Mab-PC only (not shown).

The effect of the antibody mixture on PS sensor performance was also studied (Fig. 3B), and the signal intensities with the mixture were decreased slightly, with a maximum of 4.5%. The background signal with the mixture was also increased by 10-15 pA.

These results demonstrated that the quantification of PC and PS with the mixed Cy5-2° Mabs can be as accurate as that of the respective Cy5-2° Mab, if the standard curve is also obtained with the mixed Cy5-2° Mabs. The actual cause of the signal reduction is currently being investigated.

3.3. Preliminary Study of Simultaneous PC and PS Detection

A PC sensor and a PS sensor were placed in the dual-sensing unit for simultaneous detection. Samples with the mixture of PC and PS were applied through the two sensor chambers at a flow velocity of 0.7 cm/s (the optimum for the PC sensor). Figure 4A shows the signal intensities resulting from 10 μg-PS/ml-plasma on the PC sensor and 4 μg-PC/ml-plasma on the PS sensor, respectively, in the dual-sensing unit. The response from the PC sensor is less than 1% of that from the PS sensor, indicating a minimal PS cross-reactivity with the PC sensor at the maximum PS concentration. The PS sensor also demonstrated a minimal cross-reactivity with the 4 μg/ml PC.

To investigate the interference of PS molecules to the PC sensor during simultaneous, dual factor detection, two samples were prepared: a sample of 0.5 μg-PC/ml-plasma (the lower sensing limit for PC); in another sample, PS was added at 10 μg/ml (the normal PS concentration in plasma) to 0.5 μg-PC/ml-plasma. Again, the measurements were performed on the same sensor (Fig. 4B). The signal intensity from the factor mixture was decreased by 10%, compared with the intensity of the sample with

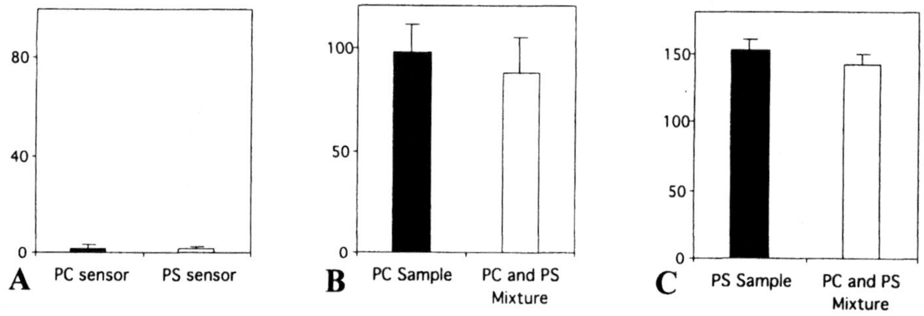

Figure 4. A: Cross-reactivity in the dual-sensing unit [10 μg/ml PS over the PC sensor and 4 μg/ml PC over the PS sensor; Cy5-2° Mab mixed]. **B**: Effect of PC and PS mixture on the PC sensing [dual-sensing unit; (0.5 μg/ml PC) and (mixture of 0.5 μg/ml PC and 10 μg/ml PS); Cy5-2° Mab mixed]. **C**: Effect of PC and PS mixture on the PS sensing [dual-sensing unit; (1.5 μg/ml PS) and (mixture of 1.5 μg/ml PS and 4 μg/ml PC); Cy5-2° Mab mixed].

PC only. Similarly, a mixture of 1.5 μg-PS/ml-plasma (the lower sensing limit for PS) and 4μg-PC/ml-plasma (the normal concentration) was tested to investigate the interference of PC molecules to the PS sensor (Fig. 4C). The signal intensity generated by the PC and PS mixture was decreased by 7%. Compared with the PC sensor, PC interference to the PS sensor appeared to be less. We speculate that the reason for this lower amount of reduction in signal intensity is that the greater PS concentration (as high as 10 μg/ml) caused more hindrance in the mass transfer for the much lower PC concentration (0.5 μg/ml).

Unlike the Cy5-2° Mabs interference, the PC molecule interference to the PS sensing (and *vice versa*) will have to be studied more carefully because some patients may have deficiencies of more than one factor.

4. CONCLUSIONS

A fiber optic PS biosensor was developed to quantify functional PS in plasma in the deficiency range, and the feasibility of fiber optic two-channel biosensing system was demonstrated for PC and PS dual-sensing. The mixture of Cy5-2° Mab-PC and Cy5-2° Mab-PS has shown a minimal cross-interference to the PC or PS sensor. Therefore, the mixture can be applied through the dual-sensing unit for PC and PS detection with an easier operation. The cross-reactivity between PC and PS sensors in the dual-sensing unit was demonstrated to be minimal (less than 1%), with 4 μg/ml PC and 10 μg/ml PS (the normal concentrations). Samples with mixed PC and PS were applied to the dual-sensing unit to test the interference caused by the other analyte. When 10 μg/ml PS was added to 0.5 μg/ml PC, a signal reduction of 10% was shown for the PC sensor. When 1.5 μg/ml PS was mixed with 4 μg/ml PC, the reduction was 7% for the PS sensor. A faster flow velocity and/or a longer sample incubation time may overcome the interference.

Other future work includes the integration of micro-electro-mechanical systems (MEMS) technology for the development of a smaller, more cost-effective, automatic, and multi-anticoagulant sensing cartridge. For the future clinical practice, since the sample volume requirement for our sensing system is less than 1 ml, micro-filtration syringes will be explored to first separate plasma from cellular components in whole blood samples of patients, and then to inject the separated plasma into the sensing unit.

5. ACKNOWLEDGEMENTS

The authors would like to thank the National Science Foundation (CAREER Award BES-9733207) and the University of Louisville (Multidisciplinary Research Incentive Grant) for the financial support. We also thank the American Red Cross for supplying the material. Special thanks are expressed to the American Heart Association for Liang Tang's predoctoral research fellowship.

REFERENCES

1. R. W. Colman, J. Hirsh, V. J. Marder, and E. W. Salzman, *Haemostasis and Thrombosis: Basic Principles and Clinical Practice* (3rd ed.; J.B. Lippincott Company, Philadelphia, 1994).
2. C. S. Kitchens, Thrombophilia and thrombosis in unusual sites, in: *Hemostasis and thrombosis: Basic Principles and Clinical Practice*, 3rd edition, edited by R. W. Colman, J. Hirsh, V. J. Marder, and E. W. Salzman, (J.B. Lippincott Company, Philadelphia, 1994), pp 1255-1263.
3. F. R. Rosendaal, Venous thrombosis: A multicausal disease, *Lancet* **353**, 1167-1173 (1999).
4. J. O. Spiker, and K. A. Kang, personal communication and unofficial documentation (2000).
5. T. Koster, F. R. Rosendaal, H. de Ronde, E. Briët, J. P. Vandenbroucke, and R. M. Bertina, Venous thrombosis due to poor anticoagulant response to activated protein C: Leiden thrombophilia study, *Lancet* **342**, 1503 (1993).
6. G. Tans, A. Nicolaes, and J. Rosing, Regulation of thrombin formation by activated protein C - effect of the factor V leiden mutation, *Semin. Hematol.* **34**(3), 244-255 (1997).
7. I. Martinelli,, P. M. Mannucci, V. de Stefano, E. Taioli, V. Rossi, F. Crosti, K. Paciaroni, G. Leone, and E. M. Faioni, Different risks of thrombosis in four coagulation defects associated with inherited thrombophilia: A study of 150 families, *Blood* **92**(7), 2353-2358 (1998).
8. J. O. Spiker, K. A. Kang, W. N. Drohan, and D. F. Bruley, Preliminary study of biosensor optimization for the detection of protein C, *Adv. Exp. Med. Biol.* **454**, 681-688 (1998).
9. H. I. Balcer, J. O. Spiker, and K. A. Kang, Sensitivity of a protein C immuno-sensor with and without human serum albumin, *Adv. Exp. Med. Biol.* **471**, 605-612 (1999).
10. J. O. Spiker, W. N. Drohan, and K. A. Kang, Reusability study of fiber optic based protein C biosensor, *Adv. Exp. Med. Biol.* **471**, 731-739 (1999).
11. J. O. Spiker and K. A. Kang, Preliminary study of real-time fiber optic based protein C biosensor, *Biotech. Bioeng.* **66**(3), 158-163 (1999).
12. H. I. Balcer, H. J. Kwon, and K. A. Kang, Assay procedure optimization of rapid, reusable protein immunosensor for physiological samples, *Ann. Biomed. Eng.* **30**(10), 141-147 (2002).
13. H. J. Kwon, H. I. Balcer, and K. A. Kang, Sensing performance of protein C immuno-biosensor for biological samples and sensor minimization, *Comp. Biochem. Phys. A* **132**, 231-238 (2002).
14. L. Tang, H. J. Kwon, and K. A. Kang, Studies on effect of sample circulation on sensing performance of protein C immunosensor, proceedings of the Second Joint EMBS/BMES Conference and 24th Annual International Conference of the Engineering in Medicine and Biology Society, Oct. 23-26, 2002, Houston, TX, pp. 1805-1806 (2002).
15. L. Tang, and K. A. Kang, Sensing improvement of protein C biosensor by sample circulation, *Adv. Exp. Med. Biol.* **540**, 177-182 (2003).

CIRCULATION TIME IN MAN FROM LUNG TO PERIPHERY AS AN INDIRECT INDEX OF CARDIAC OUTPUT

Chris B. Wolff, Sophie K. Checkley, Georgina Bhageerutty, Himanshu Bhatt, Atholl Johnston, David J. Collier, Ilias Tachtsidis, N. Garvie, M. E. Rosenberg, and Nigel Benjamin

Abstract: Circulation time (Ct) between lung and periphery may be a surrogate for cardiac output, estimated here, for the most part, as the time between taking a breath of nitrogen and peripheral detection of a desaturation pulse. Use of pulse oximetry involves an internal, instrument delay; however, using the ear, we found shortening with exercise (12.1 ± 0.37 sec, at rest; 9.1 ± 0.25 sec at 100 watts), lengthening after ß-blockade, and lengthening in patients with echocardiographic and clinical left heart failure (8 patients 16.2 ± 1.1 sec; 6 controls 12.0 ± 0.5 sec). Pulse oximetry failed, however, to discriminate heart failure from normal in several patients. In patients referred to a department of nuclear medicine for assessment of chest pain, pulse oximetry (finger and ear) showed unacceptable variability. Nuclide delays between lung and carotid artery correlated significantly with the reciprocal of gated SPECT estimated cardiac output (Q_{gs}); not so, however, for lung to finger. In normal subjects, an old Waters fast response oximeter gave short, reproducible Ct estimates and a significant correlation with the reciprocal of (indirect Fick) cardiac output (Q_{if}). The relationship for normal subjects was: $Ct = 0.28 \times 60/Q_{if} + 2.8$ sec (Q_{if} in L min.; P slope < .001).

1. INTRODUCTION

Wexler et al.[1] first measured lung to ear circulation time in 1946, well before pulse oximetry was introduced. The investigators showed that a desaturation pulse followed inspiration of a large breath of nitrogen (N_2) after a delay of 4-5 sec. Recently, circulation time, measured by pulse oximetry, has been used to assess the severity of heart failure in hospitalised patients.[2] In this study, elevation of inspired oxygen was used and the authors suggested that circulation time might be useful clinically. Pulse oximetry should bear a similar, inverse relationship to cardiac output, as expected with fast oximetry, but pulse oximetry has an extra, uncertain instrument delay between the breath of N_2 and the instrument response. The presented work has appeared previously as an abstract.[3]

2. METHODS AND MODEL

2.1. Measurements

Ideally, fast oximetry would have been used in this study because of its simple, inverse relation to cardiac output. However, pulse oximetry was readily available and clinically acceptable.

2.1.1. Oximetry

A single breath of nitrogen was taken (timed by means of a pressure pulse from the airway or, initially, with a stethograph). Pulse oximetry was measured at the ear in both normal subjects and ambulant patients with left heart failure. Six normal subjects undertook rest and exercise measurements (pulse oximetry - ear) on different days, either following placebo or 100 mg atenolol, a ß-blocker.

2.1.2. Patients with Chest Pain

Delays between arrival of nuclide at the lung and at the periphery (carotid artery and finger) were measured, as well as pulse oximetry (ear and finger). Gated SPECT[4] outlines of the ventricles were obtained, allowing estimation of cardiac output (Q_{gs}). Q_{gs} and the delays were then related.

2.1.3. Fast Oximetry and Rebreathing Cardiac Output (Normal Subjects)

The Waters oximeter (Waters Instruments Inc., Model 0-500, Rochester, MN) pre-dates pulse oximeters; it has no internal delays. Lung to ear Ct and rebreathing Q (indirect Fick, Q_{if}) were measured, as in Eq. (1)[5]:

$$Q_{if} = VCO_2 / (CvCO_2 - C_aCO_2) \quad (1)$$

The indirect Fick method depends on obtaining temporary equilibrium between mixed venous (pulmonary artery) PCO_2 ($PvCO_2$) and alveolar PCO_2 (P_ACO_2) values. This is obtained by rebreathing, initially, from an oxygen-filled 6 L anaesthetic bag until PCO_2 has risen a little above $PvCO_2$. After 2 mins, the subject re-breathes again and PCO_2 in the bag falls to the mixed venous value, resulting in a brief plateau (mixed venous PCO_2). CO_2 content values are calculated ($CvCO_2$ and C_aCO_2) from $PvCO_2$ and P_ACO_2. The subject's CO_2 production rate is measured using a Douglas bag.

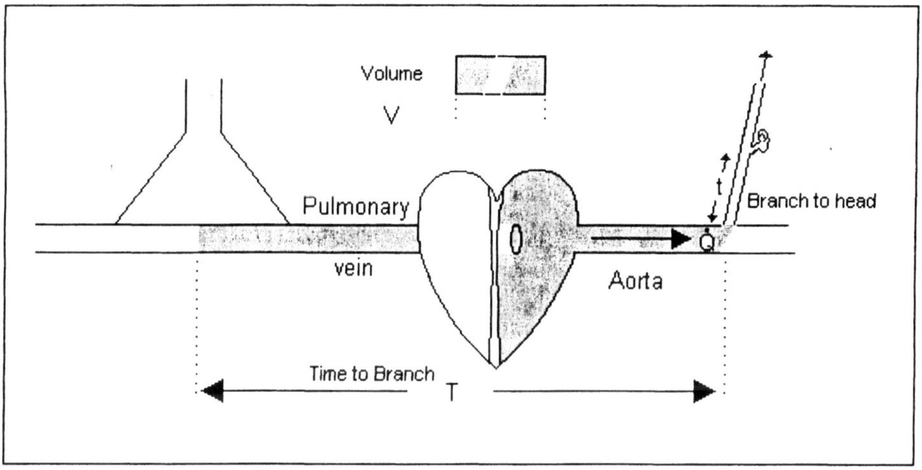

Figure 1. Schematic of lung, pulmonary vein, heart, and aorta, with vessels to the head and ear. Darker shading represents the volume of blood (V) in the circuit as far as the branch; T = the transit time; t = the delay between branching and arrival at the periphery (the ear in this case).

2.2. Model

Figure 1 shows a schematic illustrating the relevant pathway between lung and ear. Ct, the circulation time from lung to ear (as in Fig. 1), is T + t. The volume of blood in the circulation between the lung and the vascular branch from the aorta to the head, V, contains blood, which is all part of the total cardiac output. Since the time elapsing from blood passing through the lung and arriving at the branch point is T, cardiac output (Q) is V/T. Hence, we obtain T = V/Q. But Ct, circulation time, is T + t, where t is the transit time from the branch point to the oximeter on (in this case) the ear. If there is no internal oximeter instrumentation delay, then:

$$Ct = V/Q + t \qquad (2)$$

Hence, it is possible, if both circulation time and cardiac output can be measured, to obtain an estimate of V (the volume of blood in the circulatory pathway between lung and aortic branch point; this will be the slope of a line fitting a plot of Ct against 1/Q) and also of t, the transit time between the aortic branch point and the oximeter. The value t, being part of the cerebral circulation, is likely to be relatively constant.

With pulse oximeters, there is normally an internal instrument delay, but it is expected that various maneuvers will have predictable effects: exercise should shorten Ct while ß-blockade may lengthen it; heart failure, with reduced cardiac output, should also lengthen it.

2.3. Experimental Series

The first series was of 20 normal subjects undertaking measurements with pulse oximetry (ear) at rest, at 50 watts and 100 watts. Two of these same subjects plus four

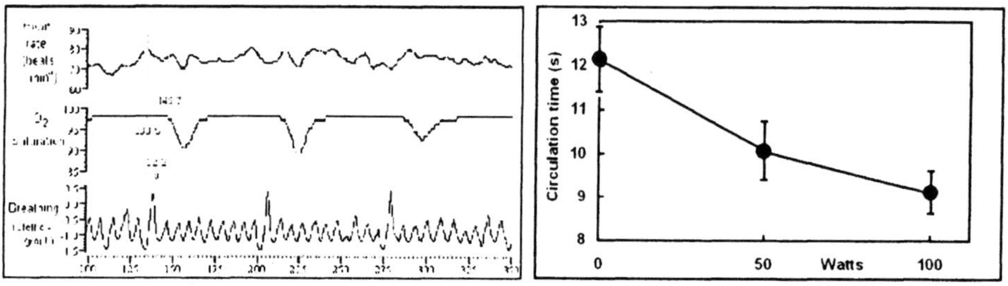

Figure 2. The left diagram shows respiration (stethograph), SaO_2, and heart rate with pulse oximetry at the ear; the right figure shows the mean values from 20 normal subjects for lung to ear estimates of circulation time, at rest, 50 watts, and 100 watts exercise.

additional subjects compared measurements at rest and at 50 watts, either on placebo or on atenolol (separate days). Eight patients with clinically confirmed left heart failure[6] were compared with 6 age-matched control subjects. Sixteen patients with chest pain had ear and finger oximetry, first pass nuclide and, separately, gated (ECG) SPECT cardiac measurements. Six normal subjects had fast (Waters) oximetry undertaken and, separately, rebreathing (CO_2) cardiac output measurements, both measured at rest and during 50 watts exercise.

3. RESULTS

3.1. Pulse Oximetry in Normal Subjects

Figure 2 (left) illustrates the effect of a moderately large breath of nitrogen on arterial oxygen saturation (SaO_2) at the ear. Mean values (n = 20) for lung to ear are shown on the right for rest and exercise (50 watts and 100 watts): rest = 12.1 ± 0.37 sec (± SEM); 50 watts exercise = 10.7 ± 0.33 sec; 100 watts exercise = 9.1 ± 0.25 sec.

3.2. Beta-Blockade and Left Heart Failure

Values for lung to ear circulation time (Ct, pulse oximetry) decreased slightly following placebo at rest, but did not fall in 50 watts exercise (Figure 3 - left and center panels). In contrast, these values increased following 100 mg atenolol, both at rest and in

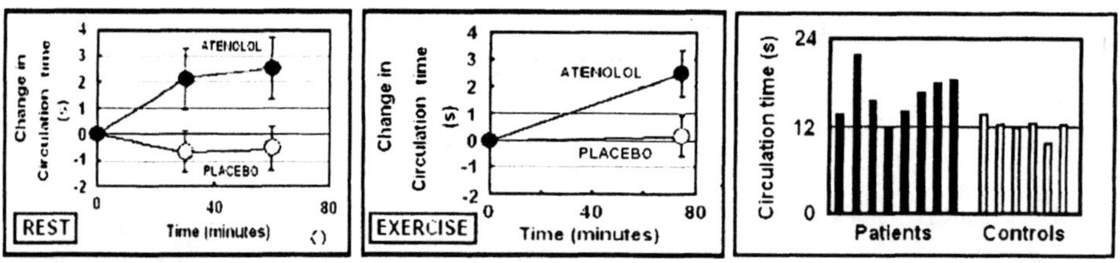

Figure 3. Mean (lung to ear) circulation time (± SEM) changes after ß-blockade (atenolol, 100 mg) or placebo are shown for 6 normal subjects: left panel - rest; center panel - 50 watts exercise. Right panel: individual mean values for 8 patients with left heart failure are shown with values for 6 age-matched control subjects.

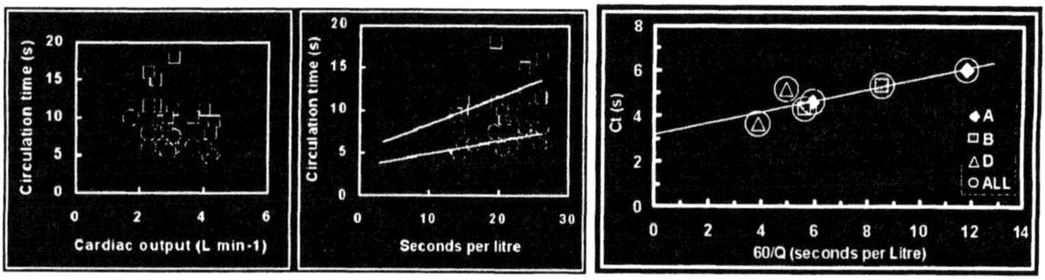

Figure 4. The left and middle panels show individual results for 16 patients referred to nuclear medicine for chest pain. The left plot is of two circulation times (delay between arrival of [1] nuclide at lung and carotid artery and [2] between lung and finger) versus cardiac output (Q_{gs}, gated SPECT estimate). The center panel shows these two circulation times plotted against $60/Q_{gs}$ (with best fit regression lines). Symbols: lung to carotid = O; lung to finger = □. The right panel shows Ct versus $60/Q_{if}$ for normal subjects (Q_{if} estimated from the CO_2 rebreathing method; only available for 3 of the 6 normal subjects, A, B, and D).

50 watts exercise. Changes were significant for rest ($P < .02$), but not for exercise ($P = .069$) due to an atypical subject result.

The use of pulse oximetry to assess left heart failure is also shown in Figure 3 (right panel). It is apparent that heart failure gives greater values, reflecting the lower cardiac output but with overlap between patients and controls.

3.3. Circulation Time and Cardiac Output

Since circulation time is Ct = V/Q + t (see Eq. 2), a linear relationship is expected between Ct and 1/Q. If cardiac output (Q) is measured in L/min, and Ct and t are measured in seconds, then the equation becomes:

$$Ct = V \times 60/Q + t \tag{3}$$

In Figure 4, the left and center panels relate to 16 patients referred to Nuclear Medicine with chest pain. They were monitored with first pass nuclide, giving circulatory delays (circulation times) between lung and carotid and lung and finger. These have been plotted against gated SPECT cardiac output estimates (Q_{gs}). The left panel shows the inverse relationship between the circulation time and Q_{gs}. The center panel shows the linear relationship between the circulatory delay and $60/Q_{gs}$ (sec/L of blood flow). The right panel shows Ct measured using the fast response (non-pulse, Waters) oximeter plotted against $1/Q_{if}$, where Q_{if} was estimated by means of the CO_2 rebreathing method[5]; valid Q_{if} values were obtained in only 3 of the six normal subjects: A, B, and D. The relationship was found to be: Ct = 0.28 x $60/Q_{if}$ + 2.8 sec; Q_{if} in L min; *P* slope < .001. Hence, V = 0.28 L, and t = 2.8 sec (from Figure 1 and Section 2.1.).

4. DISCUSSION

Exercise caused the expected trend in circulation time (a decrease) when measured with pulse oximetry. The conclusion of a direct relationship between Ct and 1/Q would be misleading from these data since there is an unknown instrumental delay. However, cardiac output is known to increase with exercise and, as expected, Ct values were reduced. Furthermore, ß-blockade is known to decrease cardiac output, so the increased Ct was also as expected. The measurements in patients with heart failure were, as a group, significantly longer than found in age-matched normal subjects. However, the potential diagnostic use of this technique was confounded by overlap, the longest values in normal subjects being greater than the shortest ones in heart failure patients.

Examining delays in the circulation and cardiac output using Nuclear Medicine methods confirmed the expected inverse relationship between cardiac output and circulatory delays across 16 individual subject rest values. Delays from lung to periphery using fast oximetry correlated well with $60/Q_{gs}$ (cardiac output in litres). Only CO_2 rebreathing cardiac output values from 3 of the 6 subjects showed adequate plateaus, but these gave a highly significant trend, supporting the theoretical Ct/Q relationship. The Waters oximeter was not suitable for clinical use, but these results on normal subjects, supported by the trends suggested by pulse oximetry and the nuclear medicine study, suggest that safe modern oximetry with a minimal instrument delay could be of clinical value as a surrogate cardiac output measuring device.

5. ACKNOWLEDGMENTS

Thanks are due to the Dunhill Trust for financial support, Mr. Ted Carter (Queen Mary and Westfield campus) for technical help, Georgina Lewis (Charterhouse Square campus) for help with patient recruitment, Steven Wilson, Cardiac Services Manager, for Echo-Cardiography services (Barts, Smithfield site), and to the technical staff in Nuclear Medicine at the Royal London Hospital. Thanks are also due to the Oxford University Physiological Laboratories for the loan of the Waters oximeter used in this study.

6. REFERENCES

1. J. Wexler, J. L. Whittenberger, and S. Himmelfarb, An objective method for determining circulation time from pulmonary to systemic capillaries by the use of an oximeter, *J. Clin. Invest.* **25**, 447-450 (1946).
2. B. Kasravi, J. P. Boehmer, and U. A. Leuenberger, A noninvasive method for estimating cardiac output using lung to finger circulation time of oxygen, *Am. J. Cardiol.* **82**, 915-917 (1998).
3. C. B. Wolff, S. K. Checkley, G. Bhageerutty, H. Bhatt, A. Johnson, D. J. Collier, N. Garvie, M. E. Rosenberg, and N. Benjamin, Circulation time in man from lung to periphery - pulse and non-pulse oximetry, *J. Physiol.* 543.1, 41P (2002).
4. M. R. Mansoor, and G. V. Heller, Gated SPECT imaging, *Semin. Nucl. Med.* **29**, 271-278 (1999).
5. N. L. Jones, *Clinical Exercise Testing*, 4th edition (WB Saunders, Philadelphia, 1997).
6. New York Heart Association, in: *Nomenclature and Criteria for Diagnosis of Diseases of the Heart and Great Vessels*, 9th edition, edited by The Criteria Committee of the New York Heart Association (Little, Brown and Co., Boston, 1994), pp. 253-256.

FACTORS CONTROLLING OXYGEN UTILIZATION

John Biaglow, Mark Dewhirst, Dennis Leeper, Randy Burd, and Steve Tuttle

Abstract: We demonstrate, theoretically, that oxygen diffusion distance is related to the metabolic rate of tumors (QO_2) as well as the oxygen tension. The difference in QO_2 rate between tumors can vary by as much as 80-fold. Inhibition of oxygen utilization by glucose or chemical inhibitors can improve the diffusion distance. Combining respiratory inhibitors with increased availability of oxygen will further improve the oxygen diffusion distance for all tumors. A simple means for inhibiting oxygen consumption is the use of glucose (the Crabtree effect). The inhibition of tumor oxygen utilization by glucose occurs in R323OAc mammary carcinoma and 9L glioma cells. However, stimulation of oxygen consumption is observed with glucose in the Q_7 hepatoma cell line. MIBG, a known inhibitor of oxygen utilization, blocks oxygen consumption in 9L, but is weakly inhibitory with the Q_7. Q_7 tumor cells demonstrate an anomalous behavior of glucose and MIBG on oxygen consumption. Our results clearly demonstrate the necessity for comparing effects of different agents on different tumor cells. Generalizations cannot be made with respect to the choice of inhibitor for *in vivo* use. Our work shows that oxygen consumption also can be inhibited with malonate and chlorosuccinate. These substrates may be effective *in vivo*, where glucose is low and glutamine is the major substrate. Our results indicate that information about individual tumor substrate-linked metabolic controls may be necessary before attempting to inhibit oxygen utilization *in vivo* for therapeutic benefit.

1. INTRODUCTION

Historically, there is a good deal of interest in decreasing the hypoxic fraction of tumors in order to improve their response to radiation and photodynamic therapies. One simple approach is to use the *Crabtree effect*, inhibition of tumor oxygen utilization by glucose.[1-3] Inhibition of oxygen utilization will increase the diffusion distance of oxygen into hypoxic cells. Glucose is low in tumors and therefore tumor oxygen consumption is more susceptible to hyperglycemia.[3] We are interested in the factors that control cellular oxygen consumption *in vitro* and ultimately *in vivo*. Historically, tumor oxygen consumption *in vivo* varies by a factor of 80-fold between rat tumors, depending on their size.[4] Biopsied human tumor tissue varies in oxygen consumption by 16-fold.[4] The huge variation in oxygen consumption between tumor cells is complicated by both endogenous and exogenous substrate availability (i.e. glucose, glutamine). We have already documented some of the factors influencing oxygen consumption.[5] It was the purpose of this study to document some of the factors influencing oxygen consumption between various tumor cell lines. We measured the effect of both glucose and glutamine on oxygen consumption. In addition, we determined the effect of various inhibitors such as Cl-succinate, malonate, and meta-iodobenzylguanidine (MIBG).

2. METHODS

Oxygen consumption was measured with the aid of the Clark electrode system, as previously described.[5] Cells were grown in culture and harvested, as previously described.[1, 5] Theoretical calculations are as previously described.[4]

3. RESULTS

Figure 1 shows the calculated diffusion distance for rat tumors, as a function of tumor size. Hypoxic fractions would occur in small tumors with a high QO_2, but they could also be present with a low QO_2 if the tumor was large enough. For Q7 hepatoma tumors, the QO_2 (nmoles/min/mg protein) varied from a high of 8.0 for the smallest tumor to a low value of 2.4 for the largest. The Walker 256 tumor shows 4.35 and 2.0 for the smallest and the largest, respectively. The DS tumor is 2.4-1.13 for the smallest and largest, respectively, and the QO_2 varies between 1.0-0.1 for the smallest vs. largest 4956 tumors. The theoretical curve demonstrates that oxygen distance could be improved in tumors with the highest respiratory rate through the use of inhibitors of oxygen consumption or by increasing oxygenation. For example, 90% inhibition of utilization with a PO_2 of 160 mm results in an increase in diffusion distance from 30 μm to nearly 1000 μm, far beyond the distance between capillaries. Note, increasing the oxygen consumption will decrease the diffusion distance.

The oxygen consumption can vary considerably, depending upon the microenvironment and the metabolic status of the tumor, as shown in Figure 1. Most recently, we have been interested in the clinical application of the Crabtree effect on tumor cell oxygen consumption *in vivo*.[1, 2, 5] The hypothesis is that inhibition of oxygen utilization by glucose would increase the diffusion distance. As predicted, there would be further increases if glucose were administered and if the tumor-bearing animal were allowed to breathe oxygen.[2] We found in the past that inhibition of R3230Ac oxygen utilization by a combination of inhibitors, glucose, and MIBG radiosensitized the tumors.[6] Furthermore, glucose sensitized tumors when given with insulin.[3]

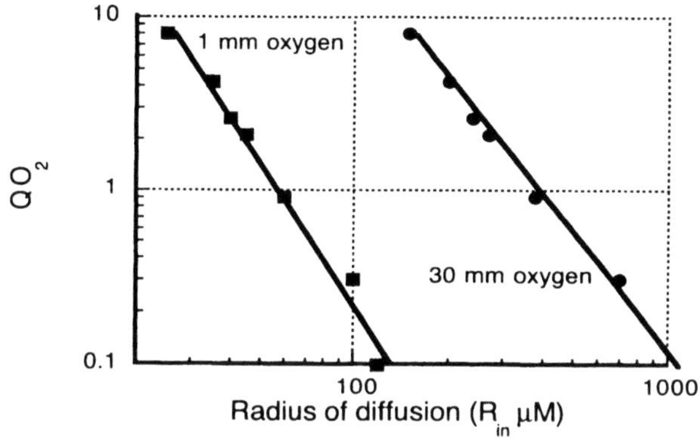

Figure 1. Theoretical calculation of diffusion distance as a function of QO_2 and oxygen tension (adapted from Biaglow, et al.[4]). QO_2 values are plotted from the highest to the lowest values.

The effect of glucose alone on oxygen consumption in R323OAc cells is shown in Figure 2. We were interested in extending this principle to other tumors; therefore, we tested the effect of glucose on Q_7 hepatoma cells. While glucose inhibits oxygen utilization in R323OAc cells, we observed stimulation of oxygen consumption for the Q_7 cells with glucose (Figure 2). Stimulation of oxygen consumption by addition of glutamine was not observed in Q_7 cells. However, maximum observed rate of oxygen utilization occurred in the presence of glutamine and succinate, as seen in Figure 3. There is a 60-fold increase in stimulation when 2 mM glutamine is followed by 2 mM succinate. The addition of 3 mM KCN totally inhibits oxygen consumption under all these conditions.

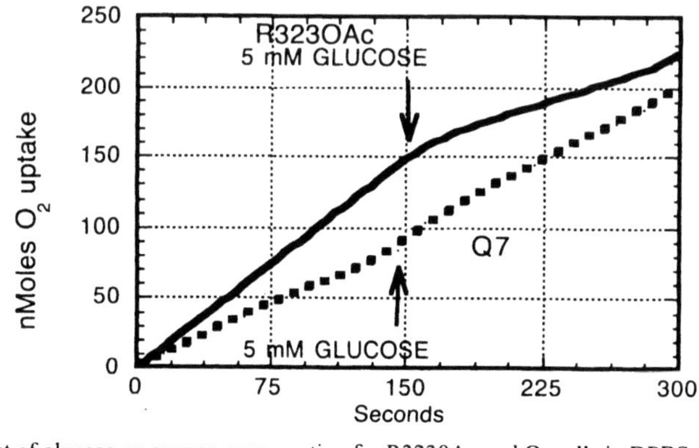

Figure 2. The effect of glucose on oxygen consumption for R3230Ac and Q_7 cells in DPBS with 2 mM $CaCl_2$ and $MgCl_2$, 10 mM KCl, 20 mM Pi, pH 7.4, and 37°C. Glucose was 5 mM when added.

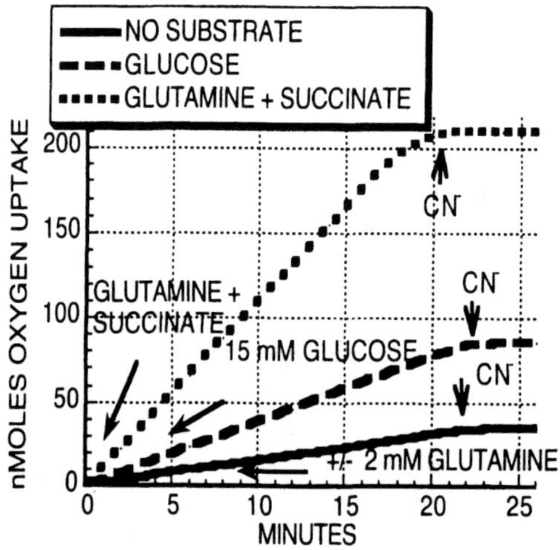

Figure 3. The effect of glucose, glutamine, and succinate on the oxygen consumption of Q_7 hepatoma cells. The reaction conditions are the same as those listed in legend of Figure 2. Final concentrations were: glucose 3 mM, glutamine 2 mM, and succinate 5 mM.

Other factors that influence oxygen consumption are shown in Figure 4. The 9L tumor cells, rapidly isolated from the animal, have a very low respiratory rate compared with cultured cells, isolated by trypsinization and suspended in DPBS. However, the maximum rate of oxygen consumption occurs when the same cells are suspended in MEM growth medium, which contains 2 mM glutamine. Oxygen uptake differs by 90-fold for these three conditions.

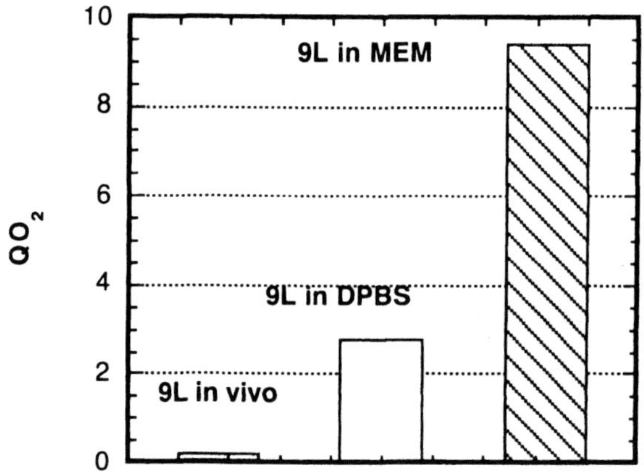

Figure 4. Comparison of oxygen consumption for cells isolated directly from the animal tumor and suspended in DPBS. Cells were also harvested from tissue culture and suspended in DPBS or MEM.

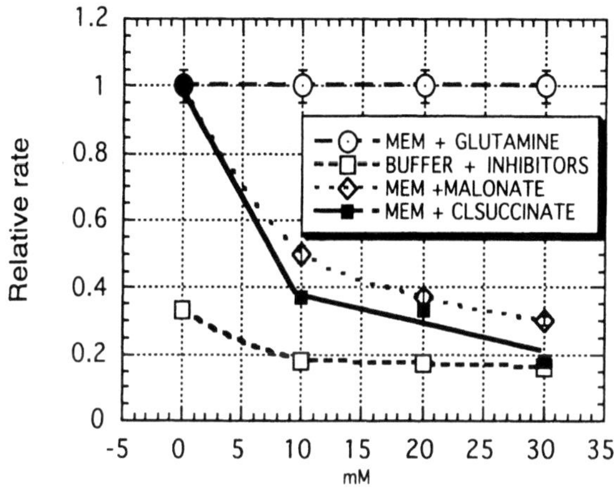

Figure 5. The effect of malonate and chlorosuccinate on cells suspended in DPBS or MEM with glutamine. The reaction conditions are the same as those listed in the legend of Figure 2.

Figure 5 shows the effect of different inhibitors on the oxygen consumption for 9L cells cultured *in vitro* and harvested by trypsinization. The maximum rate occurred in MEM. Glutamine is metabolized, after deamination, via the Krebs cycle. Classical inhibitors of the Krebs cycle activity inhibit oxygen consumption in MEM. There was an immediate inhibition of oxygen consumption observed upon the addition of 10 mM chlorosuccinate or malonate. Maximum inhibition occurs with either inhibitor at 30 mM. Without substrate (lower curve), there is little inhibition of the already decreased rate.

Figure 6 shows the difference in oxygen consumption between tumor cells in response to MIBG. We demonstrated that MIBG sensitized R3230Ac tumors *in vivo* to radiation, when used in combination with glucose.[7] We are also using MIBG to enhance the hyperthermic response.[8] Oxygen consumption in 9L cells is inhibited by 50% with 0.1 mM MIBG, and the level of inhibition increases to 80% with 0.5 mM MIBG. In contrast, Q_7 cells are relatively insensitive to MIBG at concentrations less than 0.2 mM.

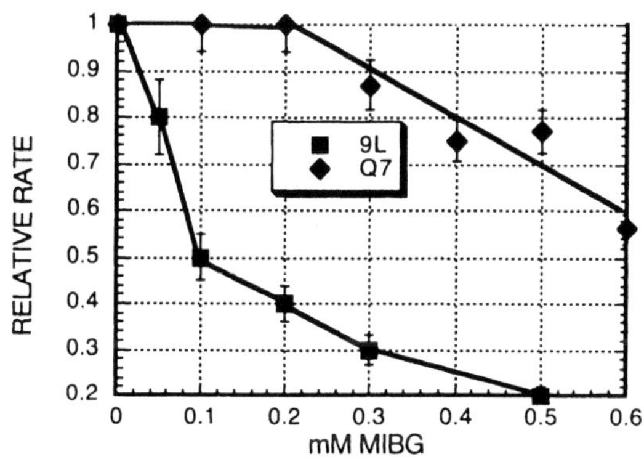

Figure 6. The effect of MIBG on the oxygen consumption of R3230Ac and the Q_7 tumor cell suspensions. The reaction conditions are the same as those listed in the legend of Figure 2.

4. DISCUSSION

Oxygen is the most efficient radiation sensitizer known. Tumors containing hypoxic fractions are considerably less sensitive to ionizing radiation. Increasing the oxygen diffusion distance, by either increasing the amount of oxygen available or inhibiting oxygen utilization, would sensitize hypoxic cells to radiation. In the latter case, when studying the effects of inhibitors, it is necessary to test these inhibitors with different tumor cells grown *in vitro* under identical conditions. We have shown a variation of nearly 80-fold in tumor cell oxygen consumption *in vivo* between 4 different tumor lines (cf. Fig 1).

Differences in rates of oxygen consumption also occur between the 9L, R323OAc, and the Q_7. The R323OAc shows a Crabtree effect, i.e. inhibition of oxygen utilization, whereas the Q_7 does not. The 9L is sensitive to MIBG, whereas the Q_7 oxygen consumption is not. The Q_7 shows stimulation of oxygen consumption with glucose, low sensitivity to MIBG, and no stimulation of oxygen consumption by glutamine. However, the Q_7 is increased with succinate plus glutamine. In conclusion, our results demonstrate that inhibition of oxygen utilization may be useful in some tumors *in vivo*, especially when used with hyperbaric oxygen.[2] It is important to stress that some tumors may not respond to respiratory inhibitors.[2] In the case of the Q_7, addition of glucose *in vivo* may actually increase the hypoxic fraction, as shown for stimulation of oxygen consumption *in vitro*.[9] Tumors that are sensitive to glutamine, in the absence of glucose, may respond favorably to inhibitors of the Krebs cycle, such as malonate and chlorosuccinate.

5. CONCLUSIONS

We demonstrate, theoretically, that oxygen diffusion distance is related to the metabolic rate of tumors (QO_2), as well as the oxygen tension. The difference in QO_2 rate between tumors can vary by as much as 80-fold. Inhibition of oxygen utilization by glucose or chemical inhibitors can improve the diffusion distance. Combining respiratory inhibitors with increased availability of oxygen will further improve the oxygen diffusion distance for all tumors. A simple means for inhibiting oxygen consumption is the use of glucose (the Crabtree effect). The inhibition of tumor oxygen utilization by glucose occurs in R323OAc mammary carcinoma and 9L glioma cells. However, stimulation of oxygen consumption is observed with glucose in the Q_7 hepatoma cell line. MIBG, a known inhibitor of oxygen utilization, blocks oxygen consumption in 9L but is weakly inhibitory with the Q_7. Q_7 tumor cells demonstrate an anomalous behavior in relation to the effect of glucose and MIBG on oxygen consumption. Our results clearly demonstrate the necessity for comparing effects of different agents on different tumor cells. Generalizations cannot be made with respect to the choice of inhibitor for *in vivo* use. Our work shows that oxygen consumption can also be inhibited with malonate and chlorosuccinate. These substrates may be effective *in vivo*, where glucose is low and glutamine is the major substrate. Our results indicate that information about individual tumor substrate-linked metabolic controls may be necessary before attempting to inhibit oxygen utilization *in vivo* for therapeutic benefit.

REFERENCES

1. R. Burd, P. R. Wachberger, J. E. Biaglow, M. I. Wahl, I. Lee, and D. R. Leeper, Absence of Crabtree effect in human melanoma cells adapted to growth at low pH: reversal by respiratory inhibitors, *Cancer Res.* **14**, 5630-5635 (2001).
2. S. A. Synder, J. L Lanzen, R. D. Braun, G. Rosner, T. W. Secomb, J. Biaglow, D. M. Brizel, and M. W. Dewhirst, Simultaneous administration of glucose and hyperbaric gas achieve greater improvement in tumor oxygenation than hyperoxic gas alone, *Int. J. Radiat. Oncol. Biol. Phys.* **51**, 494-506 (2001).
3. J. E. Biaglow, P. S. Lavik, and N. Ferencz, Jr., Modification of radiation response through glucose-controlled respiration, *Radiat. Res.* **39**, 623-633 (1969).
4. J. E. Biaglow, M. E. Varnes, B. Jacobson, and H. D. Suit, Control of oxygen utilization *in vitro* and *in vivo:* implications for radiotherapy of tumors, *Adv. Exp. Med. Biol.* **180**, 323-332 (1984).
5. J. E. Biaglow, M. E. Varnes, B. Jacobson, and C. J. Koch, Factors influencing the oxygen consumption and radiation response of cultured mammalian cells, *Adv. Exp. Med. Biol.* **159**, 347-358 (1983).
6. J. E. Biaglow, Y. Manevich, D. Leeper, B. Chance, M. W. Dewhirst, T. Jenkins, S. W. Tuttle, K. Wroblewski, J. D. Glickson, C. Stevens, and S. M. Evans, MIBG inhibits respiration: potential for radio- and hyperthermic sensitization, *Int. J. Radiat. Oncol. Biol. Phys.* **42**, 871-876 (1998).
7. I. Lee, J. D. Glickson, M. W. Dewhirst, D. B. Leeper, R. Burd, H. Poptani, L. Nadal, W. G. McKenna, and J. E. Biaglow, Effect of mild hyperglycemia +/- meta-iodo-benzylguanidine on the radiation response of R3230Ac tumors, *Adv. Exp. Med. Biol.* **530**, 177-186 (2003).
8. R. Burd, S. N. Lavorgna, C. Daskalakis, P. R. Wachsberger, M. I. Wahl, J. E. Biaglow, C. W. Stevens, and D. R. Leeper, Tumor oxygenation and acidification are increased in melanoma xerographs after exposure to hyperglycemia and meta-iodobenzylguanidine, *Radiat. Res.* **59**, 328-335 (2003).
9. J. Biaglow, and R. Durand, The effects of nitrobenzene derivatives on oxygen utilization and radiation response of an in vitro tumor model, *Radiat. Res.* **65**, 529-539 (1976).

AMPUTATION LEVEL VIABILITY IN CRITICAL LIMB ISCHAEMIA:
Setting new standards

David K. Harrison and Ian E. Hawthorn

Abstract: We have previously demonstrated the successful use of skin oxygen saturation (SO_2) measurements to predict the healing viability in lower limb amputations for critical limb ischaemia. The measurements are quick and easy to perform, but the instrument that has been used to date is now obsolete and a new, lightweight, portable instrument has recently been introduced. However, fundamental differences between the two instruments could influence the criteria used for determining amputation level viability. The purpose of this study was to compare the *in vivo* measurements using the two instruments in order to validate amputation level viability criteria using the RM200.

Skin SO_2 measurements were carried out on critically ischaemic lower limbs of patients, and on the forearms of normal volunteers during before, during and after a 5 minute period of tourniquet ischaemia. A linear correlation ($r^2 = 0.91$) was found between the values obtained from the two instruments within the range of interest (0 to 40% SO_2). Differences between the instruments lay within 1 standard deviation of the mean, demonstrating a high degree of agreement between the two methods. The RM200 is thus an acceptable replacement for the MCPD instrument for amputation level viability assessments.

1. INTRODUCTION

We have previously reported on the possibility of using skin oxygen saturation (SO_2) measurements to predict the healing viability in lower limb amputations for critical limb ischaemia.[1] The technique involves determining the degree of tissue hypoxia (DTH),[1] which is defined as the percentage of SO_2 values along the leg with a value of less than 10%. In the initial study, it was found that if the DTH was less than 15% or the mean SO_2

Table 1. Principal differences between the skin SO_2 measurement systems.

	Photal MCPD 1000	Whitland RM200
Resolution	2 nm	3 nm
Algorithm	500-586 nm : 6 wavelengths	500-586 nm: Melanin and scattering corrections, Least squares fit
Light source	Quartz halogen via 18 x 200 ∂m ∅ fibres	White LED and lens mounted in probe
Receiving fibres	12 x 200 ∂m ∅ fibres	1 x 100 ∂m ∅ fibre

at two specified sites along the line of a proposed below knee amputation (BKA) was 30% or greater, healing at the BKA level was successful.

The technique, using the above criteria, has been applied to routine clinical practice in the University Hospital of North Durham since 1999. The results of an audit of this application of the technique have been published[2] and demonstrated a 94% healing rate of BKAs was being achieved at a BKA to above knee amputation (AKA) ratio of 9:2. The measurements are quick and easy to perform, but the instrument that has been used to date, a Photal MCPD 1000 (Otsuka Electronics, Osaka) spectrophotometer, is bulky and relies on offline evaluation of the measurements. A new, lightweight, portable online system has recently been developed (RM200, Whitland Research, Whitland, UK). However, there are differences between it and the MCPD instrument that could influence the criteria used for determining amputation level viability: these are given in Table 1.

The purpose of this study is to compare the *in vivo* measurements using the two instruments in order to validate amputation level viability criteria using the RM200. During the course of the study, some interesting observations were made during tourniquet-induced ischaemia in normal forearm skin. These are also described.

2. METHODS

2.1. Patients

Ten patients with critical limb ischaemia undergoing routine amputation level viability assessments were included; in one case, both legs were assessed. The measurements were carried out as previously described[1,2] at approximately 10 mm intervals along the medial aspect of the limb, from the level of the tibial tuberosity as far as and including the big toe and at the two specified sites along the line of the proposed BKA flap. In this study, additional measurements were carried out at 10 mm intervals around the entire circumference of the proposed flap. A holder was used with the optodes of the two instruments mounted 10 mm apart. The recordings from one instrument thus lagged one measurement point behind that of the other and this was taken into account in analysing the results. Typically about 70 pairs of SO_2 values were obtained from each patient.

2.2. Normal Subjects

In order to define DTH, it was necessary to obtain sufficient data points in the region 0%-10% SO_2. A further series of measurements was therefore carried out on both forearms of 2 volunteers during a 6 minute period of brachial artery occlusion by means of a blood pressure cuff inflated to 200 mm Hg. On three occasions, a repeat measurement was carried out on the same arm. The same probe holder was used as that for the patients. The probes were therefore 10 mm apart. Recordings were made simultaneously with each instrument at 10 or 20 second intervals.

In the light of some anomalous results obtained during these measurements, a further series of 5 measurements was carried out in the forearms of the same volunteers in a matrix of 25 points covering a 40 mm x 40 mm area of skin during the 5th to 7th minute of cuff ischaemia. The RM200 instrument was used for all these measurements but the MCPD 1000 was used in addition on one occasion. In a further experiment, for reasons evident below, a "micro-lightguide" probe consisting of a single 100 μm illuminating fibre, instead of the LED and lens, was used in order to reduce the catchment volume.

3. RESULTS

3.1. Patients

Figure 1 shows that considerable variations in SO_2 values were found within very short distances, even of the order of mm, along the leg. In order to reduce the effects of these spatial heterogeneities, SO_2 values from the two probes were averaged over the same areas of skin (approximately 9 points).

No SO_2 values less than 10% (as measured with the MCPD 1000) and relatively few less than 20% were found in this group of patients. It was therefore necessary to use the tourniquet ischaemia model in normal subjects in order to complete the lower range data.

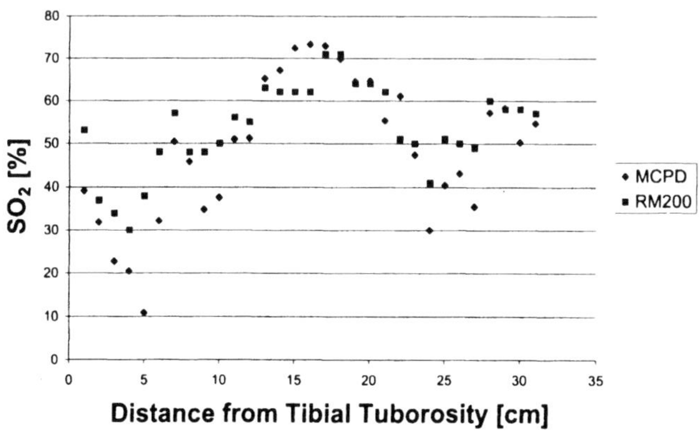

Figure 1. Typical skin SO_2 values measured with the RM200 and MCPD 1000 along the leg of a patient.

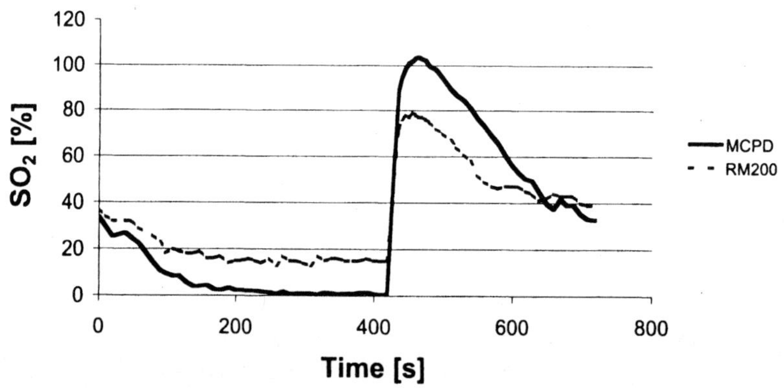

Figure 2. A typical response of SO_2 during cuff occlusion is shown by the MCPD in the figure, but that measured by the RM200 1 cm away fell only to 18%. Note that the MCPD also produced SO_2 values greater than 100% during maximal reactive hyperaemia.

3.2 Normal Subjects

It has been assumed in many previous studies that 6 minutes of tourniquet ischaemia reduces skin SO_2 values to almost 0%. We found that this is clearly not the case, as shown in Figure 2. The phenomenon was not suspected to be instrument-dependent in that values approaching 0% had already been recorded using the RM200 in the contralateral arm of the same subject. In order to try to detect heterogeneities of SO_2 during cuff occlusion, SO_2 mapping measurements were carried out.

Figures 3 to 5 show the results of preliminary experiments to map SO_2 values over small areas of forearm skin during the 5th to 7th minutes of tourniquet ischaemia. Figure 3 shows measurements on the same area of skin with a time interval of 48 hours between. A single high value can be seen at the centre of the medial edge of the measurements. The overall pattern remains very similar after 48 hours. Figure 4 demonstrates the similarity of mappings carried out with both the RM200 and MCPD 1000 instruments. On the other hand, the RM200 instrument with the "micro-lightguide" probe gives a very different pattern of values compared with that of the normal probe.

In the light of these observations, in the cuff occlusion experiments to compare the RM200 and MCPD in the normal subjects, only those in which the SO_2 values recorded by both instruments were < 10% at the end of the 6 minutes of cuff occlusion were included. These results and the patient data are combined in Figure 6 for the 0%-40% SO_2 range (as measured by the MCPD 1000).

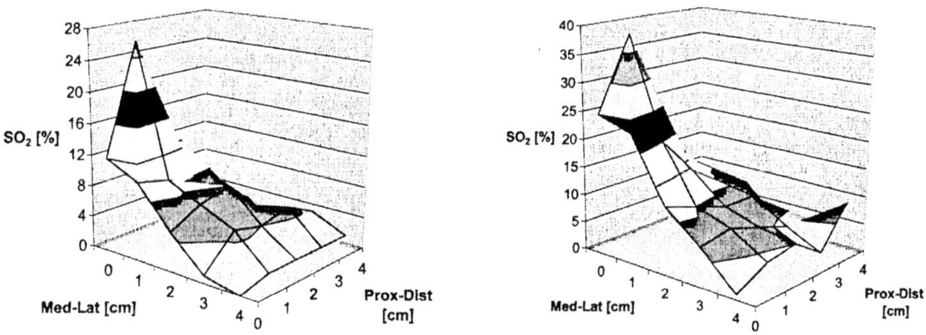

Figure 3. Left: SO_2 values in the right forearm of one volunteer after 5 minutes cuff occlusion. The axes in the x-y plane indicate measurements in the medial to lateral and proximal to distal directions. **Right:** As left, but 48 hours later.

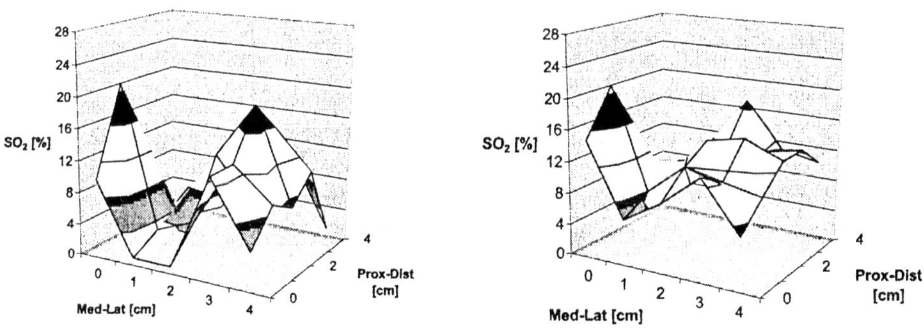

Figure 4. Left: SO_2 values in the left forearm of one volunteer after 5 minutes cuff occlusion measured with the RM200. **Right:** As left, but measured with the MCPD 1000.

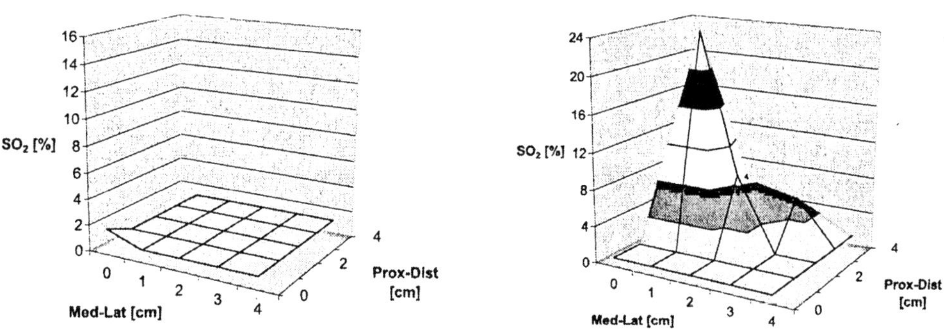

Figure 5. Left: SO_2 values in the left forearm of one volunteer after 5 minutes cuff occlusion measured with the RM200 "micro-lightguide" probe. **Right:** As left, but measured with the normal probe.

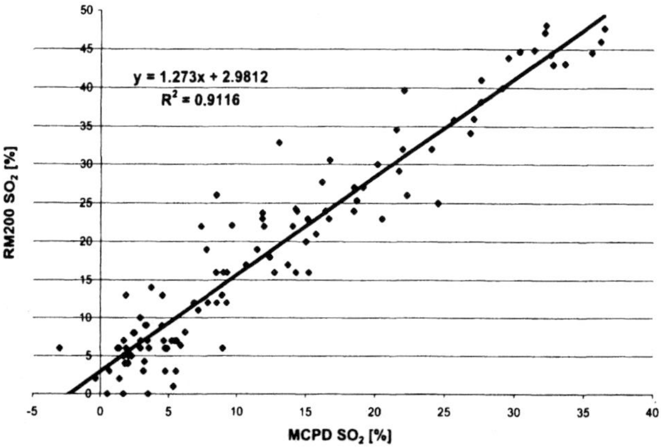

Figure 6. Skin SO_2 values measured with RM200 vs. those measured with the MCPD 1000 in normal subjects and patients for the range 0-40%.

Figure 7 shows the results of Bland and Altman analysis[3] of the differences between the values shown in Figure 6 after the RM200 had been converted to equivalent MCPD values using the inverse of the regression equation given in Figure 6. A high degree of agreement between the two systems can be seen in Figure 7, since all differences between values lay within 1 standard deviation of the mean.

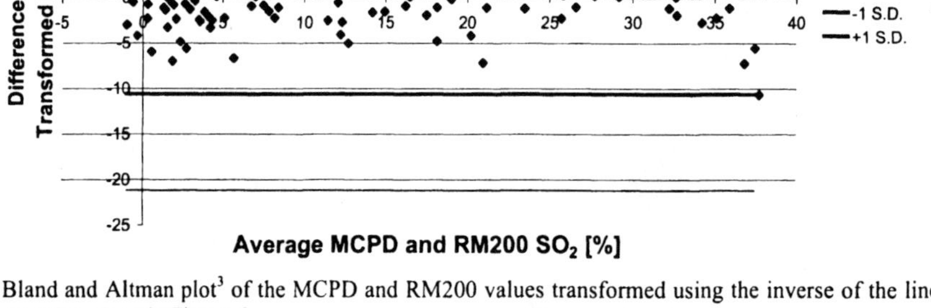

Figure 7. Bland and Altman plot[3] of the MCPD and RM200 values transformed using the inverse of the linear regression equation given in Figure 6.

4. DISCUSSION AND CONCLUSIONS

The most surprising finding arising from this study was the spatial heterogeneity of SO_2 values found in normal skin even after 6 minutes of tourniquet ischaemia. Using both the RM200 and MCPD instruments, isolated centres of high SO_2 could be found which were surrounded by values that decreased over the surrounding 20-30 mm (Figure 4). They could not be eliminated by inflating the cuff to 300 mm Hg, so were unlikely to be due to collateral arteries not occluded by the tourniquet. However, such an area was not detected with the RM200 instrument using the "micro-lightguide," which certainly had a more superficial catchment volume than the other two probes.

The question therefore arises whether the source of the high SO_2 values may be a deeper-lying arteriovenous anastomosis. Under conditions of normal flow, the anastomosis would "shunt" oxygenated arterial blood directly into the venous system. Under such circumstances, oxygenated blood may be trapped by the occlusion in the relatively impermeable capacitance vessels (the venules and small veins). While there, it could possibly remain oxygenated for several minutes and may be detected by the more deeply penetrating optodes of the RM200 and MCPD. Such anastomoses are very common in acral areas of skin, but their presence in other areas is disputed. However, a study using laser Doppler flowmetry has demonstrated the possibility of their existence in forearm skin. Further evidence was provided in this study in that the effect could be reproducibly observed in the same location on the skin (Figure 3) using the RM200 and MCPD instruments with the larger catchment volume optodes. Capillary microscope examination of the surface of the area revealed no visible abnormal anatomical features or pigmentation. Clearly, further studies will be needed to answer some of the questions raised by these few experiments. In the meantime, caution should be exerted when interpreting skin oxygenation values in a tourniquet ischaemia model.

The results shown in Figure 4 imply that the RM200 and MCPD instruments, as used for amputation level viability assessments, have similar catchment volumes. However, Figure 6 shows that the criteria for amputation level assessments using the RM200 should be modified such that the definition of degree of tissue hypoxia should become the percentage of SO_2 values along the leg with a value of less than 15%. The second criterion for predicting healing is that of a mean SO_2 value of 30% at the two specified sites. This should be amended to 40% for the RM200. With these amendments to the criteria, the RM200 should be an acceptable replacement for the MCPD 1000 instrument for amputation level viability assessments, but clearly outcomes should continue to be audited.

REFERENCES

1. D. K. Harrison, P. T. McCollum, D. J. Newton, P. Hickman, and A. S. Jain, Amputation level assessment using lightguide spectrophotometry, *Prosthet. Orthot. Int.* **19**, 139-147 (1995).
2. J. M. Hanson, D. K. Harrison, and I. E. Hawthorn, Tissue spectrophotometry and thermographic imaging applied to routine clinical prediction of amputation level viability, in: *Functional Monitoring of Drug-Tissue Interaction*, edited by M. D. Kessler, and G. J. Müller, *Proc SPIE* **4623**, 187-194 (2002).
3. J. M. Bland, and D. G. Altman, Statistical methods for assessing agreement between two methods of clinical measurement, *Lancet* **I**, 307-310 (1986).
4. C. W. Song, L. M. Chelstrom, and D. J. Haumschild, Changes in skin blood flow by hyperthermia, *Int. J. Radiat. Oncol, Biol. Phys.* **18**, 903-907 (1990).

HYPOXIA IN BREAST CANCER:

Role of blood flow, oxygen diffusion distances, and anemia in the development of oxygen depletion

Peter Vaupel, Arnulf Mayer, Susanne Briest, and Michael Höckel

Abstract: Heterogeneously distributed hypoxic areas are a characteristic property of locally advanced breast cancers. Hypoxia results from an imbalance between the supply and consumption of oxygen (O_2). Major pathogenetic mechanisms for the emergence of hypoxia are (i) structural and functional abnormalities in the tumor microvasculature, (ii) an adverse diffusion geometry, and (iii) tumor-related and therapy-induced anemia leading to a reduced O_2 transport capacity of the blood. There is pronounced intertumor variability in the extent of hypoxia, which is independent of clinical size, stage, histology and grade. Hypoxia is intensified in anemic patients, especially in tumor (areas) with low perfusion rates.

Tumor hypoxia is a therapeutic problem since it makes solid tumors resistant to sparsely ionizing radiation, some forms of chemotherapy, and photodynamic therapy. However, besides more direct mechanisms involved in the development of therapeutic resistance, there are, in addition, indirect machineries that can cause barriers to therapies. These include hypoxia-mediated alterations in gene expression, proteomic and genomic changes, and clonal selection. These in turn can drive subsequent events that are known to further increase resistance to therapy in addition to critically affecting long-term prognosis.

1. INTRODUCTION

Over the last decade, evidence has accumulated showing that 50%-60% of locally advanced solid tumors exhibit hypoxic and/or anoxic tissue areas that are heterogeneously distributed within the tumor mass. Further, pretherapeutic oxygenation status assessed in cancers of the breast,[1, 2] uterine cervix,[3-5] head and neck cancers,[6, 7] prostate cancer,[8] pancreatic cancer,[9] brain tumors,[10, 11] soft tissue sarcomas,[12, 13] and in malignant melanomas,[14] has been shown to be independent of clinical size, stage, histology, grade, and a series of other tumor characteristics or patient demographics. Hypoxia is known to directly or indirectly confer resistance to irradiation, some chemotherapeutic drugs, and other O_2-dependent therapies (e.g., photodynamic therapy) leading to treatment failure.[15,16] In addition, sustained tumor hypoxia also gives rise to adaptations which allow cells to survive and even thrive. As a consequence, a more malignant phenotype may develop due to (i) HIF-1α-mediated mechanisms favouring tumor growth and malignant progression, (ii) HIF-1α-independent up- and down-regulation of genes, and (iii) effects via genome changes (e.g., resistance to hypoxia-induced apoptosis, genomic instability leading to clonal heterogeneity, and selection of resistant and/or aggressive clonal variants).[17-22] Due to the association between tumor hypoxia, malignant progression, and treatment failure, tumor hypoxia and the expression of "hypoxia-mediated" proteins (e.g., HIF-1α, VEGF, glucose transporter GLUT 1, carbonic anhydrase IX) have been suggested to be (independent) prognostic factors for local control, and overall and disease-free survival.[23-30]

Hypoxia is predominantly caused by (a) structural and functional abnormalities of the newly and hastily formed, and thus immature, tumor microvessels, (b) a disturbed microcirculation, and (c) tumor-associated (and/or therapy-induced) anemia, which can substantially worsen tumor O_2 depletion preferentially in low-flow tumors or tumor areas. Additional pathogenetic factors leading to hypoxia are enlarged diffusion distances, an adverse diffusion geometry, carboxyhemoglobin (HbCO) formation in heavy smokers ("toxic anemia") and hypoxemia in microvessels arising from the venous side of the vascular network.[2, 15, 16] Despite the fact that breast cancer is the leading cause of cancer-related death in women, there is a striking paucity of information concerning the individual contribution (or power) of the different pathogenetic mechanisms for development of hypoxia in breast cancers. In this article, therefore, information on the role of parameters determining the oxygenation status has been compiled and reviewed.

2. VASCULARIZATION, BLOOD FLOW, AND PERFUSION-LIMITED OXYGEN DELIVERY

Major determinants of tumor tissue oxygenation are vascularization, microcirculation, and blood flow. Newly formed microvessels in most solid tumors do not conform to the normal morphology of the host tissue vasculature. Microvessels in solid tumors exhibit a series of severe structural and functional abnormalities. They are often dilated, tortuous, elongated, and saccular. There is significant arteriovenous shunt perfusion accompanied by a chaotic vascular organization, without any regulation aimed at the metabolic demand or functional status of the tissue. Excessive branching is a common finding, often coinciding with blind vascular endings. Incomplete or even missing endothelial lining and interrupted basement membranes result in an increased vascular permeability with extravasation of red blood cells and of blood plasma, expanding the interstitial fluid space and drastically increasing the hydrostatic pressure in the tumor interstitium. In solid tumors, there is a rise in viscous resistance to flow, due mainly to hemoconcentration (see Table 1). Aberrant vascular morphology and a decrease in vessel density are responsible for an increase in geometric resistance to flow, which can lead to

Table 1. Major structural and functional abnormalities of tumor microvessels.

Blood vessels	**Lymphatic vessels**
missing differentiation	commonly infiltrated by tumor cells (periphery)
loss of vessel hierarchy (disorganized vascular network)	flattened vessels without lumen (center)
increased intervessel distances	VEGF-C-induced growth at tumor margin
existence of avascular areas	inadequate lymphatic drainage
large diameter (sinusoidal) microvessels	interstitial fluid flow
elongated, tortuous (convoluted) vessels	interstitial hypertension
contour irregularities	
saccular microvessels, blind endings	
aberrant branching	
incomplete endothelial lining, fenestrations	
interrupted or absent basement membranes	
presence of lumen-less endothelial cell cords	
existence of vessel-like cavities not connected to the blood stream	
existence of tumor cell-lined vascular channels	
arterio-venous anastomoses (shunts)	
vessels originating from the venous side	
missing innervation	
lack of physiological/pharmacological receptors	
lack of smooth muscle cells	
paucity of pericytes	
absence of vasomotion	
absence of flow regulation	
increased vascular permeability, plasma leakage	
increased geometric resistance to flow	
increased viscous resistance to flow	
unstable flow velocities	
unstable direction of flow	

Conclusion

It is not only the **quantity of microvessels** that counts, but also - or even more so - the **quality of vascular function** in terms of the tumor tissue supply or drainage!

an inadequate perfusion. Substantial spatial heterogeneity in the distribution of tumor vessels and significant temporal heterogeneity in the microcirculation within a tumor may result in a considerably anisotropic distribution of tumor tissue oxygenation and other factors, which are usually closely linked and which define the so-called metabolic microenvironment. Variations in these relevant parameters between tumors are often more pronounced than differences occurring between different locations or microareas within a tumor.[31-34]

Studies on blood flow through breast cancer are scarce and therefore definite conclusions cannot be drawn. Considering the currently available data, the following (preliminary) conclusions can be drawn when flow data derived from different reports are pooled:[35-38] blood flow through breast cancers can vary considerably (0.08-0.80 ml/g/min); breast cancers can have high flow rates similar to those measured in the lactating breast; some tumors exhibit flow rates that are even lower than those of tissues with a low metabolic rate, such as postmenopausal breast; there is no association between tumor size and blood flow in breast cancers.[38-41] In contrast, in a later study using color-coded and spectral Doppler flow in breast carcinomas, tumor flow velocity showed a significant positive correlation with tumor size and lymph node metastasis, but no correlation with mean vascular density.[42] Furthermore, tumor blood flow velocity may be an independent prognostic factor of survival: breast cancer patients with a high systolic blood flow velocity had a higher risk of dying than those with a lower velocity.[43]

Perfusion-limited O_2 delivery leads to **ischemic hypoxia**, which is often transient (Fig. 1B). For this reason, this type of hypoxia is also called "acute" hypoxia, a term that does not, however, take into account the mechanisms underlying this condition.

3. DIFFUSION-LIMITED HYPOXIA

Hypoxia in tumors can also be caused by an increase in diffusion distances, so that cells far away (> 70 μm) from the nutritive blood vessel receive less oxygen (and other nutrients) than needed. This condition is termed **diffusion-limited hypoxia**, also known as "chronic" hypoxia (Fig. 1A). In addition to enlarged diffusion distances, an adverse diffusion geometry (e.g., concurrent vs. countercurrent tumor microvessels) can also cause hypoxia.

Oxygen diffusion distances in human breast cancer (xenografts) have been computed considering cell-line specific *in vivo* data.[44] Critical supply radii around a tumor microvessel are presented as a function of intravascular pO_2 (Fig. 2). From these mathematical evaluations, there is clear indication that hypoxia has to be expected at distances above 70 μm from the vessel wall. O_2 deficiency starts to develop first in tissue areas far away from a tumor microvessel and, more dramatically, at the venous end of the microvessel. If intercapillary distances exceed 140 μm, hypoxia will be present even at the arterial end of microvessels.

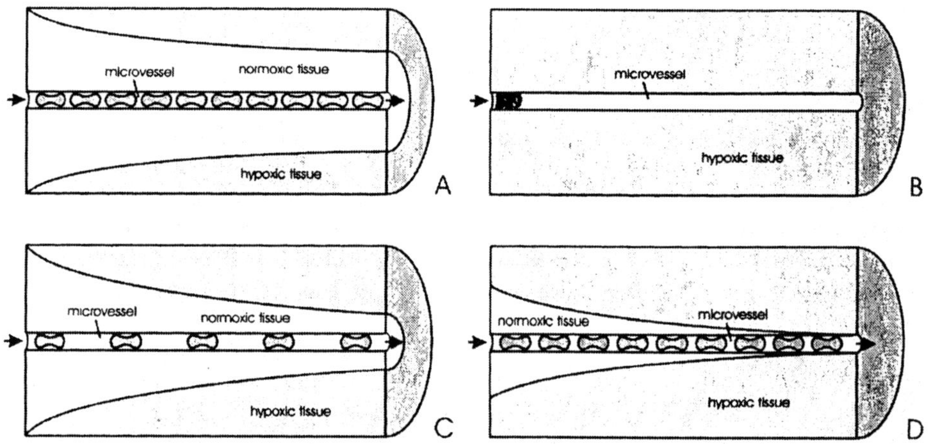

Figure 1. Schematic illustration of various types of hypoxia. Shown are longitudinal sections of tissue cylinders around a central microvessel with blood flow (erythrocyte flow) from left to right. The shaded areas represent hypoxic tissues. **A**: diffusion-limited hypoxia, **B**: ischemic hypoxia, **C**: anemic hypoxia, **D**: hypoxemic hypoxia around a microvessel arising from the venous side.

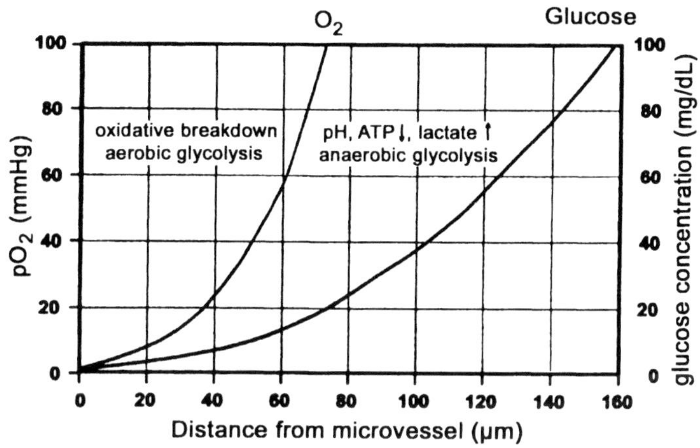

Figure 2. Critical diffusion distances for oxygen and glucose in breast cancer as a function of vascular oxygen partial pressure (pO_2) and blood glucose concentration. For computation of the critical diffusion distances the following supply parameters have been considered:[44] arterial pO_2 = 90 mm Hg, O_2 consumption rate = 20 μl O_2/g/min, Krogh's diffusion coefficient = 1.9×10^{-5} ml O_2/cm/min/atm, blood flow rate = 0.3 ml/g/min, arterial glucose concentration = 5.5 mM, glucose uptake = 1 μM/g/min, glucose diffusion coefficient (D) = 2×10^{-6} cm^2/sec. Since diffusion distances for glucose are substantially larger than those for O_2, in oxygen-depleted tissue volumes, extracellular acidosis, elevated lactate levels, and a poor ATP yield are common features due to anerobic glycolysis.

4. ANEMIC HYPOXIA

The oxygenation status of primary breast cancers critically depends on the blood hemoglobin level cHb (see Fig. 3). In moderately anemic patients (8 g/dl < cHb ≤ 10 g/dl, grade II anemia, NCI-system), all median pO_2 values were < 6 mmHg. At a mean cHb of 8.5 g/dl, the median pO_2 was 3 mm Hg. With increasing cHb values, the median pO_2 exponentially increased ($\log pO_2 = 0.114\ cHb - 0.472$, $r^2 = 0.985$), reaching a maximum pO_2 of 15 mm Hg in the cHb range above the median. The hypoxic fraction of pO_2 values ≤ 5 mm Hg significantly increased from 53% to 77% (P = .02) when breast cancers of non-anemic patients (cHb = 13.6 ± 0.2 g/dl) were compared with anemic patients (cHb = 9.4 ± 0.5 g/dl).[45] In contrast, in normal breast tissue, the median pO_2 was 52 mm Hg over the cHb range from 8.4 to 14.8 g/dl, with no cHb-related differences in pO_2 being evident. In subcutis and skeletal muscle the median pO_2 values were also substantially higher (51 mm Hg and 37 mm Hg, respectively) and were relatively constant over Hb levels from 10 to 16 g/dl. Only in moderate anemia were relatively small decreases in the median pO_2 values observed (-9% in skeletal muscle, -23% in the subcutis), which is most probably of no biological relevance.[45, 46]

In normal tissues of anemic patients (approx. 30% of breast cancer patients were anemic at the time of first presentation), the reduced O_2-carrying capacity of the blood is fully compensated by a rise in local blood flow (and a possible increase in the O_2 extraction from the blood), whereas in locally advanced breast cancers (and most probably in other solid tumors), these two adaptive (compensatory) mechanisms may not work since flow regulation is limited and O_2 extraction ratios are already high at physiological hemoglobin levels.[47]

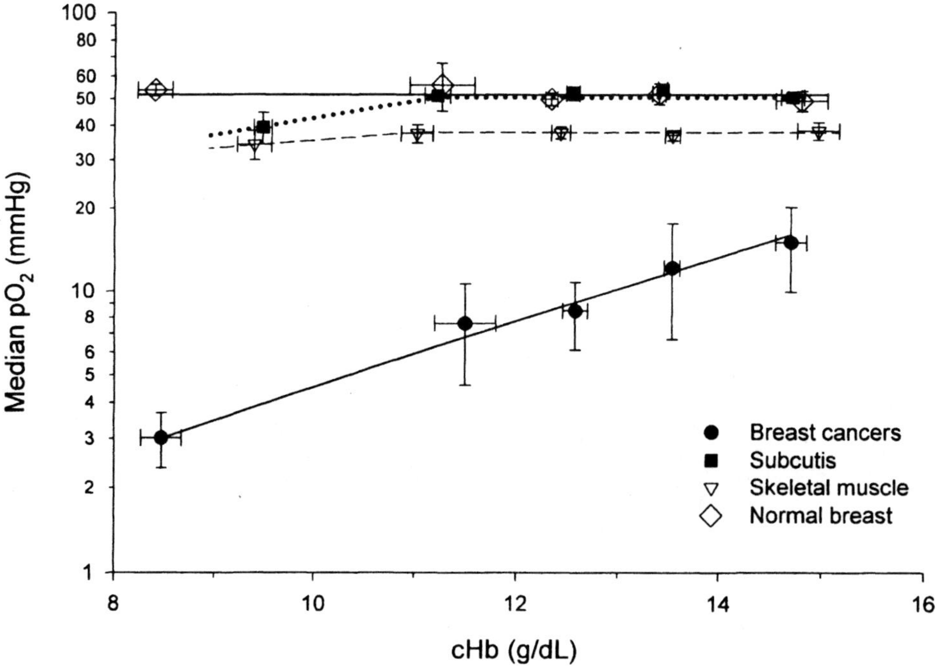

Figure 3. Median pO_2 values in breast cancers (lower curve) and normal tissues (upper curves) as a function of hemoglobin concentration (cHb). Values are means ± SEM (modified from Vaupel et al.[45]).

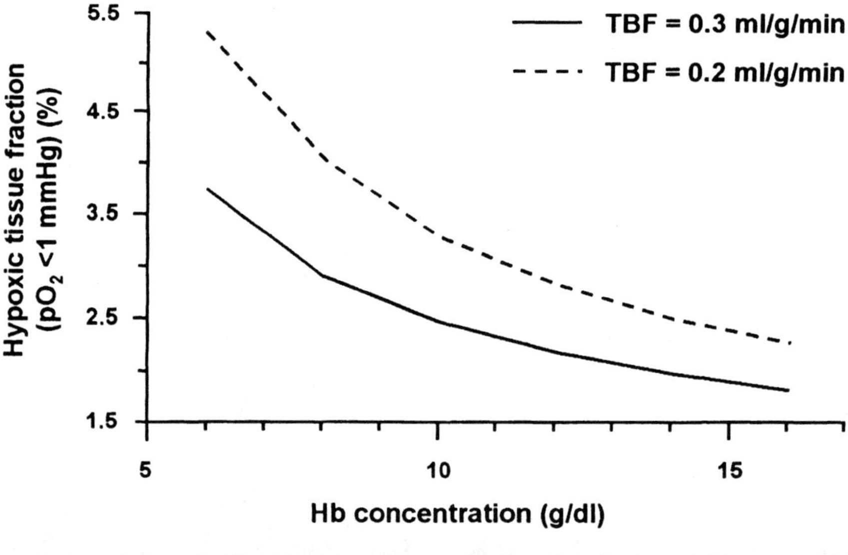

Figure 4. Hypoxic tissue fraction ($pO_2 < 1$ mmHg) as a function of varying hemoglobin concentration and tumor blood flow rate (TBF). Computed data for human breast cancer xenotransplanted into nude rats (modified from Vaupel et al.[48]).

Tumor-associated and/or therapy-induced anemia can significantly contribute to the development of hypoxia in breast cancers. Anemic hypoxia is intensified especially in tumors or tumor areas that have low perfusion rates. Falling cHb values from 14 to 8 g/dl can increase the hypoxic tissue fraction ($pO_2 < 1$ mm Hg) by a factor of approximately 4 (at a given blood flow of 0.3 ml/g/min). A reduction of blood flow to 0.2 ml/g/min leads to a further enlargement of the hypoxic fraction by a factor of 1.7 – 2 (Fig. 4).[48]

A condition similar to anemia can be caused by carboxyhemoglobin formation in heavy smokers leading to a "functional" anemia, because hemoglobin blocked by carbon monoxide (CO) can no longer transport oxygen. Very often tumor micro-vessels are perfused (at least transiently) by plasma only. In this situation, hypoxia around those vessels develops very rapidly since only a few tumor cells at the arterial end can be supplied adequately under these conditions.

5. CONSEQUENCES OF TUMOR HYPOXIA

Tumor hypoxia is a therapeutic problem as it makes breast cancers (and other solid tumors) resistant to sparsely ionizing radiation, some forms of chemotherapy, and photodynamic therapy. Besides more direct mechanisms involved in the development of therapeutic resistance, there are indirect mechanisms that can present barriers to therapies.[15, 16] These include hypoxia-driven alterations in gene expression and proteome changes (< 1% O_2, i.e., < 7 mm Hg). O_2 concentrations < 0.1% (i.e., pO_2 values < 0.7 mm Hg) can lead to genome changes (and subsequent clonal selection) and thus to "downstream" events that are known to further increase resistance to therapy and to the development of an aggressive tumor phenotype. Taken together, these alterations can critically affect long-term prognosis (Figs. 5 and 6).[21, 49, 50]

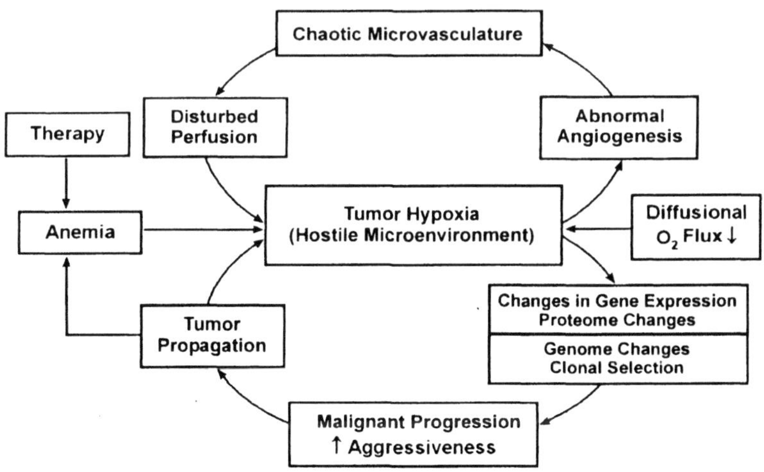

Figure 5. Schematic representation of the pivotal role of hypoxia in the development of an aggressive tumor phenotype and in malignant progression through changes in gene expression, in the proteome and genome. Hypoxia is central in two composite vicious circles that can be regarded as fundamental biological mechanisms of malignant disease (updated from Vaupel et al.[21]).

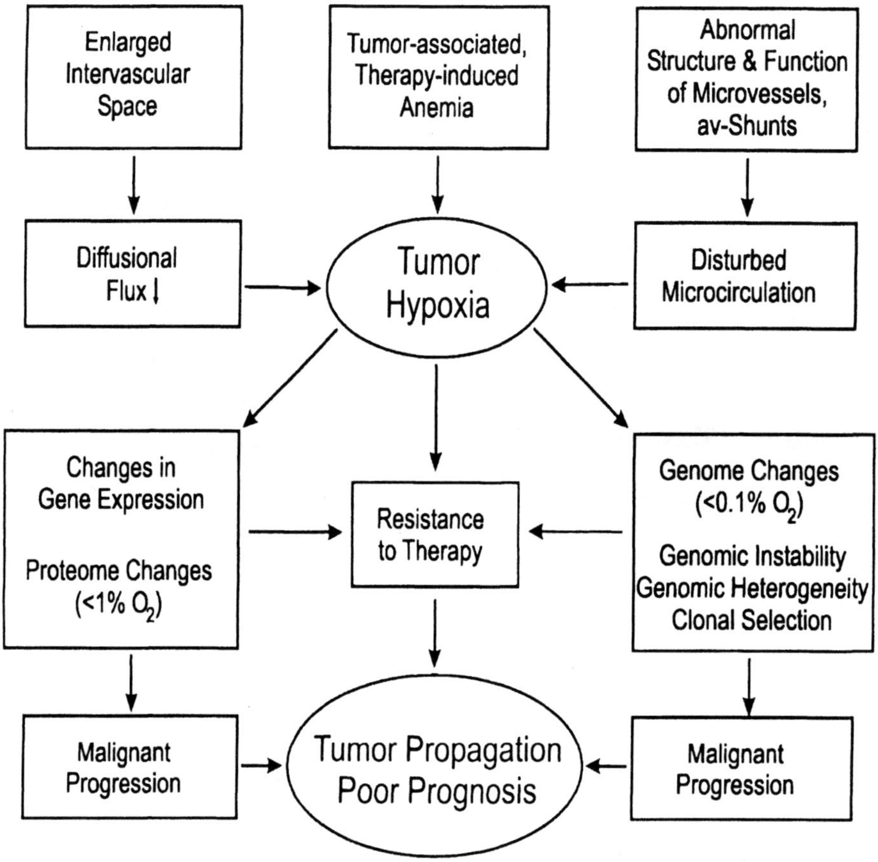

Figure 6. Schematic representation of major pathogenetic factors causing hypoxia and mechanisms leading to poor long-term prognosis.

6. ACKNOWLEDGEMENT

We thank Dr. Debra Kelleher for valuable editorial help during manuscript preparation.

7. REFERENCES

1. P. Vaupel, K. Schlenger, C. Knoop, and M. Höckel, Oxygenation of human tumors: Evaluation of tissue oxygen distribution in breast cancers by computerized O_2 tension measurements, *Cancer Res.* **51**, 3316-3322 (1991).
2. P. Vaupel, S. Briest, and M. Höckel, Hypoxia in breast cancer: Pathogenesis, characterization and biological/therapeutic implications, *Wien. Med. Wschr.* **152**, 334-342 (2002).
3. M. Hoeckel, K. Schlenger, C. Knoop, and P. Vaupel, Oxygenation of carcinomas of the uterine cervix: Evaluation of computerized O_2 tension measurements, *Cancer Res.* **51**, 6098-6102 (1991).

4. A. W. Fyles, M. Milosevic, R. Wong, M.-C. Kavanagh, M. Pintilie, A. Sun, W. Chapman, W. Levin, L. Manchul, T. J. Keane, and R. P. Hill, Oxygenation predicts radiation response and survival in patients with cervix cancer, *Radiother. Oncol.* **48**, 149-156 (1998).
5. T. H. Knocke, H.-D. Weitmann, H.-J. Feldmann, E. Selzer, and R. Pötter, Intratumoral pO_2-measurements as predictive assay in the treatment of carcinoma of the uterine cervix, *Radiother. Oncol.* **53**, 99-104 (1999).
6. M. Nordsmark, and J. Overgaard, A confirmatory prognostic study on oxygenation status and loco-regional control in advanced head and neck squamous cell carcinoma treated by radiation therapy, *Radiother. Oncol.* **57**, 39-43 (2000).
7. D. M. Brizel, G. S. Sibley, L. R. Prosnitz, R. L Scher, and M. W. Dewhirst, Tumor hypoxia adversely affects the prognosis of carcinoma of the head and neck, *Int. J. Radiat. Oncol. Biol. Phys.* **38**, 285-289 (1997).
8. B. Movsas, J. D. Chapman, E. M. Horwitz, W. H. Pinover, R. E. Greenberg, A. L. Hanlon, R. Iyer, and G. E. Hanks, Hypoxic regions exist in human prostate carcinoma, *Urology* **53**, 11-18 (1999).
9. A. C. Koong, V. K. Mehta, Q. T. Le, G. A. Fisher, D. J. Terris, J. M. Brown, A. J. Bastidas, and M. Vierra, Pancreatic tumors show high levels of hypoxia, *Int. J. Radiat. Oncol. Biol. Phys.* **48**, 919-922 (2000).
10. R. Rampling, G. Cruickshank, A. D. Lewis, S. A. Fitzsimmons, and P. Workman, Direct measurement of pO_2 distribution and bioreductive enzymes in human malignant brain tumors, *Int. J. Radiat. Oncol. Biol. Phys.* **29**, 427-432 (1994).
11. D. R. Collingridge, J. M. Piepmeier, S. Rockwell, and J. P. S. Knisely, Polarographic measurements of oxygen tension in human glioma and surrounding peritumoural brain tissue, *Radiother. Oncol.* **53**, 127-131 (1999).
12. D. M. Brizel, S. P. Scully, J. M. Harrelson, L. J. Layfield, J. M. Bean, L. R. Prosnitz, and M. W. Dewhirst, Tumor oxygenation predicts for the likelihood of distant metastases in human soft tissue sarcoma, *Cancer Res.* **56**, 941-943 (1996).
13. M. Nordsmark, J. Keller, O. S. Nielsen, E. Lundorf, and J. Overgaard, Tumour oxygenation assessed by polarographic needle electrodes and bioenergetic status measured by ^{31}P magnetic resonance spectroscopy in human soft tissue tumours, *Acta Oncol.* **36**, 565-571 (1997).
14. E. Lartigau, H. Randrianarivelo, M.-F. Avril, A. Margulis, A. Spatz, F. Eschwege, and M. Guichard, Intratumoral oxygen tension in metastatic melanoma, *Melanoma Res.* **7**, 400-406 (1997).
15. P. Vaupel, O. Thews, and M. Höckel, Treatment resistance of solid tumors: Role of hypoxia and anemia, *Med. Oncol.* **18**, 243-259 (2001).
16. P. Vaupel, and M. Höckel, Tumor hypoxia and therapeutic resistance, in: *Recombinant Human Erythropoietin (rhEPO) in Clinical Oncology*, edited by M. R. Nowrousian (Springer, Berlin, Heidelberg, New York, 2002), pp. 127-146.
17. M. Höckel, and P. Vaupel, Tumor hypoxia: Definitions and current clinical, biologic and molecular aspects, *J. Natl. Cancer Inst.* **93**, 266-276 (2001).
18. G. L. Semenza, Targeting HIF-1 for cancer therapy, *Nat. Rev. Cancer* **3**, 721-732 (2003).
19. G. L. Semenza, HIF-1 and tumor progression: pathophysiology and therapeutics, *Trends Mol. Med.* **8**, S62-S67 (2002).
20. A. L. Harris, Hypoxia – a key regulatory factor in tumour growth, *Nat. Rev. Cancer* **2**, 38-47 (2002).
21. P. Vaupel, A. Mayer, and M. Höckel, Tumor hypoxia and malignant progression, *Methods Enzymol.* **383**, 335-354 (2004).
22. T. I. Goonewardene, H. M. Sowter, and A. L. Harris, Hypoxia-induced pathways in breast cancer, *Microsc. Res. Tech.* **59**, 41-48 (2002).
23. S. M. Evans, and C. J. Koch, Prognostic significance of tumor oxygenation in humans, *Cancer Letters* **195**, 1-16 (2003).
24. G. Gasparini, Prognostic value of vascular endothelial growth factor in breast cancer, *Oncologist* **5**, 37-44 (2000).
25. M. Schindl, S. F. Schoppmann, H. Samonigg, H. Hausmaninger, W. Kwasny, M. Gnant, R. Jakesz, E. Kubista, P. Birner, and G. Oberhuber, Overexpression of hypoxia-inducible factor 1α is associated with an unfavorable prognosis in lymph node-positive breast cancer, *Clin. Cancer Res.* **8**, 1831-1837 (2002).
26. P. N. Span, J. Bussink, P. Manders, L. V. A. M. Beex, and C. G. J. Sweep, Carbonic anhydrase-9 expression levels and prognosis in human breast cancer: association with treatment outcome, *Br. J. Cancer* **89**, 271-276 (2003).
27. S. K. Chia, C. C. Wykoff, P. H. Watson, C. Han, R. D. Leek, J. Pastorek, K. C. Gatter, P. Ratcliffe, and A. L. Harris, Prognostic significance of a novel hypoxia-regulated marker, carbonic anhydrase IX, in invasive breast carcinoma, *J. Clin. Oncol.* **19**, 3660-3668 (2001).
28. R. Bos, P. v. d. Groep, A. E. Greijer, A. Shvarts, S. Meijer, H. M. Pinedo, G. L. Semenza, P. J. v. Diest, and E. v. d. Wall, Levels of hypoxia-inducible factor-1α independently predict prognosis in patients with lymph node negative breast carcinoma, *Cancer* **97**, 1573-1581 (2003).

29. L. Tomes, E. Emberley, Y. Niu, S. Troup, J. Pastorek, K. Strange, A. Harris, and P. H. Watson, Necrosis and hypoxia in invasive breast carcinoma, *Breast Cancer Res. Treat.* **81**, 61-69 (2003).
30. M. Younes, R. W. Brown, D. R. Mody, L. Fernandez, and R. Laucirica, GLUT 1 expression in human breast carcinoma: correlation with known prognostic markers, *Anticancer Res.* **15**, 2895-2898 (1995).
31. P. Vaupel, O. Thews, and M. Höckel, Tumor oxygenation: Characterization and clinical implications, in: *rhErythropoietin in Cancer Supportive Treatment,* edited by J. F. Smyth, M. A. Boogaerts, and B. R.-M. Ehmer (Marcel Dekker, New York, 1996), pp. 205-239.
32. P. Vaupel, F. Kallinowski, and P. Okunieff, Blood flow, oxygen and nutrient supply, and metabolic microenvironment of human tumors: A review, *Cancer Res.* **49**, 6449-6465 (1989).
33. P. W. Vaupel, Blood flow, oxygenation, tissue pH distribution, and bioenergetic status of tumors. Lecture 23, Berlin: Ernst Schering Research Foundation (1994).
34. P. Vaupel, and M. Höckel, Durchblutung, Oxygenierungsstatus und metabolisches Mikromilieu des Mammakarzinoms. Pathomechanismen, Charakterisierung und biologische/therapeutische Relevanz, in: *Diagnostik und Therapie des Mammakarzinoms - State of the Art,* edited by M. Untch, H. Sittek, I. Bauerfeind, G. Konecny, M. Reiser, H. Hepp (Zuckschwerdt, München, 2002), pp. 289-307.
35. R. P. Beaney, Positron emission tomography in the study of human tumors, *Semin. Nucl. Med.* **14**, 324-341 (1984).
36. R. P. Beaney, A. A. Lammertsma, T. Jones, C. G. McKenzie, and K. E. Halnan, Positron emission tomography for in-vivo measurements of regional blood flow, oxygen utilisation, and blood volume in patients with breast carcinoma, *Lancet* **1**(8369), 131-134 (1984).
37. R. Johnson, A thermodynamic method for investigation of radiation induced changes in the microcirculation of human tumors, *Int. J. Radiat. Oncol. Biol. Phys.* **1**, 659-670 (1976).
38. C. B. J. H. Wilson, A. A. Lammertsma, C. G. McKenzie, K. Sikora, and T. Jones, Measurements of blood flow and exchanging water space in breast tumors using positron emission tomography: A rapid and noninvasive dynamic method, *Cancer Res.* **52**, 1592-1597 (1992).
39. E. M. Grischke, M. Kaufmann, M. Eberlein-Gonska, T. Mattfeld, Ch. Sohn, and G. Bastert, Angiogenesis as a diagnostic factor in primary breast cancer: Microvessel quantitation by stereological methods and correlation with color Doppler sonography, *Onkologie* **17**, 35-42 (1994).
40. P. Vaupel, Vascularization, blood flow, oxygenation, tissue pH, and bioenergetic status of human breast cancer, *Adv. Exp. Med. Biol.* **411**, 243-254 (1997).
41. P. Vaupel, and M. Höckel, Blood supply, oxygenation status and metabolic micromilieu of breast cancers: Characterization and therapeutic relevance, *Int. J. Oncol.* **17**, 869-879 (2000).
42. C. Peters-Engl, M. Medl, M. Mirau, C. Wanner, S. Bilgi, P. Sevelda, and A. Obermair, Color-coded and spectral Doppler flow in breast carcinomas - Relationship with the tumor microvasculature, *Breast Cancer Res. Treat.* **47**, 83-89 (1998).
43. C. Peters-Engl, W. Frank, S. Leodolter, and M. Medl, Tumor flow in malignant breast tumors measured by Doppler ultrasound: an independent predictor of survival, *Breast Cancer Res. Treat.* **54**, 65-71 (1999).
44. P. Vaupel, Oxygenation of human tumors. *Strahlenther. Onkol.* **166**, 377-386 (1990).
45. P. Vaupel, A. Mayer, S. Briest, and M. Höckel, Oxygenation gain factor: A novel parameter characterizing the association between hemoglobin level and the oxygenation status of breast cancers, *Cancer Res.* **63**, 7634-7637 (2003).
46. A. Becker, P. Stadler, R. S. Lavey, G. Hänsgen, T. Kuhnt, C. Lautenschläger, H. J. Feldmann, M. Molls, and J. Dunst, Severe anemia is associated with poor tumor oxygenation in head and neck squamous cell carcinomas, *Int. J. Radiat. Oncol. Biol. Phys.* **46**, 459-466 (2000).
47. P. Vaupel, O. Thews, D. K. Kelleher, and M. A. Konerding, O_2 extraction is a key parameter determining the oxygenation status of malignant tumors and normal tissues, *Int. J. Oncol.* **22**, 795-798 (2003).
48. P. Vaupel, D. K. Kelleher, and M. Höckel, Oxygenation status of malignant tumors: Pathogenesis of hypoxia and significance for tumor therapy, *Semin. Oncol.* **28**, 29-35 (2001).
49. D. P. Bottaro, and L. A. Liotta, Out of air is not out of action, *Nature* **423**, 593-595 (2003).
50. M. Höckel, K. Schlenger, B. Aral, M. Mitze, U. Schäffer, and P. Vaupel, Association between tumor hypoxia and malignant progression in advanced cancer of the uterine cervix, *Cancer Res.* **56**, 4509-4515 (1996).

MONITORING METABOLITE GRADIENTS IN THE BLOOD, LIVER, AND TUMOR AFTER INDUCED HYPERGLYCEMIA IN RATS WITH R3230 FLANK TUMORS USING MICRODIALYSIS AND BIOLUMINESCENCE IMAGING

Cathryn Peltz, Thies Schroeder, and Mark W. Dewhirst

Abstract: Hypoxia is a common cause of reduced tumor response to treatment such as irradiation. The purpose of this study was to establish a method in a rat model that is clinically applicable to monitor the efficiency of glucose transport to both tumor and normal tissue following the induction of hyperglycemia. Female Fischer 344 rats bearing subcutaneous R3230 rat mammary adenocarcinomas received glucose (1 g/kg in 200 mg/ml Normosol) injected in the femoral vein with an infusion pump at a rate of 0.1 ml/min. Microdialysis sampling was performed on all animals. The perfusion marker Hoechst 33342 was injected intravenously at a dose of 5 mg/kg ten minutes prior to sacrifice. After the last blood sample was collected, the tumor and liver were removed and snap frozen for bioluminescence imaging and the rat was sacrificed. Imaging bioluminescence was performed on cryosections of the tumor and liver of the animal to monitor local metabolite gradients and concentrations of glucose in relation to the perfused vasculature, as determined by injected Hoechst 33342. Microdialysis and bioluminescence show comparable data when monitoring the changes of blood, liver, and tumor glucose concentrations as a result of induced hyperglycemia.

1. INTRODUCTION

Hypoxia is a common cause of reduced tumor response to treatment such as inrradiation.[1] Induced hyperglycemia is a promising new therapeutic tool, since it has been shown to decrease oxygen consumption in tumors by inducing the Crabtree effect (the shifting of cells from oxidative respiration to glycolysis when exposed to increased levels of glucose), thereby increasing the local oxygen partial pressure.[2] The purpose of this study was to observe the delivery of glucose and its subsequent conversion to lactate following hyperglycemia using microdialysis in blood and tumor interstitium and using bioluminescence imaging to obtain these measurements in the tumor as a whole.

2. MATERIALS AND METHODS

2.1. Animals and Procedures

All procedures were performed under a protocol approved by the Duke Institutional Animal Care and Use Committee. Female Fischer 344 rats (Charles River Laboratories, Raleigh, NC) were implanted subcutaneously in the left hind limb with 1-2 mm pieces of R3230 rat mammary adenocarcinoma tumors. When these tumors reached 1-2 cm in diameter, the rats were anesthetized with 50 mg/kg Nembutal i.p. The femoral artery and vein were cannulated for recording of blood pressure and for venous infusion of glucose and supplemental Nembutal, respectively. Rat body temperature was maintained by placing the animal on a temperature regulated, water-heated blanket (K-module, Baxter Healthcare, Valencia, CA).

2.2. Hyperglycemia

A solution of glucose in Normosol (200 mg/ml) was prepared and injected through the femoral vein cannula at a dose of 1 g/kg in the rats at time t = 0. Previous experiements by our group[3] have shown that this amount of moderate hyperglycemia does not significantly affect tumor blood flow. The solution was infused using a perfusion pump (Model MD-1001, Bioanalytical Systems, West Lafayette, IN) at a rate of 0.1 ml/min. The infusion time ranged from 8-10 minutes in these experiments.

2.3. Randomization

Ten rats were randomized to two sets of end points in these experiments. Rats were sacrificed before and at 30, 45, 75, and 120 minutes post-glucose infusion. Tumor and liver were also collected and snap frozen at these timepoints for bioluminescence imaging.

2.4. Microdialysis

2.4.1. Experimental Setup

The rats were placed on a water heated blanket and the arterial cannula was connected to a blood pressure transducer. Mean blood pressure in mmHg and heart rate were monitored throughout the experiment. Microdialysis probes with 0.5 mm diameters, 10 mm long polycarbonate membranes, and a molecular weight cutoff of 20 kD were used (CMA/20 model 830-95711, CMA Microdialysis, North Chelmsford, MA). The probe was connected to a 1 ml gas tight syringe filled with Normosol (Abbott

Laboratories, North Chicago, IL). The probe was inserted into the tumor using a needle and guiding tubing following a stab incision through the skin using a #11 scalpel blade.

The perfusate syringe was inserted into a perfusion pump that was powered by a microinfusion controller (Model MD-1020, Bioanalytical Systems, West Lafayette, IN). The microdialysis probe was perfused at a rate of 2 μL/min throughout the experiment. Samples were collected every 15 minutes using a microfraction collector (Model CMA/142, CMA Microdialysis, North Chelmsford, MA). The first 3 samples (t = -60 to t = -15 min) were considered equilibration samples and were discarded. The next sample (t = -15 to t = 0 min) was considered the baseline for each rat. Glucose was infused to induce hyperglycemia at t = 0 in all rats except those used strictly for baseline bioluminescence measurements. Microdialysis samples were then taken every 15 minutes until the animal was sacrificed.

2.4.2. Data Analysis

The microdialysis and plasma samples were analyzed for lactate, glucose, and urea concentrations with the CMA/600 Microdialysis Analyzer (CMA Microdialysis, North Chelmsford, MA). The metabolite concentrations from the microdialysis samples were used to calculate the true tumor interstitial concentrations using the no net flux method data previously determined in our laboratory.[4] The term 'recovery' was used to indicate the ratio of analyte concentration in the microdialysis sample divided by the estimated interstitial concentration.

2.5. Blood Collection

Blood samples were collected in heparanized, micro-hematocrit capillary tubes (Kimble/Kontes, Vineland, NJ) from a small tail vein incision. Plasma was collected and frozen at –80°C after centrifugation. The first blood sample was collected during the t = -15 to t = 0 time interval. While microdialysis samples represent a collection over a time of 15 minutes, blood samples represent only a single point in time. Additional blood samples were taken at the end of each microdialysis sample collection time period until the end of the experiment.

2.6. Bioluminescence Imaging

A technique previously established[5] for metabolic mapping in rapidly frozen tissue using quantitative bioluminescence and single photon imaging was used on sections of tumor and liver from the same rats used for microdialysis. Cryostat sections of the snap frozen tumor and liver were immersed into an enzyme solution using a specifically designed sandwich array, which was positioned on a thermostat-controlled microscope stage. The solution contained enzymes that link the substrate of interest, in our case glucose or lactate, to luciferase from bacteria or insects. When the temperature of the array was adjusted to 20°C, the respective metabolites in the tissue section reacted with the luciferases resulting in the emission of photons with intensity directly proportional to local metabolite content.

The light reaction was captured with a precision microscope and an ultrasensitive camera, allowing the detection of single photons including the exact location of emission.

Table 1. Baseline glucose and lactate values in blood and tumor interstitial fluid as measured by microdialysis.

	Blood Glucose (mM)	Interstitial Glu.(mM)	Blood Lactate (mM)	Interstitial Lac.(mM)
Average	7.73	1.48	0.82	11.64
Std. Dev.	0.63	0.84	0.17	2.59

Using standards in the estimated range of metabolite values, this method can be calibrated in absolute terms, resulting in 2D distributions with each pixel value representing a local metabolite content in µmole/g within each section. These distributions were displayed as color-coded images to illustrate spatial variations.

Due to the fact that different solutions had to be used for glucose and lactate measurements, and that the sections were severely affected after each measurement, the metabolites were measured in serial 20 µm sections, meaning they were determined at quasi-identical locations.

3. RESULTS

3.1. Microdialysis

Baseline results for both glucose and lactate microdialysis measurements in blood and tumor interstitial fluid are shown in Table 1. Following hyperglycemia, a peak in blood glucose was seen 15 minutes post-infusion, with the corresponding tumor glucose peak at 30 minutes post-infusion. Blood glucose increased approximately 2.5 fold over baseline, while tumor glucose increased 4 fold. A representative curve from one rat followed to 120 minutes post-glucose infusion is seen in Figure 1. In this figure we also see the blood lactate peak at 30 minutes post-glucose infusion, and the corresponding tumor lactate peak at the 60 minute time point.

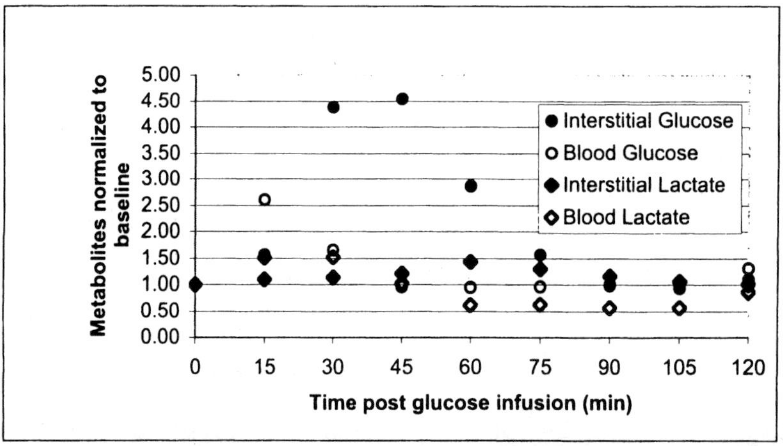

Figure 1. Representative time course of metabolites following hyperglycemia.

3.2. Bioluminescence Imaging

Bioluminescence images showed an inverse relationship between tumor glucose and lactate. Glucose tended to accumulate at the edge regions of the tumor, which are known to be the most well perfused areas, while the lactate collected in the more central regions of the tumor. This was true both for baseline images and those following the induction of hyperglycemia.

Baseline and peak results for whole tumor measurement of glucose and lactate are shown in Table 2. These values represent an average of 3 measurements from 3 serial sections from a single tumor frozen for bioluminescence imaging.

Following hyperglycemia, the whole tumor glucose peak was seen at 30 minutes post-glucose infusion, with a steep drop back to baseline levels by the next time point. Lactate levels in whole tumor tissue did not peak until the 75 minute time point and returned to normal by the next time point at 120 minutes post-glucose infusion.

4. DISCUSSION

Microdialysis and bioluminescence imaging are both valid methods to monitor changes in tumor metabolites following the induction of hyperglycemia. However, some differences in metabolite levels were found between the two measurements. The glucose microdialysis measurements showed peak tumor levels at 30 minutes post-glucose infusion, with a return to baseline at approximately 75 minutes. The bioluminescence measurements illustrated a peak at the same time point, but its return to baseline was much more dramatic, with glucose levels returning to baseline at the 45 minute time point.

These differences were also seen in the lactate measurements. The tumor lactate peak was at 45 minutes post-glucose infusion, according to the microdialysis measurements, but it was not until 75 minutes when measured with bioluminescence imaging. Future work will include a 60 minute, post-glucose infusion time point in order to ensure that 75 minutes is indeed the true whole tumor lactate peak. It is also important to note that the fold increases of tumor glucose and lactate were approximately the same in both tumor interstitium and whole tumor.

These methods allow us to follow the blood, tumor interstitial fluid, and whole tumor tissue metabolite values following the induction of hyperglycemia. Now that we have established the increases in glucose and lactate and the times of the peaks of these metabolites in the tumor following hyperglycemia, the important next step is to gauge if these increases result in an increase in tumor oxygen levels, as well as when and to what extent these oxygen increases occur.

Table 2. Whole tumor baseline and peak values as measured by bioluminescence imaging.

Baseline Glucose (t=0 min)	Peak Glucose (t=30 min)	Fold Increase
2.64 mM	15.29 mM	4.79
Baseline Lactate (t=0 min)	Peak Lactate (t=75 min)	Fold Increase
21.28 mM	35.69 mM	0.68

5. ACKNOWLEDGMENTS

This work was supported by a grant from the NIH/NCI CA40355.

REFERENCES

1. M. Höckel, and P. Vaupel, Tumor hypoxia: definitions and current clinical, biologic and molecular aspects, *J. Natl. Cancer Inst.* **93**(4), 266-276 (2001).
2. H. G. Crabtree, Observations of the carbohydrate metabolism of tumors, *Biochem. J.* **23**, 536-545 (1929).
3. S. A. Snyder, J. L. Lanzen, R. D. Braun, G. Rosner, T. W. Secomb, J. Biaglow, D. M. Brizel, and M. W. Dewhirst, Simultaneous administration of glucose and hyperoxic gas achieves greater improvement in tumor oxygenation than hyperoxic gas alone, *Int. J. Radiat. Oncol. Biol. Phys.* **51**(2), 494-506 (2001).
4. S. N. Ettinger, C. C. Poellman, N. A. Wisniewski, A. A. Gaskin, J. S. Shoemaker, J. M. Poulson, M. W. Dewhirst, and B. Klitzman, Urea as a recovery maker for quantitative assessment of tumor interstitial solutes with microdialysis, *Cancer Res.* **61**, 7964-7970 (2001).
5. S. Walenta, T. Schroeder, and W. Mueller-Klieser, Metabolic mapping with bioluminescence: basic and clinical relevance, *Biomolec. Eng.* **18**, 249-262 (2002).

POSSIBLE PROTECTIVE EFFECTS OF α-TOCOPHEROL ON ENHANCED INDUCTION OF REACTIVE OXYGEN SPECIES BY 2-METHOXYESTRADIOL IN TUMORS

Oliver Thews, Christine Lambert, Debra K. Kelleher, Hans-Konrad Biesalski, Peter Vaupel, and Jürgen Frank

Abstract: Several non-surgical tumor treatment modalities produce their cytotoxic activity by generating reactive oxygen species (ROS). Anti-oxidative enzymes such as superoxide dismutase (SOD) or exogenously supplied antioxidants may therefore reduce the efficacy of these treatments. The aim of the present study was to analyze the impact of (i) inhibiting SOD using 2-methoxyestradiol (2-ME), or (ii) application of α-tocopherol, on the cellular damage induced by hyperthermia (HT) in experimental tumors. DS-sarcoma cells grew either in culture or as solid tumors subcutaneously implanted in rats. *In vitro*, DS-cells were incubated with 2-ME, and cell proliferation, ROS formation, lipid peroxidation and apoptosis were measured. *In vivo*, DS-sarcomas were treated with a ROS-generating hyperthermia combined with 2-ME or α-tocopherol application.

Inhibition of SOD by 2-ME *in vitro* induced pronounced oxidative injury resulting in reduced proliferation. *In vivo*, ROS-generating hyperthermia led to local tumor control in 23% of the animals. The additional inhibition of SOD by 2-ME increased the control rate by approximately 50%. Application of α-tocopherol was found to have no effect on local tumor control, either in combination with ROS-generating hyperthermia or when 2-ME was additionally applied. Inhibition of SOD during ROS-generating hyperthermia results in pronounced cell injury and an improved local tumor control whereas exogenously applied vitamin E seems not to have an impact on oxidative stress.

1. INTRODUCTION

Several non-surgical tumor treatment modalities, such as ionizing radiation, chemotherapy with alkylating agents, or photodynamic therapy, produce their cytotoxic activity by generating reactive oxygen species (ROS).[1-3] The cellular damage (lipid peroxidation, protein oxidation, induction of apoptosis) is a direct function of the amount of ROS induced.[2] At the same time, however, as in normal tissues, tumor cells have effective anti-oxidative defense mechanisms that eradicate ROS. One important enzyme for the elimination of superoxide radicals is superoxide dismutase (SOD), which protects cells against oxidative injury.[4] These anti-oxidative mechanisms could, in principle, diminish the efficacy of ROS-generating treatment modalities. Huang et al.[5] demonstrated that the estrogen metabolite, 2-methoxyestradiol (2-ME), is able to effectively suppress SOD activity and proposed that ROS-generating treatments may become more effective if the anti-oxidative capacity is reduced. Cell culture experiments, on the other hand, reveal that radical scavengers such as α-tocopherol may protect cells from oxidative stress induced by anti-cancer agents.[6] These findings suggest that increased cellular levels of these antioxidants (e.g., achieved by exogenous application of vitamin E) may reduce the cytotoxic efficacy of ROS-generating treatments.[7, 8] For this reason, the aim of the present study was to analyze the impact of (i) inhibiting SOD using 2-methoxyestradiol, and/or (ii) application of α-tocopherol as an antioxidant, on the cellular damage induced by a ROS-generating treatment in experimental tumors *in vitro* and *in vivo*.

2. MATERIALS AND METHODS

2.1. Animals and Tumors

All studies were performed using the experimental DS-sarcoma of the rat. For cell culture experiments, DS-sarcoma cells were grown in RPMI medium supplemented with 10% fetal bovine serum and 2 mM glutamine. *In vivo* studies were performed using solid DS-sarcomas implanted onto the hind foot dorsum of male Sprague-Dawley rats (body weight approx. 200 g) by s.c. injection of DS-ascites cells (0.4 ml, 10^4 cells/μl). The tumor volume was determined by measurement of the three orthogonal diameters of the tumor and calculation of the volume using the formula: $V = d_1 \times d_2 \times d_3 \times \pi/6$. After ROS-generating treatment, volume was monitored for 30 days or until a volume of 3.5 ml was reached. All experimentation had previously been approved by the regional animal ethics committee and was conducted according to German federal law.

2.2. ROS-Induction

When tumors reached a volume of 0.5 to 0.75 ml *in vivo* (approx. 5 to 7 days after implantation), ROS were induced by simultaneous application of localized 44.3°C saline bath-hyperthermia (HT) to the tumor, inspiratory hyperoxia and xanthine oxidase (15 U XO/kg, i.v., 15 min prior to hyperthermia). This treatment combination is known to strongly induce ROS generation.[9]

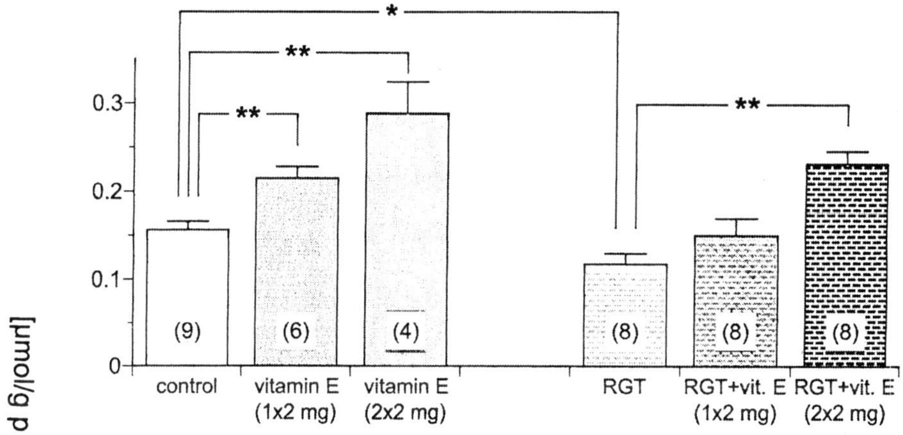

Figure 1. α-tocopherol concentration in DS-sarcomas *in vivo* after i.v. injection of vitamin E emulsion (2 mg α-tocopherol) 24 h (1 dose) or 24 h and 48 h (2 doses) prior to the measurements without or with additional ROS-generating therapy (RGT; local hyperthermia + xanthine oxidase + inspiratory hyperoxia). Bars represent means ± SEM; number of tumors in parentheses; (*) $P < .05$, (**) $P < .01$.

2.3. Methoxyestradiol (2-ME) Treatment

In vitro, DS-sarcoma cells were incubated with 2-ME (previously dissolved in ethanol to give a 20 mM solution) at a concentration of between 0.5 and 5 µM. For *in vivo* experiments, 2-ME was dissolved in ethanol at a concentration of 10 mg/ml and injected i.p. at a dose of 17.5 mg/kg 2 h prior to HT treatment. Before i.p. injection, the stock solution was diluted with 1 ml peanut oil and mixed to obtain a stable emulsion. Control animals received equivalent volumes of ethanol and peanut oil.

2.4. α-Tocopherol Treatment

One group of animals received α-tocopherol (vitamin E emulsion, Fresenius Kabi, Bad Homburg, Germany) prior to the ROS-generating therapy. Tocopherol was injected i.v. at a dose of 10 mg/kg. In order to find the optimal application course for a significant increase in tocopherol levels in tumor tissue, different schedules were tested. Figure 1 shows the results of two different treatment courses: (a) one dose of α-tocopherol; 2 mg, 24 h prior to hyperthermia and (b) two doses; 2x2 mg, 24 and 48 h prior to HT. Both schedules significantly increased the vitamin E concentration in the tumor. However, the effect of two applications was markedly greater than the single dose. Figure 1 clearly shows that the ROS-generating treatment reduced the tissue tocopherol concentration. However, with two doses of vitamin E, the tocopherol level was significantly higher than in controls. For this reason, the schedule involving the application of two 2 mg doses, 24 and 48 h prior to the HT treatment, was used in the *in vivo* experiments.

2.5. Measurement of ROS-Formation and Cellular Damage *in vitro*

ROS-formation was measured using dihydroethidium (DHE, Molecular Probes, Leiden, The Netherlands), which is oxidized by superoxide radicals and can subsequently be detected since it binds irreversibly to DNA. For staining, washed cells were incubated for 1 h with PBS containing 20 ng DHE/ml. Afterwards, cells were analyzed by flow cytometry. Lipid peroxidation induced by ROS was assessed by measurement of malondialdehyde equivalents (TBARS), as described previously.[9] In brief, cells were washed, lysed, and centrifuged. 200 µl of the supernatant was mixed with 10 µl butylated hydroxytoluene (50 mg/ml ethanol) and 200 µl orthophosphic acid (0.2 mM). Thereafter, 400 µg 2-thiobarbituric acid (in 25 µl 0.1 M NaOH) was added and incubated at 90°C for 45 min. TBARS were extracted with 1-butanol and measured using a fluorescence plate reader.

Apoptosis was assessed on the basis of DNA fragmentation and caspase-3 activation. For DNA fragmentation measurements, cells were lysed and incubated with RNase and proteinase K. After extraction, the solution was transferred to a Phase Lock Gel light tube (Eppendorf, Hamburg, Germany), centrifuged, and 3 M sodium acetate and absolute ethanol added. After drying and dissolving, DNA fragments were separated on a 1.8% agarose gel stained with ethidium bromide. For the caspase-3 activity assay, lysed and centrifuged cells were incubated with the fluorogenic caspase-3 tetrapeptide substrate, Ac-DEVD-amino-4-methylcoumarin (Calbiochem, Bad Soden, Germany), at a final concentration of 20 µM. Cleavage of the substrate was followed using a fluorescence plate reader. The number of viable cells was determined by cell counting and staining of dead cells with propidium iodide (0.2 µg/ml).

2.6. Statistical Analysis

Results are expressed as means ± standard error of the mean (SEM). Differences between the groups were assessed using the two-tailed Wilcoxon test for unpaired samples. The significance level was set at $\alpha = 5\%$. For analysis of tumor growth behavior *in vivo*, the time taken to reach a volume of 3.5 ml was determined. These time intervals were analyzed using Kaplan-Meier statistics, and differences between the probability curves were assessed using the log-rank test.

3. RESULTS

Inhibition of SOD by 2-methoxyestradiol *in vitro* induced an increased formation of ROS, even when no additional ROS-generating therapy was applied. At a dose of 5 µM 2-ME in the cell culture medium, ROS-formation was almost doubled (Fig. 2). The higher rate of ROS-formation by 2-ME *in vitro* led to a marked, dose-dependent increase in lipid peroxidation, as determined by TBARS measurements (Fig. 2). This oxidative injury resulted in a marked reduction in cell proliferation. At a 2-ME concentration of 5 µM, the number of proliferating cells decreased by 52% (24 h after 2-ME application), by 89% (48 h), and by 96% (72 h). In parallel, the fraction of dead cells increased from 3% ± 1% under control condition to 71% ± 5% when cells were incubated with 5 µM 2-ME for 72 h. This cell death was most probably the result of increased apoptosis since, at this time point, pronounced DNA-fragmentation was seen (data not shown). The caspase-3 activity (normalized to untreated controls) increased markedly within 48 h when cells were incubated with 1 µM (relative caspase-3 activity 550% ± 70%) or 5 µM (400% ± 40%) 2-ME. Since caspase-3 activity was pronouncedly increased even at a

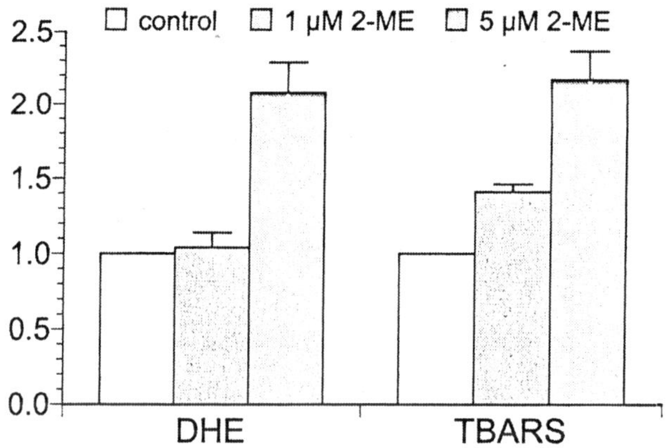

Figure 2. ROS-formation and lipid peroxidation after treatment of DS-sarcoma cells *in vitro* with different concentrations of 2-ME. ROS-formation was determined by flow cytometric measurement of dihydroethidium (DHE) staining. Lipid peroxidation was assessed by formation of thiobarbituric acid reactive substance (TBARS). Each bar represents means ± SEM of 3 experiments.

relatively low 2-ME-concentration of 1 μM (where ROS are only moderately induced, Fig. 2), it can be concluded that DS-sarcoma cells seem to be highly sensitive to oxidative stress. However, these results clearly indicate that 2-ME induces a pronounced ROS-formation leading to an oxidative cell injury resulting in apoptotic cell death.

In vivo, subcutaneously implanted DS-sarcomas grow exponentially up to a volume of 3.5 ml, with a volume doubling time of approx. 2 days, so that in untreated animals, all tumors reached the volume limit of 3.5 ml within 8 days (Fig. 3A). Without ROS-generating treatment, neither α-tocopherol nor 2-ME (or the combination of both) had any significant impact on tumor growth (Fig. 3A). The ROS-generating treatment (hyperthermia + xanthine oxidase + inspiratory hyperoxia) resulted in a growth delay of approx. 5 days and a slower growth rate in the subsequent regrowth period. In 23% of the tumors, no regrowth was seen over the whole 30 day observation period (Fig. 3B). When

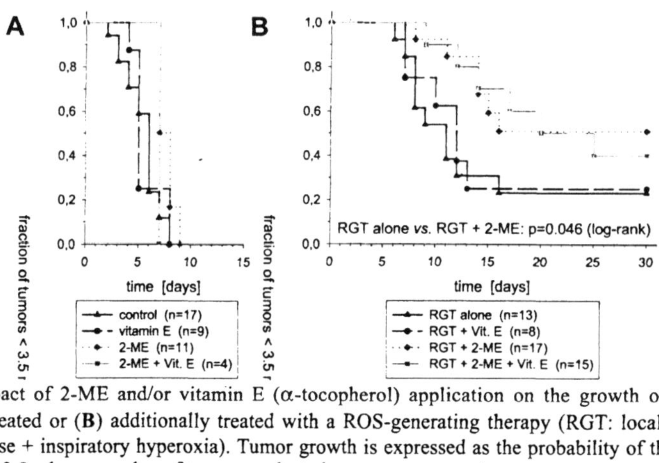

Figure 3. Impact of 2-ME and/or vitamin E (α-tocopherol) application on the growth of tumors in (**A**) otherwise untreated or (**B**) additionally treated with a ROS-generating therapy (RGT: local hyperthermia + xanthine oxidase + inspiratory hyperoxia). Tumor growth is expressed as the probability of the tumor volume being less than 3.5 ml; n = number of tumors evaluated.

the ROS-generating treatment was combined with SOD-inhibition by 2-ME, the growth delay after treatment was longer (approx. 7 days) and the local tumor control rate increased significantly to almost 51% (Fig. 3B), indicating that the 2-ME-induced ROS-formation led to an increased treatment efficacy. However, additional application of α-tocopherol had no marked impact on tumor growth, either in the group which was treated only with the ROS-generating treatment or in the group which additionally received 2-ME (Fig. 3-B).

4. DISCUSSION

On the cellular level, inhibition of superoxide dismutase by 2-ME during a ROS-generating tumor treatment resulted in an increased formation of reactive oxygen species leading to pronounced cell injury (Fig. 2). These results form the basis for the use of this agent (or other SOD inhibitors) in tumor therapy *in vivo* in order to increase the oxidative cell stress by non-surgical treatment modalities.[5] Indeed, the *in vivo* experiments in this study showed that 2-ME application, together with a ROS-generating therapy, improved local tumor control significantly (Fig. 3B). However, 2-ME alone had no impact on tumor growth *in vivo* even though in cell culture an inhibition of proliferation and induction of apoptotic cell death was seen. It seems that the anti-oxidative defense mechanisms *in vivo* are able to quench the 2-ME-induced ROS formation. In combination with a ROS-generating treatment such as hyperthermia,[9] however, the oxidative stress seems to exceed the cellular protection capacity. From these results, the question arises of whether an improvement of anti-oxidative defense (e.g., by higher concentrations of radical scavengers) may counteract increased ROS formation. In the present study, however, almost no impact of exogenously applied α-tocopherol on treatment efficacy was seen (Fig. 3B), even though the cellular tocopherol concentration was markedly increased. Several mechanisms may be responsible for this finding. Firstly, α-tocopherol is integrated into the cell membrane and protects membrane lipids from peroxidation. Since oxidative injury of the ROS-generating treatment affects not only lipids but also proteins,[9] the damage to these molecules cannot be prevented by an increased tocopherol level. In addition, even control animals already show appreciable cellular tocopherol concentrations. It is thus possible that the normal concentration of antioxidants is sufficient to prevent lipid peroxidation such that exogenously supplied vitamin E cannot further increase the membrane protection. Finally, it has been demonstrated that besides inhibiting SOD, 2-ME has a direct inhibitory impact on tumor angiogenesis and thus suppresses tumor growth.[10] Therefore, the increase in local tumor control achieved upon combination of a ROS-generating treatment with 2-ME may not only be the result of the higher rate of ROS formation, but rather of two independent anti-tumoral mechanisms. In this case, antioxidants cannot be expected to counteract possible anti-angiogenic effects of 2-ME.

In conclusion, even though inhibition of the SOD by 2-ME during a ROS-generating treatment results in an increased formation of oxygen radicals leading to an improved local tumor control *in vivo*, exogenously applied vitamin E seems not to be capable of reducing the level of oxidative stress during this treatment combination.

5. ACKNOWLEDGEMENTS

This study was supported by the Deutsche Krebshilfe (10-1798-Fr 1). The authors wish to thank Fresenius-Kabi (Bad Homburg, Germany) for the generous donation of the α-tocopherol emulsion.

6. REFERENCES

1. E. J. Hall, *Radiobiology for the Radiologist*, 5th edition (J.B. Lippincott, Philadelphia, 2000).
2. B. K. Sinha, and E. G. Mimnaugh, Free radicals and anticancer drug resistance: oxygen free radicals in the mechanisms of drug cytotoxicity and resistance by certain tumors, *Free Radical Biol. Med.* **8**, 567-581 (1990).
3. Q. Chen, Z. Huang, H. Chen, H. Shapiro, J. Beckers, and F. W. Hetzel, Improvement of tumor response by manipulation of tumor oxygenation during photodynamic therapy, *Photochem. Photobiol.* **76**, 197-203 (2002).
4. I. Fridovich, Superoxide radical and superoxide dismutases, *Ann. Rev. Biochem.* **64**, 97-112 (1995).
5. P. Huang, L. Feng, E. A. Oldham, M. J. Keating, and W. Plunkett, Superoxide dismutase as a target for the selective killing of cancer cells, *Nature* **407**, 390-395 (2000).
6. E. Germain, V. Chajes, S. Cognault, C. Lhuillery, and P. Bougnoux, Enhancement of doxorubicin cytotoxicity by polyunsaturated fatty acids in the human breast tumor cell line MDA-MB-231: relationship to lipid peroxidation, *Int. J. Cancer* **75**, 578-583 (1998).
7. J. Hannemann, and K. Baumann, Cisplatin-induced lipid peroxidation and decrease of gluconeogenesis in rat kidney cortex: different effects of antioxidants and radical scavengers, *Toxicology* **51**, 119-132 (1988).
8. C. Leonetti, A. Biroccio, C. Gabellini, M. Scarsella, V. Maresca, E. Flori, L. Bove, A. Pace, A. Stoppacciaro, G. Zupi, F. Cognetti, and M. Picardo, α-tocopherol protects against cisplatin-induced toxicity without interfering with antitumor efficacy, *Int. J. Cancer* **104**, 243-250 (2003).
9. J. Frank, D. K. Kelleher, A. Pompella, O. Thews, H. K. Biesalski, and P. Vaupel, Enhancement of oxidative cell injury and antitumor effects of localized 44°C hyperthermia upon combination with respiratory hyperoxia and xanthine oxidase, *Cancer Res.* **58**, 2693-2698 (1998).
10. T. Fotsis, Y. Zhang, M. S. Pepper, H. Adlercreutz, R. Montesano, P. P. Nawroth, and L. Schweigerer, The endogenous oestrogen metabolite 2-methoxyoestradiol inhibits angiogenesis and suppresses tumour growth, *Nature* **368**, 237-239 (1994).

SIMULTANEOUS NIR-EPR SPECTROSCOPY OF RAT BRAIN OXYGENATION

Yasuko S. Sakata, Oleg Y. Grinberg, Stalina Grinberg, Roger Springett, and Harold M. Swartz

Abstract: Changes in cerebral oxygenation were simultaneously monitored by electric paramagnetic resonance (EPR) oximetry and near-infrared spectroscopy (NIRS). The tissue oxygen tension (t-pO_2) was measured with an L-band (1.2 GHz) EPR spectrometer with an external loop resonator and the concentration of oxyhemoglobin [HbO_2] and deoxyhemoglobin [Hb] were measured with a full-spectral NIRS system. Mean cerebral hemoglobin saturation ($SmcO_2$) was calculated from the absolute [HbO_2] and [Hb]. Six adult male rats were implanted with lithium phthalocyanine (LiPc) crystals into the left cerebral cortex. The change in oxygenation of the brain was induced by altering the inspired oxygen fraction (FiO_2) in air from 0.30 at baseline to 0.0, 0.05, 0.10, and 0.15 for 1, 2, 5, and 5 minutes, respectively, followed by reoxygenation with an $FiO_2 = 0.30$. Although both t-pO_2 and $SmcO_2$ values showed a decrease during reduced FiO_2 followed by recovery on reoxygenation, it was found that $SmcO_2$ recovered more rapidly than t-pO_2 during the recovery phase. The recovery of t-pO_2 is not only related to blood oxygenation, but also to delivery, consumption, and diffusion of oxygen into the tissue from the vascular system. Further studies will be required to determine the exact mechanisms for the delay between the recovery of $SmcO_2$ and t-pO_2.

1. INTRODUCTION

The use of stable particulate oxygen sensitive materials, such as LiPc crystals, makes EPR oximetry very suitable for repetitive measurements of t-pO_2 directly *in vivo*.[1-3] On the other hand, NIRS can measure the concentration and saturation of hemoglobin in the vascular system non-invasively.[4, 5] The aims of this study were to use both types of measurements, simultaneously and continuously, in order to investigate the relationship between the value of t-pO_2 and the $SmcO_2$ during and after brief hypoxic insults in the rat brain.

2. METHODS AND MATERIALS

Six adult male rats (Fisher 344) weighing 226.1 ± 23 g were used for this study. Under anesthesia with ketamine/xylazine (80 mg/10 mg/kg), LiPc crystals (50-100μg), which were synthesized in this laboratory, were implanted seven days prior to measurements using a 25 gauge needle and a stereotactic frame. The LiPc crystals were implanted in the cerebral cortex, 1 mm behind bregma, 3 mm left of the midline, at a depth of 1.5 mm from the surface of the skull. On the measurement day, under isoflurane anesthesia (1.1%-1.3%), a tracheal tube was inserted for mechanical ventilation, and a catheter placed in the femoral artery. Blood pressure, body temperature, and arterial saturation (SpO_2; measured with a pulse oximeter) were continuously monitored during the experiment. Blood samples were withdrawn before and after the experimental procedures and analyzed with a combined blood gas analyzer and cooximeter (RapidLab 845, Bayer) to determine the stability of the rats during the experiment. An L-band (1.2 GHz) EPR spectrometer with a microwave bridge and an external loop resonator was used for the t-pO_2 measurement. Each EPR spectrum was recorded with settings carefully chosen to avoid power saturation and over-modulation. The t-PO_2 was calculated from the LiPc spectra using a previously determined calibration curve.

The NIRS system consisted of a 0.127 meter spectrograph with a charge coupled device (CCD) detector for detection and a filtered tungsten halogen light source for illumination with 1 mm optodes. Changes in [HbO_2] and [Hb] were calculated from the change in attenuation between 780-900 nm. Absolute [Hb] was calculated from a second differential spectroscopy. Absolute [HbO_2] was calculated from the change in [HbO_2] following 3 minutes of anoxia at the end of the experiment, using the assumption that [HbO_2] falls to zero during anoxia. Mean cerebral saturation ($SmcO_2$) was then calculated from absolute [HbO_2] and [Hb]. The optodes were placed 2-3 mm apart on the skull over the area in which the EPR paramagnetic material was implanted.

Hypoxic insults were induced by changing the FiO_2 in air from 0.30 at baseline to 0, 0.05, 0.10, and 0.15 for 1, 2, 5, and 5 minutes, respectively, following by reoxygenation with FiO_2 = 0.30. The interval between insults was at least 10 to 15 minutes to allow the rats to recover fully, as determined by blood pressure, SpO_2 and NIR data. Thirty EPR spectra were collected sequentially with the insult starting after the third spectrum. Each EPR spectrum consisted of a 5-second scan for the FiO_2 = 0 insult, an average of two 5-second scans for the FiO_2 = 0.05 insult, and an average of five 5-second scans for the FiO_2 = 0.10, and FiO_2 = 0.15 insults. Several EPR spectra were averaged to increase the signal-to-noise ratio. Spectral fitting by the EWVoigt program (Scientific Software Inc.)

Table 1. Body temperature and data from arterial blood samples.

	Body temperature (°C)	pH	$PaCO_2$ (mmHg)	PaO_2 (mmHg)	SaO_2 (%)	Hb (g/dL)
Pre-experiment	37. 4 ± 0.5	7.44 ± 0.04	39.8 ± 5.4	177.6 ± 7.1	99.4 ± 0.5	15.2 ± 0.5
Post-experiment	37.2 ± 0.4	7.42 ± 0.03	39.4 ± 3.5	177.5 ± 8.2	99.4 ± 0.3	14.9 ± 0.4

Each value shows the mean and standard deviation (SD) at pre- and post-study (n = 6).

was used to derive the EPR line width. The line width was transformed to cerebral pO_2 using a calibration of the LiPc that was performed at five different partial pressures of oxygen. The NIRS data were collected continuously every 0.5 seconds during baseline measurements, during the hypoxic insult, and in the recovery phase. All results are expressed as the mean and standard deviation (SD) of the group of six rats and the statistical significance of changes was determined with a paired t test.

3. RESULTS

Data from the arterial blood samples on blood gases and hemoglobin and on body weight are shown in Table 1. No statistical significance was found between pre- and post-study values in any of the measurements in Table 1 (paired t test, $P > .05$). The changes over time of t-pO_2 and $SmcO_2$ values at baseline, and during each hypoxic insult and recovery phase are shown in Figure 1. Reducing the inspired oxygen fraction resulted in a fall in both $SmcO_2$ and t-pO_2, as would be expected. These values recovered to baseline after the return to $FiO_2 = 0.30$. The changes in t-pO_2 and $SmcO_2$ for both the onset of the hypoxia and the recovery started approximately 20 to 30 seconds after the inspired gas mixture was switched. These delays in response were due to the dead space of the ventilation system, estimated to be 20 to 30 ml. $SmcO_2$ recovered promptly and reached the baseline level within 50 seconds after reoxygenation, whereas the recovery of t-pO_2 was slower. Both t-pO_2 and $SmcO_2$ reached a plateau during the exposure to $FiO_2 = 0.10$ and $FiO_2 = 0.15$.

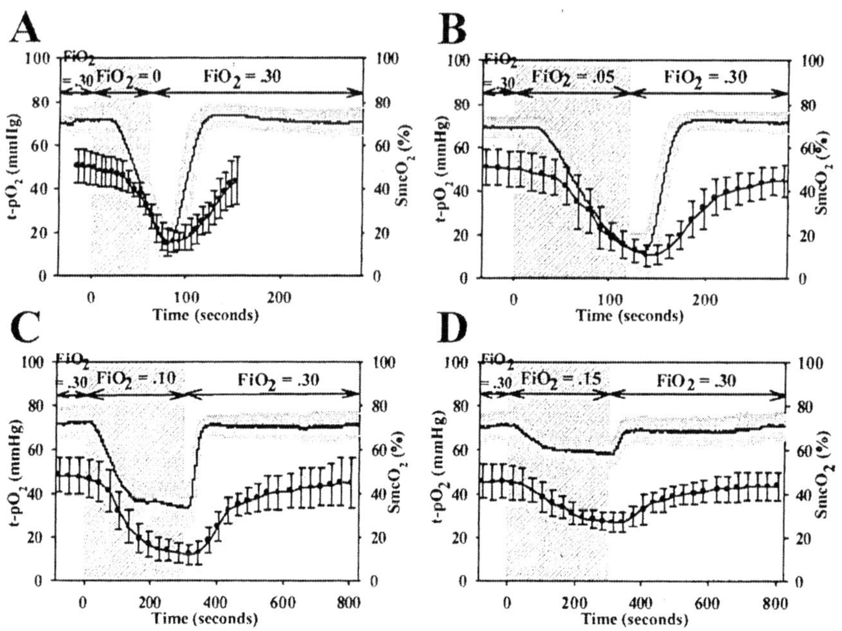

Figure 1. The mean and standard deviation (mean ± SD) of t-pO_2 and $SmcO_2$ are shown by (●) and (–), respectively (n = 6). **A**. $FiO_2 = 0$ insult for 60 seconds; **B**. $FiO_2 = 0.05$ insult for 120 seconds; **C** and **D**. $FiO_2 = 0.10$ and 0.15 insult for 5 minutes, respectively.

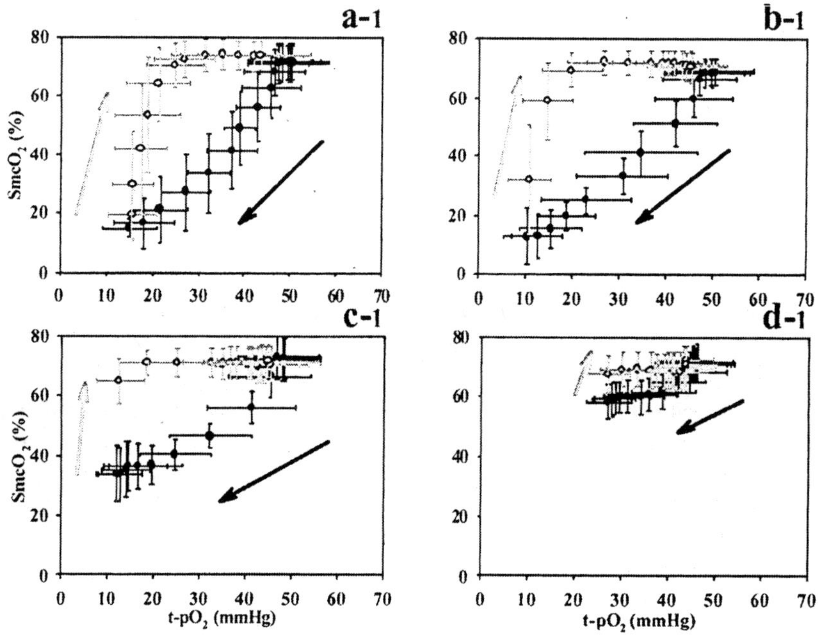

Figure 2. The X-axis represents t-pO_2 and the Y-axis represents $SmcO_2$. Each data point indicates the mean and standard deviation (n = 6). (●←) indicates the declining phase and (O→) indicates the recovery phase. Panel **a-1**, **b-1**, **c-1**, and **d-1** show FiO_2 = 0, FiO_2 = 0.05, FiO_2 = 0.10, and $Fi0_2$ = 0.15, respectively.

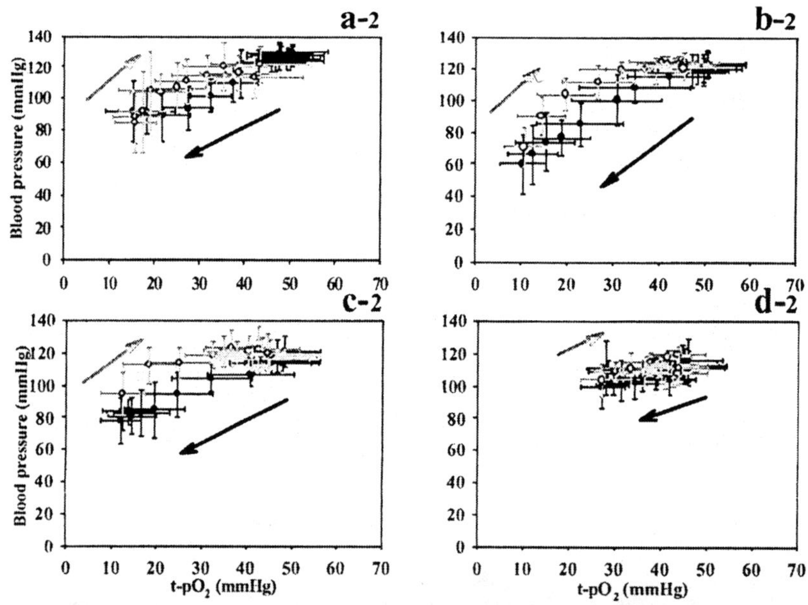

Figure 3. The X-axis represents t-pO_2 and the Y-axis represents the mean BP. Each data point indicates the mean and standard deviation (n = 6). (●←) indicates the declining phase and (O→) indicates the recovery phase. Panel **a-2**, **b-2**, **c-2**, and **d-2** show FiO_2 = 0, FiO_2 = 0.05, FiO_2 = 0.10, and $Fi0_2$ = 0.15, respectively.

Table 2. The baseline and minimum values of t-pO_2 and $SmcO_2$.

		$FiO_2 = 0.0$	$FiO_2 = 0.05$	$FiO_2 = 0.10$	$FiO_2 = 0.15$
t-pO_2 (mmHg)	Baseline	50.1 ± 7.1	50.2 ± 7.8	48.2 ± 7.6	45.8 ± 7.5
	minimum	15.2 ± 6.0**	10.4 ± 5.2**	12.1 ± 4.7**	27.2 ± 4.8**
$SmcO_2$ (%)	Baseline	70.6 ± 4.7	68.9 ± 4.3	72.2 ± 4.7	70.9 ± 5.3
	minimum	13.9 ± 6.2**	11.9 ± 8.4**	33.2 ± 8.7**	58.1 ± 6.0*

The values show the mean ± SD of baseline value and minimum value during each hypoxic insult (n = 6). *$P < .05$; **$P < .005$.

The relationship between t-pO_2 and $SmcO_2$ is shown in Figure 2. The relationship between t-pO_2 and $SmcO_2$ was not linear during the hypoxic insult and recovery phases. $SmcO_2$ declined slightly faster than the t-pO_2 value during the insult phase (●←) and $SmcO_2$ also recovered more promptly than the t-pO_2 during the recovery phase (O→). Changes in the systemic blood pressure were observed during each hypoxic insult. The systemic blood pressure was reduced to 66.5% ($P < .005$), 48.8%, 64.2%, and 83.7% ($P < .05$) of baseline during $FiO_2 = 0$, $FiO_2 = 0.05$, $FiO_2 = 0.10$, and $FiO_2 = 0.15$ insults, respectively. It recovered to baseline in all animals after reoxygenation. The relationship between t-pO_2 and the systemic blood pressure is shown in Figure 3. The changes of t-pO_2 correlated more closely with the systemic blood pressure than $SmcO_2$. The mean and standard deviation of the baseline and minimum value of t-pO_2 and $SmcO_2$ during each hypoxic insult are shown in Table 2. The largest changes in t-pO_2 and $SmcO_2$ occurred during the $FiO_2 = 0.05$ insult.

4. DISCUSSION

As expected, both t-pO_2 and $SmcO_2$ declined during each hypoxic insult, and these changes were followed by a full recovery to baseline. It was interesting and potentially important that the recovery of t-pO_2 was slower than the recovery in $SmcO_2$. These results indicate that a recovery in $SmcO_2$ after a hypoxic insult does not necessarily indicate a recovery of brain tissue oxygenation.

The minimum values of t-pO_2 and $SmcO_2$ were observed during the $FiO_2 = 0.05$ insult and not the $FiO_2 = 0$ insult. This was probably because 1 minute was not a severe enough insult compared with the $FiO_2 = 0.05$ insult for 2 minutes. However the minimum value of $SmcO_2$ during $FiO_2 = 0.10$ insult was greater than for the $FiO_2 = 0$ insult, whereas the minimum value of t-pO_2 during $FiO_2 = 0.10$ insult was less than the $FiO_2 = 0$ insult. It was possible that t-pO_2 values were not only determined by the inspired oxygen fraction, but also by the duration time of hypoxic insult or response of cellular oxygen utilization.

The recovery of t-pO_2 from hypoxic insults was found to be related not only to blood oxygen saturation, but also the systemic blood pressure. The correlation between the recovery of t-pO_2 and systemic blood pressure was greater than the correlation between the recovery of t-pO_2 and $SmcO_2$. It also is plausible that the recovery of t-pO_2 is related to cerebral blood flow, cerebral blood volume, blood CO_2 tension, tissue oxygen consumption, and oxygen diffusion from the vascular system to the tissue.[6-8] Changes in nitric oxide production occur as part of the mechanism to regulate vasodilatation, and that can lead to hypotension during the hypoxic insult.[9] The recovery of the blood pressure and oxygen consumption in the vascular wall[10, 11] might also be important factors for the recovery of t-pO_2. Further studies will be required to determine the exact mechanisms for the delay between the recovery of $SmcO_2$ and the recovery of t-pO_2. It is clear, however, that it is very useful to measure both parameters independently and simultaneously.

5. REFERENCES

1. M. Afeworki, N. R. Miller, N. Devasahayam, J. Cook, J. B. Mitchell, S. Subramanian, and M. C. Krishna, Preparation and EPR studies of lithium phthalocyanine radical as an oxymetric probe, *Free Radical Biol. Med.* **25**(1), 72-78 (1998).
2. K. J. Liu, P. Gast, M. Moussavi, S. W. Norby, N. Vahidi, T. Walczak, M. Wu, and H. M. Swartz, Lithium phthalocyanine: A probe for electron paramagnetic resonance oximetry in viable biological systems, *Proc. Natl. Acad. Sci. USA* **90**, 5438–5442 (1993).
3. N. Vahidi, R. B. Clarkson, K. J. Liu, S. W. Norby, M. Wu, and H. M. Swartz, In vivo and in vitro EPR oxymetry with fusinite: A new coal-based solid state EPR probe, *Magn. Reson. Med.* **31**, 139–146 (1994).
4. S. J. Matcher, and C. E. Cooper, Absolute quantification of deoxyhemoglobin concentration in tissue near-infrared spectroscopy, *Phys. Med. Biol.* **39**, 1295-1312 (1994).
5. S. J. Matcher, M. Cope, and D. T. Delpy, Use of the water-absorption spectrum to quantify tissue chromophore concentration changes in near-infrared spectroscopy, *Phys. Med. Biol.* **39**, 177-196 (1994).
6. M. Intaglietta, P. C. Johnson, and R. M. Winslow, Microvascular and tissue oxygen distribution, *Cardiovasc. Res.* **32**, 632-643 (1996).
7. H. Kobayashi, B. Pelster, J. Piper, and P. Scheid, Significance of Bohr effect for tissue oxygenation in a model with counter current blood flow, *Respir. Physiol.* **76**, 227-288 (1989).
8. A. G. Tsai, P. C. Johnson, and M. Intaglietta, Oxygen gradients in the microcirculation, *Physiol. Rev.* **83**(3), 933-963 (2003).
9. N. J. Edmunds, and J. M. Marshall, Vasodilatation, oxygen delivery and oxygen consumption in rat hindlimb during systemic hypoxia: roles of nitric oxide, *J. Physiol.* **532**(Pt 1), 251-259 (2001).
10. A. G. Tsai, B. Friesenecker, M. C. Mazzoni, H. Kerger, D. G. Buerk, P. C. Johnson, and M. Intaglietta, Microvascular and tissue oxygen gradients in the rat mesentery, *Proc. Natl. Acad. Sci. USA* **95**(12), 6590-6595 (1998).
11. A. Vadapalli, R. N. Pittman, and A. S. Popel, Estimating oxygen transport resistance of the microvascular wall, *Am. J. Physiol. Heart Circ. Physiol.* **279**, H657–H671 (2000).

MICROCIRCULATION AND REPERFUSION INJURY IN ORGAN TRANSPLANTATION

Giuseppe Cicco, P. C. Panzera, G. Catalano, and V. Memeo

Abstract: There are many interesting aspects regarding hemorheology and tissue oxygenation in organ transplantation (such as liver, kidney, heart, etc.). The ischemia-reperfusion injury syndrome is a very important problem. Much damage in organs appears to be induced by reperfusion injury syndrome. In fact, not only immunological etiopathogenesis but also biochemically-mediated microcirculation alterations can modulate the organ damage induced by ischemia-reperfusion injury during organ transplantation.

During ischemia-reperfusion injury, xanthine oxidase activity, the increase in oxygen free-radicals, and the activation of neuthrophils are all very important. Platelet activating factor (PAT) and LTB_4 (promoting neuthrophils adhesiveness), activated by the xanthine oxidase-derived oxidants during reperfusion, activates the final post-ischemia injury. Much research is necessary in order to gain a fuller knowledge of the microcirculation conditions and oxygenation during organ transplantation.

1. ISCHEMIA–REPERFUSION (I/R) INJURY

Organ transplantation (liver, kidney, heart, etc.) raises a number of important issues regarding *microcirculation conditions*, *hemorheology*, *tissue oxygenation*, and how they relate to ischemia-reperfusion injury. Ischemia-reperfusion (I/R) injury is a non-specific antigen-independent process that can significantly influence the outcome of organ transplantation and constitutes one of the principal risk factors for the development of long-term dysfunction of the transplanted organs.

Paradoxically, reperfusion promotes a series of very complicated pathologic events that injure tissues.[1-8] The main target of these events is the microcirculation. During

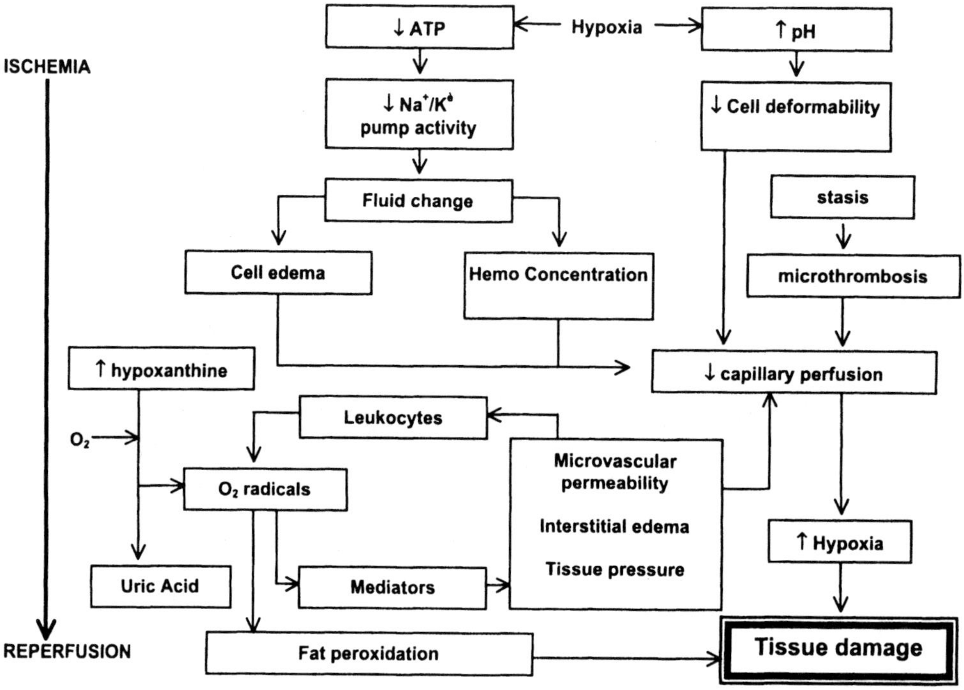

Figure 1. Ischemic-reperfusion injury.

ischemia-reperfusion injury, xanthine oxidase activity, increased oxygen free radicals (OFR), neutrophil activation, and altered adhesion, all play important roles. The neutrophils adhere[9] to the microvascular endothelium and induce microcirculatory permeability, which increases transcapillary fluid in reperfused tissues.[4, 10-15] Additionally, neutrophil adhesion causes post-ischemic capillary "no reflow" and has been implicated in reduced arteriolar sensitivity to the vasoactive substances induced by I/R injury.[2, 4, 16-20]

1.1. Ischemic Injury in Transplantation

Ischemia/Reperfusion (I/R) injury gives rise to a highly complex cascade of phenomena (Figure 1). During ischemia, hypoxia is rapidly induced, followed quickly by a drop in intracellular levels of ATP. Aerobic glycolysis necessarily converts to anaerobic glycolysis (piruvate → lactate) with reduced NADH production. With prolonged ischemia, ATP production drops and energy-dependent functions (Na^+/K^+ dependent ATPase) are arrested, inducing cellular edema and hemoconcentration. Consequently, there is a rise in capillary compression, causing reduced capillary perfusion with decreased tissue O_2. Expression of adhesion molecules increases, and these in turn increase polymorphonuclear (PMN) leukocytes adhesion to vessel walls. Hypoxia can cause decreased pH and red blood cell (RBC) deformability; this also contributes to the impairment of capillary perfusion. In such a situation, there is stasis with microthrombosis and, finally, further deterioration of capillary perfusion.

With ischemia there is an important increase of hypoxanthine. When this high concentration of hypoxanthine comes in contact with oxygen carried by the gush of blood during the early stage of reperfusion, hypoxanthine oxidase gives rise to a high quantity of oxygen-derived free radicals (ODFR) from leukocytes, and uric acid (McCord reaction). Via mediators, these ODFR could influence tissue pressure, favoring interstitial edema and microvascular permeability, thus further increasing ODFR release from leukocytes. This increase in ODFR can directly damage tissue through fat peroxidation.

1.2. Reperfusion Injury in Transplantation

The relative importance of the various mechanisms and mediators of I/R injury during reperfusion of transplanted organs are not precisely understood. Many contribute to acute or chronic organ rejection. Some etiologic factors include:

1. Ca^{2+} activation and/or alteration of intracellular calcium ion quantities
2. Depletion of intracellular ATP reserves
3. Cellular damage mediated by free radicals
4. Allo-immunization

Leukocytes can contribute to tissue dysfunction during reperfusion. A role in leukocyte adhesion is played by upregulated endothelial ICAM–1 expression during rejection in cardiac, renal, hepatic, and corneal allografts.[21-27] This suggests that inflammatory cell infiltration could play an important role in transplantation-induced tissue injury.

Moreover, I/R injury increases the risk of Delayed Graft Function (DGF) and of acute or chronic rejection. We may consider that even nowadays the possible causes of DGF are related only to immunomediated events, but it is possible that non-immunologic factors such as alterations in microcirculation could also contribute to I/R injury and to DGF. The high quantity of ODFR produced during I/R injury under the effect of xanthine oxidase could play an important role in sustaining the tissue damage.

2. CAPILLARY "NO REFLOW"

After ischemia, when the blood flow is restored in reperfusion (after liver, kidney, heart transplantation), the important phenomenon of capillary "no reflow" can arise. In fact, a large number of capillaries fail to reperfuse.[2, 4, 15-16, 28-31] The mechanisms sustaining the development of the "no reflow" phenomenon remain unclear. One hypothesis is that I/R-induced microvascular thrombus production could induce the "no reflow" phenomenon.[32] However, heparin treatment is unable to restore capillary perfusion after I/R.[33] In any case, microvascular thrombosis is unlikely to be present in post-ischemic tissues.[34-35] The most credible hypothesis is that activated neutrophils are able to induce post-ischemic capillary no reflow.[4, 16-17, 30, 36-39] Following I/R, leukocyte rolling (P–selectin-mediated) occurs followed by firm adhesion (CD18/ICAM-dependent). These phenomena can induce endothelial cell swelling and could finally induce capillary no reflow via reduction of the capillary diameter.[40-41]

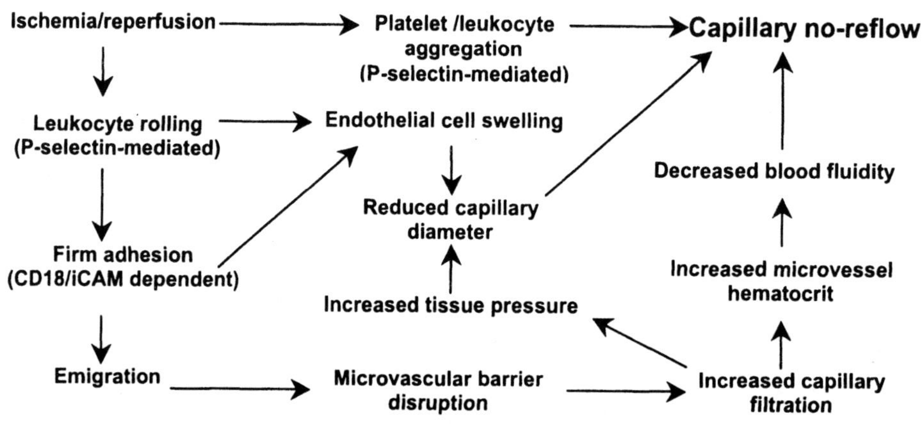

Figure 2. Mechanisms whereby leukocyte-endothelial cell adhesive interactions may contribute to the development of post ischemic capillary no reflow (modified from Akimitsu[42]).

Leukocyte adhesion to post-capillary venules contributes to the genesis of no reflow by a mechanism involving microvascular barrier disruption by activated neutrophils (Figure 2).[38-42] This induces an increase in the quantity of interstitial fluid, the formation of edema, increased capillary pressure, and extravascular connections, leading to reduced capillary diameter, favoring capillary no reflow. There could be an ensuing increase in capillary filtration, in hematocrit and blood viscosity, and, finally, the occurrence of capillary no reflow. Thus, the microhemodynamic alterations induced by neutrophil-mediated microvascular barrier disruption and altered membrane ion fluxes could contribute to capillary no reflow.

3. LIVER ISCHEMIA-REPERFUSION INJURY: PATHOPHYSIOLOGY

Reoxygenation of the tissues after prolonged ischemia can induce cell damage with microcirculation injury.[43] During early reperfusion (Figure 3), endothelial cells swell, and vasoconstriction begins; later, leukocyte entrapment, and leukostasis with firm leukocyte adhesion to endothelial cells takes place. In the liver, there can also be decreased blood flow in the sinusoids, red blood cells and platelets aggregate within the sinusoids, and hypoxia develops. Later, Kupffer cell and neutrophil activation occurs, with the release of inflammatory cytokines, ODFR, and, finally, failure of the microcirculation and increased leukocyte adhesion including the no reflow phenomenon.[44]

The causes of impaired blood flow during early liver reperfusion are:

1. The failure of active transmembrane ion transport, endothelial and Kupffer cell swelling, and capillary no reflow. It is possible to intercede in this mechanism by delivering exogenous adenosine or increasing endogenous adenosine.[45]
2. The conversion of adenosine to inosine by adenosine–deaminase; this can induce capillary no reflow, increasing vasoconstriction and stellate cell contraction, and inducing a decrease in nitric oxide (NO) production as a result

of the decrease in cofactors (O_2–NADPH). Again, it is possible to interfere in both of these mechanisms by increasing exogenous and endogenous adenosine. Nitric Oxide (NO) can be increased using an NO precursor (such as L–Arginine) or an NO donor.

3. The endothelial (ET) cell increase during ischemia could contribute to vasoconstriction and stellate cell contraction, inducing capillary no reflow.

4. CELLULAR MECHANISMS TO PREVENT ISCHEMIA-REPERFUSION INJURY

All systems operating on superoxides, nitric oxide (NO), and heme-oxygenase are responsible for the prevention, reduction, or improvement of I/R injury.

4.1. Superoxides

The superoxides are ODFRs. The ODFRs include superoxide anions, hydrogen peroxide (H_2O_2), and hydroxyl radicals (OH^-). O_2^- plays a principal role in tissue damage during I/R injury. It is produced with urates when a high quantity of oxygen (O_2) suddenly comes into contact with the tissues during reperfusion after ischemia and meets

Figure 3. Liver Ischemic-Reperfusion injury: Pathophysiology.

Early stage of Reperfusion

↓

Endothelial cell swelling

↓

Vasoconstriction

↓

Leukocyte entrapment → Leukostasis

↓

Stagnant leukocytes

↓

Decreased blood flow in sinusoids

↓

RBC and platelet aggregation within the sinusoids

↓

Hypoxic in some liver areas → Kupffer cells and neutrophils activation

↓

Inflammatory cytokines → oxygen derived free radicals (ODFR)

↓

Failure in microcirculation

a lot of hypoxanthine in the presence of xanthine–oxidase. Oxygen superoxide radicals (O_2^-) are able to abolish the beneficial function of nitric oxide (NO) on the microcirculation by scavenging NO molecules and generating highly cytotoxic peroxinitrite ($ONOO^-$). Normally, superoxide dismutase (SOD) (ZnSOD in cytoplasm and MnSOD in mitochondria), catalase, and glutathione peroxidase are able to control the production of O_2^- *in vivo.* If these systems are impaired, as during ischemia, then there is increased O_2^- and other ODFR production.

4.2. Nitric Oxide

Nitric oxide (NO), produced by levo arginine in the presence of NO synthetase (NOS), is a strong vasodilator and its effect is very important in microcirculation control. There are different NOS isoforms: inducible NOS (iNOS), constituent NOS (cNOS), endothelial NOS (eNOS), and neuronal NOS (nNOS). During organ transplantation, especially with liver, NO is able to defend the tissues by increasing tissue perfusion, inhibiting polymorphonuclear cell adhesion, and blocking ODFR damage

4.3. Heme-Oxygenase

This enzymatic system transforms the heme group with bilirubin production of biliverdin, carbon monoxide (CO), and free iron. The isoform, HO-1, is present in I/R injury and could depend on ODFR activation.

5. MEDIATORS

Many mediators are implicated in microcirculation control and play an important role in its steady state. The main vasodilators are prostacyclin (PGI_2) and nitric oxide (NO); the main vasoconstrictors are thromboxane A_2 (TxA_2) and endothelin (ET). Increased production of vasoconstrictors or decreased production of vasodilators arise due to impairment of the microcirculation after reperfusion (I/R injury). Thromboxane B_2 (the TxA_2-stable metabolite) and 6-keto-PGF-1α (the prostacyclin-stable metabolite) were both evaluated after reperfusion and raised TxB_2 was dominant in a pig liver transplantation model. A beneficial effect of donor pretreatment with a TxA_2 synthetase inhibitor has been demonstrated. Nitric oxide releasers or endothelin inhibitors could also be used during liver transplantation to both reduce mediator-related I/R injury.[46]

6. ADVANCES IN LIVER TRANSPLANTATION

Today we can make use of the Belzer liver-preserving solution from the University of Wisconsin (USA) and of ischemic preconditioning to prevent I/R injury. The University of Wisconsin Solution (Belzer, Jamieson, and Kalayoglu) used for organ (kidney 1978, liver 1988) preservation during transplantation marks a very important step forward from the Collins–Terasoky solution (1969). The presence of lactiobionate, raffinose, and hydroxyl–ethyl–starch (HES) nearly eliminates cellular edema during "cold ischemia." Glutathione and adenosine are able to increase cellular antioxidant properties and the production of high energy phosphates.

6.1. Ischemic Preconditioning

Ischemic preconditioning (IP) protects hepatic tissue from I/R injury, although the mechanisms involved are not yet fully understood. IP is carried out with short periods of ischemia followed by short periods of blood flow. This is able to reduce the amount of blood and of oxygen in direct contact with cells. As stated above, when oxygen arrives suddenly, it is metabolized to toxic ODFR by high levels of hypoxanthine (via xanthine oxidase) produced during cold ischemia.

Ischemic preconditioning exerts its protective effects via alteration of the cell inflammatory response and improvement of the microvascular reperfusion, maintaining adequate parenchymal oxygen supply, mitochondrial state, and organ function. An important study found strong evidence that the protective action of IP is based on preservation of tissue oxygenation due to the improved perfusion mediated by adenosine-associated vasodilation.

Ischemic preconditioning prevents strong Kupffer cell activation upon reperfusion after warm ischemia and confers protection via limitation of the downstream sequelae of reperfusion injury. As a result, liver perfusion with oxygen transport to the tissue is better preserved and liver function protected, as indicated by the fact that the mitochondrial redox state remains unchanged and bile flow is restricted to almost physiological conditions.

7. MATERIALS AND METHODS

In order to evaluate the transplanted liver microcirculation, we studied 5 subjects with HCV^+ liver disease (3 males and 2 females aged 46 ± 4 years), who underwent liver transplantation from cadavers. To avoid extra corporeal circulation, the "piggy back" method was used. This consists of an anastomosis between the vena cava of the liver donor and the ostium of the medial and left suprahepatic veins on the vena cava of the recipient. The ostium of the right suprahepatic vein is closed.

To measure the blood flow during liver transplantation, we used the laser–Doppler unit: Periflux 5000 (Perimed) and directly studied the liver surface in the operating room using a special probe placed on the V hepatic segment. We monitored blood flow, measured on the liver surface at time 0 (venous reperfusion), after 20 min and at arterial reperfusion.

To study the capillary morphology, we used computerized capillaroscopy (VIDEO CAP – Mitsubishi) with 200x magnification, studying the cutaneous plica of the 4th fingernail bed of the left hand. These values were detected immediately before, and 8 days after transplantation.

Considering that 34% of transplanted patients may have neurological complications, although these fortunately heal spontaneously in 80%-92% of cases, we also evaluated brain oxygenation. To measure brain perfusion and oxygenation, during all transplantation surgical phases we used a new, sophisticated, computerized instrument, the NIRO 300 (Hamamatsu). The NIRO 300 is able to evaluate non-invasively brain transcranial infrared spectroscopy, and it is able to measure ΔO_2 Hb (oxygenated hemoglobin), Δ HHb (deoxygenated hemoglobin), Δ c Hb (total hemoglobin), and Δ CytOx (oxygenated cytochrome).

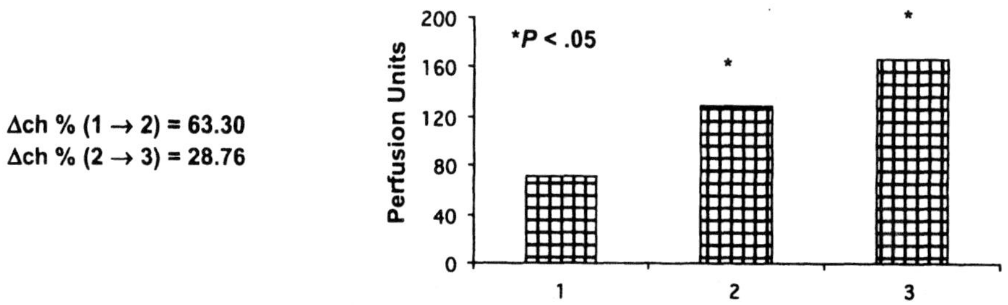

Figure 4. Laser doppler blood flow measurements.

In order to study the red blood cell deformability in these patients, we used the Laser Optical Rotational Red Cell Analyzer (LORCA). Evaluations were made at time 0, after 20 min, 90 min, 1 day, 5 days, and 8 days from transplantation. LORCA measurements were made with a sheer stress of 30 Pa. LORCA outputs include the RBC elongation index (a measure of RBC deformability).

8. RESULTS

Studying the capillary morphology, we observed a significant increase ($P < .05$) of functioning capillaries 8 days after liver transplantation compared with the values obtained before transplantation, on the basis of the capillary loop number visible in the capillaroscopy monitoring picture (6 ± 1 before versus 10 ± 2 after) ($P < .03$).

The evaluation of blood flow using the Periflux laser doppler unit yielded very interesting results. We obtained 63.30 perfusion units (PU) at baseline (time 0), 129.59 PU after 20 min, and 166.78 PU at the time of arterial reperfusion. The changes occurred after 20 min, Δ Ch% = 63.30, and continued until arterial reperfusion, Δ Ch% = 28.76 ($P < .05$) (Figure 4).

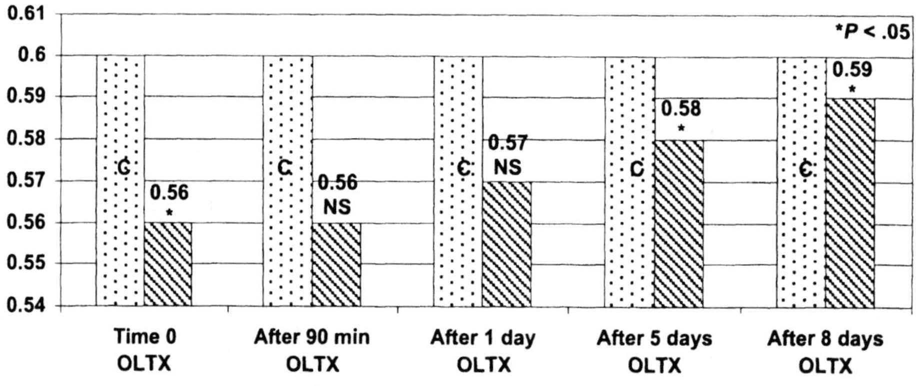

Figure 5. LORCA Elongation Index (y axis) outputs after orthotopic liver transplantation (OLTX). Measurements made with a sheer stress of 30 Pa. C = control.

Using the NIRO 300, we detected mild cerebral hypoxia only during dissection of the post-hepatic vena cava. To study the RBC deformability, we used LORCA, considering the Elongation Index (EI) (Figure 5) ($P < .05$).

9. DISCUSSION

In the transplanted patients, we observed a marked improvement of the peripheral microvasculature, studied with videocapillaroscopy and compared with the observations immediately before liver transplantation. The blood, evaluated directly on the liver surface in all patients, also revealed the absence of no reflow phenomena and an arterial reperfusion level corresponding to physiological values (an increase of approximately 20% of the total liver blood flow present at the moment of the hepatic artery connection). We found only a reversible, mild cerebral hypoxia that was resolved by increasing the inhaled oxygen. Regarding RBC deformability, it has been shown that the situation returns to normal some days later; this could be explained by resolution of the high levels of bilirubinemia that can induce RBC rigidity. This condition could also explain the low number of functioning peripheral capillaries before transplantation, due to the high blood viscosity induced by RBC rigidity, and the increase of these functioning microvessels after normalization of bilirubinemia following liver transplantation.

Further studies in transplanted patients are needed to improve the microcirculation conditions following organ transplantation, and especially liver transplantation.

REFERENCES

1. D. J. Hearse, Reperfusion of ischemic myocardium, *J. Mal. Cell. Cardiol.* **9**(8), 605-616 (1977).
2. S. N. Jerome, C. W. Smith, and R. J. Korthuis, CD–18 dependent adherence reactions play an important role in the development of the no reflow phenomenon, *Am. J. Physiol.* **264**(2 Pt 2), H479-483 (1993).
3. D. A. Parks, and D. N. Granger, Contributions of ischemia and reperfusion to mucosal lesion formation, *Am. J. Physiol.* **250**(6), G749-753 (1986).
4. D. L. Carden, J. K. Smith, and R. J. Karthuis, Neutrophils mediated microvascular disfunction in postichemic canine skeletal muscle: role of granulocyte adherence, *Circ. Res.* **66**(5), 1436-1444 (1990).
5. M. A. Perry, and S. S. Wodhwa, Gradual reintroduction of oxygen reduces perfusion injury in cat stomach, *Am. J. Physiol.* **254**(3), G366-372 (1988).
6. R. J. Korthuis, J. K. Smith, and D. L. Carden, Hypoxic reperfusion attennates post ischemic microvascular injury, *Am. J. Physiol.* **256**(2), H315-319 (1989).
7. L. O. Dahlback, and O. Rais, Morphological changes in striated muscle following ischemia: immediate post-ischemic phase, *Acta Chir. Scand.* **131**(6), 430-440 (1966).
8. J. B. Morris, U. Haglund, and G. B. Bulkey, The protection from postischemic injury by xanthine oxidase inhibition: Blockage of free radical generation or purine salvage, *Gastroenterology* **92**, 1542 (1987).
9. G. D. Dunn, N. D. Granger, and R. J. Korthuis, Leukocyte/endothelial cell adhesion and ischemia reperfusion injury, in: *Clinically Applied Microcirculation Research*, edited by J. Barker, G. L. Anderson, and M. D. Menger (CRC Press Inc., Boca Raton, FL, 1995) pp. 75-96.
10. W. K. Adkins, and A. E. Taylor, Role of xanthine oxidase and neutrophils in ischemia–reperfusion injury in rabbit lung, *J. Appl. Physiol.* **69**(6), 2012-2018 (1990).
11. L. A. Hernandez, M. B. Grisham, B. Twohig, K. E. Arfors, J. M. Harlan, and D. N. Granger, Role of neutrophils in ischemia/reperfusion induced microvascular injury, *Am. J. Physiol.* 253(3), H699-703 (1987).
12. M. J. Bishop, S. M. Kowalski, S. M. Guidotti, and J. M. Harlan, Antibody against neutrophils adhesion improves reperfusion and limits alveolar infiltrate following unilateral pulmonary artery occlusion, *J. Surg. Res.* **52**(3), 199-204 (1992).
13. M. J. Horgan, M. Ge, J. Gu, R. Rothlein, and A. B. Malik, Role of ICAM-1 in neutrophil mediated lung vascular injury after occlusion and reperfusion, *Am. J. Physiol.* **261**(5), H1578-1584 (1991).

14. M. J. Horgan, S. D. Wright, and A. B. Malik, Antibody against leukocyte integrin (CD 18) prevents reperfusion induced lung vascular injury, *Am. J. Physiol.* **259**(4), L315-319 (1990).
15. N. B. Vedder, R. K. Winn, C. L. Rice, E. Y. Chi, K. E. Arfors, and J. M. Harlan, Inhibition of leukocyte adherence by anti-CD 18 monoclonal antibody attenuates reperfusion injury in the rabbit ear, *Proc. Natl. Acad. Sci. USA.* **87**(7), 2643-2646 (1990).
16. S. N. Jerome, M. Dore, J. C. Poulson, C. W. Smith, and R. J. Korthuis, P-selectin and ICAM-1 adherence reactions: role in the genesis of postischemic no reflow, *Am. J. Physiol.* **266**(4), H1316-1321 (1994).
17. E. Mori, G. J. Del Zoppo, J. D. Chambers, B. R. Copeland, and K. E. Arfors, Inhibition of polymorphonuclear no-reflow after local cerebral ischemia in baboons, *Stroke* **23**(5), 712-718 (1992).
18. X. L. Ma, P. S. Tsao, and A. M. Lefer, Antibody to CD 18 extents endothelial and cardiac protective effects in myocardial ischemia and reperfusion, *J. Clin. Invest.* **88**(4), 1237-1243 (1991).
19. X. L. Ma, D. J. Lefer, A. M. Lefer, and R. Rothlein, Coronary endothelial and cardiac protective effects of a monoclonal antibody to intercellular adhesion molecule-1 in myocardial ischemia and reperfusion, *Circulation* **86**(3), 937-946 (1992).
20. A. S. Weyrich, X. Y. Ma, D. J. Lafer, K. H. Albertine, and A. M. Lefer, In vivo neutralization of P-selectin protects feline heart and endothelium in myocardical ischemia and reperfusion injury, *J. Clin. Invest.* **91**(6), 2620-2629 (1993).
21. D. H. Adams, S. G. Hubscher, J. Shaw, R. Rothlein, and J. M. Neuberger, Intercellular adhesion molecule-1 on liver allografts during rejection, *Lancet* **2**(8672), 1122-1125 (1989).
22. D. H. Adams, L. F. Wang, D. Burnett, R. A. Stockley, and J. M. Neuberger, Neutrophil activation--an important cause of tissue damage during liver rejection? *Transplantation* **50**(1), 86-91 (1990).
23. V. M. Elver, S. G. Elmer, M.A. Pavilack, R. F. Todd III, B. Y. Yue, and A. R. Huber, Intercellular adhesion molecule-1 in human corneal endothelium, *Am. J. Pathol.* **138**(3), 525-536 (1991).
24. S. G. Hubscher, and D. H. Adams, ICAM-1 expression in normal liver, *J. Clin. Pathol.* **44**(5), 438-439 (1991).
25. T. Omura, H. Ishikura, Y. Nakajima, J. Kiniura, K. Ito, H. Isai, T. Tomatoni, M. Miyasaka, T. Yoshiki, and J. Vehino, The expression of FA-1 ICAM-1 in liver transplantation in rats, *Transpl. Proceed.* **24**, 1618 (1992).
26. D. D. Sedmak, and C.G. Orasz, The role of vascular endothelial cells in transplantation in rats, *Transpl. Proceed.* **24**, 1237 (1992).
27. Y. Takei, I. Marzi, W. Gao, G. J. Gores, J. J. Lemasters, and R. G. Thruman, Leukocyte adhesion and cell death following orthotopic liver transplantation in the rat, *Transplantation*, **51**(5), 959-965 (1991).
28. H. A. Lehr, A. Gulhmann, D. Nolte, D. Keppler, and K. Messmer, Leukotrienes as mediators in ischemia reperfusion injury in a microcirculation model in the Hamster, *J. Clin. Invest.* **87**(6), 2036-2041 (1991).
29. A. Ames III, R. L. Wright, M. Kowada, J. M. Thurston, and G. Majno, Cerebral ischemia II. The no reflow phenomenon, *Am. J. Pathol.* **52**(2), 437-453 (1968).
30. G. W. Schmid-Schonbein, Capillary plugging by granulocytes and the no reflow phenomenon in the microcirculation, *Fed. Proceed.* **46**(7), 2397-2401 (1987).
31. G. Cicco, Hemorheology, reperfusion injury and organ transplantation, *12th ECCH, Sofia (Bulgaria) A. B.* RT1.**1**, 64 (2003).
32. W. J. Quinones–Baldrich, A. Chervu, J. J. Hernandez, M. D. Colburn, and W. S. Moore, Skeletal muscle function after ischemia "no reflow" versus reperfusion injury, *J. Surg. Res.* **51**(1), 5-12 (1991).
33. P. E. Strock, and G. M. Majno, Vascular responses to experimental tourniquet ischemia, *Surg. Gynecol. Obstet.* **129**(2), 309-318 (1969).
34. J. W. Harman, The significance of local vascular phenomena in the production of ischemia necrosis in skeletal muscle, *Am. J. Pathol.* **24**(1), 625-642 (1948).
35. U. Bagge, B. Amundson, and C. Lauritzen, White blood cell deformability and plugging of skeletal muscle capillaries in hemorrhagic shock, *Acta Physiol. Scand.* **108**(2), 159-163 (1980).
36. J. Barrosa-Aranda, G. W. Schmid-Schonbein, B. W. Zweifach, and R. L. Engler, Granulocytes and no reflow phenomenon in irreversible hemorrhagic shock, *Circ. Res.* **63**(2), 437-447 (1988).
37. R. L. Engler, G. W. Schmid–Schonbein, and R. S. Pavalec, Leucocyte capillary plugging in myocardical ischemia and reperfusion in the dog, *Am. J. Pathol.* **111**(1), 98-111 (1983).
38. S. N. Jerome, T. Akimitsu, and R. J. Korthuis, Leucocyte adhesion, edema and the development of post ischemia capillary no reflow, *Am. J.Physiol.* **267**(4), H1329-1336 (1994).
39. S. N. Jerome, S. N. Akimitsu, D. C. Gute, and R. J. Korthuis, Ischemic preconditioning alternates capillary no reflow induced by prolonged ischemia and reperfusion, *Am. J. Physiol.* **268**(5 Pt 2), H2063-2067 (1995).
40. M. C. Mazzoni, P. Borgstrom, M. Intaglietta, and K. E. Arfors, Lumenal narrowing and endothelial cell swelling in skeletal muscle capillaries during hemorrhagic shock, *Circ. Shock* **29**(1), 27-39 (1989).

41. M. C. Mazzoni, M. Intaglietta, E. J. Crogue Jr., and K. E. Arfors, Amiloride-sensitive Na+ pathways in capillary endothelial cell swelling during hemorrhagic shock, *J. Appl. Physiol.* **73**(4), 1467-1473 (1992).
42. T. Akimitsu, S. N. Jerome, D. C. Gute, and R. J. Korthuis, Reactive oxygen species, neutrophils infiltration and postischemic microvascular dysfunction, in: *Reoxygenation Injury in Skeletal Muscle*, edited by G. O. Fantini (Laudis Publication, Austin, TX, 1994), pp. 32-35.
43. D. N. Granger, Role of xanthine oxidase and granulocytes in ischemia reperfusion injury, *Am. J. Physiol.* **255**(6), H1269-1275 (1988).
44. A. Koo, H. Komatsu, G. Tao, M. Inoue, P. H. Guth, and N. Kaplowitz, Contribution of no reflow phenomenon to hepatic injury after ischemia–reperfusion: evidence for a role for superoxide anion, *Hepatol.* **15**(3), 507-514 (1992).
45. F. Serracino-Inglott, and H. Habib, Hepatic ischemia reperfusion injury, *Am. J. Surg.* **181**, 160-166 (2001).
46. T. Kurokawa, and H. Takagi, Mechanism and prevention of ischemia reperfusion injury, *Transpl. Proceed.* **31**(4), 1775-1776 (1999).

FOCUSSING ON GENOMIC AND PHENOMIC CORRELATIONS IN RESPIRATION OF NON-MELANOTIC SKIN CANCERS

David J. Maguire, Nicholas A. Lintell, Michael McCabe, and L. Griffiths

Abstract: In recent years, with the development of techniques in modern molecular biology, it has become possible to study the genetic basis of carcinogenesis down to the level of DNA sequence. Major advances have been made in our understanding of the genes involved in cell cycle control and descriptions of mutations in those genes. These developments have led to the definition of the role of specific oncogenes and tumour suppressor genes in several cancers, including, for example, colon cancers and some forms of breast cancer. Work reported from our laboratory has led to the identification of a number of candidate genes involved in the development of non-melanotic skin cancers. In this chapter, we attempt to further explain the observed (phenomic) alterations in metabolic pathways associated with oxygen consumption with the changes at the genetic level.

1. INTRODUCTION

In recent years, with the development of techniques in modern molecular biology, it has become possible to study the genetic basis of carcinogenesis down to the level of DNA sequence. Major advances have been made in our understanding of the genes involved in cell cycle control and descriptions of mutations in those genes. These developments have led to the definition of the role of specific oncogenes and tumour suppressor genes in several cancers, including colon cancers and some forms of breast cancer. Work reported from our laboratory has led to the identification of a number of candidate genes involved in the development of non-melanotic skin cancers.[1]

A previous search of the human genome revealed that there is close association between the genomic locality of key enzymes in oxygen metabolism and genetic changes identified in particular oncogene and/or tumour suppressor gene localities. Although it must be stressed that none of the genes coding key oxygen-metabolic enzymes are directly contiguous with the proposed sites of carcinogenic events, many are sufficiently close to such sites to be affected by the major chromosomal aberrations that have been observed. It was therefore proposed that the observed disruptions in oxygen metabolism at the cellular level of cancer development reflect major genome positional deletions, duplications, or insertions. In this chapter, we attempt to further explain the observed (phenomic) alterations in metabolic pathways associated with oxygen consumption with the changes at the genetic level. The approach adopted in the present work is to identify template DNA in close pairs of ETC and tumour suppressor genes or oncogenes and design PCR primers against such template regions. Using real time PCR, it has been possible to confirm our earlier proposal regarding the co-lateral disruption to genes in close proximity. It is stressed that at this stage of our investigation, there is no suggestion that there is any influence of either type of gene upon the alterations observed in the other. The approach reported here allows a much greater focus on smaller regions of the genome than has been previously reported.

2. METHODS

The samples used for this study were derived from the 1995 Nambour skin cancer trial and are composed of blood samples collected from a cohort of patients with solar keratoses (SK). The DNA was extracted using a modified salting out procedure outlined in a previous study.[2] GeneReleaser was used to further clean up the samples, albeit using a slight modification on the protocol given. 20uL of sample was added to 20uL of GeneReleaser in a 0.6mL tube and the mixture was vortexed for 5 seconds before being microwaved on high for 7 minutes. A beaker of water was placed in the microwave to redirect heat. The preparation was spun in a microfuge for 1 minute and the supernatant (~20uL) was removed and mixed with 80uL water.

The process of identifying candidate genes involved using the NCBI sequencer to pinpoint the exact locations of each of the nuclear-encoded subunits of the ETC and of oncogenes/tumour-suppressor genes associated with non-melanotic skin cancer. Using those chromosomes and locations identified in the Ashton study, nine pairs of ETC/oncogenes showing close positional relationships were identified, and two of these chosen for this stage of the project are summarised in Table 1. For primer design, a

Table 1. Summary of ETC/Oncogene 1 localities that were the focus of interest.

Cancer Gene	SMOH	PTCH
Position	7q31-q32	9q22.3
Site (Mb)	104.9	89.05
ETC Gene	NDUFA5	NDUFA8
Position	7q31	9q33.2-q34.11
Site (Mb)	114.38	113
Distance (Mb)	9.445	24.221
Genes Separation	~40	~100
Cancer	BCC	BCC

BCC: basal cell carcinoma

Table 2. Primer design for real time PCR.

			Primer Tm	Amplimer Size (Tm)
NDUFA5	Forward	GGTAATATTTTAACCTATGG	54	208 (73)
NDUFA5	Reverse	TTGGTTAAATGTTACACAAG	54	
SMOH	Forward	CCTAAGGTCACAGAATGGCC	58	244 (76)
SMOH	Reverse	CTGTACCTTCAGGTCTGGGT	58	
NDUFA8	Forward	AGCAAGGCTATGTATTTGAG	56	322 (80)
NDUFA8	Reverse	TCTGTATTTACAGAGACCCT	56	
PTCH	Forward	TTCATGGTCTCGTCTCCTAA	58	309 (82)
PTCH	Reverse	AAGTGAACGATGAATGGACA	56	

specific exon in each of the genes was targeted, partly on the basis of nucleic acid base composition and partly for its length in terms of optimal Real-Time analysis and DHPLC analysis, if required. These primer sequences are summarized in Table 2 along with their predicted Tm (melt temperature) values.

Each Real-Time run involved the simultaneous analysis of both the ETC subunit and the oncogene on 30 samples to gauge the extent of adducts or additions (if present). After completing the Real-Time run on the thermal cycler iCycler™, each of the samples was examined by electrophoresis on a 2.0% agarose gel, prepared by standard techniques, to ensure the integrity of the amplification. A standard stock of 2x BioTaq reaction buffer, 0.8 μM dNTPs, 0.5 μM primer, and 1.5 units of BioTaq polymerase were used for each of the primers, since the majority of them had been specifically selected to have similar compositions and melt temperatures. The $MgCl_2$ concentration and cycle conditions are as follows:

SMOH: MgCl2: 3.5 μM. Cycle: 95°C 1 min; 50x (95°C 30 sec, 60°C 30 sec, 72°C 45 sec).
NDUFA5: MgCl2: 3.0 μM. Cycle: 95°C 1 min; 50x (95°C 30 sec, 60°C 30 sec, 72°C 45 sec).
NDUFA8: MgCl2: 2.5 μM. Cycle: 95°C 1 min; 50x (95°C 30 sec, 56°C 30 sec, 72°C 45 sec).
PTCH: MgCl2: 3.0 μM. Cycle: 95°C 1 min; 50x (95°C 30 sec, 56°C 30 sec, 72°C 45 sec).

3. RESULTS

It was found that analysis of the samples using melt temperature in the real time thermal cycling instrument (iCycler™) in conjunction with SYBR Green involves some uncertainty. This is believed to be associated with slight inconsistencies in temperature across the 96 well plates used in the analysis. To overcome this problem, only those samples that showed an absolute temperature difference of greater than 0.5°C in at least two of the triplicate runs from the designated melt temperature were considered significant enough to warrant further analysis by DHPLC and sequencing. Samples producing amplimers that exhibit such differences in melt temperature are designated as aberrant. The term aberration in this context refers to the presence of deletions of, additions to, or significant sequence changes in specific exons in the gene sets examined.

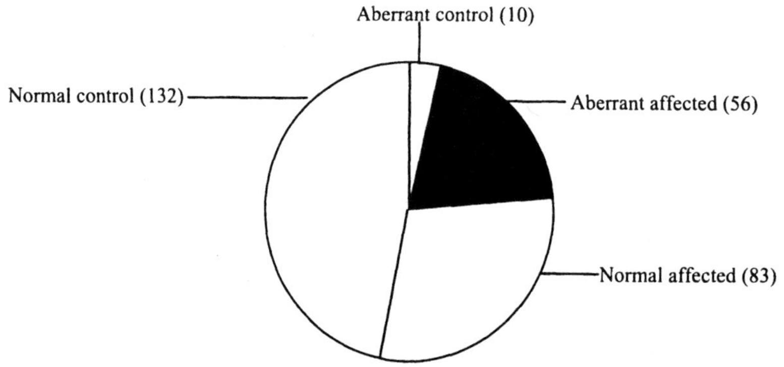

Figure 1. Pie graph summarising the distribution of aberrant samples compared to normal in an affected and control SK population, tested using NDUFA5/SMOH-targeted primers.

Aberrations of this sort on a much larger scale (i.e. significant sections on a chromosome, not just specific genes) were described in a non-melanotic skin cancer population (basal cell carcinoma, squamous cell carcinoma, and solar keratosis) utilising the Comparative Genomic Hybridisation (CGH) technique.[1] This technique can detect mega-base aberrations whilst the real-time method being implemented here is best used for anomalies between 20 base pairs up to approximately 400 base pairs.

In Figure 1, the results for the total sample population examined using the gene pair NDUFA5/SMOH are presented. It can be seen that, for this gene pair, 10 of the 142 normal samples exhibited some aberration, while 132 samples showed no aberration. By contrast, 83 of the 139 affected samples lacked aberrations in this gene pair, compared to 56 that were aberrant. In Figure 2, results for the other gene pair NDUFA8/PTCH in this work are presented. It is observed that 116 of the 142 normal samples exhibited no aberration, while 26 samples showed some aberration. By contrast, 64 of the 139 affected samples lacked aberration in this gene pair, compared with 75 that were aberrant.

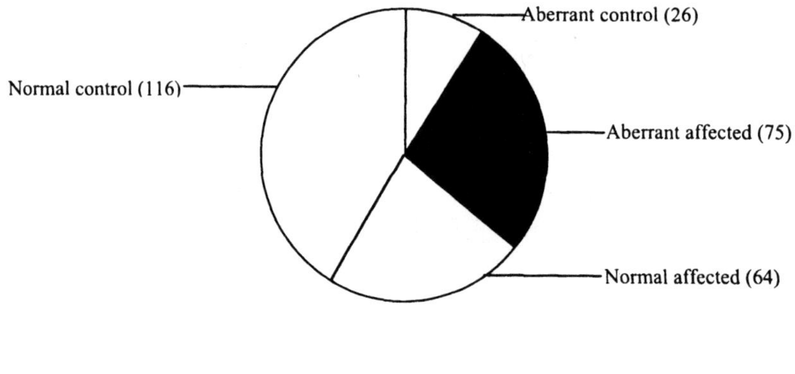

Figure 2. Pie graph summarising the distribution of aberrant samples compared to normal in an affected and control SK population, tested using NDUFA8:PTCH-targeted primers.

4. DISCUSSION

The main aim of this project was to analyse chromosomal regions containing both a nuclear-encoded electron transport chain gene and an oncogene/tumour suppressor gene associated with non-melanoma skin cancer for relatively large mutations. In order to achieve this, we used a real-time PCR melt curve analysis.

Specific exons were chosen in each of the genes for both GC content and length, since these two criteria have an enormous influence on the ability of the real-time procedure to produce a melt curve profile.[3] Samples, whether control samples or affected solar keratosis (SK) samples, were analysed simultaneously on the same run for both the ETC gene and oncogene in order to detect the presence of aberrations within the specific region. SYBR Green 1 is a dye that changes fluorophoric characteristics upon interacting with DNA. This offers a method of calculating the amount of DNA produced in a given PCR run by measuring the increase in fluorescence.[4]

The cycle number at which the level of fluorescence rises above a background threshold value is inversely proportional to the log of the initial template copies.[5] This approach was selected over several others, such as the TaqMan method, since it partly overcomes the problem of sequence heterogeneity. This means that any mutations in primer binding sequences would only hamper the first cycles(s), whereas point mutations in a probe region (like the TaqMan probe) might significantly alter the yield of signal throughout the PCR.[4] The analysis had to be performed on the same run because SYBR Green's specificity for the DNA template is directly related to the $MgCl_2$ concentration of the PCR master mix, and any slight pipetting errors can influence the results derived from real-time analysis. Thus the practise of comparing 'runs' is discouraged when utilising SYBR Green I.[3]

A disadvantage of the SYBR Green technique is that it can be incorporated into any double-stranded DNA, including unspecific products like primer-dimers.[6] Since extensive accumulation of primer-dimers is often associated with PCRs involving small amounts of target DNA or in no-template control samples, SYBR Green 1 PCR can produce false positive results.[3]

A way of accurately distinguishing non-specific amplification product from true PCR amplicons is to conduct a melting curve analysis of the amplified DNA.[3] Melting curve analysis generates a specific profile, which depicts the rate of change of fluorescence over time as a function of temperature for each sample. This profile is dependent upon the length and GC content of the PCR product.

As previously reported,[3] the assessment of DNA melt curve using SYBR Greens is not sufficiently specific to warrant the use of single occasion runs, so each of the samples was analysed in triplicate.

Eighteen affected samples produced aberration in all three runs in the NDUFA8/PTCH combination, and these were analysed via DHPLC to confirm the presence of polymorphisms. Of those eighteen, 11 were designated as being of a different polymorphic nature than the wild type, with three samples showing unique peaks. These were: a 69-year-old male with five SKs, a 66-year-old female who was part of the control group, and an affected 72-year-old female, with 122 SKs. The control female that exhibited a unique profile was the only control patient out of all the real-time runs that showed an abnormal real-time run in each of the three trials (out of 142). Sequencing of these eighteen samples will be commenced in the near future once DNA has been re-extracted from the blood samples in order to minimise degeneration peaks in the sequencing output. There were a total of 71 affected samples that expressed abnormal real-time data at least twice.

The NDUFA5/SMOH population also presented eighteen affected samples in triplicate that showed up as being aberrant in nature, though they have not as yet been completely analysed by DHPLC due to the presence of degeneration peaks in the initial runs. The DNA for these samples is currently being extracted. With regard to this pair, three control samples showed up in triplicate as being of abnormal nature.

It is not unexpected that a small proportion of the control samples should exhibit aberrations. Current theories on carcinogenesis accommodate genotype/phenotype inconsistencies on the basis that carcinogenesis involves multi-gene expression. This caveat incorporates epigenetic and environmental influences. Conversely, the identification of affected samples that did not show aberrations in this study is also not unexpected. It might be anticipated that some regions of 'normality' would be observed as a result of randomly focussing on relatively small regions of quite large genes.

Finally, when the results for both pairs of close genes are analysed, a total of 46 controls showed aberrations (36 PTCH:NDUFA8 and 10 SMOH:NDUFA5), but no control sample showed aberrations in both pairs. Conversely, for the solar keratosis samples 129 showed aberrations (75 PTCH:NDUFA8 and 57 PTCH:NDUFA5), but only 3 exhibited aberrations in both gene pair analyses.

ACKNOWLEDGEMENTS

The authors wish to thank Dr. Adele Greene of Queensland Institute of Medical Research for access to samples collected in the Nambour Skin Cancer Trial.

REFERENCES

1. K. J. Ashton, S. R. Weinstein, D. J. Maguire, and L. R. Griffiths, Molecular cytogenetic analysis of basal cell carcinoma DNA using comparative genomic hybridisation, *J. Invest. Dermatol.* **117**, 683-686 (2001).
2. R. A. Lea, S. Selvey, K. J. Ashton, J. E. Curran, P. T. Gaffney, A. C. Green, and L. R. Griffiths, The null allele of GSTM1 does not affect susceptibility to solar keratosis in the Australian white population, *J. Am. Acad. Dermatol.* **38**, 631-633 (1998).
3. J. O'Mahony, and C. Hill, A real-time PCR assay for the detection and quantitation of *Mycobacterium avium* subsp. *Paratuberculosis* using SYBR Green and the Light Cycler, *J. Microbiol. Methods* **51**, 283-293 (2002).
4. E. M. Moen, J. Sleboda, and B. Grinde, Real-time PCR methods for independent quantitation of TTV and TLMV, *J. Virol. Methods* **104**, 59-67 (2002).
5. J. S. Burgos, C. Ramirez, R. Tenorio, I. Sastre, and M. J. Bullido, Influence of reagents formulation on real-time PCR parameters, *Mol. Cell. Probes* **16**, 257-260 (2002).
6. M. Mikula, A. Dzwonek, K. Jagusztyn-Krynicka, and J. Ostrowski, Quantitative detection for low levels of *Helicobacter Pylori* infection in experimentally infected mice by real-time PCR, *J. Micro. Methods* 55(2), 351-359 (2003).

PROTEIN C PRODUCTION:

Metal ion/protein interfacial interaction in immobilized metal affinity chromatography

James J. Lee, Eileen Thiessen, and Duane F. Bruley

Abstract: Protein C (PC) is an essential blood factor in the human blood coagulation cascade. PC can help achieve blood hemostasis in many deadly disease conditions such as sepsis, cancer, HIV, etc.; reduced oxygen transport due to blood agglutination within the body can cause tissue death and organ failure as a result of low oxygen transport. Our goal is to produce large quantities of low cost zymogen PC for the treatment and prevention of blood clotting resulting from many disease states, as well as provide an effective therapy for PC deficiency.

Current studies show that Immobilized Metal Affinity Chromatography (IMAC) has high specificity and can be used for difficult separations among homologous proteins at relatively low cost compared to current methods, such as Immunoaffinity Chromatography. Thus, we are investigating the optimization of IMAC for the separation and purification of PC from Cohn fraction IV-I.

Molecular interactions within the chromatography column involve many parameters that include: the use and type of chromatographic gel and buffer solution, the pH, temperature, metal ion, chelator, and the sequence and structure of the protein itself. These parameters all influence the protein's interaction with the column. Experimental equilibrium isotherms show that PC has primary and secondary binding characteristics, demonstrating that the interaction is not just a simple process of one protein binding to one metal ion. Understanding the thermodynamics of interfacial interaction between proteins and surface-bound Cu^{2+} is essential to optimizing IMAC for PC purification, as well as for separation of other proteins in general. Hence we are undertaking theoretical and experimental studies of IDA–Cu/PC adsorption. The differences in structures of PC and other critical homologous blood factors are examined using the protein visualization program Cn3D. A better understanding of the interfacial phenomena will help determine the most effective conditions to achieve our goal.

1. BACKGROUND

1.1. Protein C

Synthesized in the liver, Protein C (PC) is a vital blood factor within the human blood coagulation cascade.[1,2] The PC used in this study was separated and purified from blood plasma fraction Cohn IV-1 provided by the American Red Cross (ARC). PC is one of several homologous vitamin K-dependent proteins, which include blood factors II (prothrombin), VII, IX, X, protein M, protein S, and protein Z.[1] Three groups can be formed of these: factors VII, IX, X, protein C, and protein Z make up the first group, factor II belongs in a second group, and protein S forms the third group.[1] They are homologous in both amino acid sequence and 3-D structure, but are differentiated into these groups by domain variations.

PC is anticoagulant,[3] anti-inflammatory,[3] and antithrombotic[4] in its active form, preventing coagulation and thrombosis in blood flow.[1] PC deficiency can lead to the formation of thrombi and embolisms, and thus disrupt normal blood hemostasis. The common method of treating PC deficiency is through coumadin. This oral anticoagulant inhibits the effects of vitamin K, which regulates several other clotting factors. This drug is effective, and though often used for long-term therapy, they are likely to cause serious complications and side effects, such as minor to persistent internal bleeding, sensitivity reactions, and in more severe cases, skin necrosis. It has also been suggested that long-term use of this oral anticoagulant could lead to tissue and organ damage.[5] Thus, PC will be used as the model molecule for the separation procedure of homologous blood factors *via* Immobilized Metal Affinity Chromatography (IMAC).

1.2. IMAC

IMAC is relatively new and was first introduced by Porath et al. in 1975.[6] IMAC has high specificity and can be used for difficult separations among homologous proteins at relatively low cost compared to current methods like Immunoaffinity Chromatography (IAC). The use of monoclonal antibodies (MAbs) in IAC yields a high affinity and specificity for a target protein, and can be generated for almost any target protein. However, the big disadvantage is cost. Since the antibodies are very sensitive to operating conditions,[7] the extensive purification of the protein for contaminants can decrease antibody activity over time. IMAC does have a lower affinity than IAC, and may leak the chelated metal ions from the affinity matrix into the product, but there is also potential of contamination by MAbs with IAC.

IMAC is promising because the relative processing cost of separation is very low in comparison to IAC. With IMAC only a single step is required for purification, the protein loading capacity is high in comparison to other affinity chromatography techniques, and the gels used in the IMAC columns can be reused without loss of its chromatographic characteristics. Metal ions can be specifically tailored to the separation of the target protein by chelating a particular metal ion to the resin using a strong chelating ligand such as iminodiacetic acid (IDA). There is a significant decrease in the risk of side effects during removal and purification of target proteins for clinical use due to the absence of biologically toxic chemicals in the process of adsorption and elution. IMAC is inexpensive and operates under mild conditions that significantly limit protein

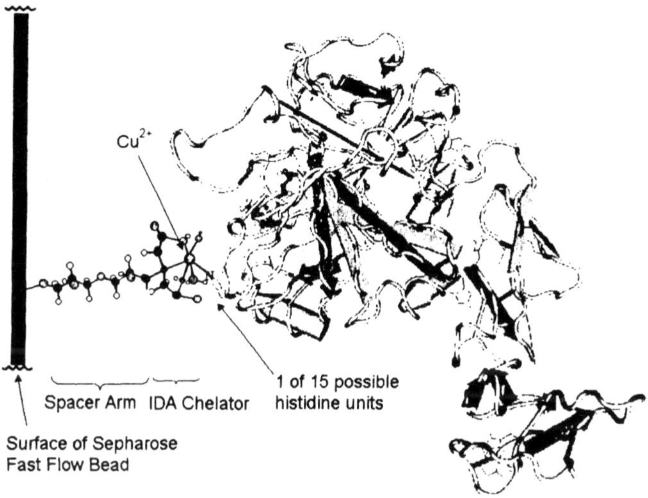

Figure 1. Illustration of the interface between PC and Cu^{2+} in IMAC.[8]

denaturing. Unlike IAC, there is no immunogenic contamination or possibility for bacterial growth. IMAC is also useful in the concentration of dilute protein solutions.[9]

The matrix or polymeric support (resin) of the chromatographic structure needs certain physicochemical characteristics in order to be suitable for IMAC.[9] The role of the chelator is to hold the metal ion to the resin *via* a spacer arm, allowing unimpeded access of the protein to the ligand, and limiting any steric hindrance.[9] Our research uses a Sepharose Fast Flow matrix, an ether chain spacer arm, an IDA chelator, and Cu^{2+} metal ion as illustrated in Figure 1.[5] This combination has been effective in the purification of PC from the blood plasma fraction Cohn IV-1.

2. THEORY

2.1. Lewis-Acid/Base Theory

The most commonly used metals are the first row transition metal ions, Co^{2+}, Zn^{2+}, Cu^{2+}, and Ni^{2+}. The principals of hard and soft acids and bases (HSAB) can explain the affinity of the proteins to the metal ions The theory states that when two atoms bond, one atom acts as a Lewis acid and the other as a Lewis base.[10] The strength of the bond is determined by the intrinsic "hardness" or "softness" ratings of the atoms involved. Bonds between the atoms with a similar rating are the strongest and *vice versa.* Co^{2+}, Zn^{2+}, Cu^{2+}, and Ni^{2+} are classified as borderline acids. The study of IMAC by Porath et al. used these metal ions because of the nature of their d-electron orbitals.[6, 11] These metal ions exhibit a preference for non-bonding lone pair electrons of nitrogen atoms in aromatic or aliphatic ligands. They show variations in affinity toward proteins that can be predicted through the HSAB theory. Histidine and tryptophan are borderline Lewis bases and therefore associate well with borderline acids such as Cu^{2+}.[10]

2.2. Histidine & Imidazole

Common metals used for IMAC are Cu^{2+}, Ni^{2+}, Zn^{2+}, Co^{2+}, Al^{3+}, and Fe^{3+}. The Cu^{2+}, Ni^{2+}, Zn^{2+}, and Co^{2+} metals have a preference for nitrogen-containing amino acid side chains. The functional groups with the highest retention strength for metal ions in IMAC are histidine and cysteine.[9] Cysteine can be ignored because in the case of most proteins it is found in its oxidized form: a disulfide bond with another cysteine. The imidazole ring of histidine units of PC gives it affinity to Cu^{2+} metal ion in IMAC.

IMAC has high specificity among the homologous vitamin K-dependent proteins because the adsorption of the protein to the metal ion is based on the interaction between an electron-donating group of the protein surface and the metal ion. The amount of histidine amino acids in a protein typically averages only 2% of the entire globular protein, of which only about half are exposed on the protein surface. So although only a small number of histidine amino acids are exposed on the protein, they are the major players in the adsorption of those proteins onto the metal ion.[9] PC has eighteen histidine units,[1] which is the highest among the vitamin K-dependent proteins looked at in this study using Cn3D: PC, factors II, VII, IXa, and Xa. Factor IX and X were found in its active form because only those available protein crystalline structures could be examined using Cn3D. Of the eighteen-histidine units, fifteen of these are theoretically accessible, allowing for interaction and adsorption with the chelated metal ions in the IMAC column.[9]

The protein bound to the column gel can be released by reducing the binding affinity between the metal ion and the adsorbed protein, by displacing the protein with a competitive agent that is similar to the amino acid involved in the metal binding.[9] Imidazole is typically used to displace the bound protein from the metal ion. It is effective in eluting proteins that have equilibrium binding affinities that are orders of magnitude stronger. It is evident that the concentration of imidazole required for elution is dependent on the number of surface histidine units on a protein.[12] An increasing concentration of imidazole is necessary to elute proteins with increasing number of histidine units. This is a critical factor in the separation of homologous vitamin K-dependent blood proteins with different number of theoretically exposed histidine units. By analysis of 3D structures with visualization software like Cn3D, the proteins in decreasing order of surface histidines are: PC (15), factor VII (9), factor IXa (8), factor Xa (5) and factor II (4). In vivo, the half-life in hours of these proteins in decreasing order is: factor II (100), factor X (65), factor IX (20), PC (6-7), and factor VII (5).[13] Thus, if PC and factor II cannot be completely separated, factor II may accumulate in vivo because of its long half-life relative to PC.

3. THERMODYNAMICS

Three major factors influence protein adsorption: surface properties, protein sequence and conformation, and solvent conditions. There is little understanding of how these factors influence the adsorption process at the molecular level. The process of adsorption is complex because the adsorbing particles are relatively large, and may undergo conformational changes upon adsorption. These conformational changes have allosteric effects on the protein. Protein adsorption thermodynamics is dependent on the intra- and intermolecular interactions between the protein molecules, the environment they are in, and the surface the proteins adsorb to.

Protein adsorption isotherms are used to view the thermodynamic properties because the data represent conditions at equilibrium. Using Eq. (1) the adsorption enthalpy (ΔH) and adsorption entropy (ΔS) from the isotherms generated by calorimetry results can be used to calculate the free energy (ΔG), or it can be calculated from the equilibrium constant of association (K_a) using Eq. (2):

$$\Delta G = \Delta H - T\Delta S \quad (1)$$

$$\Delta G = -RT \ln K_a \quad (2)$$

A negative value for free energy indicates favorable binding or adsorption of the protein to the metal ion, while a positive value indicates unfavorable or no adsorption. ΔH depends on interaction between the adsorbed proteins and the arrangement of the metal ions on the matrix or support.[14] More enthalpic energy will be required to separate an adsorbed protein that has strong interactions because of an arrangement that allows increased interaction between the histidine and metal ion. ΔH is generally negative when proteins adsorb to surfaces, suggesting the dominance of attractive forces between the adsorbing protein and the surface. Endothermic adsorptions occur when the adsorption is entropically driven, which is mainly attributed to protein conformational changes.[15]

3.1. Parameters

The exposure of the histidine units on a protein's surface greatly influences the retention of that protein to the metal ion, but this does not mean that adsorption is only available to that site of the protein.[9] The metal ion affinity to the protein can also be considered as a domain rather than a single amino acid site in which the surrounding hydrophobic residues can assist the histidine in its affinity of metal ions. So, histidine alone may not have complete control of adsorption. There are other factors that aid or hinder the adsorption of the protein to the metal ion.

Molecular interactions within the chromatography column involve several essential elements: the use and type of chromatographic gel, buffer,[16] salts,[16, 17] pH,[16, 17] temperature,[16] metal ion,[17] chelator,[18] and the protein itself.[18] These parameters all affect the protein's interaction with the metal ion.

3.2. Protein Adsorption Isotherms

The protein adsorption isotherm relates the amount of adsorbed protein to the solution concentration of protein measured at equilibrium. The Langmuir isotherm characterizes the adsorption with a basic saturation type of kinetics. The basic saturation shows a rate of adsorption that is proportional to some function of the concentration of the protein, and at a high enough concentration it reaches a saturation limit where no more protein can adsorb. The adsorption of proteins on the metal-ion all cannot be model to exactly fit Langmuir isotherm characteristics. Figure 2 exposes this point by demonstrating that PC continues to adsorb after it has reached a concentration of saturation as indicated with the arrows. Figure 3 shows how multiple sites of a protein

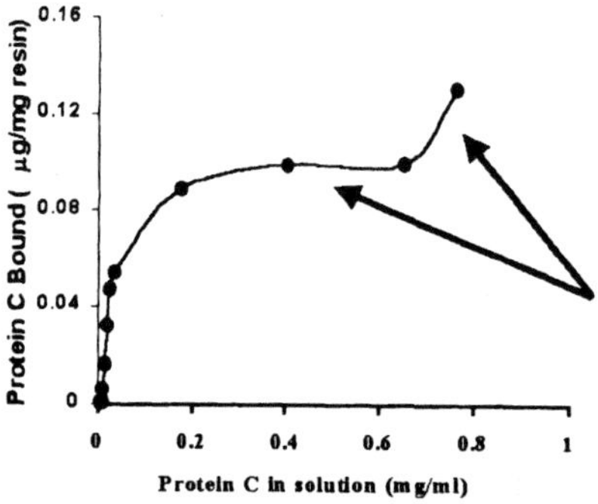

Figure 2. Adsorption Isotherm PC on IDA-Cu(II) gel.[19]

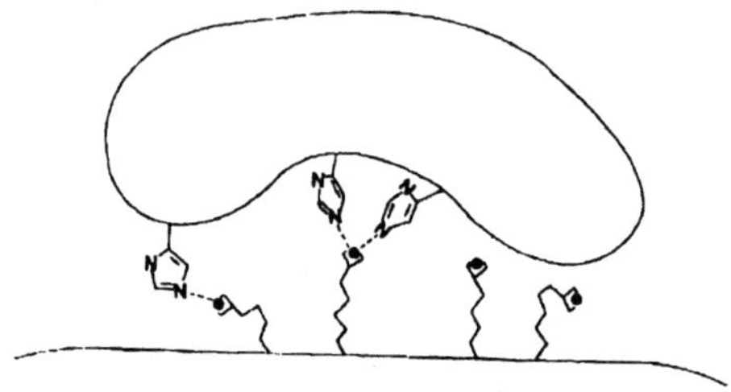

Figure 3. Multipoint attachment of protein to metal support.[10]

molecule may bind to a single metal ion. The adsorption isotherm (Figure 2) of PC in IMAC suggests a possibility that the adsorption of the metal ions is due not to one but to multiple histidine units.[10, 19]

4. CONCLUSION

PC has promising clinical potential for patients with a genetic deficiency of PC, septic shock, hip and knee replacement patients, coumadin-induced necrosis patients, heparin-induced thrombo-cytopenia, and patients undergoing fibrinolytic therapy or angioplasty, or suffering from unstable angina.[5]

Understanding the interfacial interaction between PC and Cu^{2+} is crucial to the optimization of IMAC for PC purification as well as for other protein separation processes. It is important to consider the use and type of chromatographic gel, buffer, salts, pH, temperature, metal ion, chelator, and the protein itself in studying the adsorption of proteins in IMAC. By taking measurements of the adsorption energies with respect to the different parameters and its changes in operation, it will provide a way to examine the relationship of how these parameters affect the adsorption energies and quantify them. From this it can be possible to identify thermodynamic values and relationships within IMAC. Identification of how the parameters affect the thermodynamic interfacial interactions will assist the optimization of IMAC for operation at low cost and production of high purity of PC and other important proteins.

5. ACKNOWLEDGEMENTS

Research was made possible by funding through NSF-Grants CTS-0090749 and CV523345, and material and technical support by the American Red Cross. The Whitaker Foundation Special Opportunity Award made initial research possible in biomedical engineering. Thanks to the Hematology Engineering Lab research group, and Paul Thiessen, with his expertise in Cn3D, for assistance with this project.

REFERENCES

1. D. F. Bruley, and W. N. Drohan, *Advances in Applied Biotechnology Series; Protein C and Related Anticoagulants*, Portfolio Publishing Co, Woodlands, TX, **11**, (1990).
2. R. M. Bertina, *Protein C and Related Proteins; Biochemical and Clinical Aspects*, Churchill Livingstone, Edinburgh London, (1998).
3. C. T. Esmon, The Anticoagulant and Anti-Inflammatory Roles of the Protein C Anticoagulant Pathway, *J. Autoimmun.* **15**, 113-116 (2000).
4. C. T. Esmon, Protein C anticoagulant pathway and its role in controlling microvascular thrombosis and inflammation, *Crit. Care Med.* **29**(7 Suppl), S48-51 (2001).
5. H. Wu, and D. F. Bruley, Chelator, metal ion and buffer studies for protein C separation, *Comparative Biochemistry and Physiology Part A*, **132** (2002) 213-220.
6. J. Porath, J. Carlsson, I. Olsson, and G. Belfrage, Metal chelate affinity chromatography, a new approach to protein fractionation, *Nature*, **258**(5536), 598-599 (1975).
7. L. K. Korah, D. G. Ahn, and K. A. Kang, Development of an Economic MiniAntibody Production Process for the Purification of Protein C (Anti-coagulant/Anti-thrombotic), *Adv. Exp. Med. Biol.* **510**, 127-131 (2003).
8. E. E. Thiessen, and D. F. Bruley, Theoretical Studies of IMAC Interfacial Phenomena for the Purification of Protein C, *Adv. Exp. Med. Biol.* **540**, 183-190 (2003).
9. E. K. M. Ueda, P. W. Goutt, and L. Moranti, Rev. of "Current and prospective application of metal ion-protein binding," *J. Chromatog. A*, **988**, 1-23 (2003).
10. V. Gaberc-Porekar, and V. Menart, Rev. of "Perspectives of immobilized-metal affinity chromatography," *J. Biochem. Biophys. Methods*, **49**, 335-360 (2001).
11. H. Chaouk, and M. T. W. Hearn, New ligand, N-(2-pyridylmethyl)aminoacetate, for use in the immobilised metal ion affinity chromatographic separation of proteins, *J. Chromatog. A*, **852**, 105-115 (1999).
12. R. D. Johnson, R. J. Todd, and F. H. Arnold, Multipoint binding in metal-affinity chromatography II. Effect of pH and imidazole on chromatographic retention of engineered histidine-containing cytochromes *c*, *J. Chromatog. A*, **725**, 225-235 (1996).
13. D. A. Barrow, Helena Laboratories, Hemostasis Pathways Poster, PO BOX 752, Beaumont, TX 77704.
14. R. D. Johnson, Z.-G. Wang, and F. H. Arnold, Surface site heterogeneity and lateral interactions in multipoint protein adsorption, *J. Phys. Chem*, **100**, 5134-5139 (1996).
15. M. E. Thrash, and N. G. Pinto, Characterization of enthalpic events in overloaded ion-exchange chromatography, *J. Chromatog. A*, **944**, 61-68 (2002).
16. H.-I. Jung, S. J. Bowden, A. Cooper, and R. N. Perham, Thermodynamic analysis of the binding of component enzymes in the assembly of the pyruvate dehydrogenase multienzyme complex of *Bacillus stearothemophilus*, *Protein Sci.* **11**, 1091-1100 (2002).
17. F.-Y. Lin, W.-Y. Chen, and L.-C. Sang, Microcalorimetric Studies of the Interactions of Lysozyme with Immobilized Metal Ions: Effects of Ion, pH Value, and Salt Concentration, *J. Colloid Interface Sci.* **214**, 373-379 (1999).
18. F.-Y. Lin, W.-Y. Chen, R.-C. Ruaan, and H.-M. Huang, Microcalorimetric studies of interactions between proteins and hydrophobic ligands in hydrophobic interaction chromatography: effects of ligand chain length, density and the amount of bound protein, *J. Chromatog. A*, **238**, 333-339 (2001).
19. R. Nandakumar, and H. Afshari, Personal communication and non-published data (2002).

MODELING BEHAVIOR OF PROTEIN C DURING AND AFTER SUBCUTANEOUS ADMINISTRATION

Mahesh V. Chaubal, Ladislav Dedík, Mária Durisová, and Duane F. Bruley

Abstract: Protein C is an important blood factor protein that regulates the blood coagulation process. Deficiency of protein C can lead to excessive coagulation that results in lack of tissue oxygenation, causing conditions such as deep vein thrombosis, pulmonary embolism, and stroke. Human protein C has been approved as a treatment for congenital protein C deficiency; however, the therapy requires frequent injections, due to the short residence time of the protein. Subcutaneous administration has been examined as an alternative to increase residence time and decrease injection frequency, thereby creating a more patient-friendly dosing regimen. In order to design an efficient injection or infusion protocol for subcutaneously administered proteins, it is important to accurately model the behavior (absorption, distribution, elimination) of these proteins in the body. However, several factors involved in a subcutaneous injection of the protein make modeling this behavior a challenging task. For example, absorption of the drug from the subcutaneous site into the blood stream can be variable depending on the site of injection, physical activity of the patient, etc. Furthermore, degradation of the protein can occur at the site of injection and further modify its absorption. The objective of this work was to demonstrate the utility of frequency response modeling as an alternative method to analyze the behavior of subcutaneously administered protein C. The results of our study indicate that if the dose range yielding the constant clearance of protein C is identified for the patient, models of that type, as presented in our study, can be used to adjust optimal dosing of protein C necessary to reach prescribed levels of the protein in this patient at desired time points, both specified by treatment requirements.

1. INTRODUCTION

Protein C is a vitamin K-dependent protein normally present in the human blood plasma at a concentration of 4 μg/ml. The protein exists as a zymogen precursor of a serine protease, which, when activated, regulates the coagulation cascade preventing abnormal blood clots, thereby maintaining a regular flow of blood and oxygen to tissues. Protein C deficiency, which affects up to 1 in 300 people in the general population,[1] can potentially cause a lack of regulation of the natural blood coagulation pathway and could lead to massive thrombosis (clogging of blood vessels due to clots). Such clogging of blood vessels can further cause tissue oxygen deprivation resulting in tissue damage.[2] Serious examples of this damage include stroke, heart attack, pulmonary embolism, tissue necrosis, and other complications that can result in amputation or death.

Coumadin and heparin are the two most commonly used drugs for the treatment of symptoms caused by protein C deficiency. Coumadin acts by inhibiting the action of vitamin K, which in turn inhibits activity of coagulation proteins such as Factor VIII and Factor IX. Though often used for long-term therapy, coumadin is likely to cause complications, such as minor to persistent bleeding, sensitivity reactions, and, in more severe cases, skin necrosis and thrombocytopenia (a condition leading to a decrease in platelet count). Reports also suggest that long-term use of this oral anticoagulant could lead to tissue and organ damage.[3] Similarly heparin has associated side effects, the most severe of them being hemorrhage, which has been observed in up to 1%-5% of patients. The other severe complication is heparin-induced thrombocytopenia, which may occur in up to 1% of the patients receiving this treatment.

Protein C has been approved as a drug for the treatment of protein C deficiency.[4, 5] Although the drug is reported to be safe and efficacious, it suffers from a drawback of short half-life, leading to a frequent (typically once-a-day) injection regimen. The long-term goal of our work is to evaluate alternate routes of administration to deliver protein C in a more patient friendly dosing regimen. In the past, we and other groups have demonstrated the utility of polymeric microspheres for encapsulation and sustained release of protein C.[6, 7] For sustained release technologies to be useful, the protein should have some bioavailability through the extravascular routes (subcutaneous or intra-muscular). In this report, we evaluate the utility of subcutaneous administration of protein C via alternate mathematical modeling methodologies aimed at elucidating the mechanism and pharmacokinetics of subcutaneously administered protein C.

Subcutaneous administration offers a number of advantages over intravenous injections, some of which are listed below:

1. Injections can be facilitated in a home setting by the patient (eliminating the need for expensive visits to the physician's office);
2. There is a lesser chance for side-effects such as phlebitis, commonly associated with frequent intravenous injections;[8]
3. Frequency of injections can be reduced.

Subcutaneous injection/infusion involves introducing a liquid formulation or a suspension into the subcutaneous layers. The drug slowly diffuses into the surrounding blood capillaries and lymphatic vessels. Recent studies have suggested the importance of lymphatic transport in the pharmacokinetics of subcutaneously administered proteins.[9] In order to design an efficient injection/infusion protocol for subcutaneously administered

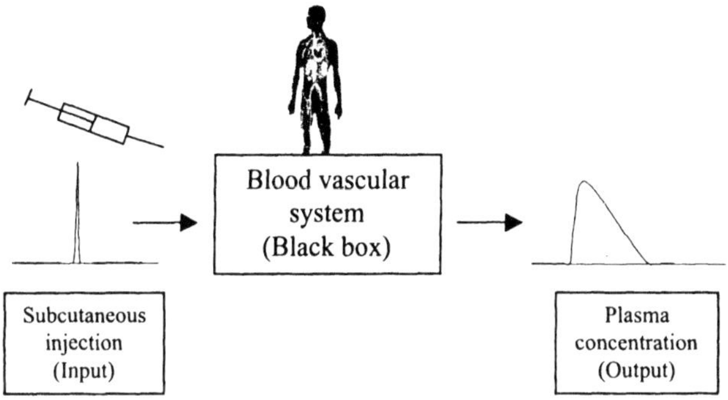

Figure 1. The "black box" approach used by Frequency Response technique to model the output (plasma concentration of the drug) for a known input (subcutaneous injection of a dose with known concentration). The fate of the drug during its journey from the subcutaneous tissues to the blood vessels is considered as the black box.

proteins, it is important to accurately model the pharmacokinetics of plasma concentration of the protein. The absorption of the drug from the subcutaneous site into the blood stream can be variable depending on the site of injection, physical activity of the patient, etc. Also degradation may occur at the site of injection, further modifying the kinetics of drug absorption. These factors make conventional approaches of pharmacokinetic modeling difficult and at times inaccurate. In this report, we have examined two novel approaches for pharmacokinetic modeling as alternative methods for subcutaneously administered proteins: a physiological transport model and a mathematical model utilizing frequency response analysis.

2. MATERIALS AND METHODS

The clinical data for subcutaneous administration of protein C, reported by Minford and coworkers,[3] was used for this study. Briefly, protein C was infused subcutaneously into the patient at a constant rate for 2 hours. The data reported by the authors measured the time from the end of infusion. In our analysis, the same data were used after adjusting the time scale from the start of infusion. A frequency response based modeling methodology was adopted. As depicted in the schematic in Figure 1, this technique treats the system as a 'black box' involving a known, well-defined input and an output, which depends on the characteristics of the system. The time data are converted into frequency domain, using Laplace transforms.[10] The ratio of the output to input in the frequency domain is called a transfer function for that system, as presented in Eq. (1) below:

$$G(\omega) = \frac{Y(\omega)}{X(\omega)} = \frac{\int_0^{Ty} y(t)\, e^{-j\omega t} dt}{\int_0^{Tx} x(t)\, e^{-j\omega t} dt} \tag{1}$$

where Tx and Ty represent the times when the input and output values become zero, respectively; x(t) and y(t) are the time domain input and output responses, i.e. the subcutaneous dose and the plasma concentrations, respectively; and w, t, and j denote the forcing frequency, time, and imaginary number, respectively. $X(\omega)$ and $Y(\omega)$ are the transformed input and output in the frequency domain, respectively. The subcutaneous infusion of the protein x(t) was approximated by Eq. (2):

$$\begin{aligned} x(t) &= 0 \quad \text{for } t \leq 0 \\ x(t) &= D/T \text{ for } T \leq t > 0 \\ x(t) &= 0 \quad \text{for } t > T \end{aligned} \tag{2}$$

where D and T are the dose and duration of protein C infusion, respectively.

All calculations for the frequency response technique were carried out using the software package CXT (CompleX Tools for Linear Dynamic System Analysis), developed by Dedik and Durisova.[10] The mathematical functions used for the calculation of pharmacokinetic parameters (such as Mean Residence Time, Clearance, etc.) have been previously described by Durisova and coworkers.[11] The Area Under the Curve (AUC) is represented by an estimate of the area under the curve for the Output response from time t = 0 to infinity. Other pharmacokinetic parameters were calculated from the frequency response model as follows:

$$\text{Gain (G)} = \text{AUC/D, where D was the total dose level} \tag{3}$$

$$\text{Clearance Cl} = 1/G \tag{4}$$

The maximum concentration, Cmax, and time of maximum concentration, tmax, were estimated from the maxima of the output curve. The elimination half-life was estimated as the time at which the concentration of the drug in the plasma fell to half of the total input. The Mean Residence Time (MRT) was calculated based on the linear extrapolation constants in the frequency domain, as explained by Durisova and coworkers.[11]

In order to compare the performance of frequency response-based pharmacokinetic modeling with a compartmental approach, the same clinical data were fitted to a classical, one-compartmental disposition model. The WinNonLin Pro 3.0 (Pharsight Corp, CA) software was used for the calculation of different pharmacokinetic parameters for the compartmental model. Absorption of the protein from the subcutaneous site into the vascular compartment was assumed to be a first order process. The performance of the two models (frequency response model and compartmental model) was compared using the Root Mean Squared Error (RMSE).

3. RESULTS AND DISCUSSIONS

Three sets of clinical data obtained from a previously reported dose ranging study were used for pharmacokinetic modeling.[3] The data set for 1500 Units dose was not used during the modeling studies due to the insufficient number of sampling points. The results of the pharmacokinetic modeling are presented in Figure 2 for the 3000, 4000, and

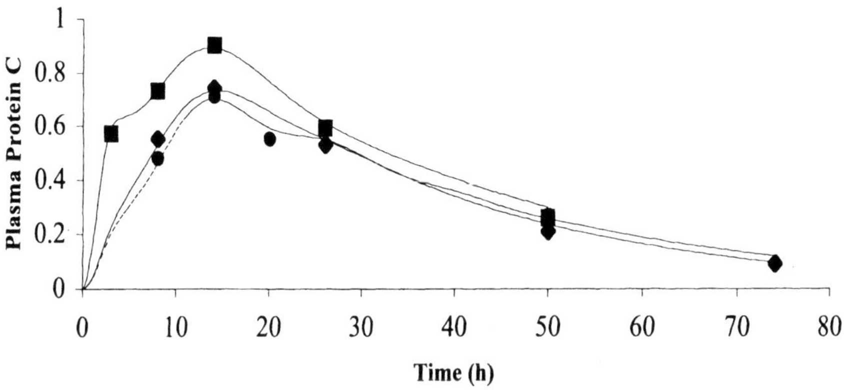

Figure 2. Typical pharmacokinetic curves for subcutaneously administered protein C, corresponding to a dose ranging study: ● 3000 U, ◆ 4000 U, and ■ 5000 U. Symbols represent actual measured data (protein C plasma concentration), and the solid line represents modeling results using frequency response analysis.

5000 Units dose levels. The clinical data suggest the presence of multiple peaks at high dose levels (5000 ppm). This behavior of protein C has been observed by other researchers and cannot be explained by a simple, first order disposition model. The frequency response method, on the other hand, provides a good fit for this data set, as can be seen in Figure 2.

Results from the frequency response analysis and compartmental modeling are reported in Table 1. As can be seen from pharmacokinetic modeling studies, the elimination half-life of protein C for subcutaneous administration was approximately 21 hours. This represents a 3-fold increase in the half-life, as compared with intravenous injections, which has a reported half-life of 6-8 hours. The peak concentration is reached at approximately 14 hours after initiation of dose. This matches well with the *in vivo* observations made by Sans-Rodriguez and coworkers.[12]

The values of the MRT of protein C were similar for all the doses and close to 40 h, as seen in Table 1. If, in the study by Minford et al.,[3] the concentration of protein C in plasma had been measured at time points of 0.5T and T after the beginning of the

Table 1. Pharmacokinetic parameters derived from Frequency Response (FR) and Compartmental Modeling (CM) approaches.

Parameter	**3000 U**	**4000 U**	**5000 U**
G (h/ml) (FR)	0.0126	0.0072	0.0059
Cl (ml/h) (FR)	78.8	139.38	170.12
C_{max} (U/ml) (FR)	0.89	0.73	0.7
t_{max} (h) (FR)	13.55	14.3	13.95
$t_{1/2el}$ (h) (FR)	24.92	19.4	20.5
MRT (h) (FR)	37.07	34.04	38.25
AUC (U.h/ml) (FR)	38.07	28.7	29.39
AUC (U.h/ml) (CM)	35.37	27.785	28.91
t_{max} (h) (CM)	10.96	14.51	15.56
C_{max} (U/ml) (CM)	0.854	0.6826	0.6225

Table 2. Goodness of fit, as determined by RMSE for Frequency Response (FR) and Compartment Modeling (CM) approaches to model the protein C pharmacokinetics data.

Dose level	% RMSE (FR)	% RMSE (CM
3000 Units	2.3%	4.3%
4000 Units	1.6%	3.4%
5000 Units	2.3%	5.8%
Average	**2.1%**	**4.5%**

subcutaneous infusion of the protein, then this would have yielded better estimates of the AUC because the duration of the infusion T was approximately 5% of the MRT in that patient.

The comparison of the two models is presented in Table 2. Root Mean Square Error (RMSE) is often used as a criterion for the selection of predictive models in forecasting studies. It can be seen that the RMSE for the frequency response model was half of that for the compartmental model. The difference in the RMSE values demonstrates the improvement that frequency response models can offer over conventional compartmental models for protein C pharmacokinetics.

4. CONCLUSIONS

This work has demonstrated the potential for frequency response techniques for the modeling of protein C behavior during and after subcutaneous administration. It was seen that subcutaneously administered protein C has an elimination half-life of approximately 21 hours, representing a 3-fold increase over the conventional intravenous treatment. Furthermore, it was seen that increasing the dose of protein C from 3000 Units to 5000 Units did not decrease the frequency of injections because of a dose-dependent clearance of protein C, as was observed from our modeling studies (Table 1).

If the dose range yielding the constant clearance of protein C is identified, models of that type, as presented in our study, can be used to adjust optimal dosing of protein C necessary to reach prescribed levels of the protein in the patient at desired time points, both specified by treatment requirements. It was seen that such a dose range of protein C was seemingly 2800 U - 3100 U, with the minimum value of the clearance for the dose of 3000 U. As shown in Figure 1, this dose was sufficient to reach the concentration of protein C of 0.3 U/ml in the patient at 48 h after the beginning of the infusion. The dose range yielding the constant clearance of protein C is very probably patient specific. It follows then that, for subcutaneous administration of this protein, the exact identification of this range for an individual patient is of crucial importance.

5. ACKNOWLEDGMENTS

This work was partially supported through a grant from the National Science Foundation (CTS-0090749).

REFERENCES

1. F. A. Nizzi, and H. S. Kaplan, Protein C and S deficiency, *Sem. Thromb. Hemostas.* **25**, 265-277 (1999).
2. K. G. Mann, and E. G. Bovill, Protein C deficiency and thrombotic risk, in: *Advances in Applied Biotechnology Series 11: Protein C and Related Anticoagulants*, edited by D. F. Bruley and W. N. Drohan (Portfolio Publications Co., Texas, 1990).
3. A. M. Minford, L. A. Parapia, C. Stainforth, and D. Lee, Treatment of homozygous protein C deficiency with subcutaneous protein C concentrate, *Br. J. Haematol.* **93**, 215-216 (1996).
4. D. F. Bruley, and W. N. Drohan, Protein C and related anticoagulants, in: *Advances in Applied Biotechnology Series 11: Protein C and Related Anticoagulants*, edited by D. F. Bruley and W. N. Drohan (Portfolio Publications Co., Texas, 1990).
5. H. P. Schwarz, W. Schramm, and M. Dreyfus, Monoclonal antibody purified protein C concentrate: Initial clinical experience, in: *Advances in Applied Biotechnology Series 11: Protein C and Related Anticoagulants*, edited by D. F. Bruley and W. N. Drohan (Portfolio Publications Co., Texas, 1990).
6. M. V. Chaubal, Z. Zhao, and D. F. Bruley, Novel injectable gels for the sustained release of protein C, *Adv. Exp. Med. Biol.* **530**, 147-155 (2003).
7. M. F. Zambaux, F. Bonneaux, R. Gref, E. Dellacherie, and C. Vigneron, Preparation and characterization of protein C-loaded PLA nanoparticles, *J. Controlled Release* **60**, 179-188 (1999).
8. S. H. Yalkowsky, J. F. Krzyzaniak, and G. H. Ward, Formulation-related problems associated with intravenous drug delivery, *J. Pharm. Sci.* **87**, 787-796 (1998).
9. C. J. Porter, and S. A. Charman, Lymphatic transport of proteins after subcutaneous administration, *J. Pharm. Sci.* **89**, 297-310 (2000).
10. L. Dedik, and M. Durisova, Frequency response method in pharmacokinetics, *J. Pharmacokinet. Biopharm.* **22**, 293-307 (1994).
11. M. Durisova, L. Dedik, A. Batorova, A. Sakalova, and J. Hedera, Pharmacokinetics of factor VIII in hemophilia A patients assessed by frequency response method, *Method. Find. Exp. Clin. Pharmacol.* **20**, 217-226 (1998).
12. C. Sanz-Rodriguez, J. J. Gil-Fernandez, P. Zapater, I. Pinilla, E. Granados, V. Gomez-G de Soria, J. Cano, N. Sala, J. M. Fernandez-Ranada, and N. Gomez-Gomez, Long-term management of homozygous protein C deficiency: replacement therapy with subcutaneous purified protein C concentrate, *Thromb. Haemost.* **81**, 887-890 (1999).

MECHANISMS FOR VASOCONSTRICTION AND DECREASED BLOOD FLOW FOLLOWING INTRAVENOUS ADMINISTRATION OF CELL-FREE NATIVE HEMOGLOBIN SOLUTIONS

Hae W. Kim, and A. Gerson Greenburg

Abstract: Acellular free hemoglobin-based oxygen carriers (HBOC) are being developed as red cell substitutes. However, following intravenous administration of some HBOC, decreased systemic blood flow and decreased functional capillary density have been observed. In isolated blood vessels, hemoglobin (Hb) in solution free of erythrocyte membranes has been shown to elicit vascular contraction. Therefore, the decreased blood flow and functional capillary density may be due to inherent vasoactive property of native Hb. There are two plausible mechanisms for the Hb-mediated vasoconstriction: nitrosylation of heme-irons and S-nitrosation of reactive β-chain cysteines (Cys93β). In this study, we investigated whether Hb Cys93β thiols play a role in Hb-mediated vascular contraction using functional bioassays with isolated rat thoracic aorta. To better define the roles of globin thiols and heme-iron, Hbs modified at the heme-iron and/or Cys93β sites were prepared and their vasoactivities tested. In addition, vasoactivities of natural heme proteins with heme and/or cysteine sites unavailable for NO reaction were also examined.

1. INTRODUCTION

Acellular free hemoglobin-based oxygen carriers (HBOC) are being developed as red cell substitutes. However, following intravenous administration of some HBOC, decreased systemic blood flow and decreased functional capillary density have been observed.[1, 2] In isolated blood vessels, Hb in solution free of erythrocyte membranes has been shown to elicit vascular contraction.[3-5] Therefore, the decreased blood flow and functional capillary density may be due to vasoconstriction. There are two plausible mechanisms for the Hb-mediated vasoconstriction: nitrosylation of heme-irons and S-nitrosation of reactive β-chain cysteines (Cys93β).

First, Hb-mediated vascular contraction could occur through heme-iron inactivation of endogenous nitric oxide (NO), a potent, smooth muscle relaxation factor constitutively synthesized and released by the vascular endothelium. NO has a high affinity for ferrous Hb.[6]

Second, reactive globin thiols of Hb are proposed to play a key role in regulating vascular tone and blood flow through reversible nitrosation reactions with endogenous NO.[7] In this theory, Cys93β thiols are proposed to serve as either NO scavengers or donors depending on the local oxygen tension. In the pulmonary venous circulation where PO_2 is high, Hb is oxygenated (HbO_2) taking the "relaxed" conformational state (R-state). In this state, Hb Cys93β thiol's reactivity for NO is reported to be high, allowing NO to be "loaded" onto Hb by forming S-nitrosothiol Hb. In the blood vessels of actively metabolizing tissues where O_2 is consumed (thus lower PO_2), the conformational structure of Hb is switched to "tense" state (T-state) as it is being deoxygenated. In the T-conformation, Hb Cys93β thiols' affinity for NO is also lowered, allowing NO to be released. In response, the vascular smooth muscle relaxes, increasing blood flow and O_2 supply to the affected tissues.[7-9] This is an intriguing theory since it assigns erythrocytic Hb a new function as a blood flow regulator through oxygen dependent NO delivery to metabolically active tissues.

There are, however, many questions that remain to be answered before this theory can be widely accepted. In this study, we investigated whether Hb Cys93β thiols play a role in Hb-mediated vascular contraction using functional bioassays with isolated rat thoracic aorta. To better define the roles of globin thiols and heme iron, Hbs modified at the heme-iron and/or Cys93β sites were prepared and their vasoactivities tested. In addition, vasoactivities of natural heme proteins with heme and/or cysteine sites unavailable for NO reaction were also examined.

2. METHODS

2.1. Preparation of Test Hbs

Based on site specific reaction chemistries, test Hbs were prepared by reacting purified human Hb with sodium nitrite, potassium ferricyanide, and/or N-ethyl maleiimide (NEM, a cysteine-specific blocking reagent). Three types of test Hbs were prepared: Hbs with low heme-iron reactivity to NO (HbNO, ferric Hb), a Hb with masked cysteine residues (*NEM-HbO_2*), and Hbs with low NO reactivity for both heme and cysteine sites (NEM-HbNO, NEM-HbCN).

- Heme-iron blocked Hb:
  deoxyHb + NO → HbNO
  oxyHb (HbO_2) + KCN/$K_3Fe(CN)_6$ → Hb^+CN (ferric heme)
- Cysteine masked Hb:
  HbO_2-$Cys_{\beta 93}$-SH + N-ethyl maleiimide (NEM) → NEM-HbO_2
- Both heme-iron and cysteine masked:
  HbNO + NEM → NEM-HbNO
  Hb^+CN + NEM → NEM-Hb^+CN

2.2. Isometric Vascular Tension Measurements

The isolated rat thoracic aortic ring model has been described previously.[5] Briefly, male Sprague Dawley rats of 250-350 g body weight were anesthetized with methoxy flurane. Through a midline incision, the thoracic aorta was carefully excised out avoiding endothelial damage. Following clearing of connective and other non-vascular tissues, the vessels were cut transversely into 2-3 mm ring segments and mounted in a tissue bath for isometric tension measurements. The tissue bath was filled with a Krebs buffer maintained at 37°C. The buffer was continuously aerated with a 95% O_2-5% CO_2 gas mixture. In hypoxic experiments, the oxygen supply was interrupted or bubbled with a gas mixture of 95% N_2-5% CO_2. The vessel rings were initially set at imposed tension of 1.5-2 g and allowed to relax for 1 hour. After the equilibration, vessel ring tensions were adjusted to 0.5 g prior to experimental procedure.

The responsiveness of vessel rings was first assessed by treatment with 50 nM norepinephrine (NE); vessel rings that developed less than 0.5 g of contraction were considered nonviable and excluded. Functional integrity of the endothelium was then assessed by treating the vessels with 10-50 μM acetylcholine (Ach). If vessel rings relaxed less than 10% of pretreatment tension, functional endothelium was considered absent and excluded from further testing.

2.3. Data Representation and Analysis

Data are presented as mean ± 1 standard deviation (SD). Statistical significance was determined by Student paired (before and after treatment comparisons) or unpaired t tests (between group comparisons) at $P = .05$ level.

3. RESULTS

3.1. Vascular Responses to Hb with Modified Heme and/or Cysteines

In NE tone-enhanced endothelium-intact vessel rings, treatment with control ferrous Hb (HbO_2) elicited a significant contraction; 0.4 μM Hb caused 58.1% ± 19.7% (mean ± SD) increase in vessel ring tension over the pretreatment value ($P < .01$, N = 8, Student t test). Ferrous oxyHb with Cys93β masked with NEM (NEM-Hb, 0.4 μM) also caused a significant contraction (50.7% ± 26.8%, $P < .01$, N = 8); vasoactivity was not attenuated when compared with control Hb. In contrast, HbCN, a ferric Hb with its heme-irons liganded with CN, did not produce a notable contraction (-3.7% ± 11.8%, N = 7, $P > .05$). Masking both the heme-iron and cysteine residues (NEM-HbCN) had no further effect. When deoxyHb was preliganded with NO (HbNO), it did not elicit any significant contraction (-5.5% ± 15.9%, N = 7, $P > .05$). Similarly, masking cysteines (NEM-HbNO) did not alter its lack of vasoactivity (3.2% ± 3.0%, N = 3, $P > .05$).

3.2. Vessel Ring Responses to Hb Under Lower PO_2 Conditions

In preliminary experiments, vessel rings subjected to a 15 min or longer period of hypoxia (stopped O_2 supply) showed lower and sluggish contractile responses than oxygenated controls in response to an equivalent NE dose (20-50 nM). In these vessel rings, a standard dose of Ach (30-50 μM) caused little or no relaxation. Subsequent treatment with 2-4 μM HbO_2 failed to elicit a significant contraction. In these vessel rings, 30 min reoxygenation did not restore normal vascular responses to Ach or Hb.

3.3. Vascular Responses to Heme Proteins with Naturally Blocked or Absent Heme-Iron or Cysteine Ligand Binding Sites

Although human and sperm whale myoglobins (Mb) are reported to contain one cysteine residue per molecule, they are not known to be reactive. Ferrous sperm whale myoglobin (Mb, 0.4 μM) elicited a significant contraction (55.1% ± 41.8%, N = 10, $P <$ .01) while ferric Mb (metMb) at a comparable concentration did not (-0.8% ± 7.6%, N = 6, $P >$ .05). Finally, equine cytochrome C (cyto-C) has two cysteine residues per molecules, but they are crosslinked to the propionic groups of the heme making them unreactive. In addition, the 6th coordinate of cyto-C heme iron (ligand binding site for Hb) is linked to methionine of the polypeptide chain, leaving no binding site for NO. Vessel rings treated with 0.2 μM cyto-C did not show notable contraction at all (-6.8% ± 13.5%, N = 10, $P >$ .05). Of note, methionines in these heme proteins do not normally participate in nitrosation reactions.

4. DISCUSSION

In this study, ferrous Hb with alkylated Cys93β thiols (NEM-HbO_2) did not significantly attenuate vascular contractile potency of unmodified control Hb. This lack of attenuation by NEM-Hb appears to indicate that globin thiols are not key players in the Hb-mediated vasoactivity, at least in this *in vitro* vessel model. To further test this hypothesis, vasoactivity of NEM-Hb preliganded with NO and CN^- (i.e., NEM-HbNO and NEM-Hb^+CN) were tested. Ferrous NEM-Hbs preliganded with NO or ferric heme NEM-Hb did not elicit contraction, but ferrous NEM-HbO_2 maintained vascular contractile activity. In addition, ferrous myoglobin (Mb), which has no reactive thiols, elicited contraction further supporting that heme-irons, not globin-thiols, play a primary role in the Hb-mediated vascular contraction. Relative reactivity of heme-irons toward the endothelial NO appears to be a primary controlling factor.

The proposition that Hb plays a key role in blood flow regulation through reversible interaction of Hb thiols with NO is intriguing. The theory postulates preferential NO binding to Cys93β thiols of Hb in the oxy- or R-conformational state, producing SNO-HbO_2, thereby preserving NO bioactivity.[7-9] Because the tissue bath was oxygenated with an enriched O_2 gas, the predominant conformational state of Hb in the bath should be the R-state (oxyHb). Therefore, it is expected that vessel rings were exposed to a substantial amount of SNO-Hb. The vessel rings, however, failed to relax when subjected to lower PO_2 conditions, created by turning off the O_2 supply for an extended period or switching to 95% N_2-5% CO_2 gas. According to the theory, SNO-Hb formed under higher PO_2 should have released NO when PO_2 was lowered, thus relaxing the vessel rings. Further, although transient hypertension was often observed, incidences of hypotension have not been reported following intravenous administration of free Hb to animal or human subjects. However, erythrocytic Hb exists in a milieu that contains a high level of glutathione (GSH), which may play a crucial role in SNO-Hb formation and subsequent NO release. Whether GSH- or thiol-containing proteins in plasma have any effect on Hb-mediated hypertension has not been investigated and requires further study.

Finally, because Cys93βs are generally conserved in many mammalian Hbs, they have been implicated to play an essential function, such as S-nitroso thiol-mediated blood flow regulation.[4] However, while they generally show an increased O_2 affinity, human Hb variants of Cys93β do not appear to cause serious hemodynamic abnormalities, suggesting that globin thiols are not essential for normal vascular homeostasis.

5. CONCLUSION

In the isolated thoracic aorta, Hbs with low heme-iron reactivity for NO elicited no or significantly reduced vascular contraction. In contrast, Hbs with masked globin thiols did not significantly alter the Hb-mediated vasoactivity. These results indicate that a primary mechanism for the Hb-mediated vasoconstriction observed following intravenous administration of some acellular Hb solutions appears to be heme-iron scavenging of endothelial NO. Therefore, Hb-based oxygen carriers with low heme-iron reactivity for NO but with normal oxygen binding/release characteristics would be desirable as a clinically useful red cell substitute.

REFERENCES

1. A. G. Tsai, B. Friesenecker, R. M. Winslow, and M. Intaglietta, Functional capillary density changes during blood substitution with ααHb and dextran 70: influence on oxygen delivery, *Art. Cells Blood Subs. Immob. Biotech.* **22**, 841-847 (1994).
2. A. Barve, A. P. Sen, P. R. Saxema, and A. Gulati, Dose response effect of diaspirin crosslinked hemoglobin (DCLHb) on systemic hemodynamics and regional blood circulation in rats, *Art. Cells Blood Subs. Immob. Biotech.* **25**, 121-133 (1997).
3. W. Martin, J. A. Smith, and D. G. White, The mechanisms by which hemoglobin inhibits the relaxation of rabbit aorta induced by nitrovasodilators, nitric oxide, or bovine retractor penis inhibitory factor, *J. Pharmacol.* **89**, 563-571 (1986).
4. R. F. Furchgott, Role of endothelium in responses of vascular smooth muscle, *Circ. Res.* **53**, 557-573 (1983).
5. H. W. Kim, and A.G. Greenburg, Ferrous hemoglobin scavenging of endothelium derived nitric oxide is a principal mechanism for hemoglobin mediated vasoactivities in isolated rat thoracic aorta, *Art. Cells Blood Subs. Immob. Biotech.* **25**, 121-133 (1997).
6. Q. H. Gibson, and F. J. W. Roughton, The kinetics and equilibria of the reactions of nitric oxide with sheep hemoglobin, *J. Physiol.* **136**, 507-526 (1957).
7. L. Jia, C. Bonaventura, J. Bonaventura, and J. S. Stamler, S-nitrosohemoglobin: a dynamic activity of blood involved in vascular control, *Nature* **380**, 221-226 (1996).
8. A. J. Gow, and J. S. Stamler, Reactions between nitric oxide and haemoglobin under physiological conditions, *Nature* **391**, 169-173 (1998).
9. T. J. McMahon, A. E. Stone, J. Bonaventura, D. J. Singel, and J. S. Stamler, Functional coupling of oxygen binding and vasoactivity in S-nitrosohemoglobin, *J. Biol. Chem.* **275**, 16738-16745 (2000).

AUTHOR INDEX

KEYWORD INDEX